International Union of Theoretical and Applied Mechanics

Dynamics of Rotors

Symposium Lyngby/Denmark
August 12-16, 1974

Editor F. I. Niordson

Springer-Verlag
Berlin Heidelberg New York 1975

Professor Ph. D. FRITHIOF I. NIORDSON

Technical University of Denmark

With 195 Figures

ISBN-13: 978-3-642-48786-6 e-ISBN-13: 978-3-642-48784-2
DOI: 10.1007/ 978-3-642-48784-2

Library of Congress Cataloging in Publication Data
Main entry under title: Dynamics of rotors.
At head of title: International Union of Theoretical and Applied Mechanics.
Bibliography: p. Includes index.
1. Rotors–Dynamics–Congresses. I. Niordson, Frithiof I. N., 1922–
II. International Union of Theoretical and Applied Mechanics.
TJ1058.D9 621.8'2 75-25649

Offsetprinting: fotokop wilhelm weihert kg, Darmstadt · Bookbinding: Konrad Triltsch, Würzburg

Preface

Requirements on high efficiency of rotating machinery forces engineers to deal with problems of self-induced vibrations and balancing of rotors in a serious and scientific manner. The last decade has shown a leap towards the solution of advanced problems of this kind as indicated by the large amount of scientific and engineering papers on rotor dynamics published in international journals.

In appreciation of this situation the General Assembly of the International Union of Theoretical and Applied Mechanics (IUTAM) at its meeting in Moscow in 1972, acting on a proposal by the Danish Adhering Organization, decided to sponsor an international symposium on the "Dynamics of Rotors".

The scientific planning of the symposium was handed by a scientific committee, consisting of Prof. R.E.D. BISHOP, UK, Prof. S.H. CRANDALL, USA, Prof. F.M. DIMENTBERG, USSR, Dr. A. TONDL, CSSR, Prof. H. ZIEGLER, Switzerland and Prof. F.I. NIORDSON, Denmark, (Chairman). This committee was charged with the selection and invitation of a small number of outstanding scientists in the field, whose contributions to the symposium are presented in this volume. Participation was restricted to invited scientists and there were 87 active participants from 18 countries.

Apart from a financial contribution from IUTAM to support travel and subsistence of invited participants, generous contributions were received from the following Danish Companies: Burmeister & Wain, Danfoss, The Danish Sugarcorporation, DISA, Elsinore Shipbuilding Engineer, Kyndbyværket, Niro Atomizer, and the Swedish Company STAL-LAVAL. The contributions are gratefully acknowledged.

An index of the scientific literature on rotor dynamics summing up the most important contributions during recent years was prepared especially for the symposium. It covers some 1200 references by author and by Key Work In Context (KWIC) and contains over 200 pages. A small edition was printed and remaining copies are available at Solidmech, TU, 2800 Lyngby, Denmark.

Local arrangements before and during the symposium were handled by the staff of the Department of Solid Mechanics at the Technical University of Denmark. The main burden of the organizing work was in the hands of Dr. VIGGO TVERGAARD, Dr. PAULI PEDERSEN, Mr. JARL JENSEN, Mr. R. ZETTERLUND, and Mr. E. DAM RAVN.

The social program included a reception for all participants and accompanying persons by the Mayor of Copenhagen at the City Hall, a banquet in the Tivoli Gardens, a tour to North Zealand, and a Ladies' program.

In order to present a nice, uniform volume, all manuscripts were retyped according to the same standards by the secretaries of the Department, Mrs. EDITH CHRISTENSEN and Mrs. BENTE ANDERSEN. Their untiring effort and the high quality of their work is documented by these pages.

In the editorial work Dr. TVERGAARD was most helpful.

I express my sincere thanks to all coworkers at the Department of Solid Mechanics, especially to those mentioned above for their contribution to a successful meeting. I also thank the Springer-Verlag for their kind and efficient cooperation in producing this volume.

Lyngby, June 1975.

Frithiof Niordson

Participants

(Authors are identified by an asterisk)

ATZORI, B., Politecnico di Torino, Istituto di Costruzione di Macchine, Corso Duca degli Abruzzi, 24, 10100 Torino, Italy

*BADGLEY, R.H., Mechanical Technology Incorporated, 968 Albany-Shaker Road, Latham, New York 12110, USA

*BALDA, M., Dvorakova 25, 320 07 Plzen, CSSR

BERRINO, A., Ansaldo Meccanico Nucleare - Via Pacinotti 20, 16151 - Genova Sampierdarena, Italy

BIGRET, R., G.E.T.T. - Centre Local/La Courneuve, 141 rue Rateau, 93123 La Courneuve, France

BISHOP, R.E.D., Department of Mechanical Engineering, University College London, Torrington Place, London WC1E 7JE, UK

*BLACK, H.F., Department of Mechanical Engineering, Heriot-Watt University, Chambers Street, Edinburgh EH1 1HX, UK

BOFFEY, D.A., Department of Mechanical Engineering, University of Edinburgh, Maryfield Road, Edinburgh EH9 3JL, UK

BOOKER, J.F., Sibley School of Mechanical and Aerospace Engineering, Cornell University, Ithaca, N.Y. 14850, USA

BOURGAIN, L.M., DGI/SET, Cen. Saclay, B.P. No. 2, 91190 Gif sur Yvette, France

*BROMMUNDT, E., Lehrstuhl A für Mechanik, Technische Universität, 33 Braunschweig, Postfach 3329, BRD

*BURDESS, J.S., Department of Mechanical Engineering, University of Newcastle Upon Tyne, Stephenson Building, Claremont Road, Newcastle Upon Tyne, NE1 7RU, UK

*CAPRIZ, G., Consiglio Nazionale delle Ricerche, Istituto di Elaborazione della Informazione, Via S. Maria, 46, 56100 Pisa, Italy

CARNEWAL, G., Université Libre de Bruxelles, Avenue F.D. Roosevelt, 50, 1050 Bruxelles, Belgium

CARRIER, A., Kellogg International Corporation, 62/72 Chiltern Street, London W1M 2AD, UK

CAZIN, M., 292 Rue Saint-Martin, 75003 Paris, France

CHRISTOPHERSEN, M., Kyndbyværket, 3640 Krogstrup, Denmark

CLAPIS, A., CISE, Casella Postale 3986, 20100 Milano, Italy

CRANDALL, S.H., Department of Mechanical Engineering, Massachusetts Institute of Technology, 77 Massachusetts Avenue, Cambridge, Mass. 02139, USA

*DONE, G.T.S., Postgraduate School of Applied Dynamics, Department of Mechanical Engineering, University of Edinburgh, Mayfield Road, Edinburgh EH9 3JL, UK

DRECHSLER, J., Technische Universität Berlin, Institut für Luft- und Raumfahrt, 1 Berlin 10, Salzufer 17/19, BRD

ESHLEMAN, R.L., Engineering Mechanics Division, IIT Research Institute, 10 West 35th Street, Chicago, Illinois 60616, USA

EWINS, D.J., Department of Mechanical Engineering, Imperial College, Exhibition Road, London SW7 2AZ, UK

FEDERN, K.G., Technische Universität Berlin, D 1000 Berlin 12, Strasse des 17. juni 135, BRD

GASCH, R., Technische Universität Berlin, Institut für Luft- und Raumfahrt, 1 Berlin 10, Salzufer 17/19, BRD

GRABOWSKI, B., Technische Universität Hannover, Lehrstuhl B für Mechanik, Appelstrasse 24 B, 3000 Hannover, BRD

*GUNTER, E.J., Department of Mechanical Engineering, University of Virginia, Charlottesville, VA. 22903, USA

GUPTA, K.K., Structures, Dynamic and Materials Section, Jet Propulsation Laboratory, Mail Stop 138/310, 4800 Oak Grove Drive, Pasadena, California 911031, USA

HOLMBERG, B.R., Avdelningen för mekanik, Chalmers tekniska Högskola, Fack 402 20 Göteborg 5, Sweden

*HOLMES, R., 12 Greenbank Ave, Brighton BN2 8GS, Sussex, UK

*HUSEYIN, K., Department of Systems Design, University of Waterloo, Waterloo, Ontario, N2L 3G1, Canada

INOUE, J., Division of Mechanical Engineering, Kyushu Institute of Technology, 1-1 Sensui-cho Tobata-ku, Kitakyushu-shi, Japan

IWATSUBO, T., Institut B für Mechanik, Technische Universität München, D-8000 München 2, Arcisstrasse 21, BRD

IYENGAR, N.G.R., Department of Aeronautical Engineering, Indian Institute of Technology, 11 T Post Office, Kanpur - 208016, India

JAECKER, M., Technische Universität Berlin, Institut für Luft- und Raumfahrt, 1 Berlin 10, Salzufer 17/19, BRD

*KANKI, H., Vibration Research Laboratory, Kobe Technical Institute, Mitsubishi Heavy Industries Ltd., Sinhama Arai-Cho, Takasago 676, Japan

KAWAI, R., Technical Faculty, Kobe University, Kobeshi Nadaku Rokkodai 1, Kobe, Japan

KAZIMIERSKI, Z., Technical University, Gdanska 155, 90-924 Lodz, Poland

*KELLENBERGER, W., Brown Boveri & Cie, Ch-5242 Birr, Switzerland

*KIRK, R.G., Pratt and Whitney Co., 400 Main Street EB3S2, East Hartford, Connecticut 06108, USA

KRÄMER, E., Technische Hochschule Darmstadt, Lehrstuhl für Machinendynamik, 61 Darmstadt, BRD

*LALANNE, M., INSA de Lyon, Bat 113, 20 Avenue Albert-Einstein, 69621 Villeurbanne, France

LARSSON, L.-O., ASEA, OCPB, 721 83 Västerås, Sweden

LEKKERKERKER, J.G., Laboratorium voor Technische Mechanica, Mekelweg 2, Delft 2208, Netherlands

LINDHARDT, P., A/S Burmeister & Wain's Motor- og Maskinfabrik af 1971, Torvegade 2, 1440 København K, Denmark

LINNETT, J.A., Postgraduate School of Applied Dynamics, Department of Mechanical Engineering, University of Edinburgh, Mayfield Road, Edinburgh EH9 3JL, UK

*LUND, J.W., Afdelingen for Maskinelementer, Bygning 403, Danmarks tekniske Højskole, 2800 Lyngby, Denmark

MAYES, I.W., Central Electricity Generating Board, Scientific Services Department, Ratcliffe-on-Soar, Nottingham, NG11 0EE, UK

MEINKE, P., MAN Werk München, Neue Technologie, 8 München 50, Dachauer Str. 667, BRD

MICHELINI, R.C., Universita di Genova, Istituto di Meccanica Applicata alle Machine, 16145 Genova, via All'Opera Pia, 11, Italy

MOLIN, S.E., ASEA, Sektion för Likströms- och Stora Växelströmsmaskiner, 721 83 Västerås, Sweden

MOORE, L.S., 5, Old Lodge Way, Stanmore, Middlesex, UK

MOREL, J.N.A., 3 Allee des Troenes, 92330 - SCEAUX, France

*MORTON, P.G., G.E.C. Power Engineering Ltd., Stafford Laboratory, Lichfield Road, Stafford, UK

*MUSZYNSKA, A., Institute of Fundamental Technological Research, Polish Academy of Sciences, ul Swietokrzyska 21, 00-049 Warszawa, Poland

MÄGI, M., Chalmers Tekniska Högskola, Inst. für Maskinelement, Fack, 402 20 Göteborg 5, Sweden

MÅRTENSSON, S., SIKOB AB, Ankdammsgatan 5H, S-171 43 Solna, Sweden

NILSSON, J.E.V., Inst. för Strömningsmaskinteknik, Chalmers Tekniska Högskola, 402 20 Göteborg 5, Sweden

NIORDSON, F.I., Danmarks tekniske Højskole, Afdelingen for Faststofmekanik, Bygning 404, 2800 Lyngby, Denmark

NORDMANN, R., TH Darmstadt, Fachgebiet Maschinendynamik, 61 Darmstadt, Petersenstrasse 18, BRD

OLSSON, K.-O., STAL-LAVAL Turbin AB, 612 20 Finspång, Sweden

*PARKINSON, A.G., Department of Mechanical Engineering, University College London, Torrington Place, London WC1E 6BT, UK

PARSZWESKI, Z., ul. Wierzbowa 38m34, 90-245 Lodz, Poland

PEDERSEN, P., Danmarks tekniske Højskole, Afdelingen for Faststofmekanik, Bygning 404, 2800 Lyngby, Denmark

PEDERSEN, P.T., Danmarks tekniske Højskole, Skibsbygningsafdelingen, Bygning 101D, 2800 Lyngby, Denmark
POSSA, G., CISE, Casella Postale 3986, 20100 Milano, Italy
RAO, J.S., Department of Mechanical Engineering, Indian Institute of Technology, Kharagpur - 721302, India
RESHEF, S., Ministry of Defence/Scientific Department, P.O.B. 7063, Tel Aviv, Israel
*RIEGER, N.F., Department of Engineering, Rochester Institute of Technology, Rochester, N.Y. 14623, USA
ROACH, C.D., US Army Scientific & Technology Information Team, I.G. Farben Hochhaus Zm 740, Grünebergweg, 6000 Frankfurt am Main, BRD
SCHIEHLEN, W., Institut B für Mechanik, Technische Universität München, D-8000 München 2, Arcisstrasse 21, BRD
*SCHWEITZER, G., Institut B für Mechanik, Technische Universität München, D-8000 München 2, Arcisstrasse 21, BRD
*SMITH, D.M., Flowermead, Winton Road, Bowdon, Cheshire, UK
SPRINGER, H., Institut für Maschinendynamik und Messtechnik, Technische Hochschule Wien, 1040 Wien, Karlsplatz 13, Austria
SQUARZONI, A., Universita di Genova, Istituto di Meccanica Applicata alle Machine, 16145 Genova, Via All'/pera Pia, 11, Italy
STECCO, S., Universita Degli Studi di Roma, Istituto Di Macchine e Technologie, Via Eudossiana 18, 00184 Roma, Italy
STEHLIN, P., Flygtekniska försögsanstalten, Box 11021, 161 11 Bromma 11, Sweden
STRAARUP, O., A/S Niro Atomizer, Gladsaxevej 305, 2860 Søborg, Denmark
TVERGAARD, V., Danmarks tekniske Højskole, Afdelingen for Faststofmekanik, Bygning 404, 2800 Lyngby, Denmark
VIND, T., Skördev. 11, S-612 00 Finspång, Sweden

VOGT, H.-J., Technische Universität Hannover, Lehrstuhl B für Mechanik, Appelstrasse 24 B, 3 Hannover, BRD

WALCZYK, J., Polska Akademia Nauk, Instytut Maszyn Przeplywowych, Skrytka pocztowa 621, 80-952 Gdansk, Poland

*WEHRLI, C., Eidgenössische Technische Hochschule, Rämistrasse 101, 8006 Zürich, Switzerland

WESTPHAL, G., Helsingør Værft A/S, Turbodepartment, 3000 Helsingør, Denmark

WOLFMAN, D., Nuclear Research Center, Nahal Soreq, Yavne, Israel

YHLAND, E.M., SKF European Research Centre B.V., Plettenburgerweg, Postbus 50, Jutphass, Netherlands

ZIEGLER, H., Eidgenössische Technische Hochschule, Zürich, Leonhardstrasse 33, Switzerland

Contents

Recent Developments in Multiplane-Multispeed Balancing of Flexible Rotors in the United States

Robert H. Badgley

Mechanical Technology Incorporated

Latham, New York, USA

ABSTRACT - This paper describes current developments in the evolution of a computer-implemented balancing procedure which permits flexible rotors to be precisely balanced in a cost-effective manner. Corrections in virtually any reasonable number of planes, computed by the procedure using signals from vibration sensors at critical locations, permit rotor operation over any design speed range. Steady-state operation at undamped critical speeds has been demonstrated. Results of recent test efforts indicate that the procedure can be applied with equal effectiveness to rotors of any size. Manufacturing and overhaul cost reductions are expected to flow from its adoption, together with performance advantages from operation in hitherto restricted dynamic regimes.

INTRODUCTION

Recent experiences in high-speed machinery development, particularly in the gas turbine industry, have confirmed the existence of a technology barrier identified ten years ago: the inability to balance rotor-bearings systems in a cost-effective manner for operation above their bending critical speeds.

Technology development and verification programs undertaken to surmount that barrier have now begun to bear fruit, and the resulting Multiplane-Multispeed Balancing capabilities are now being applied for reduced vibration operation of both existing and new rotating systems of many kinds.

More recently, an additional benefit accompanying the use of this technology has been identified: the ability to use lower-precision components in a high-precision system, with associated cost reductions. This benefit is only now beginning to be explored.

The most recent results of the balancing technology efforts have also indicated clearly that significant improvements in system performance-to-weight ratios are now possible. In effect, the demonstrated capabilities of Multiplane-Multispeed Balancing remove a major artificial restriction on the designer: the requirement to design for subcritical operation. Whole new realms of performance are available to the designer who accepts the challenge of the supercritical system.

It is the purpose of this paper to highlight the most important aspects of the newly-emerging balancing technology, through a brief description of its applications to date, and to describe the benefits which are beginning to flow from its adoption. This paper does not attempt a detailed comparison of the procedure with others developed for particular classes of machinery.

MACHINERY DESIGN TRENDS AND RESULTING NEEDS FOR NEW BALANCING CAPABILITIES

New rotating system designs, as they evolve, often involve rotational speeds which exceed one or more shaft-bearing system critical speeds at which bending of the shaft elastic axia is predicted. Such designs can be very sensitive to unbalance, particularly when rolling-element bearings are employed, because of low system damping. When such a design occurs, considerable design effort is usually expended to modify the system's configuration, not only to incorporate damping, but also to raise the critical speeds. Precise manufacturing tolerances, within the limits of manufacturing methods, are often specified to get as well-balanced a component as possible via the machining process. The intent is to perform standard low-speed balancing on these components and assemble them into the final installation without further attention.

Unfortunately, significant compromises in cost and performance often accompany achievement of the desired dynamic regime of operation (i.e., elevation of all the system bending critical speeds above the rotor speed). For instance, specification of larger shaft diameters produces larger bearing diameters, which may in turn require reduced rotor speeds to limit bearing DN values. This can lead to larger diameter aero-components for given power levels. An alternative is the addition of one or more bearings, which adds weight and complexity, and reduces overall system reliability.

The choice which must be made at each step in such a process is whether to expend further effort to force all of the rotor's bending critical speeds to lie sufficiently far above the operating speed (i.e., to require the first bending critical speed to exceed approximately 130 percent of design speed), or to accept rotor speeds above bending critical speeds and provide methods for controlling the dynamics of the resulting designs. A number of development programs are presently being conducted in quest of such methods. In these programs, efforts are being made to achieve tighter tolerances at lower cost, to eliminate rotor joints, to incorporate permanently welded subassemblies, such as compressor stages, to achieve higher bearing DN values, and to develop new kinds of intershaft bearings. At the same time, those problems are being examined which will have to be faced when in particular cases it becomes impossible to design for subcritical rotor-dynamic operation. The radical performance improvements which will flow from this approach are now being examined by several machinery manufacturers, particularly with regard to aircraft engines.

There are two procedures which, properly applied, permit supercritical rotor operation. These are Multiplane-Multispeed Balancing, which can be used to limit input vibration energy, and bearing resilient support damping, which dissipates existing vibration energy. Both energy management methods are necessary for proper supercritical rotor operation. However, since unbalance forces grow with the square of the speed, while damping forces increase only linearly with speed, it is apparent that balancing will become relatively more important as speeds increase.

Multiplane-Multispeed Balancing can, by itself, yield balanced rotor assemblies which can operate at and above bending critical speeds without bearing resilient support damping, although such designs may not always be desirable. On the other hand, damping alone may not be sufficient for initial operation of a supercritical rotor. Balancing often must be employed as well, and this balancing must properly treat the axial distribution of unbalance in multiple correction planes and at multiple speeds.

The dynamic implications of these trends may be most clearly demonstrated on the traditional "critical speed map", in which calculated lateral critical speeds are plotted as functions of rotor support (e.g., bearing) stiffness. Such a map, which is obtained without consideration of the actual damping in the system, is an intermediate step in a full rotor-dynamic analysis. It generally contains a series of curves similar to those in Figure 1. It may readily be seen that two critical speeds (the rigid-body criticals) decay to zero as stiffness tends toward zero. The straight-line nature of these two curves (on log-log paper) shows clearly that the resonances are functions of the square root of support stiffness in this low stiffness region. The third critical speed curve, and indeed all higher critical speed curves, become asymptotic at low stiffness values. These are the "free-free" criticals. The third critical speed always involves bending of the rotor's elastic axis regardless of the support stiffness, and if such a critical speed lies within 130 percent of running speed, multiplane

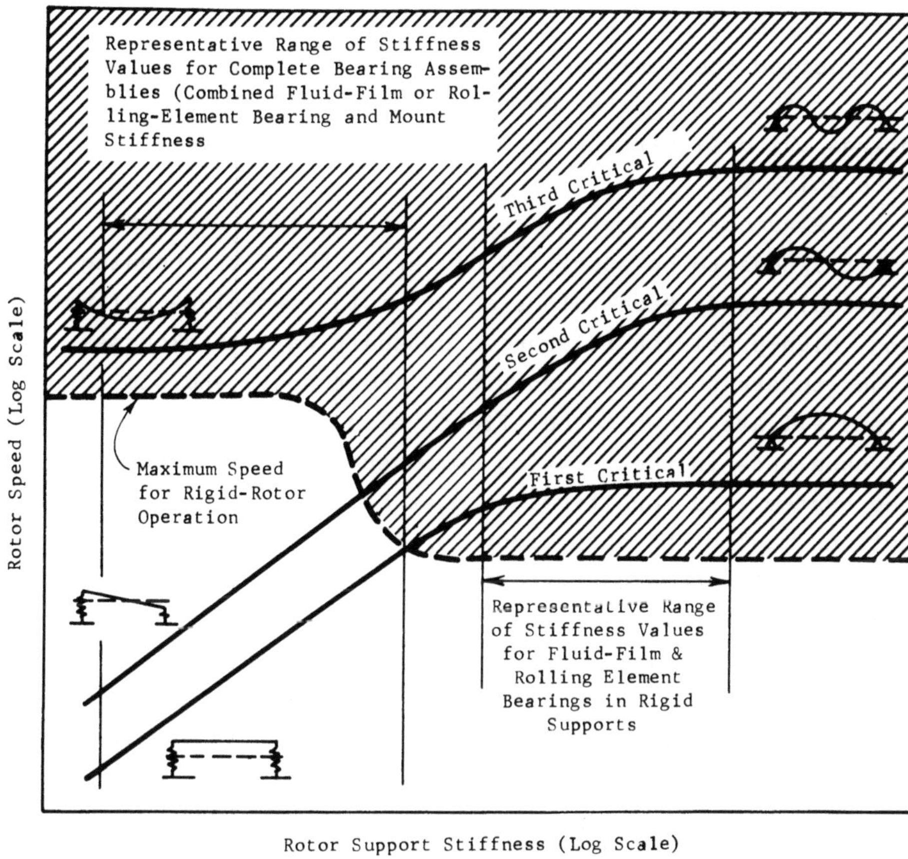

Fig. 1

balancing techniques become almost mandatory.

It should also be noted that both the first and second critical speeds involve elastic axis bending at higher support stiffness levels. Application of the "130 percent rule" to the first critical speed in this regime drastically reduces permissible operating speeds. Thus, the area above the broken line in Figure 1 is effectively denied to the designer unless he adopts Multiplane Balancing or adds heavy damping.

While many machinery designs operate in the rigid-body regime, recent trends in some classes of machinery are toward less rigid rotors. This trend tends to shift the critical speed curves downward and to the left, further restricting the designer. Any attempt to stiffen the bearings or their supports, or to operate at higher speeds, brings the designer into the restricted region that much more quickly.

A number of design actions can be taken to permit rotor speeds to be specified closer to the first bending critical speed. Such things as precision two-plane balancing, high-precision tolerances on shaft wall thicknesses, tight tolerances on concentricity and runout, and similar items all are known to be effective in reducing the level of unbalance in the rotating component, and thereby making the system less sensitive. In most cases, however, these actions may not be sufficient for design speeds which lie above the 130-percent-rule limiting speed, and reliance upon them above will drastically increase component costs. The Multiplane Balancing procedure may also be cost-effectively applied below the 130-percent-rule speed in the two-plane or one-plane options. The key, at the present time, to judging cost effectiveness in such an application is the cost of the Multiplane Balancing operation, since manufacturing costs can be fairly well quantified.

BALANCING PHILOSOPHY

Present-day metal removal procedures, in spite of their high-precision nature, all leave some distributed residual unbalance in components manufactured for use as part of rotating assemblies. This residual unbalance is essentially as random in size and location as are production tolerances. Two-plane low-speed component balancing can be used very effectively to counter this residual unbalance, yielding smooth component operation in the balancing machine. Smooth operation can sometimes be achieved in the final installation with the assembly of balance components when rigid elastic axis operation exists. Unfortunately, assembly tolerances will often result in re-introduction of unbalance at the assembly joints during final assembly. Moreover, the final assembly may operate with a flexible axis in spite of the fact that each component's elastic axis may experience no bending. Such behavior can be caused, for instance, by couplings or splines, or may be inherent in the design. The result can be considerable trial-and-error effort to find a satisfactory balance in the trim-balance stage, or greatly increased efforts toward tighter tolerances and more stringent assembly procedures, all of which can be very costly.

A generalization of the conventionally used balancing process involves more than two balancing planes and more than one balancing speed. Such a generalization is defined as Multiplane-Multispeed Balancing, and is the process of identifying and applying a single set of correction weights simultaneously in more than two planes on a rotating shaft, so as to achieve low vibration levels at each of a number of measurement locations along the shaft and at each of a number of shaft speeds. This procedure is also known as flexible-rotor or flexible-shaft balancing. Although multiple-plane balancing has been practiced for a number of years by builders and operators of various kinds of rotating machinery, particularly steam turbines, the procedures they have used are approximate only, and often depend on accumulated historical information about the particular type of machinery. Moreover, such procedures are often performed with the assistance of personnel (e.g., analysts) not commonly found in the factory-floor environment and often depend upon calculated dynamic properties of the rotor-bearing system. These characteristics have contributed to the unworkability of the earlier procedures both on systems which do not possess the dynamically well-understood properties of the steam turbine and also on systems which operate above several or more bending critical speeds. The cost of implementing these procedures has made their use prohibitively expensive for moderate- to high-volume production of machines such as gas turbines.

Flexible rotor balancing is essentially an information-gathering procedure. Thus, and regardless of how the correction weights are eventually arrived at, there exists an absolute requirement for an initial startup of the rotor to be balanced in order to obtain "as is" vibration data. If the balancing engineer has designed the rotor supports properly (or is lucky otherwise) or has chosen to balance *in situ*, the vibration modes of the rotor in the balancing apparatus will be reasonably like those to be experienced in service, and a satisfactory balance can result. In effect, the vibration behavior produced in this initial run will be caused by those components of unbalance which must be removed for smooth operation in the final installation.

It is sometimes the case that the rotor is well-enough balanced after manufacturing that it can immediately be run up to operating speed. In this (desirable) situation, the rotor need not be low-speed balanced. This behavior, however, depends upon manufacturing tolerances and, perhaps more importantly, on system damping. It is not to be expected that rotors of widely different sizes and classes can be operated to full speed during the first startup. Should full speed not be achievable on the first startup, the system damping must either be increased or an intermediate-speed balancing step must be applied. The former alternative may be difficult unless the balancing setup has been designed with this in mind.

Subsequent to the initial rotor startup (with vibration data recorded either up to desired top speed or at a speed limit set by vibration levels), the balancing engineer possesses several options by which he may determine the balancing corrections. For instance, in certain classes of machinery it is most desirable that no further startups be made in the unbalance condition because of concern for machine integrity or human safety. If no further operation of the rotor can be permitted in order to better define the unbalance distribution, then all further steps must be taken based upon calculated system dynamic properties. A large computer, time, and an analyst immediately enter the picture, and the accuracy of the corrections will depend upon the accuracy of the dynamic model constructed by the analyst. While the analyst can model most kinds of rotors, and the stiffnesses of certain kinds of bearings reasonably accurately, he will, in all likelihood, have only estimates of damping and structural stiffness. Unfortunately, the damped mode shapes are the ones of importance, rather than the undamped mode shapes, and required balance correction accuracies are presently tighter than calculated damping accuracies for many systems. Unfortunately, low-damping systems require very precise balance corrections, which in turn demand ultra-precise values of damping for use in the calculations.

An alternative approach is to calculate the damped unbalance-vibration behavior to be experienced at the balancing speeds with "trial weights" analytically attached to the rotor in each balancing plane, one at a time, and then make additional test runs with the trial weight physically attached to the rotor in each of the same balancing plane, one at a time. The calculations can be conducted in such a manner that direct comparison between predicted and measured trial weight vibration amplitudes will yield "improvements" to the remaining calculated values, which may then be used to calculate the desired balance weights. A large computer, time, and an analyst are still involved.

Finally, it is possible to dispense with the calculations entirely, and gather the required rotor behavior data experimentally. In such an undertaking, the large computer and the analyst leave the picture, and elapsed time is reduced to that required to operate the rotor. No assumptions are required as to the rotor model or as to inputs such as structural stiffness or system damping. In effect, the rotor tells the entire story through its sensors. This method is, in essence, an extension of the standard low-speed balancing procedures, and is very well suited to production.

Should either of the options involving calculations of rotor behavior be selected, it is a straightforward task to calculate the vibration amplitudes and phase angles all along the rotor's elastic axis for a unit unbalance in each selected balancing plane, and to utilize those calculated values which are found at the sensor locations. General unbalance response computer programs have become widely available during the past several years, and have proven fully satisfactory for this purpose. Separate calculation of the properties of each mode of vibration is also possible, with later superposition of results.

Such a process, however, introduces needless analytical steps and approximations into the procedure, which may be avoided through use of one of the more general response methods. Vibrations calculated by means of the total unbalance response technique are rapidly melded with the experimental data and used to determine balance corrections. Indeed, this process has been developed at the Author's company into a general simulation tool for optimizing a rotor-bearing system design for implementation of minimum time/cost balancing.

The vibration information gathered in any and all of the foregoing options is used during the balancing process as criteria for balance quality. It is vitally important, therefore, that this vibration data be meaningful to the individual rotating equipment. For example, it is common to utilize the motion of a bearing or bearing mounting surface as the balancing criterion. Alternatively, the force transmitted to the bearing support may be used. Such sensor outputs may indeed be useless if the rotor elastic axis nodes lie at the bearings, which is often the case with high-stiffness bearings. Large mid-span vibration levels may or may not accompany such a condition. The balancing engineer must understand the damped modes he faces, and must take steps to deal with them. As an additional example, consider the case of a large-diameter bull gear with short stub shafts mounted in journal bearings. If the pitch diameter is non-concentric to the bearings, as is the case with normal manufacturing tolerances, which criterion should be used for balancing: the journal dynamic motion or transmitted force, or pitch circle runout? The answer will depend upon the needs of the application, and may prove to be crucial in attaining a good-running system.

ANALYTICAL BASIS OF MULTIPLANE-MULTISPEED BALANCING SYSTEM

The Multiplane-Multispeed Balancing System makes use of linear influence coefficients for relating both calculated and measured vibration amplitudes to rotor unbalance. As such, it requires only that the rotor behavior be repeatable at the balancing speeds (e.g., that the rotor behavior not change because of loose parts or the like. Such changes indicate the presence of an engineering design problem rather than a balancing problem). The general analytical description of an influence coefficient procedure was presented in [5], and is reviewed briefly herein for convenience.

Rotor unbalance and resulting vibration amplitudes are related by means of influence coefficients, α :

$$(1) \qquad \{\omega_R\} = [\alpha] \cdot \{U_p\}$$

The α_{jp} are the linear coefficients which relate lateral rotor deflection at location j to unbalance at location p .

The displacement quantities ω_R are complex, and for any plane p ; the α_{jp} are complex as well:

(2) $$\alpha_{jp} = \alpha_{c,jp} + i\alpha_{s,jp}$$

The unbalances U_p may also be resolved into their components

(3) $$U_p = U_{c,p} + iU_{s,p}$$

where subscripts c and s denote the real and imaginary components, respectively.

The order of $[\alpha]$ will be ν by n , where ν is the total number of displacement elements in the ω_R column matrix and n is the number of balancing planes. The number of displacement elements in turn is given by

(4) $$\nu = mN$$

where m = number of axial locations along the rotor at which displacements are measured, and N = number of balancing speeds at which m measurements are taken.

There are, thus, $\nu = mN$ rows and n columns in the general $[\alpha]$ matrix. As will be discussed later, this matrix must be inverted for solution of equation (1). The $[\alpha]$ matrix must, therefore, be a square matrix. That is, ν must equal n .

The foregoing matrix equation may be written in long form as follows, where the number of the balancing speed is indicated by a superscript:

$$
(5)\quad
\begin{Bmatrix}
\omega_1^{(1)} \\ \omega_2^{(1)} \\ \cdot \\ \omega_m^{(1)} \\ \cdot \\ \cdot \\ \cdot \\ \omega_1^{(2)} \\ \omega_2^{(2)} \\ \cdot \\ \omega_m^{(2)} \\ \cdot \\ \cdot \\ \cdot \\ \omega_1^{(N)} \\ \omega_2^{(N)} \\ \cdot \\ \omega_m^{(N)}
\end{Bmatrix}
=
\begin{bmatrix}
\alpha_{11}^{(1)} & \alpha_{12}^{(1)} & \cdot & \cdot & \cdot & \alpha_{1n}^{(1)} \\
\alpha_{21}^{(1)} & \alpha_{22}^{(1)} & \cdot & \cdot & \cdot & \alpha_{2n}^{(1)} \\
\cdot & \cdot & \cdot & \cdot & \cdot & \cdot \\
\alpha_{m1}^{(1)} & \alpha_{m2}^{(1)} & \cdot & \cdot & \cdot & \alpha_{mn}^{(1)} \\
\cdot & \cdot & \cdot & \cdot & \cdot & \cdot \\
\cdot & \cdot & \cdot & \cdot & \cdot & \cdot \\
\cdot & \cdot & \cdot & \cdot & \cdot & \cdot \\
\alpha_{11}^{(2)} & \alpha_{12}^{(2)} & \cdot & \cdot & \cdot & \alpha_{1n}^{(2)} \\
\alpha_{21}^{(2)} & \alpha_{22}^{(2)} & \cdot & \cdot & \cdot & \alpha_{2n}^{(2)} \\
\cdot & \cdot & \cdot & \cdot & \cdot & \cdot \\
\alpha_{m1}^{(2)} & \alpha_{m2}^{(2)} & \cdot & \cdot & \cdot & \alpha_{mn}^{(2)} \\
\cdot & \cdot & \cdot & \cdot & \cdot & \cdot \\
\cdot & \cdot & \cdot & \cdot & \cdot & \cdot \\
\cdot & \cdot & \cdot & \cdot & \cdot & \cdot \\
\alpha_{11}^{(N)} & \alpha_{12}^{(N)} & \cdot & \cdot & \cdot & \alpha_{1n}^{(N)} \\
\alpha_{21}^{(N)} & \alpha_{22}^{(N)} & \cdot & \cdot & \cdot & \alpha_{2n}^{(N)} \\
\cdot & \cdot & \cdot & \cdot & \cdot & \cdot \\
\alpha_{m1}^{(N)} & \alpha_{m2}^{(N)} & \cdot & \cdot & \cdot & \alpha_{mn}^{(N)}
\end{bmatrix}
\begin{Bmatrix}
U_1 \\ U_2 \\ \\ U_3 \\ \cdot \\ \cdot \\ \cdot \\ \cdot \\ \cdot \\ \cdot \\ \cdot \\ \cdot \\ \cdot \\ \cdot \\ \cdot \\ U_{n-1} \\ \\ U_n
\end{Bmatrix}
$$

In equation (5), the subscripts on the elements of the displacement vector $\{\omega\}$ denote measurement locations along the rotor, these being the locations at which low vibration amplitudes are desired. The subscripts on the U vector indicate balancing planes in which trial weights are to be located during the balancing process and in which final correction weights will be placed.

The balancing procedure involves solution of equation (5) for $U_1, U_2, \ldots, U_n$, which are the components of residual unbalance in the "n" balancing planes. The process involves measurement of amplitudes and phase angles, calculation of the influence coefficients α, and finally inversion of the α matrix for solution of the U quantities.

An example of the application of this procedure is presented in [5].

DEMONSTRATED CAPABILITIES IN BALANCING VARIOUS TYPES OF HARDWARE

A practical, cost effective procedure for performing Multiplane-Multispeed balancing has been evaluated and proven at the laboratory level through a number of contract efforts undertaken by the Author's company between 1967 and the present time. The Multiplane-Multispeed Balancing procedure has now been demonstrated on a number of rotor-bearing systems and has been found to be suitable for application to a wide range of types of power-generation and power-transmission systems. A detailed discussion of the laboratory application of the balancing procedure was presented in ref. [1] and summarized in ref. [2], which described its potential impact upon the turbine engine designer. A brief description of the results to date is presented herein for ready reference.

During the period 1967-1970, an analytical method for calculating balance correction weights in multiple planes was developed and evaluated experimentally [3 , 4]. As part of these efforts, TESSARZIK, BADGLEY, and ANDERSON [5], were able to balance the two-bearing rotor used by LUND and ORCUTT [6] for sustained operation precisely at an undamped rotor-bearing system lateral critical speed in which significant shaft bending occured. This successful result verified prior analytical studies, and set in motion a series of increasingly more complex experimental evaluations which continues today.

In 1971, a considerably advanced version of the original balancing procedure became available. This advanced version was designed to permit the balancing operator to utilize more vibration data than that required to balance a rotor for theoretically zero amplitudes at each of the balancing speeds. In effect, the operator was permitted to utilize a practical number of balancing speeds and balancing data sensor locations which exceeded the maximum number permitted by the earlier program. TESSARZIK and BADGLEY [7] evaluated both the earlier and improved versions of the procedure for several different initial unbalance configurations, and found that the new procedure was equal in effectiveness to the earlier version for the direct comparisons made.

Demonstration of the system's capabilities on a rotor-bearing system which operated through several or more bending critical speeds became the next objective, together with an investigation of the quality of the results which could be obtained with various combinations of vibration sensors and balancing planes [3]. The rotor employed in this investigation is shown in Fig. 2(a). Results of these tests, an example of which is shown in Fig. 2(b), were very positive, in that it was possible to balance the (deliberately) severely unbalanced rotor quite rapidly for operation through four bending critical speeds, of which two exhibited severe resonances.

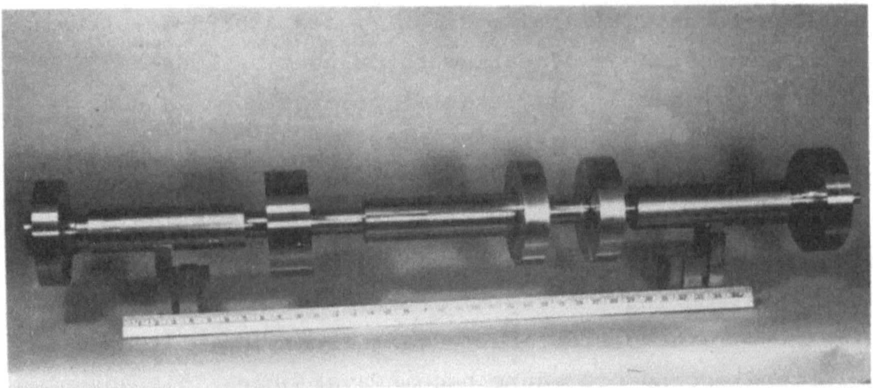

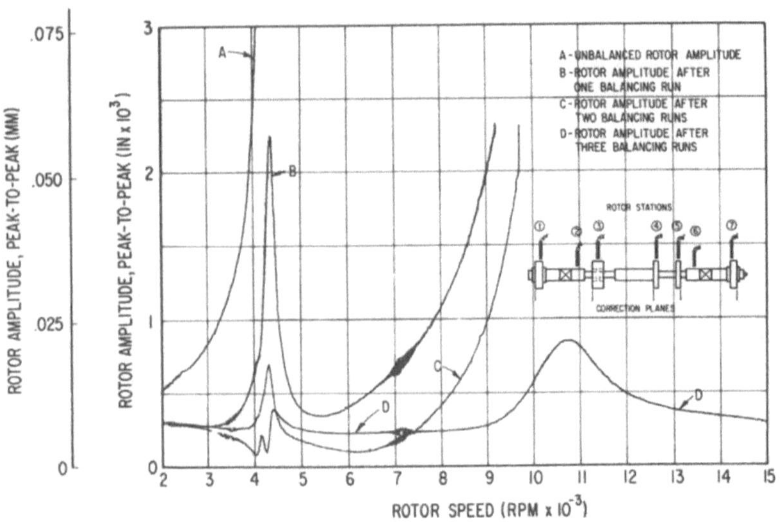

A - UNBALANCED ROTOR AMPLITUDE
B - ROTOR AMPLITUDE AFTER ONE BALANCING RUN
C - ROTOR AMPLITUDE AFTER TWO BALANCING RUNS
D - ROTOR AMPLITUDE AFTER THREE BALANCING RUNS
ROTOR STATIONS
CORRECTION PLANES
ROTOR AMPLITUDE, PEAK-TO-PEAK (MM)
ROTOR AMPLITUDE, PEAK-TO-PEAK (IN x 10^{3})
ROTOR SPEED (RPM x 10^{-3})

Fig. 2

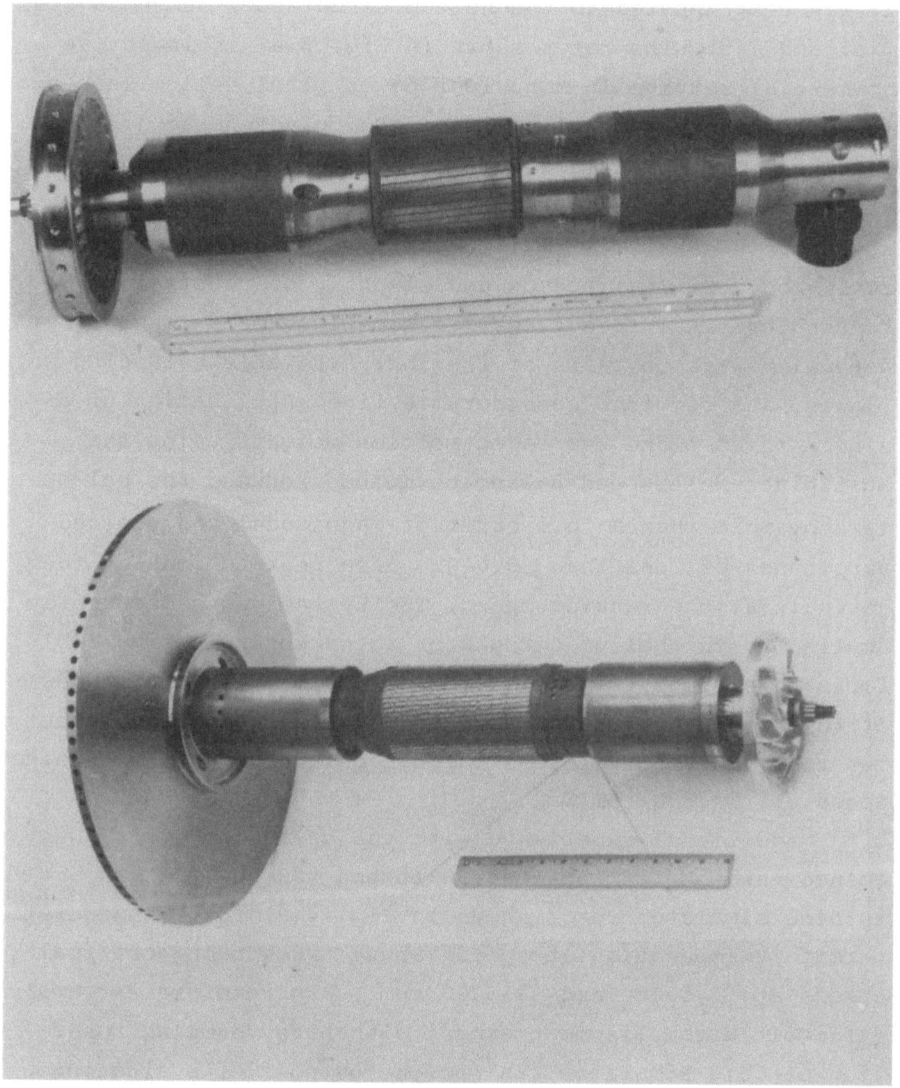

Fig. 3

To gain further experience with the balancing system, it was next applied to the helium circulators shown in Figs. 3(a) and 3(b). The rotor shown in Fig. 3(a) is in active commercial service in a nuclear power plant helium loop and is supported on two helium-lubricated pivoted-pad journal bearings. The shaft is driven by a centrally-mounted armature, and carries a single wheel. Normal operating speed of the shaft is 25,000 rpm, which lies above two rigid-body critical speeds and slightly below its first bending critical speed. Earlier two-plane balancing was done with various pairs of the four available balancing planes, and required considerable time. Application of a single three-plane set of correction weights using the Multiplane-Multispeed Balancing method reduced the balancing time by more than a full order of magnitude, and yielded amplitudes of less than 0.0001 inch peak-to-peak at the journals at the running speed. The system was subsequently applied to the helium circulator rotor shown in Fig. 3(b) for single-plane trim balancing. The vibration amplitudes of this rigid-body rotor were reduced by a factor of about two to satisfactory levels through balancing at the running speed of 14,500 rpm.

Additional experience with the balancing system was gained through its application to the 30,000 rpm gas turbine simulator rotor shown in Fig. 4(a) [8], in order to permit its operation at speeds above two bending critical speeds at 7,5000 and 21,000 rpm, a heretofore extremely difficult accomplishment except with heavy damping. Operation of this simulator was for the purpose of evaluating the Hydresil™ intershaft gas-lubricated foil bearing. The key to a meaningful comparison of dynamic behavior with and without the intershaft foil bearing was the ability of the simulator to produce repeatable results. This, in turn, required that once achieved, the simulator's balance never be disturbed. This, together with the requirement for reasonable amplitudes over the speed range of the simulator,

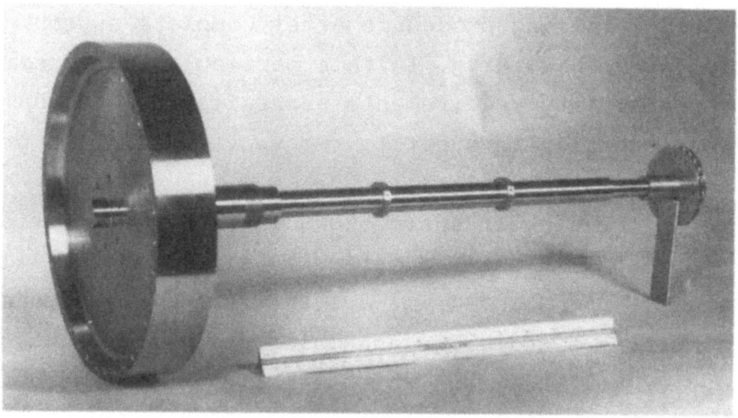

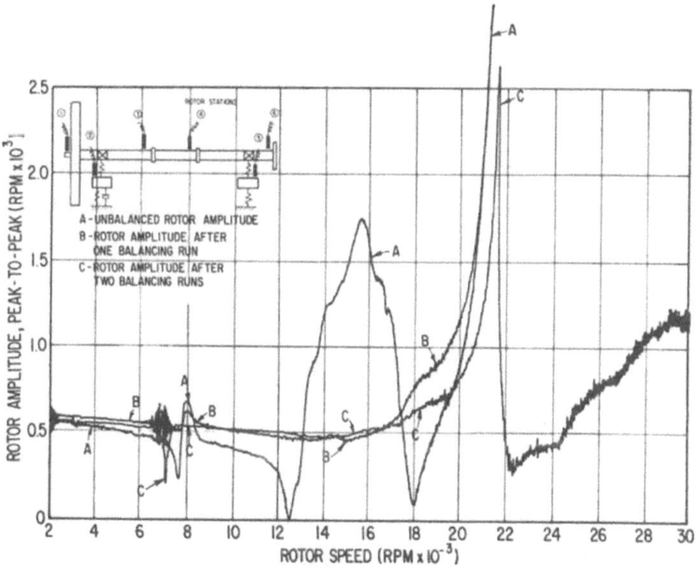

ROTOR STATIONS
A - UNBALANCED ROTOR AMPLITUDE
B - ROTOR AMPLITUDE AFTER ONE BALANCING RUN
C - ROTOR AMPLITUDE AFTER TWO BALANCING RUNS
ROTOR AMPLITUDE, PEAK-TO-PEAK (RPM x 10^3)
ROTOR SPEED (RPM x 10^-3)
0
0.5
1.0
1.5
2.0
2.5
2 4 6 8 10 12 14 16 18 20 22 24 26 28 30
A
B
C

Fig. 4

was easily achieved through the use of the Multiplane-Multispeed Balancing procedure after complete assembly, with no low-speed balancing. In this particular case, simulator repeatability requirements necessitated a "no-damping" configuration in order to remove variabilities produced by damper fluid heating.

The slender shaft shown in Fig. 4(a) carries a massive overhung disk which dynamically simulates two axial turbine stages, and a small air-drive turbine which dynamically simulates a gear. The assembled unit is supported by two duplex ball bearings which are flexibly-mounted so as to properly simulate engine structural impedance. An oil squeeze-film damper is located outboard of the bearing on the disk end. Five balance planes are provided, two on the disk, two on the shaft (raised areas) and one on the drive turbine. Excursions of the shaft are monitored by inductance proximity probes in four planes. Excursions of the flexibly-supported bearing housings are monitored by inductance probes in two locations.

Results of this balancing are shown in Fig. 4(b). It can be seen that one five-plane balancing step completely suppressed the first critical speed and an apparent disk-end pedestal resonance at about 15,000 rpm, and that the second five-plane correction proved adequate for the rotor to pass through the second critical speed, which was allowed to remain for purposes of the foil-bearing tests.

Present basic technology efforts [3] have recently demonstrated the ability of the balancing system to balance the rotor shown in Fig. 2(a) with its bearing housings supported in flexures having dissimilar orthogonal stiffnesses. This condition, which produces distinctly elliptical orbits, was treated without difficulty by the system. Additional technology-level efforts now underway include (1) balancing of a rotor having dissimilar bending stiffness in mutually orthogonal planes; (2) balancing with sensor mount resonances; and (3) preliminary efforts directed at identification of methods suitable for optimization of machinery during the design process to fully utilize Multiplane-Multispeed Balancing.

In addition to the foregoing technology efforts, the Multispeed-Multiplane Balancing System has also entered the application technology phase, and is now being evaluated for application to several new and current-inventory U.S. Army and civilian aircraft turbine engines. These programs are directed at achievement of cost benefits, as well as performance improvements.

In addition to these turbine engine applications, the Balancing System has recently been applied for _in situ_ balancing of a remotely-located rotor-bearing system. This system included both a driver and a driven unit, which were connected by a flexible coupling. Because of the time element and of the location of the rotor system, vibration data recorded on-site was transmitted to the Author's company via telephone and used to calculate a three-plane, single-speed correction, which was similarly returned. Total turnaround time was less than one hour, and the single correction produced a satisfactory balance. No trial and error was involved.

CURRENT APPLICATIONS

The capabilities demonstrated to date by the Multiplane-Multispeed Balancing System have led to its preliminaty integration into the rotating system design procedure by several organizations.

Analytical Response-Balancing Optimization. Selection of an optimum rotor-bearing configuration from the balancing standpoint can be readily performed through the use of existing, integrated computer program systems for response-balancing calculations. Such calculations can be used to evaluate the penalties associated with off-optimum design, such as location of sensors or balancing planes at points which are not ideal. In effect, the analytical tools simulate the balancing procedure and permit the designer to identify the most effective locations for sensors and balancing planes.

Application Techniques. Several alternative procedures exist for applying Multiplane-Multispeed Balancing to the needs of particular rotating systems:

(1) Adaptation to an existing balancing apparatus or machine, to improve the accuracy or precision of the process through the use of additional sensors and new, precise vibration-signal-handling electronics. This adaptation may produce a more rapid one-plane or two-plane balance, as well as a better balance where more than two planes are required;

(2) Use of a customized, low-cost fixture (which may range from simple mounts to an engine shell) which provides for representative bearings, good access to the rotor, and a practical drive method;

(3) Addition of a multiplane trim-balance procedure which can be applied in a test area to provide an alternative to lengthy trial-and-error sequences and rotor-bearing system disassembly;

(4) Addition of a field-portable multiplane trim-balance procedure which can be used to correct balance degradations without removing the rotor-bearing system from its working environment.

These options span the entire range of options open to the balancing engineer, and give him the maximum flexibility required for elimination of certain manufacturing and assembly steps and reduction of costs. Each, within certain limits, is capable of achieving good performance (low vibration levels). Each has its own set of hardware and operational requirements and restrictions.

The major concept common to each of these options is that provision for balancing must be considered as an integral part of the design process, rather than as something which is added as an afterthought or relegated to the manufacturing operation. The designer must incorporate balancing provisions into the design, thereby achieving a totally-integrated rotor-bearing system from the manufacturing, assembly, maintenance, and overhaul standpoints.

CONCLUSIONS

Several important conclusions may be drawn concerning Multiplane-Multispeed Balancing as a result of its applications to date:

1. Balance corrections may be made simultaneously in virtually any number of planes. A single set of correction weights satisfies at once the requirement for low vibration levels at each of a number of sensor locations and at each of a number of speeds.
2. Either existing or new vibration sensors may be used, depending upon the application. The sensors need not be all of the same type.
3. Corrections may be made in any or all accessible balancing planes, depending upon the application.
4. The balancing process need know nothing about the rotor's construction, and thus may be applied with equal effectiveness to rotor of any size.
5. Multiplane-Multispeed Balancing offers viable alternatives to the traditional approach to rotor bearing system manufacture and assembly. These alternatives include, but

are not limited to: provision of additional balancing planes on the rotor to take advantage of the ability to make axially-distributed balance corrections; designing the rotor-bearing system as a module which can be multiplane-balanced at elevated speeds in an inexpensive fixture after final assembly; providing access to the rotor in its final installation for multiplane trim balancing; and relaxing balance-related fabrication tolerances in anticipation of final-assembly balancing. These several alternatives may yield significant cost and performance benefits for hardware ranging in size from the small turbocharger to the large gas turbine used for electric power production or naval ship propulsion.

6. Multiplane-Multispeed Balancing may be implemented in a number of ways and at a number of points in the manufacturing-assembly process. In effect, the balancing process need no longer be relegated solely to the final step in manufacturing. Not only can the multiplane technique provide a better, more rapid balance in a traditional balancing machine for components which require more than two balancing planes, but the balancing operation may now be applied with equal assurance of success during or after subsystem assembly and during or after final installation of the rotating components.

7. Multiplane-Multispeed Balancing provides the designer with a firm base from which he may advance with assurance of success into the flexible rotor dynamic regime. The existence of a proven, practical method for balancing a rotor simultaneously in multiple planes permits rotor speeds to be deliberately specified above bending critical speeds.

ACKNOWLEDGEMENTS - The capabilities of the Multiplane-Multispeed Balancing System have been demonstrated experimentally by Mr. Juergen M. Tessarzik, Senior Development Engineer, with the assistance of Mr. Walter Spodnewski, Dynamics Test Specialist, and Mrs. Frieda Gillham, Engineering Analyst, all of MTI. Major portions of these efforts were performed under National Aeronautics and Space Administration Contract NAS3-14420 under the technical cognizance of Mr. William J. Anderson, Chief, and Dr. David Fleming, Project Manager, Bearings and Mechanical Power Transfer Branch, NASA-Lewis Research Center, Cleveland, Ohio. Additional portions of the efforts described herein were performed under U.S. Air Force Contract F33615-72-C-1801 under the technical cognizance of Mr. Everett A. Lake, Project Engineer, Air Force Aero Propulsion Laboratory, Wright-Patterson Air Force Base, Dayton, Ohio.

REFERENCES

[1] BADGLEY, R.H.: "The Potential Impact of Multiplane-Multispeed Balancing on Gas Turbine Production and Overhaul Costs", ASME Paper No. 74-GT-94.

[2] BADGLEY, R.H.: "Implications of Multiplane-Multispeed Balancing for Future Turbine Engine Design and Cost", Submitted for presentation at the 1974 Aerospace Engineering & Manufacturing Meeting of the SAE, September 30 - October 4, 1974, San Diego, California.

[3] "Development of High-Speed Flexible Rotor and Shafting Balancing Techniques", National Aeronautics and Space Administration Contract NAS3-14420 with Mechanical Technology Incorporated, Latham, New York, March 1971.

[4] "Balancing Techniques for High-Speed Flexible Rotors", National Aeronautics and Space Administration Contract NAS3-18520 with Mechanical Technology Incorporated, Latham, New York, May 1974.

[5] TESSARZIK, J.M., BADGLEY, R.H., and ANDERSON, W.J.: "Flexible Rotor Balancing by the Exact Point-Speed Influence Coefficient Method", ASME Paper No. 71-Vibr-91.

[6] LUND, J.W., and ORCUTT, F.K.: "Calculations and Experiments on the Unbalance Response of a Flexible Rotor", ASME Paper No. 67-Vibr-27.

[7] TESSARZIK, J.M., and BADGLEY, R.H.: "Experimental Evaluation of the Exact Point-Speed and Least Squares Procedures for Flexible Rotor Balancing by the Influence Coefficient Method", ASME Paper No. 73-DET-115.

[8] CUNDIFF, R.A., BADGLEY, R.H., and REDDECLIFF, J.: "Pneumomechanical Critical Speed Control for Gas Turbine Engine Shafts", Air Force Aero Propulsion Laboratory Technical Report 73-101, November 1973. Prepared by Mechanical Technology Incorporated under USAF Contract F33615-72-C-1801.

Dynamic Properties of Turboset Rotors

Miroslav Balda

Research Institute of the SKODA Works
C.S.S.R.

1. INTRODUCTION

Problems of rotor dynamics are known to be complex problems and by no means the problems of rotors alone, and that is why they were approached as such in the SKODA Works. Although in complete generality the phenomena to be dealt with are nonlinear, linear procedures have been found to work to satisfaction for turboset rotors provided the processes being studied are taking place at low vibration amplitudes.

2. ALLIED PROBLEMS OF TURBOSET ROTOR DYNAMICS

Before we turn to the dynamic properties of turboset rotors, let us note some of the problems which - though seemingly only marginally connected with rotor dynamics - must be well under control before any attempt is made at its successful solution.

2.1. Effect of stiffness and damping of supports on resonant vibration. It is generally known that the finite stiffness of supports reduces the critical speed of rotors. That it is so is also proved by the dependence of the root λ of the frequency equation of the uniform homogeneous shaft mounted in two bearings with different stiffnesses, K_1 and K_2 according to Fig. 1 where

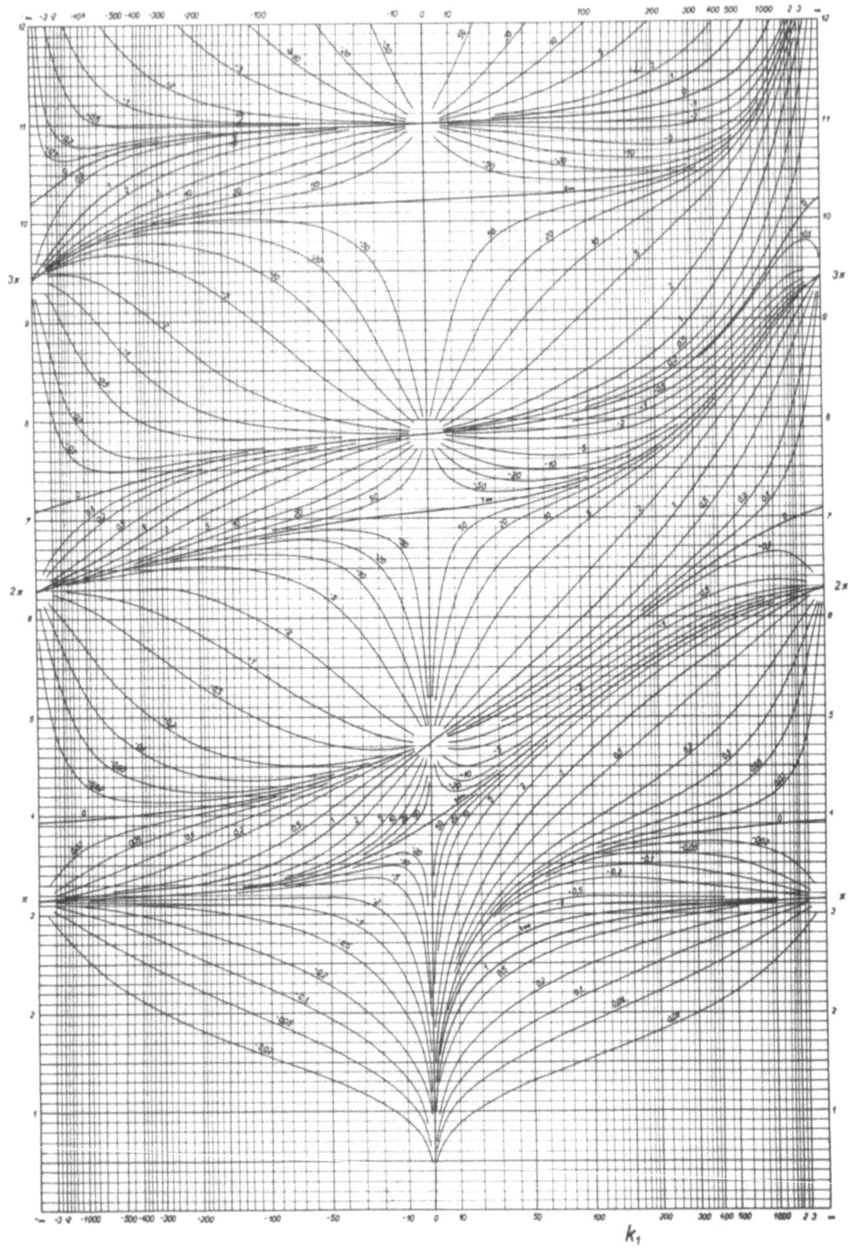

Fig. 1. Root λ of the frequency equation of a uniform shaft supported in two bearings, as a function of dimensionless stiffness of one support, $k_1 = K_1 \ell^3/(EJ)$, and of the ratio of stiffnesses of supports, $p = K_2/K_1$. The natural frequency of the shaft is given by the expression

$$\Omega = \frac{\lambda^2}{\ell^2}\sqrt{\frac{EJ}{\rho A}}$$

$$k_1 = \lambda^3 \frac{(1+p)B \pm \sqrt{(1+p)^2 B^2 + 4pS_1 D}}{2pS_1} \quad ;$$

$$k_1 = K_1 \ell^3 / EJ$$

$$p = K_2 / K_1$$

$$B = \cosh\lambda \sin\lambda - \sinh\lambda \cos\lambda$$

$$D = \cosh\lambda \cos\lambda - 1$$

$$S_1 = 2\sinh\lambda \sin\lambda$$

The effect of bearing damping is of equal import. As pointed out earlier by SKYBA the effect of support damping on the magnitude of resonant vibration is not of monotonic character. It can be shown that the resonant vibration of a rotor with stiffness $C = m\Omega_\infty^2$ mounted in two damped bearings of complex stiffness $Z = K + iH$, with mass m at the centre of the span reaches its minimum for the ratio $\zeta = (H/K)^2$ equal to

$$\zeta_m = 1 + \frac{C}{2K}$$

These two parameters enable us to establish also the relations defining the modulus of relative response, y, of mass m in the middle of the rotor, to unbalance μr in the same location, known as the Q-factor

$$Q = \left|\frac{my}{\mu r}\right| = \sqrt{1 + \frac{(\zeta_m + \zeta)^2}{\zeta(\zeta_m - 1)^2}}$$

and the relative frequency at which the vibration amplitudes reach their maximum

$$\left(\frac{\omega}{\Omega_\infty}\right)^2 = \frac{\zeta_m + \zeta}{\zeta_m^2 + \zeta}\left[1 + \frac{\zeta(\zeta_m - 1)^2}{(\zeta_m + 1)^2}\right]$$

In the above Ω_∞ is the frequency of the shaft considered mounted on absolutely rigid supports, i.e. $\Omega_\infty = \sqrt{C/m}$.

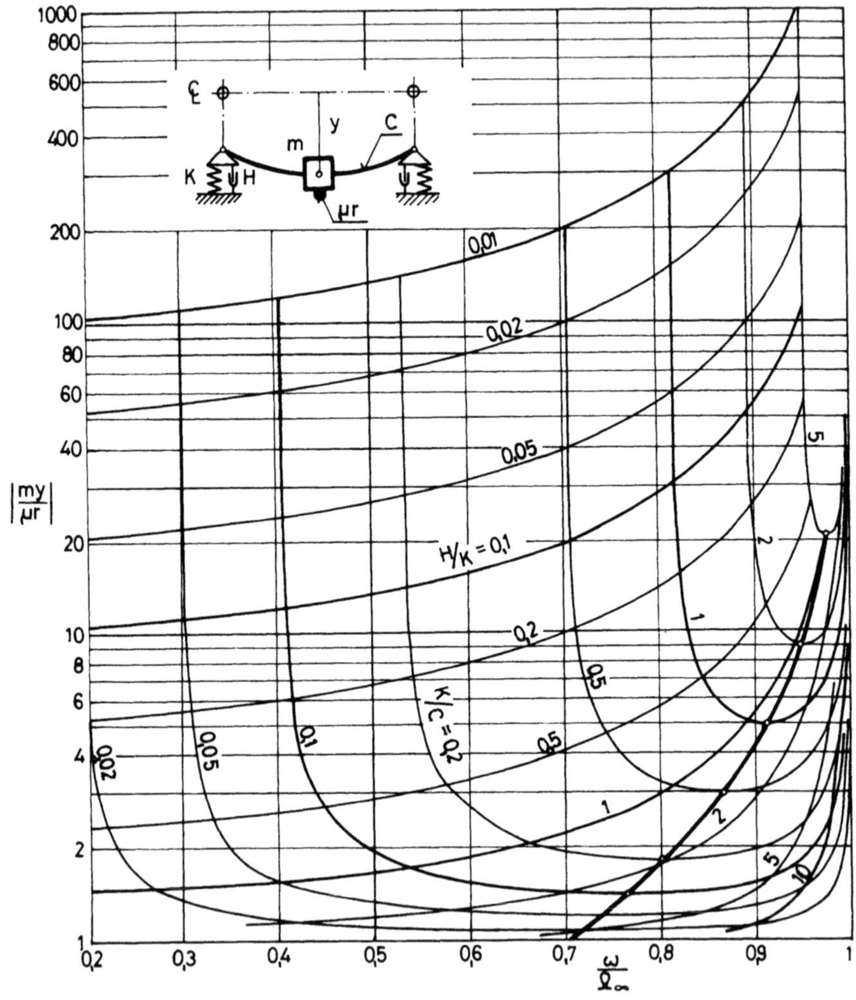

Fig. 2. Dependence of Q-factor and ratio between actual critical speed ω and critical speed on rigid supports Ω_∞ , on ratios H/K and K/C .

In Fig. 2 these relationships are summarized in a diagram showing the effect of stiffness as well as of damping on the peak response of the rotor, and the frequency at which this response occurs and the magnitude of the optimum ratio between damping and stiffness of supports.

Too small a damping and too large a damping are both equally dangerous. For the optimum stiffness-damping ratio, on the other hand, the relative response of peak vibration turns out to be minimum, i.e.

$$Q_m = \left|\frac{my}{\mu r}\right| = \frac{\zeta_m + 1}{\zeta_m - 1} \quad \text{for} \quad \frac{\omega_m}{\Omega_\infty} = \sqrt{\frac{\zeta_m + 1}{2\zeta_m}}$$

It is clear from this simple example the significance of a thorough knowledge of the dynamic properties of the supporting structure of rotor mounting.

<u>2.2. Dynamic properties of supporting structure</u>. In the simpliest case when the individual supports - bearing stands and housings - are mounted on a rigid foundation, the deflections of the r-th bearing in normal space (Fig. 3)

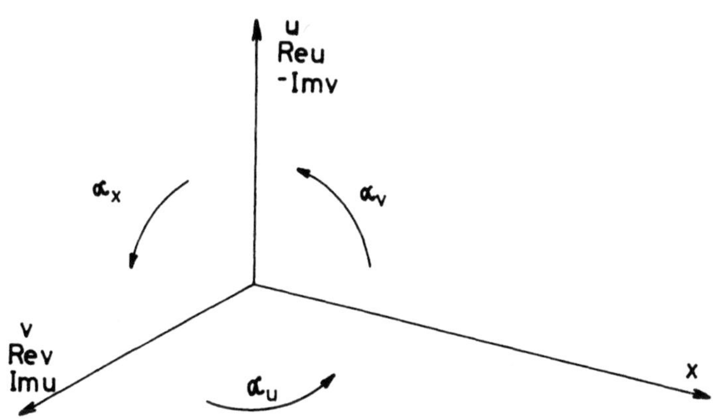

Fig. 3.

are defined by the relation

(2.1) $$q_r = G^s_{rr} Q_r \quad , \quad r = 1,\ldots,R$$

where

(2.2) $$q_r = \left[\alpha_u \, , \alpha_r \, , \alpha_x \, , u \, , v \, , x\right]^T_r$$

is the vector of generalized deflections,

(2.3) $$Q_r = \left[M_u \, , M_v \, , M_x \, , F_u \, , F_v \, , F_x\right]^T_r = [Q_j]_r$$

the vector of generalized forces, and G^s_{rr} a square matrix with 6×6 elements - complex flexibilities - receptances.

For the considered system of isolated supports, all deflections in dependence on excitation may be described by the equation

(2.4) $$\begin{bmatrix} q_1 \\ q_2 \\ \vdots \\ q_R \end{bmatrix} = \text{diag}[G_{11} \, , G_{22} \, , \ldots \, , G_{RR}]^s \begin{bmatrix} Q_1 \\ Q_2 \\ \vdots \\ Q_R \end{bmatrix}$$

or in shorthand notation

(2.5) $$q = G^s Q$$

In the general case of supports mutually coupled across a common elastic damped foundation, the matrix G^s is no longer quasidiagonal. In general all of its off-diagonal submatrices G^s_{qr} are non-zero and frequency dependent. The determination of the elements of matrix G^s can be done in two ways, namely by calculation or by in-situ experiments in the power station.

The lubricating oil film in journal bearings, though of virtually negligible dimensions, exercises a dominant effect on rotor dynamics by its action. It is usually the source not only of a major portion of mounting flexibility but of damping of the whole system as well. It is therefore only natural that we must deal also with the properties of the oil film.

The first information published in the fifties by HAGG and SANKEY gave impetus to our own research in this field. Our first experiments made on a small test equipment with a 75-mm dia.horizontal journal, which we called OF1 , were followed by those conducted on equipment OF2 (160mm dia.-vertical) and OF3 (430mm dia.-horizontal).

The linearized properties of journal bearing oil film are defined by the frequency-dependent matrix G^F_{rr} of complex dynamic flexibilities, having the same structure as the compliance matrix of the mechanical support. They will, therefore, be evidently amenable to investigation by similar methods, namely by successive excitation of the object by harmonic forces. That is why it was used already on the earliest equipment OF1 .

As is well known, the geometries being equal, the working condition of a bearing is characterized by the dimensionless parameter - the Sommerfeld number

$$S_o = \frac{Q}{LD} \cdot \frac{(\Delta D/D)^2}{\eta\omega} \tag{2.6}$$

where are Q - reaction to journal , L , D - length and diameter of bushing, η - dynamic viscosity of oil, and ω - angular frequency of journal rotation.

On this parameter depend both the static and the dynamic properties of oil film.

Since an analysis has revealed the elements of the compliance matrix corresponding to rotational motion (tilting of the journal) to be of no special significance for short bearings, the tests were restricted, as in the case of supports, to four elements of matrix G^F_{rr} corresponding to transverse displacements and excitation, that

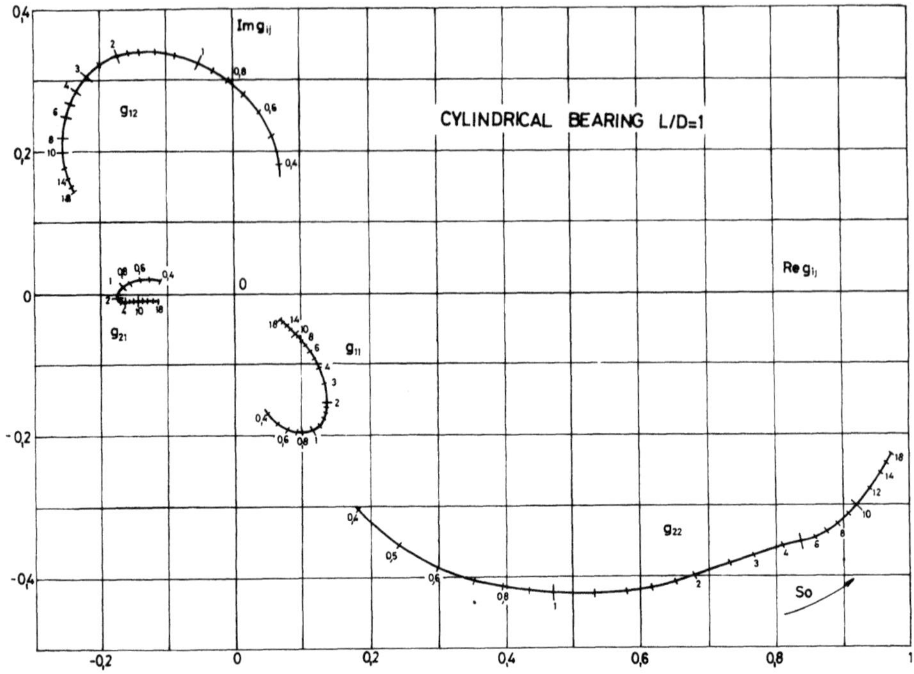

Fig. 4. Dimensionless complex flexibilities of oil film of a cylindrical (L/D = 1) journal bearing as a function of the Sommerfeld number. Subscripts i and j refer to axes u and v .

is to say, to the u- and v-th elements of submatrix $[G_{yF}]^F_{rr}$. Fig. 4 shows one of the many diagrams plotting the complex dimensionless elements g_{ij} of the oil film compliance matrix as functions of So for a cylindrical bearing with the L/D ratio equal to 1 (the limiting case of an elliptical bearing with $\Delta D_u/\Delta D_v = 1$) . Using these we can set up a simplified matrix of the oil film, $\hat{G}^F$, corresponding only to the displacements in the direction of axes u and v of the form

$$(2.7) \qquad \hat{G}^F = \frac{\Delta D}{2Q}[\mathrm{Re}\, g + i\, \mathrm{Im}\, g]$$

Consulting the figure we see that $g_{12} \neq g_{21}$ and that therefore matrix $\hat{G}^F$ is a skew matrix. As it is well known this fact is apt to lead to unstable running of the rotor. So long as the frequency of journal whirl Ω in the bearing differs from the working frequency ω , matrix $\hat{G}^F$ must be corrected to

$$(2.8) \qquad G^F = \left[\mathrm{Re}\,(\hat{G}^F)^{-1} + i\,\frac{\Omega}{\omega}\,\mathrm{Im}\,(\hat{G}^F)^{-1} \right]^{-1}$$

In calculations one usually operates with the stiffness matrix

$$Z = \left[G^S + G^F \right]^{-1}$$

3. LINEAR BENDING VIBRATION OF ROTORS IN SPACE

If a rotor is placed in the fixed Cartesian coordinate system u, v, x (Fig. 3) in a way that makes its longitudinal centreline coincides with axis x , the motion of rotor elements transverse to axis x can be described by a pair of complex coupled partial differential equations which have been reported in literature. In order that we might obtain a ready solution to the problem yet not make it markedly affected by flagrantly neglecting the secondary effects, we introduce the following mathematical model of the rotor:

a) the rotor is composed of prismatic homogeneous segments

b) changes in cross-sectional and material quantities occur solely on the interfaces between the sections

c) all additional effects such as
   - the effect of the surroundings (supports, medium) on the rotor
   - the gyroscopic effect of the sections as well as of added parts
   - the effect of shear force on the deflection
   - the effect of point mass
   - external disturbances - excitation

act on the interfaces between the sections.

This hybrid rotor model is advantageous not only in that it closely resembles actual rotors but also in that its deformations are described by pairs of very simple differential equations

(3.1)
$$u^{(IV)} - \lambda^4 u = f_u(x)$$
$$v^{(IV)} - \lambda^4 v = f_v(x)$$
$$\lambda^4 = \frac{\rho A}{EJ}\Omega^2$$

We are looking for the solution $y = (u, v)^T$ in the form of the sum of solutions to a homogeneous and a nonhomogeneous problem

(3.2)
$$y = y^H + y^N$$

All quantities may be complex.

Since the equations between the interfaces of section i are homogeneous, we know that each solution is of the form

(3.3)
$$y_i^{HT} = [u, v]_i = [e^+, e^-, c, s]_i [a_u, a_v]_i^H$$

where

a_{ui}^H, a_{vi}^H - vectors of the integration constants of the solution to the homogeneous problem, and

e^+, e^-, c, s - symbols of functions $e^{\lambda x}$, $e^{-\lambda x}$, $\cos\lambda x$ and $\sin\lambda x$.

On denoting the derivatives of the deflections divided by the respective power by

$$U_{j+1,i} = u_i^{(j)}/\lambda_i^j \quad \text{and} \quad V_{j+1,i} = v_i^{(j)}/\lambda_i^j \quad ; \quad j = 1, 2, 3, 4$$

we can use them in the setting-up of the transformed state vector

$$(3.4) \qquad y_{x,i} = [u^T, v^T]^T = A_{xi} a_i^H = \begin{bmatrix} H, & 0 \\ 0, & H \end{bmatrix}_{xi} \begin{bmatrix} a_u \\ a_v \end{bmatrix}_i^H$$

where

$$(3.5) \qquad H_{xi} = \begin{bmatrix} e^+, & e^-, & c, & s \\ e^+, & -e^-, & -s, & c \\ e+, & e^-, & -c, & -s \\ e^+, & -e^-, & s, & -c \end{bmatrix}_{xi}$$

is the matrix of the functions of the solution. The vector of the integration constants is established from the initial state, using equation (3.4) for $x = 0$, viz.

$$(3.6) \qquad a_i = A_o^{-1} y_{o,i}$$

By joining the last equation with equation (3.4) we can deduce that the states at the end of rotor will be

$$(3.7) \qquad y_{\ell i} = A_{xi} A_o^{-1} y_{o,i} = B_{Hi} y_{o,i}$$

Matrix B_{Hi} is the transfer matrix between the states at the beginning and the end of a homogeneous segment of the rotor. The transformed states at the beginning of the next section, $i+1$ can be expressed as follows:

$$(3.8) \qquad {}_o y_{i+1} = T_{i+1}^{-1} B_{z,i} B_{TI,i} T_i y_{\ell,i}$$

This expression includes the backward transition to the physical quantities of deformation, angular rotation, moment and shear forces, which is effected by the matrix

$$(3.9) \qquad T_i = \mathrm{diag}[1\ ,\ \lambda\ ,\ \lambda^2 EJ\ ,\ \lambda^3 EJ\ ,\ 1\ ,\ \lambda\ ,\ \lambda^2 EJ\ ,\ \lambda^3 EJ]_i$$

The transfer across the gyrational effect and the effect of the shear force is concentrated in the respective transfer matrix

$$(3.10) \qquad B_{TI,i} = \begin{bmatrix} 1 & & & -\frac{\ell}{A_s G} & & & & \\ & 1 & & & & & & \\ & -\Omega^2 I_\alpha & 1 & & & i\omega\Omega I_p & & \\ & & & 1 & & & & \\ & & & & 1 & & & -\frac{\ell}{A_s G} \\ & & & & & 1 & & \\ & -i\omega\Omega I_p & & & & -\Omega^2 I_\alpha & 1 & \\ & & & & & & & 1 \end{bmatrix}$$

The transition across the point mass, the r-th support or the effect of the surroundings is ensured by the matrix

$$(3.11) \qquad B_{z,i\to r} = \begin{bmatrix} 1 & & & & & & & \\ & 1 & & & & & & \\ z_{24} & z_{22} & 1 & & z_{25} & -z_{21} & & \\ m\Omega^2 - z_{44} & -z_{42} & & 1 & -z_{45} & z_{41} & & \\ & & & & 1 & & & \\ & & & & & 1 & & \\ -z_{14} & -z_{12} & & & -z_{15} & z_{11} & 1 & \\ -z_{54} & -z_{52} & & & m\Omega^2 - z_{55} & z_{51} & & 1 \end{bmatrix}_{i\to r}$$

This form of transfer matrix has been found convenient for the purpose of respecting all the effects of the oil film on journals, as well as of steam on rotor wheels and labyrinth seals. In this case elements Z_{jk} are the elements of the quasidiagonal matrix G_{rr}^{-1} to which we referred earlier. Matrix T_{i+1}^{-1} produces the backward transition from the physical quantities to the transformed states.

By combining equations (3.7) and (3.8) we get

$$ {}_{o}y_{i+1} = B_i y_{o,i} \tag{3.12} $$

where

$$ B_i = T_{i+1}^{-1} B_{z,i} B_{TI,i} T_i B_{Hi} $$

is the transfer matrix of section i with interface on its end.

Method of transfer matrices. It is assumed that $y_{o,i+1} = {}_{o}y_{i+1}$ at the beginning of each new section so that at the end of the rotor (at the beginning of a fictitious section added to the end of the rotor) it is

$$ y^H_{o,n+1} = \prod_{i=1}^{n} B_i y^H_{o,1} = B y^H_{o,1} \tag{3.13} $$

From the resultant transfer matrix B and the state vector from excitation $y^N_{o,n+1}$ calculated parallel to it, we choose a system of four equations for four end conditions and determine the four unknown initial elements of vector $y^H_{o,1}$.

Method of linear combination. This is also a method of transfer but the initial vector $y^H_{o,1}$ is chosen in the form of a linear combination of four independent vectors $y^{(j)}_{o,1}$ (j = I , II , III , IV) . Then it follows from equation (3.13) that

$$(3.14)\qquad y^{H}_{o,n+1} = \prod_{i=1}^{n} B_i [y^{(j)}]_{o,1} [\alpha^{(j)}]$$

This method is roughly twice as fast as the method of transfer matrices.

New method. Laments over ill-conditionality of the numerical process of the transfer methods have already appeared many a time in professional literature. It can be proved by an analysis of the problem that the source of difficulties lies in function $e^{\lambda x}$ which - as its argument grows - causes the effect of the remaining functions of the solution to be rapidly suppressed.

A new method which does away with these difficulties proceeds from the method of linear combination, but in contradistinction to it let us prescribe that the vectors of the integration constants, as the columns of the matrix, should be of the form

$$(3.15)\quad [a^{(j)}]_i = [a^{I}, a^{II}, a^{III}, a^{IV}]_i = \begin{bmatrix} 1 & 0 & 0 & 0 \\ a_2^{I} & a_2^{II} & a_2^{III} & a_2^{IV} \\ a_3^{I} & a_3^{II} & a_3^{III} & a_3^{IV} \\ a_4^{I} & a_4^{II} & a_4^{III} & a_4^{IV} \\ 0 & 1 & 0 & 0 \\ a_6^{I} & a_6^{II} & a_6^{III} & a_6^{IV} \\ a_7^{I} & a_7^{II} & a_7^{III} & a_7^{IV} \\ a_8^{I} & a_8^{II} & a_8^{III} & a_8^{IV} \end{bmatrix}$$

This will ensure that the partial vectors y^{I} and y^{II} determined from equation (3.4) will contain the contributions of the positive exponential from a given section alone while y^{III} and y^{IV} will be free of it. Since the resultant vector of the integration constants is a linear combination of partial vectors, it will be of the form

$$(3.16)\quad \alpha_i^H = \left[\alpha^{(j)}\right]_i \left[\alpha^{(j)}\right]_i = \left[\alpha^I , a_2 , a_3 , a_4 , \alpha^{II} , a_6 , a_7 , a_8\right]_i^{HT}$$

At the beginning of the (i+1) - th section there must hold the equivalence between the resultant solution ${}_oy_{i+1}^H$ derived from the preceding section, and the solution $y_{o,i+1}^H$ initial for the subsequent solution, i.e.

$$(3.17)\quad {}_o\left[y^{(j)}\right]_{i+1}\left[\alpha^{(j)}\right]_i = \left[y^{(j)}\right]_{o,i+1}\left[\alpha^{(j)}\right]_{i+1} = A_o\alpha_{i+1}^H$$

and from there

$$(3.18)\quad \alpha_{i+1}^H = A_{oo}^{-1}\left[y^{(j)}\right]_{i+1}\left[\alpha^{(j)}\right]_i$$

From the first and the fifth equation of (3.18) and from a comparison with (3.16) it follows after some manipulations that the relation between the vectors of the constants of the linear combination $\alpha = [\alpha^{AT} , \alpha^{BT}]^T = [\alpha^{(j)}]$ in sections i and i+1 is

$$(3.19)\quad \alpha_i = \begin{bmatrix}\alpha^A \\ \alpha^B\end{bmatrix}_i = \begin{bmatrix} E^A , & E^B \\ 0 , & I\end{bmatrix}_{i+1} \begin{bmatrix}\alpha^A \\ \alpha^B\end{bmatrix}_{i+1} = E_{i+1}\alpha_{i+1}$$

where

$$\alpha_i^A = \left[\alpha^I , \alpha^{II}\right]_i^T \quad ; \quad \alpha_i^B = \left[\alpha^{III} , \alpha^{IV}\right]_i^T$$

and

$$E_{i+1}^A = 4\ {}_o\begin{bmatrix} S_u^I , & S_u^{II} \\ S_v^I , & S_v^{II}\end{bmatrix}_{i+1} \quad \text{and} \quad E_{i+1}^B = -\ E_{i+1}^A\ {}_o\begin{bmatrix} S_u^{III} , & S_u^{IV} \\ S_v^{III} , & S_v^{IV}\end{bmatrix}_{i+1}$$

Symbol ${}_oS_{u,i+1}^{(j)} = \sum_{k=1}^{4} {}_oU_{k,i+1}^{(j)}$ and ${}_oS_{v,i+1}^{(j)} = \sum_{k=1}^{4} {}_oV_{k,i+1}^{(j)}$

Thus in contradistinction to the method of linear combination, in which the coefficients $[\alpha^{(j)}]$ continue to be constant across the whole rotor, here only the elements α^B are constant whereas α^A is a function of the sections.

The substitution of expression (3.19) into (3.17) gives the relation for the new initial vector in section i+1 , viz.

$$[y^{(j)}]_{o,i+1} = {}_o[y^{(j)}]_{i+1}E_{i+1} \tag{3.20}$$

This product will cause a re-distribution of information between the vectors and "clearing" of vectors y^{III} and y^{IV} from the average element. Let us also express the whole transition from the beginning of the section:

$$[y^{(j)}]_{o,i+1} = B_i[y^{(j)}]_{o,i}E_{i+1} \tag{3.21}$$

Evidently, the complete transition across a section with an interface will be effected by multiplying the matrix of initial vectors $[y^{(j)}]_{o,i}$ by the transfer matrix B_i from the left and then by the "stabilization" matrix E_{i+1} from the right.

The particular solution y^N from external disturbances will be found in the conventional manner, as in the method of transfer matrices. From the recurrent relation (3.19) there is then set up - with the aid of the end conditions F^A_{n+1} , F^B_{n+1} and F^N_{n+1} - the following system of equations for the sought α_i :

$$
(3.22)\qquad
\left[\begin{array}{cccccc|cc}
-I & E_2^A & & & & & & E_2^B \\
 & -I & E_3^A & & & & & E_3^B \\
 & & \ddots & \ddots & & & & \vdots \\
 & & & -I & E_{i+1}^A & & & E_{i+1}^B \\
 & & & & \ddots & \ddots & & \vdots \\
 & & & & & -I & E_{n+1}^A & E_{n+1}^B \\
\hline
 & & & & & & F_{n+1}^A & F_{n+1}^B \\
 & & & & & & &
\end{array}\right]
\left[\begin{array}{c}
\alpha_1^A \\ \alpha_2^A \\ \vdots \\ \alpha_i^A \\ \vdots \\ \alpha_n^A \\ \alpha_{n+1}^A \\ \alpha^B
\end{array}\right]
=
\left[\begin{array}{c}
\\ \\ \\ 0 \\ \\ \\
\hline
-F_{n+1}^N \\
\end{array}\right]
$$

Unlike the methods referred to in the foregoing, our new method yields absolutely stable results, and this fact can be proved both analytically and numerically.

Using the similar approach, the method has been extended even to the case of coupled supports.

The calculation of *forced vibration* proceeds exactly in accordance with the analysis just presented. The excitation is assumed to be capable of acting anywhere. It has an origin in production technology and is introduced into the calculation in the form of discontinuities in the elements of the state vectors. Accordingly, four kinds of discontinuities are recognized: in the deflection (displacement of the axis), in the slope (break in the axis), in the moment (obliquely fitted wheel and couplings) and in the shear force (unbalance). All types of excitation may be combined.

If quantity δ_r with components δ_{ru} and δ_{rv} is the discontinuity in the real plane u, v, the complex discontinuities

$$
(3.23)\qquad \delta_u = \delta_{ru} + i\delta_{rv} \quad \text{and} \quad \delta_v = -\, i\delta_u
$$

are used in the calculation of the particular solution. For each excitation frequency equal to the speed frequency the problem is solved from the beginning of the rotor to the end (forward solution), while the rotor deformations are obtained by backward solution after the systems of equations (3.22) have been solved for α . The program is prepared in FORTRAN IV language. Fig. 5 shows an example of such a calculation of forced vibration of rotors of a fairly small turboset with a subjectively chosen spatial unbalance at each point of the rotor. The diagrams plot the trajectories of the various points of the rotor. Symbols + and x on these trajectories denote the direction of precession, + the positive , x - the negative. At the same time they mark the points at which the rotor was at time $t = 0$; the line connecting these points represents the deformation line of the rotor centre line.

Calculation of *free vibrations of damped rotors* is a far more complicated matter than the determination of forced vibration. It is required to find such complex frequencies Ω_k for which all boundary conditions are satisfied even without external excitation. On denoting by Δ the determinants of the systems of equations (3.22), we find that this state occurs whenever $\Delta = 0$. Since Δ is a complex quantity, it must hold simultaneously that $\mathrm{Re}\Delta(\omega,\Omega) = \mathrm{Im}\Delta(\omega,\Omega) = 0$.

What we actually have here is a system of two non-linear equations with two unknowns. Eventhough the closed mathematical formula of these equations is not known, we can determine the values of $\mathrm{Re}\Delta$ and $\mathrm{Im}\Delta$ numerically in each iteration cycle.

After many tests of several methods we came to the conclusion that the most efficient of all is the solution of the system of nonlinear equations by the method of least squares. We have used Fletcher's modification of the Marquardt algorithm which has proved to be highly efficient.

The free vibrations of rotor is obtained when the real part of the whirl frequency, i.e. $\mathrm{Re}\Omega$ is chosen for the working frequency of the rotor, ω . The solution of our

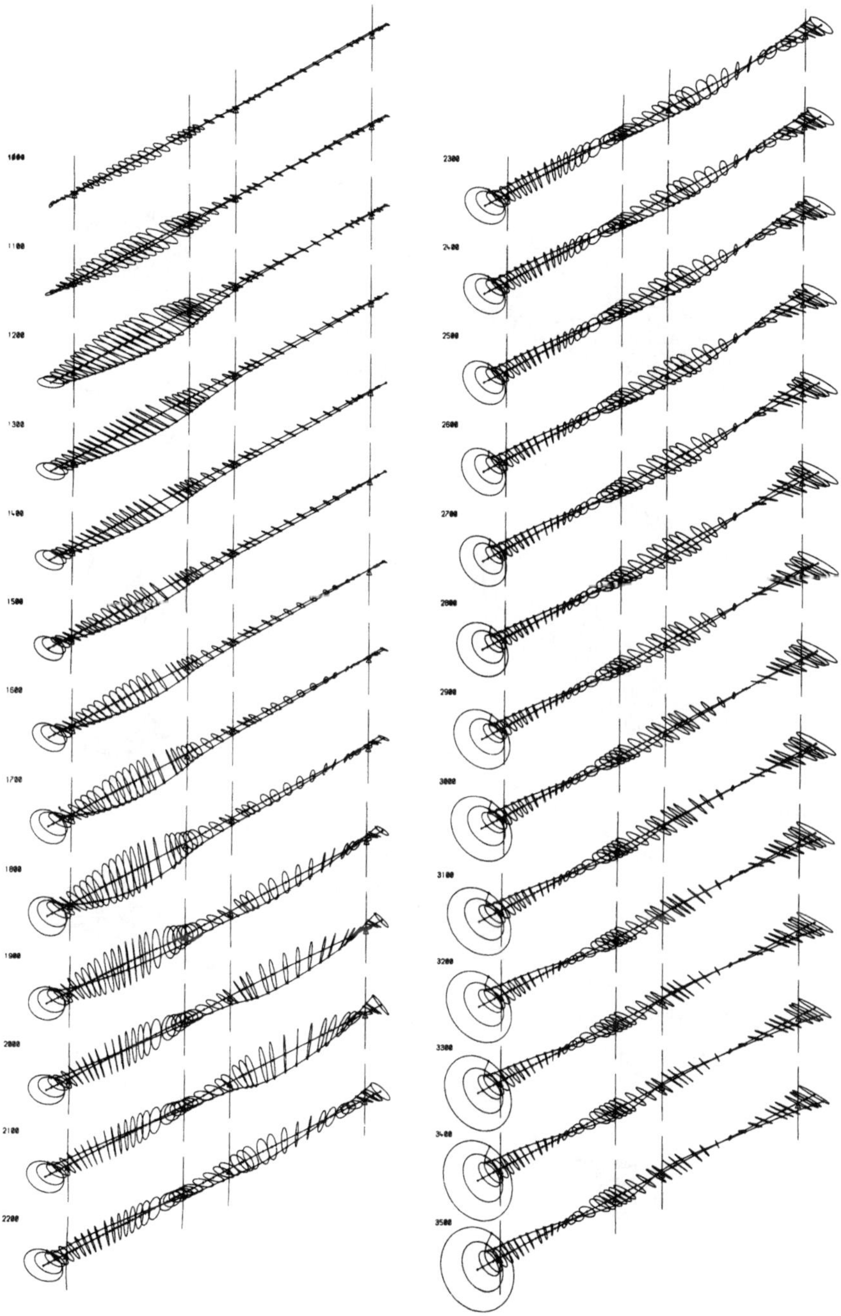

Fig. 5. Forced vibration of turbine and generator rotors in speed interval 1000 to 3,500 rpm .

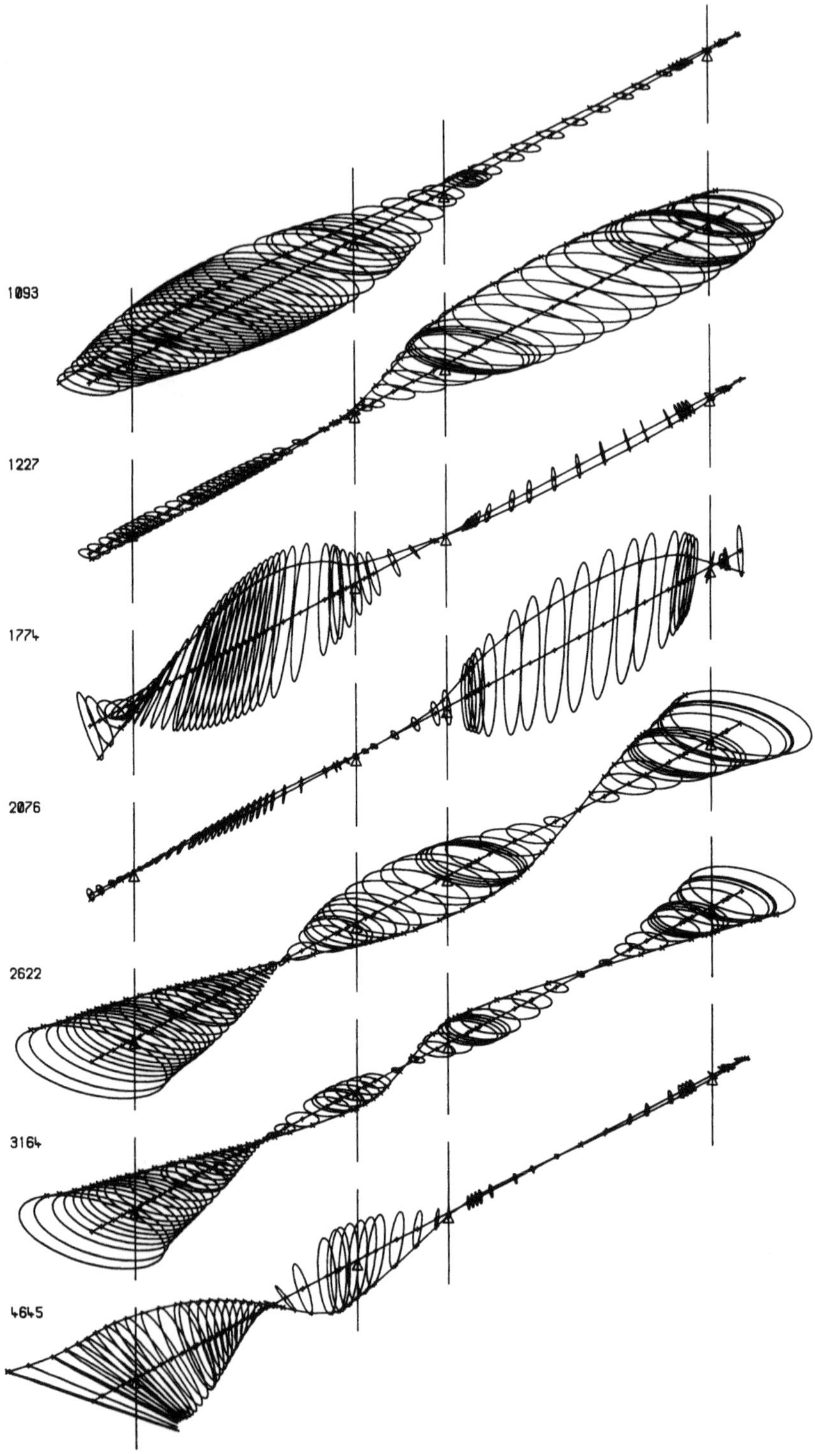

Fig. 6. Vibration modes at natural frequencies of rotor.

system of nonlinear equations will yield the complex natural frequency $\mathrm{Re}\Omega_k + i\mathrm{Im}\Omega_k$. To each natural frequency Ω_k belongs the natural vibration mode $\phi_k(x) = [\phi_u(x) , \phi_v(x)]_k^T$ composed of the transverse rotor deflections $u(x , \Omega_k) = \phi_{uk}^{(x)}$ and $v(x , \Omega_k) = \phi_{vk}^{(x)}$.

Fig. 6 shows examples of the natural vibration modes with the effect of $\mathrm{Im}\Omega_k$ omitted, calculated by our programme and drawn by an automatic plotter.

Since we are dealing with vibration in natural modes we can determine the relative damping of the system in the k-th mode as

$$\kappa_k = \frac{\mathrm{Im}\Omega_k}{|\Omega_k|} \tag{3.24}$$

and hence the sensitivity of the system in the k-th resonance - the Q-factor

$$Q_k = \frac{1}{2\kappa_k} = \frac{|\Omega_k|}{2\mathrm{Im}\Omega_k} \tag{3.25}$$

The *limit of stability* is a working frequency ω_s at which one of the natural frequencies Ω_k is *real*.

The limit of instability may be sought by the same methods as the free vibration of rotors, the only exception being that now it is ω and $\mathrm{Re}\Omega_k$ for $\mathrm{Im}\Omega_k = 0$ that are the unknown parameters.

4. BALANCING OF FLEXIBLE ROTORS

A great deal of human effort has already been devoted to this field of rotor dynamics. This statement is borne out by innumerable treatises by outstanding experts the world over. Yet the question remains forever alive. Since we have also been engaged in providing an answer to it, some of the results of our research follow.

To a general rotor mounted in a general anisotropic support system belongs a set of eigenvalues $\{\Omega\} = \{\Omega_I\} \cup \{\Omega_{II}\}$ and a set of natural modes $\{\phi(x)\} = \{\phi_I(x)\} \cup \{\phi_{II}(x)\}$ corresponding to it, for which there are satisfied all boundary conditions without external excitation. Any element of subscript k of the set of natural modes is formed by a vector with two elements $\phi_{jk} = [\phi_u(x) , \phi_v(x)]^T_{jk}$, $j=I,II$. Assume that for each ω the dynamic properties of the rotor mounting are characterized by matrix $G(\omega)$ with $(2R , 2R)$ elements at least, those elements being the dynamic flexibilities. It has been said already that for a system of journal bearings mounted on a common elastic structure, matrix $G(\omega)$ is full, complex and skew.

As it is well known, the problem with the transpose of the compliance matrix of the mounting, that is to say with matrix $\overset{T}{G}(\omega)$, has the same set of eigenvalues but a different set of natural modes $\{\Phi(x)\} = \{\Phi_I(x)\} \cup \{\Phi_{II}(x)\}$.

To each vibration order 'i' belongs generally a couple of frequencies Ω_{ij} and the modes $\phi_{ij}(x)$ and $\Phi_{ij}(x)$, $j=I,II$ corresponding to it. The two frequencies and modes become one in isotropic rotor mounting. Between any two natural modes of the two sets applies the condition of generalized orthogonality

$$S\left\{\Phi^T_m(x)\, m(x)\, \phi_n(x)\right\}_j = \delta_{mn} \quad , \tag{4.1}$$

where symbol $S\{\ \}$ is the Stieltjes integral over the whole length of the rotor, δ_{mn} the Kronecker delta, and $m(x)$ the mass of the rotor.

The general real spatial unbalance with components $n_u(x)$ and $n_v(x)$ from the (u , v) plane can also be thought of as a complex function in plane $(\mathrm{Re}\, u , \mathrm{Im}\, u)$, as $n(x) = n_u(x) + in_v(x)$. With a view to equation (3.23) the complex vector of unbalance will be of the form

$$n(x) = n(x)[1 , -i]^T \tag{4.2}$$

The relation between the two formulations of unbalance is clearly $n(x) = \frac{1}{2}[1\ ,\ i]\,n(x)$. Complex unbalance can then be expressed by the expansion into natural modes of one of the subsets

$$(4.3) \qquad n(x) = \frac{1}{2}[1\ ,\ i] \sum_{k=1}^{\infty} m(x)\,\phi_{jk}(x)\,c_{jk}^{N} \quad ; \quad (j = I \text{ or } II)$$

There thus exist two systems of generally complex, so-called generalized Fourier coefficients of the linear combination of the products of natural modes and masses of which we can compose the original unbalance.

The disadvantage of the "point" balancing, sometimes used for a solution, lies in that the individual weights possess generally contributions to all vibration modes thus making the matrix of influence coefficients ill-conditioned or at least sensitive to errors of measurement.

4.1. Modal balancing. The shortcomings just mentioned can be done away with completely by using - in place of point test balancing weights $w(x)$ - groups of such weights $W_k(x)$. On expanding them, as we have done with the unbalance, into natural modes of one of the sets ($j = I$ or II , this notation being omitted hereinafter) we get

$$(4.4) \qquad W(x) = \sum_{k=1}^{\infty} W_k(x) = \sum_{k=1}^{\infty} m(x)\,\phi_k(x)\,c_k^{W}$$

From here we can determine the Fourier coefficient of the group, viz.

$$(4.5) \qquad c_k^{W} = S\left\{\phi_k^T(x)\,W(x)\right\} = \sum_{s=1}^{S} \Phi_{sk}^T w_s$$

where $\Phi_{sk} = [\Phi_u\ ,\ \Phi_v]_{sk}^T$ are vectors $\Phi(x)$ at points $x = s$, and w_s is the vector of the components of a partial weight from group $W(x)$ at point s . Should the balancing weights $W(x)$ contain only one component $W_k(x)$ affinitive to the single k-th natural mode, it will produce only this mode, as implied by equation (4.1).

Introducing the new complex variables $\Phi_{sk} = [\Phi_u - i\Phi_v]_{sk}$ and $w_{sk} = \frac{1}{2}[1\,,\, i]\omega_{sk}$ we may determine the elements of the test weights from the written-out relation (4.5) as

$$\begin{bmatrix} \Phi_{11}\,, \Phi_{12}\,, \ldots, \Phi_{1k} \\ \Phi_{21}\,, \Phi_{22}\,, \ldots, \Phi_{2k} \\ \vdots \quad\; \vdots \qquad\quad \vdots \\ \Phi_{s1}\,, \Phi_{s2}\,, \ldots, \Phi_{sk} \end{bmatrix}^T \begin{bmatrix} w_{11}\,, w_{12}\,, \ldots, w_{1k} \\ w_{21}\,, w_{22}\,, \ldots, w_{2k} \\ \vdots \quad\; \vdots \qquad\quad \vdots \\ w_{s1}\,, w_{s2}\,, \ldots, w_{sk} \end{bmatrix} = \begin{bmatrix} c_1^W & & & \\ & c_2^W & & \\ & & \ddots & \\ & & & c_k^W \end{bmatrix} \tag{4.6}$$

or

$$\Phi^T \omega = c^W \tag{4.7}$$

Since we are dealing with test weights we can replace the matrix of the multiplicative constants c^W by the unit matrix I . It is convenient to make the number of balancing planes equal to the number of balancing speeds i.e. $S = K$. Then the matrix of test counterweights is directly $\omega = \Phi^{-T}$. In this way we would determine the groups of weights ω_k - the coulmns of matrix ω .

The determination of matrix Φ remains naturally a problem. Fortunately, it can be determined approximately, either numerically or for weakly non-self-adjoint systems it can be replaced directly by the matrix ϕ of the natural modes of the original problem, with a certain error, of course.

What is achieved by the balancing process is that the Fourier coefficients of the resultant unbalance $c_k^{NW} = c_k^N + c_k^W$ are zero or small at least, for several first orders K .

The balancing groups ω_k produce changes of the vector of the original vibration, y_o , on balancing speeds.

The vectors of these changes form the columns of the complex matrix of group influence coefficients, A . A linear combination of the balancing groups results in the same linear combination of influence coefficients. And that is why in balancing we endeavour to have

$$A\alpha + y_o = 0 \quad , \tag{4.8}$$

where α is the vector of the coefficients of the linear combination of the groups of weights.

4.2. Optimization of the balancing process. If the test counterweights are expressed by the expansion into a set of functions $\{\Phi_j\}$ ($j = I$ or II), the balancing process will well eliminate the manifestation of unbalance merely in a definite set of frequencies while in the other set of frequencies it will mainfest itself differently provided the system is not isotropic.

It follows from the above that one group of weights, say that for k-th order, is incapable of simultaneous balancing a pair of natural modes ϕ_{Ik} and ϕ_{IIk} . A certain optimum may be achieved, however. If we choose for the balancing speed a speed from all resonance regions and if we measure simultaneously in two directions, usually at right angles to one another, we get an overdetermined system of $2RK$ linear algebraic equations (4.8) with S unknown coefficients of vector α . Its approximate solution may be obtained, for example, by the method of least squares which leads to the minimization of Euclid's norm of vector of the residual vibration Δy

$$A\alpha + y_o = \Delta y \tag{4.9}$$

If we are interested in emphasizing some of the speed regions, we can multiply the whole of system (4.9) by a certain diagonal matrix of weights.

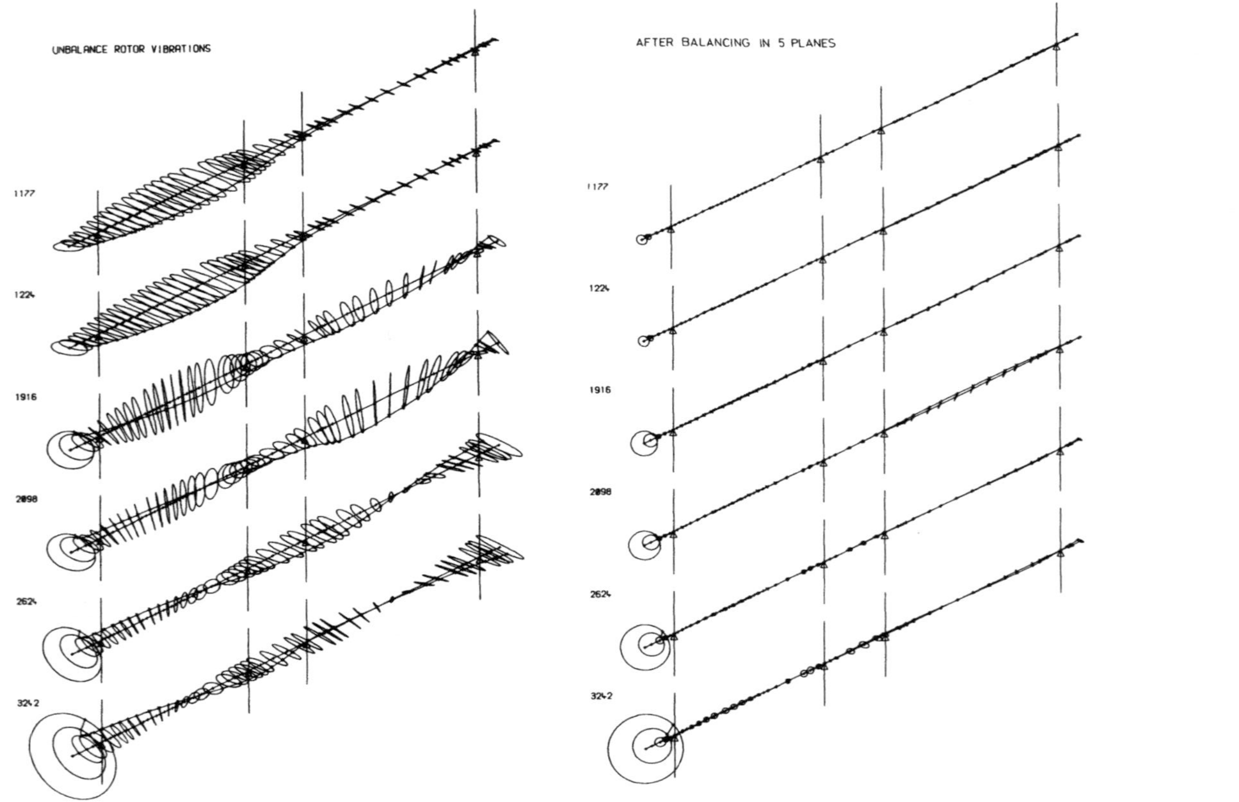

Fig. 7. Simulated run of an unbalanced turboset rotor before and after optimized balancing by means of five groups of weights in five balancing planes.

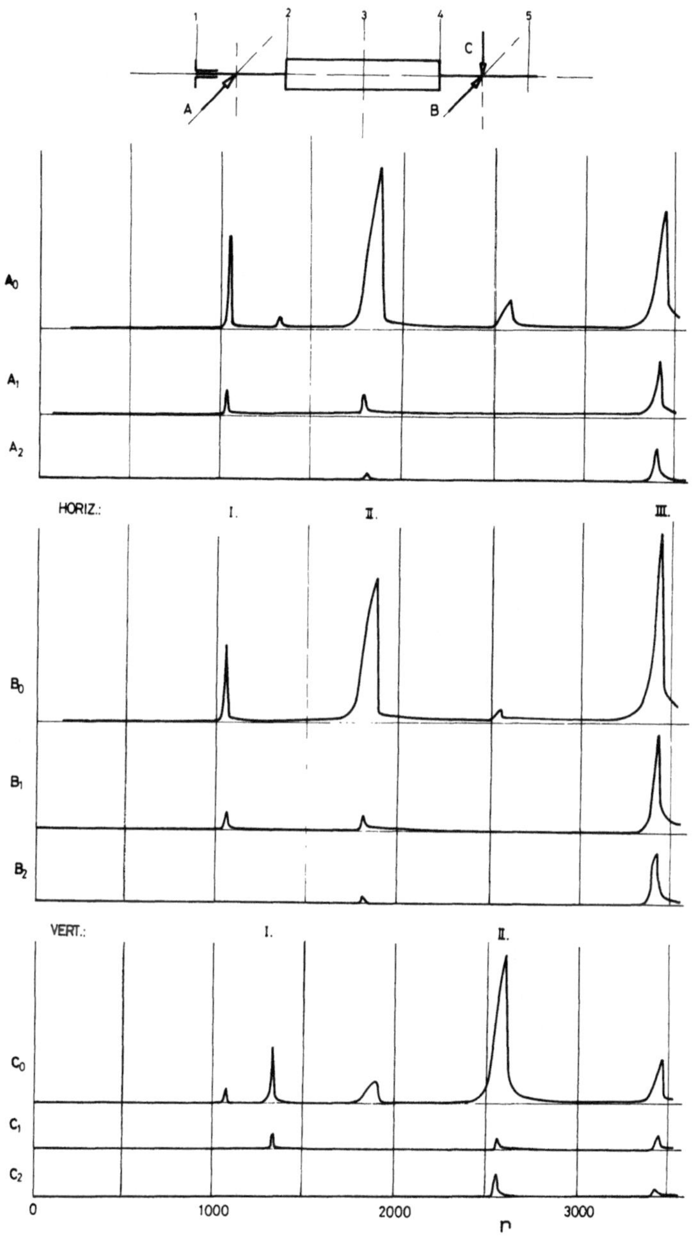

Fig. 8. Vibration of a generator rotor model before balancing (A_o, B_o, C_o), after a series of 3 balancing runs (A_1, B_1, C_1), and following a final correction (A_2, B_2, C_2). Balancing process carried out in five balancing planes.

A subjectively chosen unbalance distributed along the rotor produced forced spatial vibration calculated by the afore-mentioned method. The calculation of forced vibration from the test groups of five weights on the balancing speeds gave the rectangular matrix of influence coefficients. The coefficients of the linear combination of the test groups were determined by optimization. On adding the resultant group of weights thus established to the original unbalance we obtained the state shown in Fig. 7. Consulting it we see no marked peak in the range 1000 to 3000 rpm and conclude that the balancing was successful.

The procedure outlined above was also used in the balancing of a mechanical model of the generator rotor mounted on slightly nonlinear supports. Fig. 8 shows the vibration amplitudes measured at three points when running through the speed interval prior to balancing, after the first stage of balancing in three test runs and following the correction for imperfect convergence brought about by the nonlinearities.

Evidently, so long as we know the matrix of influence coefficients, say either from calculation or from preceding balancing of identical rotors, the process of optimum balancing can theoretically be effected in one run. Actually, however, minute deviations in installation produced by various effects (such as the oil temperature, bushing geometry, rotor alignment and possibly nonlinearities) will cause this information to be of use at best for the improvement of the initial state.

CONCLUSIONS

This paper has presented a cross-section of some research studies in the field of rotor dynamics performed at the Central Research Institute of the SKODA Works and, in part, also in some collaborating laboratories.

Theoretical as well as numerical analyses have revealed the importance of having available reliable data, especially on the stiffness and damping of full-scale bearings, and on the dynamic properties of foundations. The methods outlined in the paper enable the designer to calculate without undue difficulties both the free and the forced vibration of linear statically indeterminate rotors mounted in linear supports of any kind, and to establish the limits of stability of their motion.

It has been found that in consequence of anisotropy of stiffnesses and dampings, and of the different properties of rotor supports, the rotors vibrate in complicated spatial modes, frequently with different directions of precession in different parts of the rotor simultaneously. As this fact implies, it is hard to choose uniform precession in simplified rotor calculations.

Anisotropy of supports also affects the convergence of the balancing process. By means of simulated problems and models we have proved the method of optimized modal balancing to be successful even in the case of slightly nonlinear problems.

ACKNOWLEDGMENT - The author wishes to express his thanks to all his colleagues at the Central Research Institute of the SKODA Works, especially to Mr. V. Horák for the many years of cooperation in the field of rotor dynamics, and to the management of the SKODA Works for their kind permission to publish some of the results of our long-term studies. Thanks are also due to all the authors of references quoted herein.

Lateral Stability and Vibrations of High Speed Centrifugal Pump Rotors

H.F. Black

Department of Mechanical Engineering
Heriot-Watt University, Edinburgh, UK

NOTATION

B	axial width of impeller outlet
c	annular clearance
F_y, F_z	force components in y, z directions
j	$\sqrt{-1}$
k	hydrodynamic bearing force factor (equation (1))
K	diffuser force coefficient (equation (11))
M	concentrated rotor mass
n	frequency ratio of running speed
p	ratio of cross-coupling to radial stiffness coefficients in seals
p_d	ratio of cross-coupling to radial stiffness coefficients for diffuser-impeller forces
p_i	impeller pressure head
P	pump pressure head
q	ratio of radial to cross-coupling stiffness coefficients in hydrodynamic bearings (equation (1))
Q	ratio of flow to optimum flow rates through pump
r_e	instantaneous shaft displacement at station e (y_e+jz_e)
R_e	whirl radius at station
R_i	impeller radius
s	shaft stiffness in simple bearings
s_1	radial stiffness of seals at nominal speed ω_0
s_d	radial negative stiffness of diffuser forces at nominal speed ω_0
t	time
0_{yz}	reference frame for lateral shaft displacements

α_d diffuser vane angle

Δ mass unbalance eccentricity

ξ "damping ratio"

ρ fluid density

Ω asynchronous precession speed

ω shaft speed

ω_n $\sqrt{s/M}$

ω_{n1} $\sqrt{s_1/M}$

ω_{nd} $\sqrt{s_d/M}$

Other symbols are defined in the text.

INTRODUCTION

This discussion deals principally with two phenomena, contributing to lateral vibrations of rotors, which are peculiar to high speed centrifugal pump rotors. Since the present purpose is to clarify effects, the simplest possible representation of the rotor system suffices: that is, a single concentrated mass supported on a flexible shaft which may be in simple, or circular hydrodynamic bearings.

Figure 1 indicates the two features peculiar to this problem. One is the hybrid bearing forces arising from annular internal clearances under pressure, e.g. the neck-ring bushes indicated in figure 1. The purely radial stiffness of these was investigated theoretically by LOMAKIN [1]; experimental studies revealing indirect evidence of hybrid bearing effects were carried out by KARINCEV and MARZINKOWSKI [2]; BLACK, JENSSEN, MURRAY and COCHRANE, [3], [4], [5], [6] carried out theoretical and direct experimental studies of the hybrid bearing effects. A second feature arises from interaction between the impeller and diffuser (figure 1),

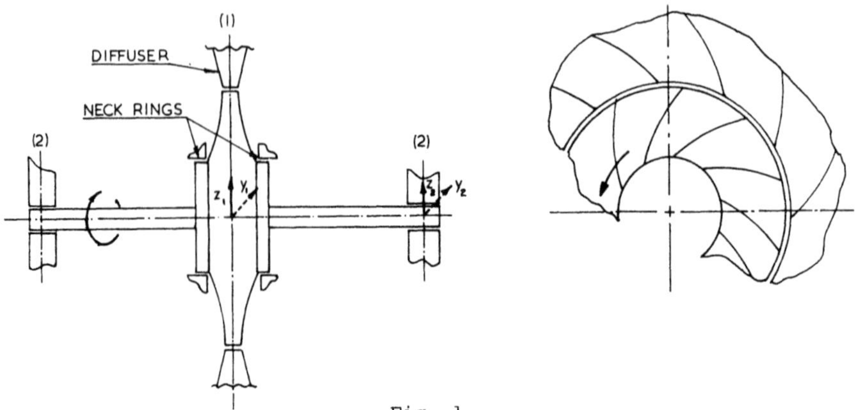

Fig. 1.

particularly at flow rates below the ideal design value for a given running speed. The diffuser vanes are then loaded hydrofoils, and consequently large lateral forces act on the impeller when it moves out of concentricity with the diffuser. Some experimental evidence of these forces has been demonstrated by HERGT and KRIEGER [7]; this discussion contributes a semi-empirical theoretical treatment which is compared with some of the experimental results from [7].

The effects of forces from annular clearance seals under pressure and from diffuser-impeller interaction show markedly in the predicted stability and forced response to mass unbalance. The implications with regard to linear theory are discussed in the following paragraphs. Some non-linear behaviour consequent to forced or unstable whirling across the entire internal clearance space is also discussed.

REPRESENTATION OF CROSS-COUPLING FORCES

(i) Circular hydrodynamic bearings. Centrifugal pump rotors are light in relation to journal bearings of a reasonable size in relation to the shafts. It is therefore inevitable that journal bearings of high speed rotors will be lightly loaded, i.e., they will run at low eccentricity ratio. The journal dynamics when linearised can then be approximated closely by

(1) $$\begin{bmatrix} F_y \\ F_z \end{bmatrix} = - k\omega \begin{bmatrix} q & 1 \\ -1 & q \end{bmatrix} \begin{bmatrix} y \\ z \end{bmatrix} - 2k \begin{bmatrix} 1 & 0 \\ 0 & 1 \end{bmatrix} \begin{bmatrix} \dot{y} \\ \dot{z} \end{bmatrix}$$

The ratio q is small $(<<1)$ within the range of this approximation. When a range of speeds with constant bearing load is considered, q diminishes as speed increases: for short sleeve bearings

(2) $$q = \mu/\omega^4$$

where μ is a constant, is a good approximation for eccentricity ratios less than 0.25 .

(ii) Hybrid bearing forces of annular clearance seals. Taking into account that the pressure generated in a centrifugal pump is very nearly proportional to the square of speed, the hybrid bearing forces of seals can be closely approximated by

(3) $$\begin{bmatrix} F_y \\ F_z \end{bmatrix} = - s_1 \left(\frac{\omega}{\omega_0}\right)^2 \begin{bmatrix} 1 & p \\ -p & 1 \end{bmatrix} \begin{bmatrix} y \\ z \end{bmatrix} - 2s_1 p \left(\frac{\omega}{\omega_0^2}\right) \begin{bmatrix} 1 & 0 \\ 0 & 1 \end{bmatrix} \begin{bmatrix} \dot{y} \\ \dot{z} \end{bmatrix}$$

This expression follows from [3], [4], omitting only some

relatively unimportant contributions due to fluid inertia effects in the clearance space.

Comparing the results equations (3) and (1), the salient differences are:

(a) in equations (3), radial stiffness is dominant unless the seals are exceptionally long: in equations (1), the cross-coupling stiffness is always dominant;

(b) the forces of equation (3) are proportional to square of speed; those of equation (1) are directly proportional to speed.

In the case of flexible shaft systems, designed to run above the first critical speed, it is normally the case that s_1 is of the same order of magnitude as the shaft stiffness s .

(iii) Diffuser - Impeller inter-action. Figure 2 represents an impeller having a displacement and velocity in the 0_y direction. If α represents the mean outflow angle with the impeller centred, it is assumed that at part flow ratio, Q , the effect of loading the diffuser vanes alters this angle locally when the impeller is displaced in direct proportion to local change of clearance c' , such that at location θ (figure 2) the change in outflow angle is,

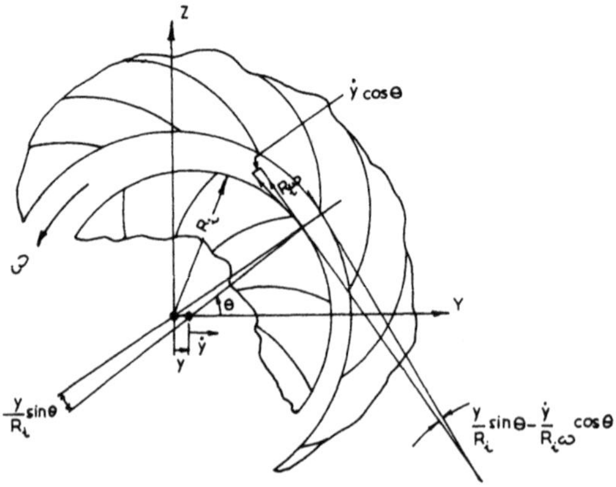

Fig. 2. Impeller-diffuser configuration.

with $c' = y\cos\theta$,

(4) $$\delta = \frac{y}{R_i}\sin\theta + (\alpha_d - \alpha)\frac{y}{c}\cos\theta - \frac{\dot{y}}{R\omega}\cos\theta$$

Treating the impeller outlet angle β as small, a semi-empirical expression for outlet whirl velocity is

(5) $$v_w \simeq 0.8R_i\omega - u_m/\beta$$

the factor 0.8 being a slip coefficient taking into account secondary flows within the impeller. If α is also treated as a small angle, the radial outflow velocity

(6) $$u_m \simeq \alpha v_w$$

The pressure at the impeller outlet is given by

(7) $$p_i = \rho R_i \omega v_w - \frac{1}{2}\rho u^2$$

neglecting the inflow kinetic energy. When α is changed to $\alpha + \delta$, and it is assumed that $u \simeq v_w$ (consistent with the assumption of a small outlet angle α) , a first order perturbation applied to equations (5), (6) and (7) yields a local pressure change

$$\Delta p_i = - p_i\left[\frac{\alpha+0.2\beta}{(\alpha+0.6\beta)(\alpha+\beta)}\right]\cdot\delta$$

or

(8) $$\Delta p_i = - p_i f(\alpha,\beta)\cdot\delta$$

Upon incorporating the value of δ given by equation (4), integrating for force components around the impeller, and considering a similar impeller displacement in the 0_z direction, the inter-action force components are found to be

$$\begin{bmatrix} F_y \\ F_z \end{bmatrix} = \begin{bmatrix} \lambda_d & -\lambda_c \\ \lambda_c & \lambda_d \end{bmatrix} \begin{bmatrix} y \\ z \end{bmatrix} - \begin{bmatrix} \frac{\lambda_c}{\omega} & 0 \\ 0 & \frac{\lambda_c}{\omega} \end{bmatrix} \begin{bmatrix} \dot{y} \\ \dot{z} \end{bmatrix} \tag{9}$$

where

$$\begin{cases} \lambda_d = \pi B p_i \left(\frac{R_i}{c}\right) f(\alpha,\beta)(1-Q)\alpha_d \\ \\ \lambda_c = \pi B p_i f(\alpha,\beta) \end{cases} \tag{10}$$

with the further approximation $\alpha = Q\alpha_d$.

Static force measurements by HERGT and KRIEGER [7] give a good indication of the direct radial static force coefficient λ_d ; the cross-coupling force characterised by λ_c cannot be quantitatively discerned due to scatter, but is clearly present. In [7], direct radial force measurements are presented in terms of a dimensionless coefficient

$$K = \frac{\text{Force}}{2BR_i P} \tag{11}$$

where P is total pressure head over the impeller and diffuser. Figure 3 gives equivalent coefficients K_d and K_c

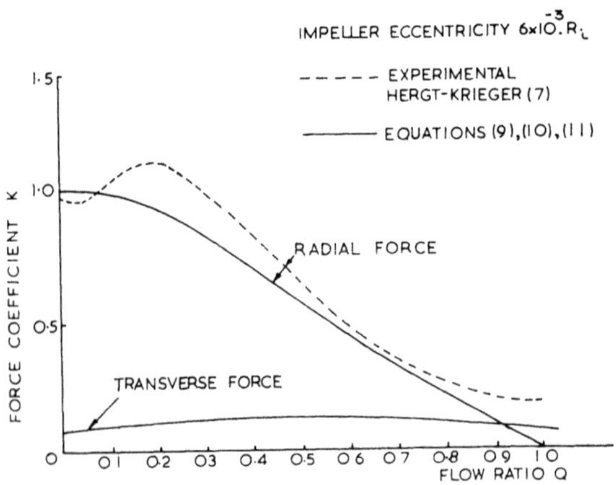

Fig. 3. Diffuser-impeller inter-action.

for radial and cross-coupling forces respectively as defined by equations (9), (10) corresponding to the measurements reported in [7]; it has been assumed that $p_i = 0.7\,P$. Quantitative comparison with the measured experimental results for the radial component of force is encouraging.

Bearing in mind that normally pressure is nearly proportional to square of speed, equation (9) may be restated in the form

$$(12) \qquad \begin{bmatrix} F_y \\ F_z \end{bmatrix} = s_d\left(\frac{\omega}{\omega_0}\right)^2 \begin{bmatrix} 1 & -p_d \\ p_d & 1 \end{bmatrix} \begin{bmatrix} y \\ z \end{bmatrix} = -s_d p_d \left(\frac{\omega}{\omega_0^2}\right) \begin{bmatrix} 1 & 0 \\ 0 & 1 \end{bmatrix} \begin{bmatrix} \dot{y} \\ \dot{z} \end{bmatrix}$$

The following points are emphasised:

(a) the radial force component is wholly destablising when the flow rate is less than the ideal $(Q < 1)$;

(b) the cross-coupling force provides a forward whirling drive;

(c) s_d and p_d depend on the pump delivery, i.e. the flow ratio Q .

STABILITY OF ROTOR INCLUDING BEARING AND PRESSURE SEAL FORCES

Assuming the rotor system figure 1 to be symmetric, equations (1) and (3) may be taken to relate to the bearings and seals respectively lumped together. Then, with

$$(13) \qquad r = y + jz$$

the free motion equations excluding impeller diffuser interaction become

$$\ddot{r}_1 + 2p\left(\frac{\omega_{n1}}{\omega_0}\right)^2 \omega\dot{r}_1 + \left[\omega_n^2 + \left(\frac{\omega_{n1}}{\omega_0}\right)^2 \omega^2 - jp\left(\frac{\omega_{n1}}{\omega_0}\right)^2 \omega^2\right] r_1 - \omega_n^2 r_2 = 0$$

(14)

$$\omega_n^2 r_1 + 4\xi\omega_n \dot{r}_2 + \left[\omega_n^2 + 2q\xi\omega_n\omega - j2\xi\omega_n\omega\right] r_2 = 0$$

Seeking the stability boundary with the quasi-stable substitution

(15) $$r_e = R_e e^{j\Omega t}$$

yields

(16) $$\Omega = \frac{1}{2}\omega$$

i.e. a half-speed whirl; the onset speed is given by

(17) $$\frac{\omega}{\omega_n} = \sqrt{\frac{4}{1-4\left(\frac{\omega_{n1}}{\omega_0}\right)^2} + \left(\frac{1}{4q\xi}\right)^2} - \frac{1}{4q\xi}$$

It is first of all note-worthy that if

(18) $$\frac{\omega_{n1}}{\omega_0} > \frac{1}{2}$$

unstable whirling is suppressed at all speeds.

Figure 4 shows plots of onset speed ratio ω/ω_n against $q\xi$, for various values of ω_{n1}/ω_0. Since ξ is a measure of the relative stiffness of bearings and shaft,

$\xi \to \infty$ denotes a flexible shaft

system and

$\xi \to 0$ denotes a very stiff shaft.

Bearing in mind the relation equation (2), this implies

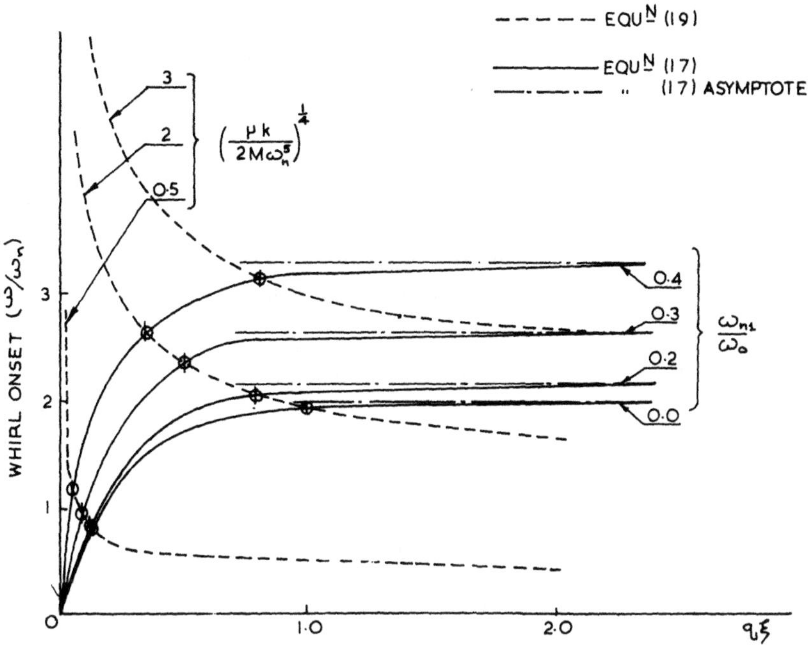

Fig. 4. Unstable whirl onset with seals and hydrodynamic bearings.

(19) $$\frac{\omega}{\omega_n} = \left(\frac{\mu k}{2M\omega_n^5}\right)^{1/4}\left(\frac{1}{q\xi}\right)^{1/4}$$

for a particular set of bearings. Three possible curves of operating point according to equation (19), for varying relative shaft flexibility, are also shown in figure 4. The intersection of the curves for equation (17) and equation (19) marks the unstable whirl onset speed of a particular assembly.

For very flexible shafts, $q\xi \to \infty$, so that the onset speed of unstable whirling becomes

(20) $$\frac{\omega}{\omega_n} = \frac{2}{\sqrt{1-4\left(\frac{\omega_{n1}}{\omega_0}\right)^2}}$$

and is now determined wholly by the bearing properties of the annular seals. It appears therefore that a flexible shaft system in hydrodynamic bearings may be treated as though the bearings were simple supports.

STABILITY OF FLEXIBLE ROTOR INCLUDING PRESSURE SEAL AND IMPELLER-DIFFUSER FORCES

If the hydrodynamic bearings are treated as infinitely stiff, then with equations (13), (3) and (12) a single free motion equation results:

$$\text{(21)} \qquad \ddot{r}_1 + \left[2p\left(\frac{\omega_{n1}}{\omega_0}\right)^2 + p_d\left(\frac{\omega_{nd}}{\omega_0}\right)^2\right]\dot{r}_1 + \left[\omega_n^2 + \left(\frac{\omega_{n1}^2-\omega_{nd}^2}{\omega_0^2}\right)\omega^2 - j\left(\frac{p\omega_{n1}^2+p_d\omega_{nd}^2}{\omega_0^2}\right)\right]r_1 = 0$$

The substitution

$$\text{(22)} \qquad r_1 = R_1 e^{j\Omega t}$$

yields at the stability boundary

$$\text{(23)} \qquad \Omega = \left[\frac{1 + \left(\frac{\omega_{nd}}{\omega_{n1}}\right)^2 \frac{p_d}{p}}{2 + \left(\frac{\omega_{nd}}{\omega_{n1}}\right)^2 \frac{p_d}{p}}\right]\omega$$

the onset speed being given by

$$\text{(24)} \qquad \frac{\omega}{\omega_n} = \left[\frac{\left(2 + \left(\frac{\omega_{nd}}{\omega_{n1}}\right)^2 \frac{p_d}{p}\right)^2}{\left(1 + \left(\frac{\omega_{nd}}{\omega_{n1}}\right)^2 \frac{p_d}{p}\right)^2 - \left(\frac{\omega_{n1}^2-\omega_{nd}^2}{\omega_0^2}\right)\left(2 + \left(\frac{\omega_{nd}}{\omega_{n1}}\right)^2 \frac{p_d}{p}\right)^2}\right]^{\frac{1}{2}}$$

It is first noted that when the diffuser forces vanish, as expected, equations (23) and (24) are identical with equations (16) and (20) respectively.

Equation (23) shows that the precession speed of unstable whirling becomes nearly synchronous when diffuser forces dominates; equation (24) indicates that the whirl onset speed also falls, approaching

$$\frac{\omega}{\omega_n} = \frac{1}{1 + \left(\frac{\omega_{nd}}{\omega_0}\right)^2} \tag{25}$$

Thus, in situations where the restraining forces of seals are weak or negligible, nearly synchronous unstable whirling may commence at speeds close to or even below the first critical speed of the rotor running dry.

Plots of equations (23) and (24) for varying ratio of diffuser forces to seal forces are shown in figure 5.

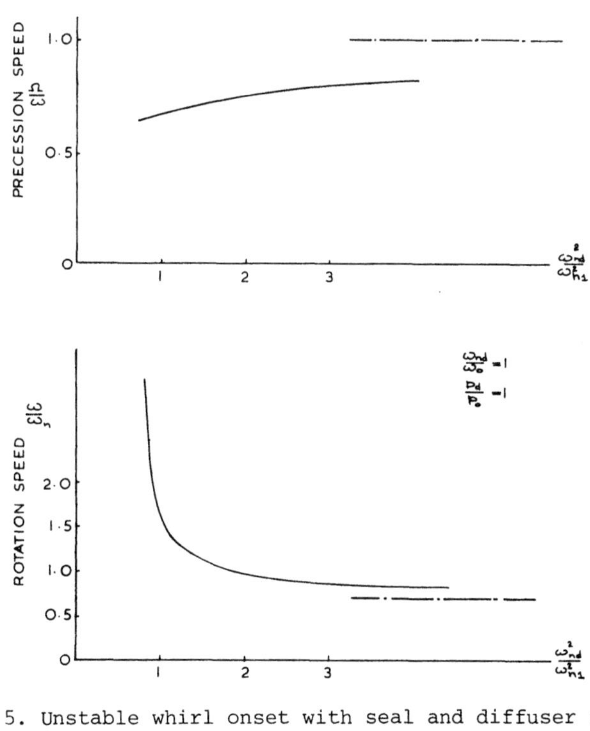

Fig. 5. Unstable whirl onset with seal and diffuser forces.

SYNCHRONOUS FORCED WHIRLING OF FLEXIBLE SHAFT

The response to mass unbalance of the flexible shaft system considered in the preceding paragraph is given by putting the right-hand side of equation (21) equal to $\Delta\omega^2 e^{j\omega t}$ and substituting

$$r_1 = R_1 e^{j\omega t} \tag{26}$$

The resulting expression for the amplitude of forced synchronous whirling is

$$\frac{|R_1|}{\Delta} = \frac{n^2}{\sqrt{\left[1 - \left\{1 - \frac{(\omega_{n1}^2 - \omega_{nd}^2)}{\omega_0^2}\right\} n^2\right]^2 + p^2\left(\frac{\omega_{n1}}{\omega_0}\right)^4 n^4}} \tag{27}$$

A maximum in amplitude occurs when

$$n = \frac{1}{\sqrt{1 - \frac{(\omega_{n1}^2 - \omega_{nd}^2)}{\omega_0^2}}} \tag{28}$$

this denotes the critical whirling speed.

Then

$$\frac{|R_1|_{max}}{\Delta} = \frac{1}{p\left(\frac{\omega_{n1}}{\omega_0}\right)^2} \tag{29}$$

In high speed running, the synchronous whirl amplitude becomes asymptotic to

$$\frac{|R_1|}{\Delta} = \frac{1}{\sqrt{\left[1 - \frac{(\omega_{n1}^2 - \omega_{nd}^2)}{\omega_0^2}\right]^2 + p^2\left(\frac{\omega_{n1}}{\omega_0}\right)^4}} \tag{30}$$

When

$$\frac{\omega_{n1}^2 - \omega_{nd}^2}{\omega_0^2} \geq 1 ,$$

there is no "critical whirling speed"; the whirl amplitude then merely rises gradually with speed, eventually approaching the asymptote equation (30). In parameter ranges where critical speeds exist, these may be lower than the dry running critical speed: this is likely to be the case when the diffuser vanes are heavily loaded, that is, when the pump is operating at reduced flows.

Figure 6 indicates the effect of varying the bearing strength of seals on maximum forced whirl amplitude and critical speed.

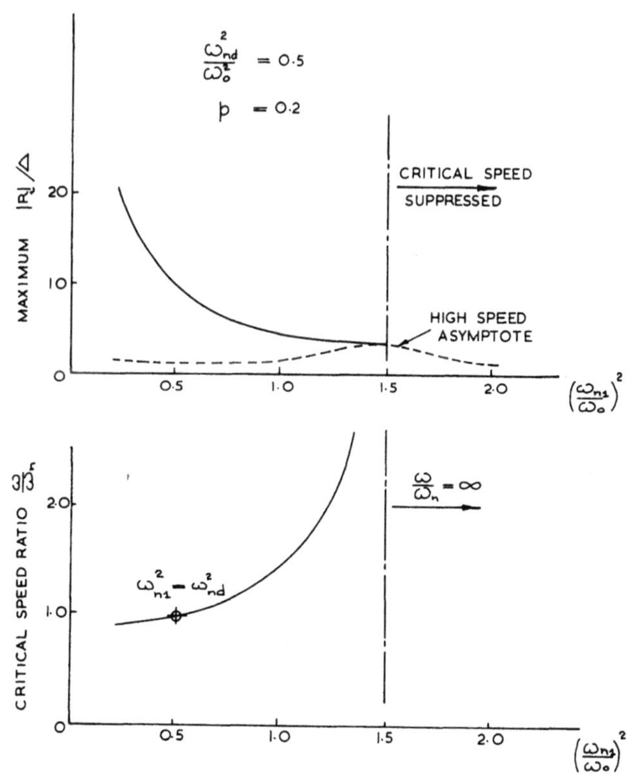

Fig. 6. Maximum whirl radius and critical speed with varying seal stiffness.

If sufficient mass unbalance is present to cause the whirl amplitude given by equation (27) to exceed the available internal clearance c at the seals, synchronous forced whirling with rubbing at the seal surface may continue over a wide or indefinite speed range above the critical speed. If whirling in this state is possible, it terminates when the tangential component of mass unbalance force can no longer sustain the friction drag on the seal surface. The terminal equilibrium conditions are

$$(31) \qquad \begin{cases} F + \left[s + (s_1 - s_d)\left(\dfrac{\omega}{\omega_0}\right)^2\right]c = M\omega^2 c \\[2ex] fF + s_1 p\left(\dfrac{\omega}{\omega_0}\right)^2 c = M\omega^2 \Delta \end{cases}$$

where F is the radial force between shaft and seal, and f is the coefficient of friction. It has been assumed that seal and diffuser forces remain linear; this is certainly justifiable for the latter; for the former, the non-linearities at high eccentricity are slight compared with those of hydrodynamic bearings [3].

Then, the terminal speed for whirling in this condition is given by

$$(32) \qquad \left(\frac{\omega}{\omega_n}\right)^2 = \frac{1}{1 - \left(\dfrac{\omega_{nl}^2 - \omega_{nd}^2}{\omega_0^2}\right) - \dfrac{1}{f}\left[\dfrac{\Delta}{c} - p\left(\dfrac{\omega_{nl}}{\omega_0}\right)^2\right]}$$

Assuming that a critical speed is not entirely suppressed from equation (29), this type of whirling can only occur when

$$(33) \qquad \frac{\Delta}{c} > p\left(\frac{\omega_{nl}}{\omega_0}\right)^2$$

If the inequality (33) is satisfied, then whirling with rubbing at the seal can go on indefinitely if

$$(34) \qquad \frac{\Delta}{c} > f\left[1 - \left(\frac{\omega_{n1}^2 - \omega_{nd}^2}{\omega_0^2}\right)\right] + p\left(\frac{\omega_{n1}}{\omega_0}\right)^2$$

Figure 7 indicates the mass unbalance needed to trigger and sustain this type of whirling. It is seen that quite small mass unbalance can result in a wide speed range, above the critical speed, of whirling with rubbing at the seals.

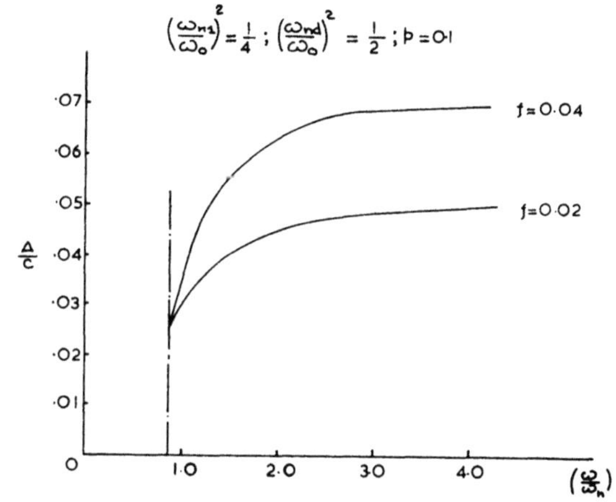

Fig. 7. Terminal speed of synchronous whirl with rubbing at seals.

NON-LINEAR ASYNCHRONOUS WHIRLING

If unstable whirling due to seal and diffuser forces occurs, this will ultimately be limited by rubbing at the seals. Unstable whirling may then continue, at a precession speed determined by the friction coefficient at the rubbing surface. Retaining the assumption of linearity for the seal and diffuser forces, the equilibrium of a whirl with rubbing at precession speed Ω is governed by

$$F + \left[s + (s_1 - s_d)\left(\frac{\omega}{\omega_0}\right)^2\right]c = M\Omega^2 c \tag{35}$$

and

$$fF + \left[s_1\left(2\frac{\Omega}{\omega} - 1\right)p\left(\frac{\omega}{\omega_0}\right)^2 + s_d\left(\frac{\Omega}{\omega} - 1\right)p_d\left(\frac{\omega}{\omega_0}\right)^2\right]c = 0 \tag{36}$$

The precession speed is then given by

$$\frac{\Omega}{\omega_n} = \sqrt{1 + n^2\left[\left(\frac{\omega_{n1}^2 p + \omega_{nd}^2 p_d}{f\omega_0^2}\right) + \frac{1}{4}\left(\frac{2\omega_{n1}^2 p + \omega_{nd}^2 p_d}{f\omega_0^2}\right)^2 - \left(\frac{\omega_{nd}^2 - \omega_{n1}^2}{\omega_0^2}\right)\right]} - \frac{1}{2}n\left(\frac{2\omega_{n1}^2 p + \omega_{nd}^2 p_d}{f\omega_0^2}\right) \tag{37}$$

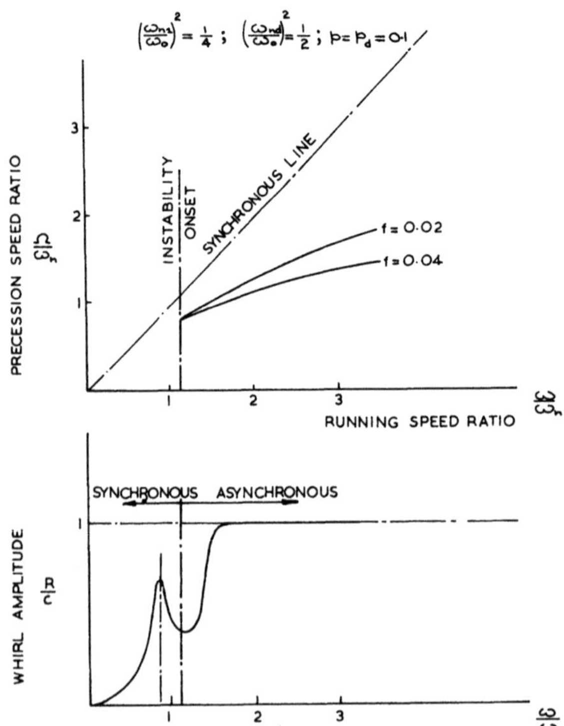

Fig. 8. Asynchronous whirling with rubbing at seals.

subject to the condition of force closure at the rubbing face,

$$\frac{\Omega}{\omega_n} > \sqrt{1 - n^2\left(\frac{\omega_{nd}^2 - \omega_{nl}^2}{\omega_0^2}\right)} \tag{38}$$

A plot of possible whirl precession speeds in this regime is shown in figure 8. A plot of probable whirl amplitudes with small mass unbalance is also shown.

CONCLUSIONS

The lateral stability and response to mass unbalance of high speed centrifugal pump rotors is seen to depend radically on fluid forces in internal clearance spaces and, through hydraulic forces arising in diffusers, on the pump delivery rate. Thus, for example, in boiler feed pumps, the character of rotor vibrations may change considerably for different settings of the feed regulation valve.

Apart from the factors considered in this discussion, pump rotors can also be subjected to large very low frequency lateral forces arising from stall and consequent reverse flow in some diffuser passages. This phenomenon occurs at very low flow rates, for example when a pump is on stand-by duty with shut discharge valve. The forces in question usually take the form of a fairly regular slowly rotating radial thrust, with a precession speed of approximately one-tenth of running speed [7]. In applications where pumps have to operate in this condition for some proportion of their working life, progressive wear of internal clearance is inevitable. Eventually, such machines then develop unacceptable synchronous or asynchronous whirls at more normal running conditions, the latter being directly attributable to the diffuser impeller inter-action characterised by equation (9). The asynchronous whirls in question have a precession rate well above half the rotational speed and are sometimes only just below the rotational speed: such whirls have occured in pumps with rotors in rolling element bearings.

REFERENCES

[1] LOMAKIN, A.A. and BEDCER, F.S.: "Determination of Critical Speed of a Pump with Reference to Forces Arising in the Seal Gaps", Mash. Giz. 12, 5, 1957.

[2] KARINCEV, J.B. and MARZINKOWSKI, W.A.: "Influence of Seal Gaps on the Critical Speeds of Rotors", Energomashinostroenie, 7, 4.

[3] BLACK, H.F. and JENSSEN, D.N.: "Dynamic Hybrid Bearing Properties of Annular Pressure Seals", Proc. I. Mech. E. 184, 3N, 1970.

[4] BLACK, H.F. and JENSSEN, D.N.: "Effects of High Pressure on Pump Rotor Vibrations", ASME 71 - WA/FE-38, 1971.

[5] BLACK, H.F. and MURRAY, J.L.: "The Hydrostatic and Hybrid Bearing Properties of Annular Pressure Seals in Centrifugal Pumps", BHRA RR1026, 1969.

[6] BLACK, H.F. and COCHRANE, E.A.: "Leakage and Hybrid Bearing Properties of Serrated Seals in Centrifugal Pumps", 6th Int. Conf. on Fluid Sealing, 1973.

[7] HERGT, P. and KRIEGER, P.: "Radial Forces in Centrifugal Pumps with Guide Vanes", Proc. I. Mech. E, 184, 3N, 1970.

On the Numerical Investigation of Nonlinear Periodic Rotor Vibrations

E. Brommundt

Technische Universität Braunschweig, Germany

1. INTRODUCTION

Rotor vibrations can be strongly nonlinear because of the nonlinear force deflection relations of journal bearings. But the literature (cf. [12]) contains few papers which apply the notions and ideas of the theory of strongly nonlinear oscillations (e.g. [10], [23], [25]) to these problems.

Valuable findings about the nonlinear mechanisms of various phenomena were achieved by the averaging method (e. g. [16], [19], [24], [34]), and by numerical or analog computer investigations of initial value problems (e.g. [8], [18],[26]). Disadvantages of these procedures are, e.g., the necessity to introduce a small parameter (a weak nonlinearity) in the first case, and the difficulty to obtain a general view, in the second case, when several solutions of different characteristics exist.

In a series of papers (cf. [35], [36], [38]) URABE and REITER have shown that high order Fourier approximations for solutions of nonlinear differential equations can be efficiently computed by a suitably adapted Galerkin procedure. WARNCKE ([40]) was the first to apply this method to the computation of rotor vibrations. The present paper adapts these procedures to a systematic study of such problems. A numerical stability investigation is included. The investigation of bifurcations is possible on the basis of the Fourier expansions alone. This permits, to the author's opinion, a systematic study of rotor motions.

A rather general model is chosen which is assumed to

be linear except for the oilfilm forces, other nonlinea-ities can be treated similarly, cf. [1]. A single disc rotor mounted on elastically supported bearings serves as an example.

2. ROTOR MODEL

Fig. 2.1 Reference configuration

Fig. 2.1 shows the front section of a rotor composed of the shaft ① which carries discs ② and is mounted on sleeve bearings ③ , ④ . All parts are specified by numbers: $j = 1,\ldots, J$ specifies bearings, $k = 1,\ldots K$ discs, $\ell = 1,\ldots, L$ marks points of the shaft where either bearings or discs are placed, $(L = J + K)$. Primes discriminate different parameters of the same family, e.g. ℓ', ℓ'' .

Coordinates for horizontal and vertical deflections are distinguished by the (first) subscripts $i = 1, 2$, respectively, see Fig. 2.1. The deflection of the shaft at the place ℓ is given by the deflection vector

$$(2.1) \quad \underset{\sim}{x}_\ell(t) := \underset{\sim}{x}_\ell := \{x_{1\ell}, x_{2\ell}\} := (x_{1\ell}, x_{2\ell})^T .$$

The superscript T denotes transposition, $\{. , .\}$ is equivalent to $(. , .)^T$. The index ℓ is dropped if it is clear from the context; $\dot{\underset{\sim}{x}} := d\underset{\sim}{x}/dt$, etc.

Frequently complex notation is convenient $(i := \sqrt{-1})$:

$$(2.2) \quad \underset{\sim}{x} = x_1 + i\, x_2 = x \cos\phi + i\, x \sin\phi = x\, e^{i\phi} , \quad x := |\underset{\sim}{x}| .$$

A rotation of $\underset{\sim}{x}$ by an angle α in the positive direction (cf. Ω in Fig. 2.1) is given by $e^{i\alpha}\underset{\sim}{x}$. For the computations the complex products must be replaced by matrix products:

$$(2.3/4) \quad e^{i\alpha}\underset{\sim}{x} \sim \begin{vmatrix} \cos\alpha & -\sin\alpha \\ \sin\alpha & \cos\alpha \end{vmatrix} \begin{vmatrix} x_1 \\ x_2 \end{vmatrix} , \quad i \sim \underset{\sim}{E} := \begin{vmatrix} 0 & -1 \\ 1 & 0 \end{vmatrix} .$$

All deflections and forces are assumed to be parallel to the 1- , 2- plane. This implies that inertia effects due to the inclination of the shaft as well as effects caused by a misalignment of the journal are neglected.

<u>The bearing housings</u> are assumed to be point masses m_j , in zero position placed on the centerline s and acted upon by linearly viscoelastic support forces plus oilfilm forces. The equations of motion are

$$(2.5) \quad m_j\, \ddot{\underset{\sim}{u}}_j + \underset{\sim}{B}_j\, \dot{\underset{\sim}{u}}_j + \underset{\sim}{C}_j\, \underset{\sim}{u} + \underset{\sim}{f}_j(\underset{\sim}{v}_j) = 0 , \quad j = 1,\dots, J .$$

Notations: $\underset{\sim}{u}$ deflection of the bearing; $\underset{\sim}{B}$, $\underset{\sim}{C}$ damping and stiffness matrices of the support,

$$(2.6) \quad \underset{\sim}{B} := \begin{vmatrix} b_{11} & b_{12} \\ b_{21} & b_{22} \end{vmatrix} , \quad \underset{\sim}{C} := \begin{vmatrix} c_{11} & c_{12} \\ c_{21} & c_{22} \end{vmatrix} ;$$

$\underset{\sim}{f}(\underset{\sim}{v}) := \underset{\sim}{f}(\ddot{\underset{\sim}{v}}, \dot{\underset{\sim}{v}}, \underset{\sim}{v})$ oilfilm force, positive when acting on the journal, cf. below; $\underset{\sim}{v}$ journal position with respect to its bush.

The discs are considered as point masses m_k attached at the eccentricities $\underset{\sim}{a}_k^e = a_k^e e^{i\alpha_k}$ to the shaft, cf. Fig. 2.2, and acted upon by the shaft forces, the weight, a linear external damping and linear "sealings and leakage" forces (cf. [21], [30]). Equations of motion:

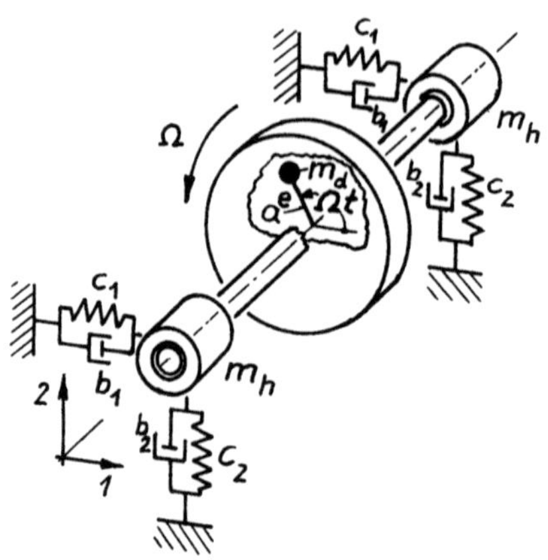

Fig. 2.2 Single disc rotor

$$(2.7)\quad m_k \ddot{\underset{\sim}{x}}_k + b_k^e \dot{\underset{\sim}{x}}_k - \kappa_k^l\, i\, \underset{\sim}{x}_k = -m_k\, ig + m_k\, \Omega^2\, \underset{\sim}{a}_k^e\, e^{i\Omega t} - \underset{\sim}{p}_k .$$

Notations: $\underset{\sim}{x}$ deflection of the shaft at the disc; b^e external damping; κ^l leakage force (positive for a turbine, negative for a turbocompressor); g gravitational acceleration; Ω constant angular velocity of the shaft; a^e, α magnitude and phase of the eccentricity; $\underset{\sim}{p}$ shaft force, cf. below.

The shaft is assumed to be massless and isotropically linearly viscoelastic. Let $c_{\ell'\ell''}$ and $b_{\ell'\ell''}$ be the stiffness and the damping coeffocients, respectively, of the nonrotating shaft bent in a plane. The (hyper-) vectors of the deflections and the forces acting upon the shaft are denoted by

$$(2.8)\qquad \underline{x} := \{\underset{\sim}{x}_1, \ldots, \underset{\sim}{x}_L\} \quad , \quad \underline{p} := \{\underset{\sim}{p}_1, \ldots, \underset{\sim}{p}_L\} \quad .$$

For the nonrotating shaft holds the force deflection relation

$$p = C\,x + B\,\dot{x}\,, \tag{2.9}$$

where the elements of C and B are given by $C_{\ell'\ell''} := I\, c_{\ell'\ell''}$ and $B_{\ell'\ell''} := I\, b_{\ell'\ell''}$, respectively. A frame transformation leads to the force deflection relation for the rotating shaft:

$$p = (C - \Omega B i)\, x + B\,\dot{x} = (C - \Omega BE)\, x + B\,\dot{x}\,. \tag{2.10}$$

Oilfilm forces (cf. [3], [22], [8]): We choose OTT's equation for a "long bearing" with suppressed regions of negative pressure (cf. [20]). In complex notation it can be written as

$$f(\dot{v},v) = -\frac{6}{\nu} P\, e^{i\phi} \left\{ \frac{2W\varepsilon}{(2+\varepsilon^2)(1-\varepsilon^2)^{3/2}} + \right.$$

$$+ \pi \left[1 + \frac{3\varepsilon\varepsilon'}{W+3|\varepsilon\varepsilon'|}\right] \left[\frac{\varepsilon'}{(1-\varepsilon^2)^{3/2}} + \right. \tag{2.11}$$

$$\left.\left. + i\, \frac{(2\phi'-\nu)\varepsilon}{(2+\varepsilon^2)\sqrt{1-\varepsilon^2}}\right]\right\},$$

where $W := \sqrt{1-\varepsilon^2}\,\sqrt{\varepsilon^2(2\phi'-\nu)^2 + 4\varepsilon'^2}$.

Notations: f force on the journal; r journal radius; Δr radial clearance; $v = v(t)$ journal position; $\bar{v} := v/\Delta r$; $\varepsilon := |\bar{v}|$; $\phi := \operatorname{arc} \bar{v}$; $\tau := \Omega t/\nu$; $\varepsilon' := d\varepsilon/d\tau$, $\phi' = d\varepsilon/d\tau$; $\nu \equiv 1/\omega$ (nondimensional) order of a subharmonic, relative reciprocal frequency (see sect. 4) ; $P := \eta\Omega(r/\Delta r)^2\, 2rL$; η absolute viscosity; $2rL$ bearing area.

3. ROTOR EQUATIONS

We rearange the elements of the vectors $\underline{x}$ and $\underline{p}$, cf. (2.8), in such a way that the first J elements are adjoined to bearings and the following K elements to discs. The corresponding (sub-)vectors are denoted by $\underline{x}_b$, $\underline{p}_b$ and $\underline{x}_d$, $\underline{p}_d$, respectively:

$$\underline{x} = \{\underline{x}_b , \underline{x}_d\} \quad , \quad \underline{p} = \{\underline{p}_b , \underline{p}_d\} \ . \tag{3.1}$$

The pertinent decompensation of (2.10) is

$$\begin{vmatrix} \underline{p}_b \\ \\ \underline{p}_d \end{vmatrix} = \begin{vmatrix} \underline{B}_{bb} & \underline{B}_{bd} \\ \\ \underline{B}_{db} & \underline{B}_{dd} \end{vmatrix} \begin{vmatrix} \dot{\underline{x}}_b \\ \\ \dot{\underline{x}}_d \end{vmatrix} + \begin{vmatrix} \underline{C}_{bb} & -\Omega\underline{B}_{bb}i & \underline{C}_{bd} & -\Omega\underline{B}_{bd}i \\ \\ \underline{C}_{db} & -\Omega\underline{B}_{db}i & \underline{C}_{dd} & -\Omega\underline{B}_{dd}i \end{vmatrix} \begin{vmatrix} \underline{x}_b \\ \\ \underline{x}_d \end{vmatrix} \ . \tag{3.2}$$

The equations (2.5) and (2.7) get the forms

$$\underline{M}_b \, \ddot{\underline{u}} + \underline{B}_b \, \dot{\underline{u}} + \underline{C}_b \, \underline{u} + \underline{f}(\dot{\underline{v}},\underline{v}) = 0 \ , \tag{3.3}$$

$$\underline{M}_d\ddot{\underline{x}}_d + \underline{B}_d\dot{\underline{x}}_d - \underline{K}_d i\underline{x}_d + \underline{p}_d = - \underline{M}_d\underline{g} + \Omega^2\underline{M}_d\underline{a}^e e^{i\Omega t} \ . \tag{3.4}$$

Notations:

$$\begin{cases} \underline{u}:=\{\underset{\sim}{u}_1,\ldots,\underset{\sim}{u}_J\} \ , \ \underline{v}:=\{\underset{\sim}{v}_1,\ldots,v_J\} \ , \ \underline{f}:=\{\underset{\sim}{f}_1,\ldots,\underset{\sim}{f}_J\} \ , \\ \\ \underline{g}:=\{ig,\ldots,ig\} \ , \ \underline{a}^e:=\{\underset{\sim}{a}_1^e,\ldots,\underset{\sim}{a}_K^e\} \ , \end{cases} \tag{3.5}$$

$$\begin{cases} \underline{M}_b:=\mathrm{diag}(\underset{\sim}{I}\ m_j) \ , \ \underline{B}_b:=\mathrm{diag}(\underset{\sim}{B}_j) \ , \ \underline{C}_b:= \mathrm{diag}(\underset{\sim}{C}_j) \ , \\ \\ \underline{M}_d:=\mathrm{diag}(\underset{\sim}{I}\ m_k) \ , \ \underline{B}_d:=\mathrm{diag}(\underset{\sim}{I}\ b_k^e) \ , \ \underline{K}_d:=\mathrm{diag}(\underset{\sim}{I}\ \kappa_k^1) \ . \end{cases} \tag{3.6}$$

Geometrical compatibility at the bearings:

$$\underline{x}_b = \underline{u} + \underline{v} \ . \tag{3.7}$$

By elimination of $\underline{p}_d$, $\underline{x}_b$ and putting $\underline{p}_b = \underline{f}$ we obtain three sets of equations of motion for our rotor system. They have the general form

(3.8)
$$\underline{A}_{11}\,\ddot{\underline{x}} + \underline{B}_{11}\,\dot{\underline{x}} + \underline{C}_{11}\,\underline{x} + \underline{B}_{12}\,\underline{u} + \underline{C}_{12}\,\underline{u} + \underline{B}_{13}\,\dot{\underline{v}} + \underline{C}_{13}\,\underline{v} = \underline{c}^1 + \underline{q}^1(t)\ ,$$

(3.9)
$$\underline{A}_{22}\,\ddot{\underline{u}} + \underline{B}_{22}\,\dot{\underline{u}} + \underline{C}_{22}\,\underline{u} + \underline{B}_{21}\,\dot{\underline{x}} + \underline{C}_{21}\,\underline{x} + \underline{B}_{23}\,\dot{\underline{v}} + \underline{C}_{23}\,\underline{v} = 0\ ,$$

(3.10)
$$\underline{A}_{22}\,\ddot{\underline{u}} + \underline{B}_{32}\,\dot{\underline{u}} + \underline{C}_{32}\,\underline{u} + \underline{f}(\dot{\underline{v}},\underline{v}) = 0\ .$$

Here $\underline{x} := \underline{x}_b$, the constant vector $\underline{c}^1$ contains the weight, the vector $\underline{q}^1(t)$ represents the unbalance forces; it has the (least) period T ,

(3.11)
$$\underline{q}^1(t + T) = \underline{q}^1(t)\ , \quad \text{where} \quad T := 2\pi/\Omega\ .$$

The details of the notations in the $M := 2\times(K+J+J)$ equations (3.8/9/10) are easily seen by comparison with the preceding equations. The equations (3.8), (3.9) are linear, the equation (3.10) is nonlinear (with respect to $\underline{v}$). For convenience we introduce the M-dimensional configuration vector (cf. [29])

(3.12)
$$\underline{w}(t) := \{\underline{x},\ \underline{u},\ \underline{v}\} =: \{w_1, \ldots, w_m, \ldots, w_M\}$$

and abbreviate the three equations above by

(3.13)
$$\underline{g}(\ddot{\underline{w}},\dot{\underline{w}},\underline{w},t) = 0 \quad \text{for short:} \quad \underline{g}(D\underline{w},t) = 0\ .$$

The equation (3.13) is nonautonomous (time-dependent) and T-periodic,

(3.14)
$$\underline{g}(D\underline{w},\ t + T) = \underline{g}(D\underline{w},\ t)\ ,$$

if the unbalance is not zero, $\underline{q}^1(t) \not\equiv 0$. It is autonomous (independent of the time) if $\underline{q}^1(t) \equiv 0$. In the latter case we replace (3.13) by

$$(3.15) \qquad \underline{g}(\ddot{\underline{w}},\dot{\underline{w}},\underline{w}) = 0 \ , \quad \text{for short:} \quad \underline{g}(D\underline{w}) = 0 \ .$$

For general discussions and stability investigation it is advisable to introduce the state vector

$$(3.16) \qquad \underline{y}(t) := \{\dot{\underline{x}},\ \underline{x},\ \dot{\underline{u}},\ \underline{u},\ \underline{v}\} =: \{\underline{y}_1,\ \underline{y}_2,\ \underline{y}_3,\ \underline{y}_4,\ \underline{y}_5\}$$

having the elements y_s and the dimension $S:=2\times(K+K+J+J+J)$.

The equations of motion (3.8/9/10) are supplemented by $\dot{\underline{y}}_2 = \underline{y}_1$, $\dot{\underline{y}}_4 = \underline{y}_3$ and abbreviatedly written as

$$(3.17) \qquad \underline{h}(\dot{\underline{y}},\ \underline{y}) = \underline{c} + \underline{q}(t) \quad ,$$

where

$$(3.18) \quad \underline{c} := \{\underline{c}^1,\ 0,\ 0,\ 0,\ 0\} \ , \ \underline{q} := \{\underline{q}^1(t),\ 0,\ 0,\ 0,\ 0\} \ .$$

The equation of motion (3.17) is T-periodic or autonomous under the same conditions as (3.13/15).

4. NUMERICAL CALCULATION OF PERIODIC OSCILLATIONS

We adapt URABE/REITER's numerical version [36], [38] of Galerkin's procedure to our problem.

<u>4.1 Investigation of a nonlinear nonautonomous system</u>. It is advisable to normalize the period $T_s = kT$ of a k-th order subharmonic solution $\underline{p}(t)$ to 2π by introducing the nondimensional time

$$(4.1) \qquad \tau := \frac{2\pi}{\nu T}\, t = t\ \Omega/\nu \ , \quad \nu := k \ , \quad ' = d/d\tau$$

into the equation of motion (3.13):

$$(4.2) \qquad \underline{g}(D\underline{w},\ \tau) = 0 \ ;$$

the equation (4.2) is $2\pi/\nu$-periodic with respect to τ .

We approximate $\underline{p}(\tau + 2\pi) = \underline{p}(\tau)$ by a trigonometric polynomial, a "truncated Fourier series", of the order N :

$$(4.3) \qquad \tilde{\underline{p}}(\tau) := \underline{a}_o + \sum_{n=1}^{N} (\underline{a}_{2n-1} \sin n\tau + \underline{a}_{2n} \cos n\tau) \ .$$

The (approximate) Fourier coefficients $\underline{a}_n$ are M-vectors. They establish the $L := (2N+1)\times M$-dimensional "Fourier vector" $\underline{a}$,

$$(4.4) \qquad \underline{a} := \{\underline{a}_o, \underline{a}_1, \ldots, \underline{a}_{2N}\} =: \{a_1, \ldots, a_\ell, \ldots, a_L\} \ .$$

For any set of numbers $\underline{a}$ we get from (4.3) a function

$$(4.5) \qquad \tilde{\underline{p}} = \tilde{\underline{p}}(\tau, \underline{a}) = p(\tau+2\pi, \underline{a}) \ .$$

When we put this $\tilde{\underline{p}}$ into the equation (4.2), we obtain

$$(4.6) \qquad \tilde{\underline{g}}(\tau, \underline{a}) := \underline{g}(D\tilde{\underline{p}}, \tau)$$

which is 2π-periodic again and can be expanded into a Fourier series:

$$(4.7) \qquad \tilde{\underline{g}}(\tau, \underline{a}) = \underline{r}_o + \sum_{n=1}^{\infty} (\underline{r}_{2n-1} \sin n\tau + \underline{r}_{2n} \cos n\tau) \ .$$

The M-dimensional Fourier coefficients $\underline{r}_n$ are given by

$$(4.8) \quad \underline{r}_o = \frac{1}{2\pi} \int_0^{2\pi} \tilde{\underline{g}}(\tau, \underline{a})\, d\tau \ , \qquad \left.\begin{matrix} \underline{r}_{2n-1} \\ \\ \underline{r}_{2n} \end{matrix}\right\} = \frac{1}{\pi} \int_0^{2\pi} \tilde{\underline{g}}(\tau, \underline{a}) \left\{\begin{matrix} \sin n\tau \\ \\ \sin n\tau \end{matrix}\right\} d\tau.$$

If $\tilde{\underline{p}}$ is a solution of (4.2), the function $\tilde{\underline{g}}$ vanishes, $\tilde{\underline{g}}(\tau, \underline{a}) \equiv 0$, consequently the Fourier coefficients $\underline{r}_n(\underline{a})$ vanish. For an approximate solution $\tilde{\underline{p}}$ we truncate (4.7) after the N-th order terms and demand

$$(4.9) \qquad \underline{r}(\underline{a}) = 0 \ ,$$

where

$$(4.10) \quad \underline{r} := \{\underline{r}_o, \underline{r}_1, \ldots, \underline{r}_{2N}\} =: \{r_1, \ldots, r_\ell, \ldots r_L\} \ .$$

The condition (4.9) constitutes a set of L nonlinear equations for the L unknowns a_ℓ. (See [35], [36] for the possibility of such approximations and for error estimates).

The equation (4.9) is solved by Newton's iterative method (cf. [13], [36], [37]): Let $\underline{a}^{(\rho)}$ be the ρ-th approximate solution of (4.9) and $\underline{r}^{(\rho)} := \underline{r}(\underline{a}^{(\rho)})$ be the ρ-th residual vector. The $(\rho + 1)$-st approximation has the form

$$(4.11)\qquad \underline{a}^{(\rho+1)} := \underline{a}^{(\rho)} - \left[\underline{R}^{(\rho)}\right]^{-1} \underline{r}^{(\rho)} \quad ,$$

where

$$(4.12)\quad \underline{R}^{(\rho)} := \underline{R}(\underline{a}^{(\rho)}) := \frac{\partial \underline{r}}{\partial \underline{a}} := \frac{\partial(r_1, \ldots, r_\ell, \ldots, r_L)}{\partial(a_1, \ldots, a_\ell, \ldots, a_L)}$$

is the Jacobian matrix of (4.9) taken at $\underline{a} = \underline{a}^{(\rho)}$.

Newton's method converges to a solution $\underline{a}$ of (4.9) if the initial approximation $\underline{a}^{(o)}$ is close enough to $\underline{a}$, and $\det \underline{R} \neq 0$ for a sufficiently large neighbourhood of $\underline{a}$ (cf. [13]). As a starting vector $\underline{a}^{(o)}$ can serve either the solution $\underline{a}$ of a neighbour problem (with slightly different parameters), an extrapolated value, cf. sect. 5.4, or, at bifurcation points, a separately calculated vector, cf. sect. 5.5.

Since the equations (3.8) and (3.9) are linear, the Fourier terms corresponding to $\underline{x}$ and $\underline{u}$ can be eliminated formally from (4.9) which reduces the size of $\underline{a}$ but can damage the structure of $\underline{R}$.

4.2. Investigation of the nonlinear autonomous system.

The period T_s of a periodic solution $\underline{p}(t)$ of the autonomous equation of motion (3.15) is a function of the parameters of the system. Parallel to (4.1) we normalize that period to 2π by introducing $\omega := 1/\nu$ as an unknown nondimensional relative frequency into (4.1) which enters the transformed equation of motion as a parameter

$$(4.13)\qquad \underline{g}(D\underline{w}, \omega) = 0 \quad ,$$

and has to be determined from it. Because of $\nu = 1/\omega$ all results from the section 4.1 can be used. Since $\underline{g}(D\underline{w}, \omega)$ does not depend explicitly on τ , an arbitrary shift τ_s of the periodic solution $\underline{p}(\tau)$ is possible. Therefore, τ_s can be chosen such that a certain (reference) Fourier term of the expansion (4.3) vanishes, say $a_{\ell'} = 0$. The Fourier vector $\underline{a}$, cf. (4.4), is modified by exchanging $a_{\ell'}$ for ω :

$$\underline{a} := \{a_1, \ldots, a_{\ell'-1}, \omega, a_{\ell'+1}, \ldots, a_L\} .$$

The following calculations run parallel to the nonautonomous case.

4.3 Constant solutions. For a vanishing unbalance vector $\underline{q}^1(\tau)$, cf. (3.8), we obtain the constant (static) deflection $\underline{a}_o$, see (4.3). A simplified version of the nonlinear equation (4.9) must be solved by Newton's method.

5. VARIATIONAL EQUATION, STABILITY, BIFURCATIONS

In this section we discuss some effects of (small) parameter changes on periodic solutions. We linearize the equation of motion (3.17), or the equation (4.9) for the Fourier coefficients, with respect to "variations" of parameters and solutions and apply to the variational equations the theory of linear equations. This supplies insight into the neighborhood of the known solution and the character of branching off solutions at bifurcation points, it permits to trace the influences of parameter changes (to investigate the stability), or to calculate the changes necessary to accomplish a desired variation (cf. the "sensitivity coefficients" in [31], [32]).

5.1 Variation of the differential equation. Let the vector $\underline{\xi}$ comprise the "essential" parameters of the motion $\underline{y} = \underline{y}(\tau,\underline{\xi})$. It may include the initial conditions and, in an autonomous problem, the relative frequency ω . The equation (3.17) gets the form

$$\underline{h}(\underline{y}', \underline{y}, \underline{\xi}) = \underline{c}(\underline{\xi}) + \underline{q}(\tau,\underline{\xi}) . \qquad (5.1)$$

Let

$$(5.2) \qquad \underline{p}(\tau, \underline{\xi}_o) \equiv \underline{y}(\tau, \underline{\xi}_o)$$

denote a known 2π-periodic (possibly subharmonic) solution of (5.1), the "undisturbed motion". A small deviation (disturbance, variation) $\Delta\underline{\xi}$ of $\underline{\xi}$ from $\underline{\xi}_o$ causes a variation $\underline{z}$ of the motion $\underline{y}(\tau)$, which, in linear approximation, is given by

$$(5.3) \qquad \underline{z} = \underline{y}_{\underline{\xi}} \, \Delta\underline{\xi} \quad .$$

For $\underline{z}(t)$ holds the (nonhomogeneous) linear differential equation with periodic coefficients, the variational equation

$$(5.4) \qquad \underline{z}' = \underline{H} \, \underline{z} + \underline{f}(\tau)$$

where

$$(5.5) \qquad \underline{H}(\tau) = \underline{H}(\tau + 2\pi) := - \underline{h}_{\underline{y}'}^{-1} \, \underline{h}_{\underline{y}} \ ,$$

$$(5.6) \qquad \underline{f}(\tau) := \underline{h}_{\underline{y}'}^{-1} \, (\underline{c}_{\underline{\xi}} + \underline{q}_{\underline{\xi}} - \underline{h}_{\underline{\xi}}) \Delta\underline{\xi} \ .$$

The Jacobians $\underline{y}_{\underline{\xi}} := \partial\underline{y}/\partial\underline{\xi}$, $\underline{h}_{\underline{y}'} := \partial\underline{h}/\partial\underline{y}'$, etc. are taken at $\underline{\xi} = \underline{\xi}_o$, $\underline{y} = \underline{p}$; $\det \underline{h}_{\underline{y}'} \neq 0$.

For the theory of linear differential equations with periodic coefficients see [5], [9], [10], [11] .

<u>5.3 Stability</u>. For the stability investigation the homogeneous (matrix) variational equation, cf. (5.4),

$$(5.7) \qquad \underline{Z}' = \underline{H} \, \underline{Z} \ , \quad \underline{Z}(\tau) := (\underline{z}_1, \ldots, \underline{z}_S)$$

has to be solved numerically for the initial condition $\underline{Z}(0) = \underline{I}$. Its characteristic multipliers σ_s are obtained from

$$(5.8) \qquad \det(\underline{Z}(2\pi) - \sigma_s \underline{I}) = 0 \ .$$

The solution is stable, if $|\sigma_s| < 1$ for $s = 1, \ldots, S$. (On some general discussions see [4], [15]).

The condition above is too strict for a periodic oscillation $\underline{p}(\tau)$ of a completely balanced rotor where the equation of motion (3.17) is autonomous and has a multiplier $\sigma_1 = 1$. By Andronov-Witt's theorem (cf. [17]) the solution is stable if $|\sigma_s| < 1$ for $s = 2,\dots, S$. A stable periodic solution of an autonomous equation of motion is a limit cycle (cf. [10], [23], [25]).

<u>5.4 Variation of the Fourier coefficients</u>. The (approximate) Fourier coefficients $\underline{a}(\underline{\xi}_o)$ of the 2π-periodic solution $\underline{p}(\tau, \underline{\xi}_o)$ are calculated from the nonlinear equation (4.9),

$$(5.9) \qquad \underline{r}(\underline{a}, \underline{\xi}) = 0 \quad \text{with} \quad \underline{\xi} = \underline{\xi}_o \ .$$

For a variation $\Delta\underline{\xi}$ we obtain by linearization of $\underline{r}$ with respect to $\Delta\underline{a}$ and $\Delta\underline{\xi}$

$$(5.10) \qquad \Delta\underline{a} = - \underline{R}^{-1}\, r_{\underline{\xi}}\Delta\underline{\xi} \quad \text{if} \quad \det \underline{R} \neq 0 \ , \quad \text{cf. (4.12)} \ .$$

This $\Delta\underline{a}$ represents the Fourier coefficients of the *periodic* variation corresponding to the nonhomogeneous equation 5.4 if no Fourier coefficients a_ℓ are exchanged for other parameters. It can be used in Newton's method to "extrapolate" starting vectors $\underline{a}^{(o)} := \underline{a} + \Delta\underline{a}$ for changed parameters. Subharmonic variations (and subharmonic resonances) must be investigated separately with proper adaptions of ν in (4.1) and in the trigonometric polynomial (4.3). A small $|\det \underline{R}|$ indicates the closeness of resonance (cf. [7] chapter V for a discussion of related problems), the corresponding variation $\Delta\underline{a}$ will be large (in general). At resonance, i.e. at $\det \underline{R} = 0$, the homogeneous equation $\underline{R}\Delta\underline{a} = 0$ has the eigenvector $\underline{a}^h$,

$$(5.11) \qquad \underline{R}\, \underline{a}^h = 0 \ ,$$

which represents the Fourier coefficients of a 2π-periodic solution for the homogeneous differential equation (5.4) .

<u>5.5 Bifurcations</u>. We discuss bifurcation (or branching) points as a heuristic means to detect "hidden" periodic solutions, to construct appropriate starting vectors for the

Newton iterations of section 4. The survey article [27] gives a brief introduction and provides examples which demonstrate various phenomena and problems. More comprehensive treatments can be found in [13], [39] .

The equation (5.9) is an implicit representation of the vector $\underline{a}_o := \underline{a}(\xi_o)$ where $\xi = \xi_o$ is a scalar parameter. By the theorem of implicit functions, cf. e.g. [13], the equations are uniquely solvable in the beighborhood of a solution $(\underline{a}_o, \xi_o)$ if $\det \underline{R} \neq 0$, the relation (5.10) holds (in linear approximation) for the variations $\Delta\underline{a}, \Delta\xi$. A small $|\det \underline{R}|$, a large $\Delta\underline{a}$ at a resonance can simply be the consequence of a rapid change of $\underline{a}(\xi)$ in the vicinity of a vertical tangent, cf. Fig. 5.1. (An exchange of ξ and (the fastest varying) a_ℓ , removes such difficulties). On the other hand a bifurcation, cf. Fig. 5.2, can occur.

If $\det \underline{R}(\xi)$ has a simple zero at $\xi = \xi_o$, it can be shown (cf. [13], [27]) that, under certain conditions, the branching off solution has the tangent $\underline{a}^h$ at the branching point, cf. Fig. 5.2. This can be easily exploited in numerical calculations by choosing a number θ and assuming a starting value $\underline{a}^{(o)}$ for Newton's method, $\underline{a}^{(o)} = \underline{a}_o + \theta\underline{a}^h$. The possible convergence of $\underline{a}^{(\rho)}$ to the bifurcation point is prevented by an exchange of ξ for a suitably chosen Fourier coefficient $a_{\ell'}$. The arrows in Fig. 5.2 indicate the procedure ("N" direction of the convergence in Newton's method).

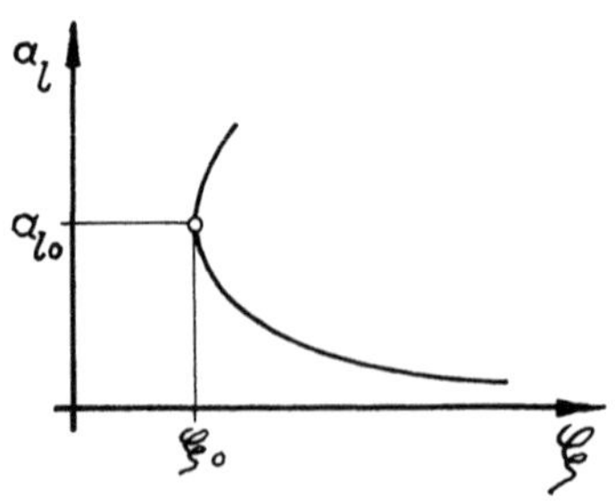

Fig. 5.1 Vertical tangent

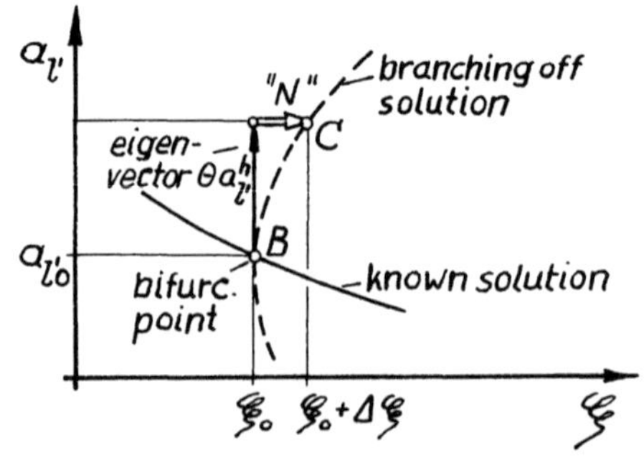

Fig. 5.2 Bifurcation point

There are some numerical problems: The solutions $\underline{a}_o$ are easily computed if the branching off solution is a subharmonic. Because of $\det \underline{R} \neq 0$ no difficulties arise in Newton's method, the Jacobian $\underline{R}^k$ for the k-th order subharmonic is calculated separately. (A branching off *harmonic* solution reveals its existence beforehand by irregular convergence of Newton's method). Since the resonance point ξ_o can not be computed exactly, $\det \underline{R}^k = 0$ is only approximately satisfied. The difficulty can be overcome by a Gauss algorithm with "complete pivoting", cf. [41]. In the resulting triangularized form of $\underline{R}^k$ the last diagonal element is the smallest one (if ξ_o is sufficiently good); it is canceled, the eigenvector $\underline{a}^h$ can be calculated, cf. (5.11).

<u>5.6 Close-by points</u>. Calculations show that curves $a_\ell(\xi)$ representing different (types of) solutions happen to approach each other to a certain degree, there exist "close by" points, cf. points B' , B" in Fig. 5.3. Such couples of points can be interpreted as a prior bifurcation point which was dissolved by parameter changes. Consequently, $\Delta^k(\xi) := |\det \underline{R}^k(\xi)|$ has a relative minimum at $\xi = \xi_{B'}$. This characteristic permits to calculate $\xi_{B'}$. The complete pivoting triangularization of $\underline{R}^k(\xi_{B'})$, cf. above, with a subsequent canceling of the last diagonal element yields an artificial eigenvector $\underline{a}^a$. This vector can be used to arrive at the close-by solution by a Newton method similarly as $\underline{a}^h$ was used in sect. 5.5 for the detection of the branching off solution. The procedure is pointed out by the arrows in Fig. 5.3.

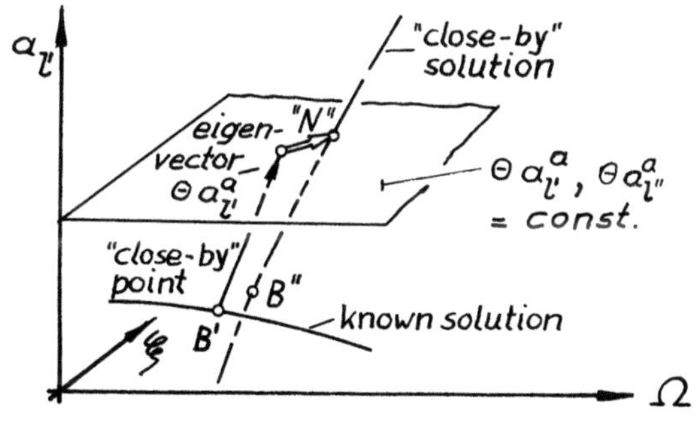

Fig. 5.3 "Close-by" point

6. INVESTIGATION OF A SINGLE DISC ROTOR

We apply the general procedures of the previous sections to investigate symmetrical motions of the symmetrical rotor of a turbine shown in Fig. 2.2.

6.1 Equations of motion. The nondimensionalized equations of motion (3.8/9/10) are

$$\frac{\Omega^2}{\nu^2}\,\underset{\sim}{x}'' + \frac{\Omega}{\nu}\,(b+b^e)\,\underset{\sim}{x}' + (c-(\Omega b+\kappa)\,\underset{\sim}{E})\,\underset{\sim}{x}$$

$$(6.1)\qquad - \frac{\Omega}{\nu}\,b(\underset{\sim}{u}+\underset{\sim}{v})' - (c-\Omega b\underset{\sim}{E})\,(\underset{\sim}{u}+\underset{\sim}{v}) = -\,i + \Omega^2\,a^e\,e^{i\nu\tau}$$

$$\frac{\Omega^2}{\nu^2}\,m\,\underset{\sim}{u}'' + \frac{\Omega}{\nu}\,(\underset{\sim}{B}+b)\,\underset{\sim}{u}' + (\underset{\sim}{C}+c-\Omega b\underset{\sim}{E})\,\underset{\sim}{u}$$

$$(6.2)\qquad - \frac{\Omega}{\nu}\,b(\underset{\sim}{x}-\underset{\sim}{v})' - (c-\Omega b\underset{\sim}{E})\,(\underset{\sim}{x}-\underset{\sim}{v}) = 0\quad ,$$

$$(6.3)\qquad \frac{\Omega^2}{\nu^2}\,m\,\underset{\sim}{u}'' + \frac{\Omega}{\nu}\,\underset{\sim}{B}\,\underset{\sim}{u}' + \underset{\sim}{C}\,\underset{\sim}{u} + \frac{\Omega}{S}\,\underset{\sim}{f}(\underset{\sim}{v}',\,\underset{\sim}{v}) = 0\quad ,$$

where

$$\underset{\sim}{B} := \begin{vmatrix} b_1 & 0 \\ 0 & b_2 \end{vmatrix}\,,\quad \underset{\sim}{C} := \begin{vmatrix} c_1 & 0 \\ 0 & c_2 \end{vmatrix}\,,\quad m := 2m_h/m_d\quad .$$

We restrict our investigations to the parameters $m = 0.2$, $S = 1.0$, $c = c_1 = 0.4$, $c_2 = 0.8$, $b^e = b_1 = b_2 = \kappa = 1\cdot 10^{-3}$, $b = 2\cdot 10^{-3}$. Variable is the frequency: $0.3 \le \Omega \le 6$, the eccentricity is chosen: $a^e = 0$, $a^e = 1$.

6.2 Motions of a shaft on a rigid oilfilm. The motions of a shaft on a rigid oilfilm, $\underset{\sim}{v} \equiv 0$, are governed by the linear equations (6.1), (6.2). We solve those equations to have reference motions for the nonlinear problem.

Natural frequencies: When we neglect all nonconservative terms in (6.1/2), we obtain four natural frequencies $(\omega\Omega)_j = (\Omega/\nu)_j$, cf. (4.1) and section 4.2, the modal vectors are shown in Table 6.1. The arrows (to be multiplied by the scale factors 1 or 1/10) indicate the relative magnitudes and directions of the modal vibrations.

j	1	2	3	4
$(\omega\Omega)$	0.436	0.511	2.052	2.478
x_1	1	0	-0.105	0
x_2	0	1	0	-0.070
u_1	0.475	0	1	0
u_2	0	0.348	0	1
$\underset{\sim}{x}$	×1	×1	×1/10	×1/10
$\underset{\sim}{u}$	×1	×1	×1	×1

Table 6.1 Natural frequencies, modal vectors; $\underset{\sim}{v} \equiv 0$

Forced vibrations: For $a^e = 1$ and $\omega = 1/\nu = 1$ we get the oscillations

$$x_i = a_{i1}^1 \sin\tau + a_{i2}^1 \cos\tau = Q_i^1 \cos(\tau - \psi_i^1) \quad ,$$

(6.4)

$$u_i = a_{i1}^2 \sin\tau + a_{i2}^2 \cos\tau = Q_i^2 \cos(\tau - \psi_i^2) \quad .$$

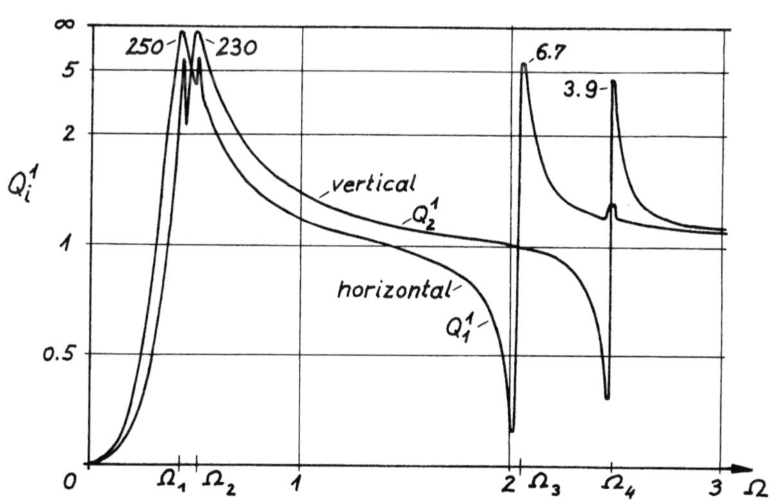

Fig. 6.2 Amplitudes of the disc; $\underset{\sim}{v} \equiv 0$

The amplitudes Q_i^1 are shown in Fig. 6.2. (The vertical axis is divided reciprocally proportional to Q for $Q > 1$). The resonant peaks at the critical speeds Ω_j are very sharp because of the small damping terms.

6.3 General remarks on the calculation on the nonlinear oscillations. For the evaluation of the equations (6.1/2/3) the eliminations mentioned at the end of section 4.1 were carried out formally to save computer time; there are $4N+2$ nonlinear equations of the form (4.9). Equally, derivatives corresponding to (4.12) were calculated formally as far as possible. Derivatives with respect to parameters were obtained numerically. For Fourier analysis see [28]. Standard programs of the "Rechenzentrum der TU Braunschweig" were used for linear equations, and the NAG-library ("Nottingham Algorithms Group") for step by step integrations and eigenvalue problems (sect. 5.3). A special subroutine was written for complete pivoting (sect. 5.5/6)

In most cases a reasonably chosen arbitrary starting vector $\underline{a}^{(o)}$ converged to a solution of (4.9) when - in the initial steps of Newton's method - the correction term in (4.11) was multiplied by a factor $\Lambda < 1$. Subsequent variations of the interesting parameters or Fourier coefficients yielded families of solutions, bifurcations led into the neighborhood. Single term Fourier approximations $(N = 1)$ were sufficient for exploratory calculations, three term approximations were used in the vicinity of (harmonic) resonances, up to six and more were necessary for subharmonics.

6.4 Constant solutions. Fig. 6.3 shows the static deflections of the journal in a v_1- , v_2- , Ω -space , and in the v_1- , v_2-plane for $0.3 \leq \Omega < 6$. The stability is investigated by (5.8). The state vector $\underline{z}$ has 10 elements, cf. (3.16), the matrix $\underline{H}$ is constant. By $\underline{z} = \underline{a}\, e^{\lambda\tau}$ we get the eigenvalue problem

$$(6.5) \qquad (\lambda \underline{I} - \underline{H})\underline{a} = 0 \ .$$

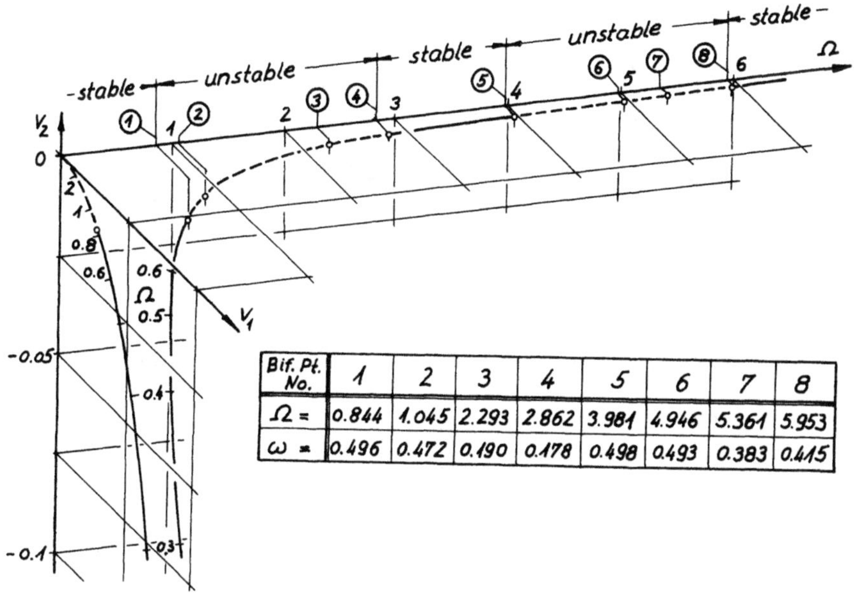

Bif. Pt. No.	1	2	3	4	5	6	7	8
Ω =	0.844	1.045	2.293	2.862	3.981	4.946	5.361	5.953
ω =	0.496	0.472	0.190	0.178	0.498	0.493	0.383	0.415

Fig. 6.3 Constant deflections of the journal

It has five complex conjugate eigenvalues

(6.6) $$\lambda_s = \delta_s + i\omega_s \;, \quad s = 1,\ldots,5 \;.$$

The corresponding solutions can be written in the form

(6.7) $$\underline{z}_s^F(\tau) = A_s e^{\delta_s(\tau-\tau_s)}\{\cos\omega_s(\tau-\tau_s)\,\mathrm{Re}\,\underline{a}_s - \sin\omega_s(\tau-\tau_s)\,\mathrm{Im}\,\underline{a}_s\}$$

where A_s (amplitude), τ_s (time shift) are the constants of integration. The expression in the braces can be interpreted as a rotary (elliptical) modal vibration, cf. Fig. 6.5.

Fig. 6.4 shows the five δ_s . (The δ-axis is divided $\sim\sqrt[3]{\delta}$ to reach vertical intersections with the zero axis). The zeros of δ_s are enumerated ①, ..., ⑧, Fig. 6.3 contains a list with the corresponding values of Ω and ω . At the -/+ crossing of δ_s the stationary deflection becomes unstable, cf. Fig. 6.3.

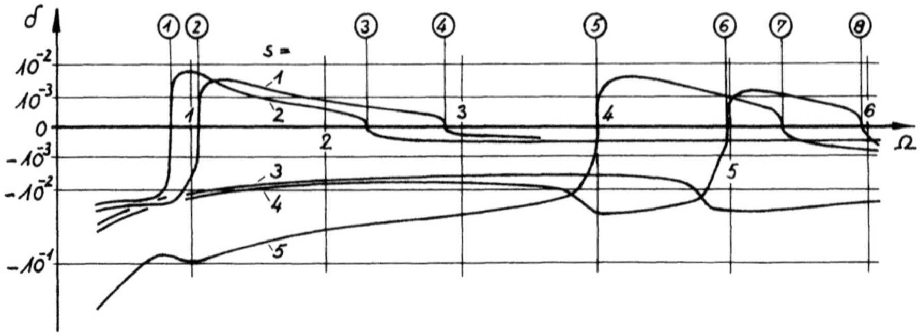

Fig. 6.4 Real parts δ_s for the constant solution

The five frequencies ω_s (as far as they are less then 0.5) are shown in the lower part of Fig. 6.6 by the lines partly full, partly chaindotted. The dotted lines in that figure present the four natural frequencies ω_j calculated in section 6.2 for a rotor on rigid oilfilm, cf. Table 6.1. With growing Ω the ω_s converge from below to four ω_j , except in the vicinity of the points ⑤ and ⑥ where they switch partners. The Fig. 6.5 presents the modal

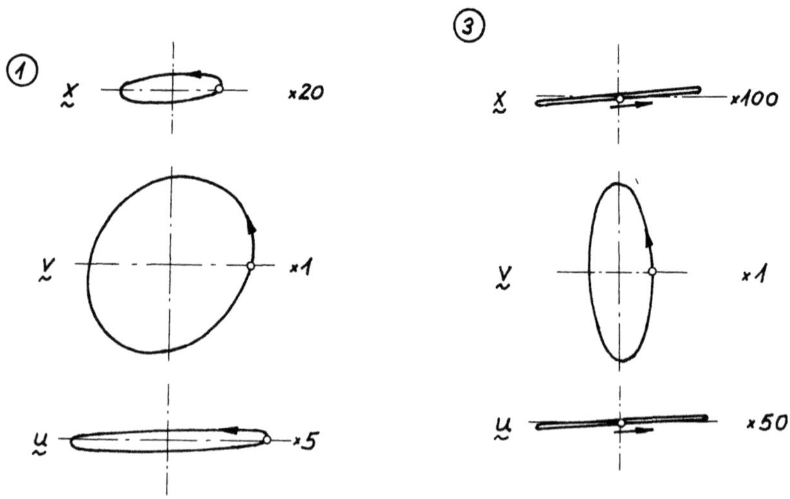

Fig. 6.5 Modal vibrations at the points ① and ③

vibrations at the stability limits ① and ③ ($\underset{\sim}{x}$ disc, $\underset{\sim}{u}$ bearing, $\underset{\sim}{v}$ journal).

A look at the oilfilm force (2.11) reveals a bit of the mechanism underlying the oscillations at the stability limits as well as the self-excitations, treated in the next section: The journal motion is roughly circular, cf. 6.5 ,

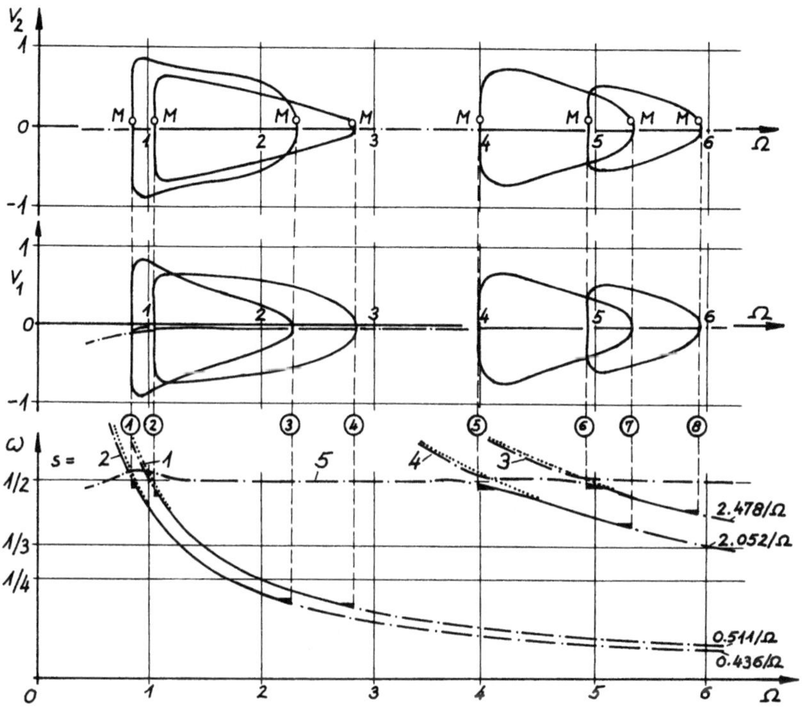

Fig. 6.6 Self-excited oscillations, envelopes and frequences

$\phi' \approx 1$ and $\nu = 1/\omega > 2$. The tangential term of the oilfilm force (containing the imaginary unit i in (2.11)) pushes the journal in the Ω -direction. The transition from ① to ③ is characterized by the fact that the frequencies $\omega\Omega$ of the oscillations change from 0.419 , which is somewhat lower than the natural frequency $(\omega\Omega)_1$ listed in Table 6.1, to 0.436 , which is very close to the natural frequency. The oscillations at the latter point can be understood as an "internal resonance": The amplitudes of disc and bearings have greatly increased, (cf. the scale factors), there is a 90° phase difference between the journal motion and the motions of the other parts and an energy balance between the exciting and the damping forces (cf. [2]). For larger values Ω the oilfilm is no longer able to sustain motions similar to the considered normal modes because the damping forces grow too fast. Similar explanations hold for the other transition points.

6.5 Self-excited vibrations. By the methods of point mappings it can be shown that $2\pi/(\omega_s\Omega)$- periodic (self-excited) oscillations branch off from the constant solutions at the stability limits ① ,..., ⑧ . (More complicated situations can arise at multiple eigenvalues, cf. [27]). Accordingly, the modal vibrations (6.7) together with the corresponding ω_s are the eigenvectors $\underline{a}^h$ of equation (5.11). The branching off solution can be calculated, as outlined in section 5.5, by the Fourier approximations for autonomous systems, cf. section 4.2, we get the "half-frequency" whirl.

Fig. 6.6 shows, as functions of Ω , the constant deflections (chaindotted), the mean values of the oscillating solutions of the journal (very close to the Ω-axis), the envelopes of the motions (solid lines), and the frequencies ω . In most cases only one-term Fourier approximations were used. No careful search for sharp peaks, cf. Fig. 6.2, or for secondary branchings was done. At the points marked by the letter M three-term Fourier expansions were computed and the stability investigated. All eight eigenvalue problems are ill-conditioned. Only the two solutions belonging to ④ and ⑧ are stable. But even for them some of the multipliers are so close to the unit circle that further phenomena (secondary branchings?) are to be expected.

6.6 Forced vibrations. Starting vectors $\underline{a}^{(o)}$ for forced vibrations are easy to guess at frequencies Ω where small oscillations occur. A stepwise variation of Ω permits to investigate the whole interval. The convergence of Newton's method may be poor near resonant peaks. Only the journal motions exhibit appreciable second and third order harmonics. Except for the resonant peaks, which are reduced by 25 - 50%,

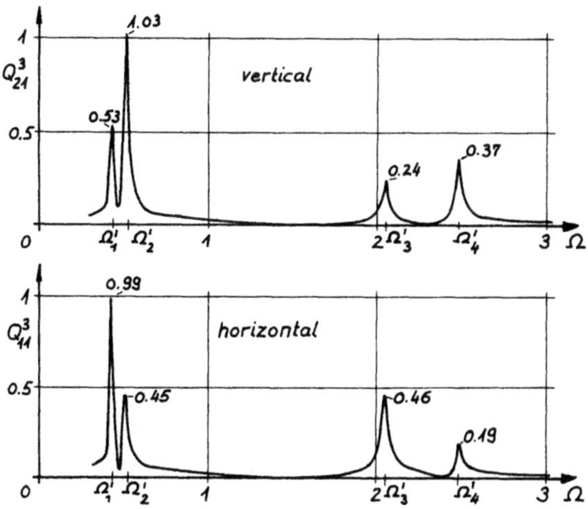

Fig. 6.7 Forced journal oscillations, amplitudes of first harmonics

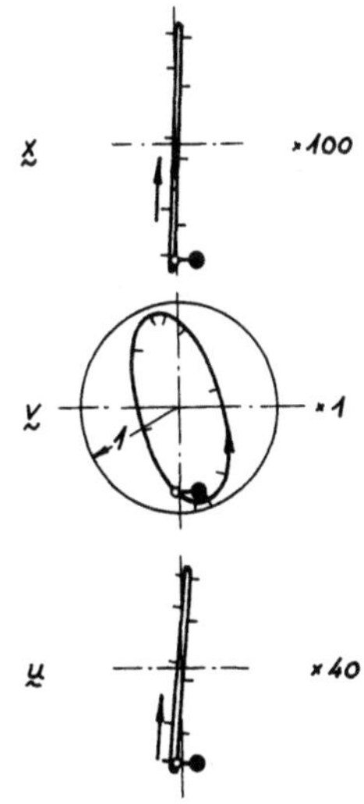

Fig. 6.8 Forced oscillations at Ω_2'

the ocsillations of disc and bearings are almost identical with the linear ones shown in Fig. 6.2. The amplitudes of the first harmomics of the journal oscillations, Q_{il}^3, are presented in Fig. 6.7. The higher order harmonics ensure that the journal does not touch the bush; Fig. 6.8 shows the motions at the resonance frequency $\Omega_2' = 0.509$. The little "pendula" mark the relative phase of the unbalance, the dashes at the orbits mark time intervals of $2\pi/10$ duration.

Stability investigations at several frequencies of the abovementioned interval indicate that the calculated forced vibrations have about the same unstable regions as the stationary deflections studied in section 6.4.

6.7 Bifurcations. In the vicinity of the stability limit of the constant solution, $\Omega_{①} = 0.844$ (cf. Fig. 6.3), the forced vibration gets unstable. Fig. 6.9 shows det $\underline{R}^2$ for a second order subharmonic in that Ω-range. There are two single zeros at $\Omega_{①}' \simeq 0.827$ and $\Omega_{①}'' \simeq 0.843$, two subharmonic solutions branch off. A calculation in the direction of the eigenvectors $\underline{a}^h$, cf. sect. 5.5, leads into the branches characterized in Fig. 6.10 by the response curves for the initial values $v_1(0)$. Fig. 6.11 shows a stable second order subharmonic oscillation.

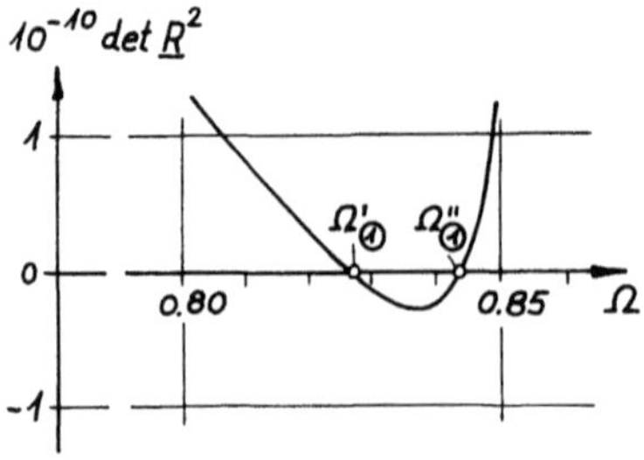

Fig. 6.9 Singular points of $\underline{R}^2$

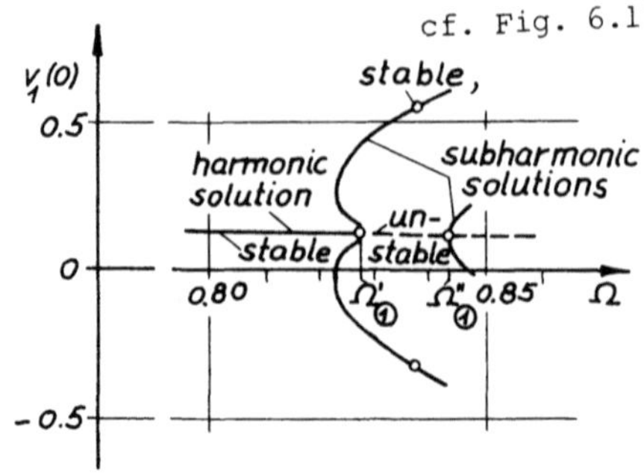

Fig. 6.10 Bifurcation points

6.8 A close-by solution. As an example for the application of the "close-by" procedure outlined in section 5.6 the Jacobian $\underline{R}^3$ for third order subharmonics was tested along the forced vibrations calculated in section 6.6. At about Ω=1.51 the determinant det $\underline{R}^3$ has a minimum, the "artificial" eigenvector $\underline{a}^a$ has (subharmonic) terms equivalent to

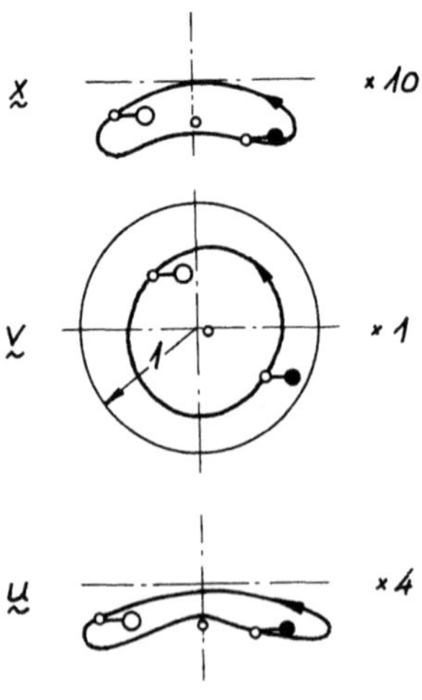

Fig. 6.11 Stable second order subharmonic solution, Ω = 0.8376

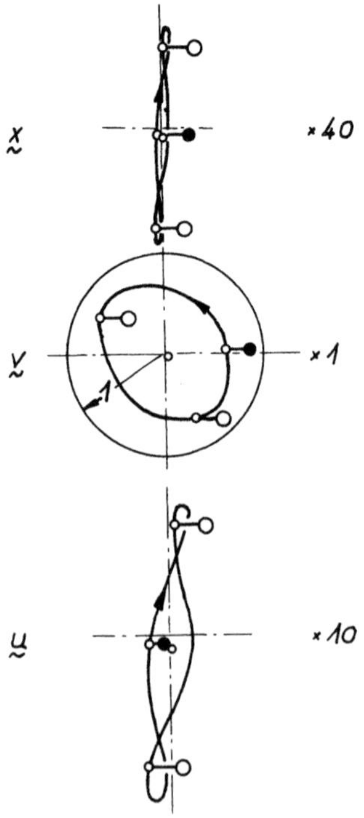

Fig. 6.12 Unstable third order subharmonic solution, Ω = 1.52...

$$(6.8)\quad \begin{cases} v_1 \sim -\,0.64\,\sin\tau + 0.26\,\cos\tau \;, \\ v_2 \sim +\,0.59\,\sin\tau + 0.42\,\cos\tau \;. \end{cases}$$

A calculation with fixed Fourier coefficients for the subharmonics of v_2, a free adaptable frequency Ω and an additional free component of the unbalance force led to a solution of the sought type. We arrive at $\Omega = 1.52...$ and the subharmonic terms

$$(6.9)\quad \begin{cases} v_1 = -\,0.50\,\sin\tau + 0.51\,\cos\tau + \ldots, \\ v_2 = \,0.60\,\sin\tau + 0.19\,\cos\tau + \ldots \;. \end{cases}$$

This demonstrates the practicability of the procedure. The corresponding subharmonic motion is shown in Fig. 6.12, it is unstable.

REFERENCES

[1] BROMMUNDT, E.: Eindimensionale Kontinua; Untersuchung nichtlinearer Schwingungen, (Habilitation) Techn. Hochschule Darmstadt: 1968.

[2] BROMMUNDT, E.: Stabilitätsaussagen über Turborotoren mit Hilfe von Energieausdrücken, Ing.-Archiv 40 (1971) 159-167.

[3] CAMERON, A.: The Principles of Lubrication, London: Longmans 1966.

[4] CESARI, L.: Asymptotic Behavior and Stability Problems in Ordinary Differential Equations, Berlin/Göttingen/ Heidelberg: Springer 1963.

[5] Coddington, E.A. & LEVINSON, N.: Theory of Ordinary differential Equations, New York/ Toronto/London: McGraw-Hill 1955.

[6] GANTMACHER, F.R.: The Theory of Matrices, Vols. 1, 2, New York: Chelsea 1959.

[7] GIACAGLIA, G.E.O.: Perturbation Methods in Non-Linear Systems, New York/Heidelberg/Berlin: Springer 1972.

[8] GUNTER, E.J. & KIRK, R.G.: Transient Journal Bearing Analysis, NASA Contractor Report CR-1549, Washington: NASA 1970

[9] HALANAY, A.: Differential Equations; Stability, Oscillations, Time Lags, New York/London: Academic Press 1966.

[10] HALE, J.K.: Ordinary Differential Equations, New York/ London/Sydney/Toronto: Wiley-Interscience 1969.

[11] IAKUBOVICH, V.A. & STARZHINSKII, V.M.: Linear Differential Equations With Periodic Coefficients and Their Applications (in Russian), Moscow: Nauka 1972.

[12] JENSEN, J., NIORDSON, F. & PEDERSEN, P.: Author & KWIC Index on Rotor Dynamics, Solidmech, Techn. Univ. of Denmark, 2800 Lyngby: 1974.

[13] KRASNOSEL'SKII, M.A., VAINIKKO, G.M., ZABREIKO, P.P., RUTITSKII, Ya.B. & STETSENKO, V.Ya.: Approximate Solution of Operator Equations, Groningen: Wolters-Noordhoff Publishing 1972.

[14] KUSHUL', M.Ia.: Auto-oscillations of Rotors, Moscow: Nauka 1963.

[15] La SALLE, J. & LEFSCHETZ, S.: Stability by Liapunov's Direct Method, New York/London: Academic Press 1961.

[16] LUND, J.W. & SAIBEL, E.: Oil Whip Whirl Orbits of a Rotor in Sleeve Bearings, J. Engng. Ind., Trans. ASME 89 B (1967) 813-823.

[17] MALKIN, J.G.: Theorie der Stabilität einer Bewegung, Berlin: Akademie-Verlag 1959.

[18] MITCHELL, J.R., HOLMES, R. & BYRNE, J.: Oil Whirl of a Rigid Rotor in 360° Journal Bearings: Further Characteristics, Proc. Instn. Mech. Engrs. 180 (1965-66) 593-610.

[19] MUSZYŃSKA, A.: Vibrations Nonlinéaires des Arbres Tournants, Conf. Fasc. 101, Warszawa: Państwowe Wydawnictwo Naukowe 1973.

[20] OTT, H.H.: Zylindrische Gleitlager bei instationärer Belastung, (Dissertation) ETH Zürich: 1948.

[21] PILTZ, E.: Belastungen von Dampfturbinenlaufrädern bei Auslenkungen der Welle, Konstruktion 24 (1972) 141-148.

- : Einfluss von Dampfbelastungen auf die Stabilitätsgrenze einer einstufigen Modellturbine, Konstruktion 25 (1973) 130-135.

[22] PINKUS, O. & STERNLICHT, B.: Theory of Hydrodynamic Lubrication, New York/Toronto/London: McGraw-Hill 1961.

[23] PLISS, V.A.: Nonlocal Problems of the Theory of Oscillations, New York/London: Academic Press 1966.

[24] POZNIAK, E.L.: Nonlinear Oscillations of Rotors on Sleeve Bearings; in Dynamics of Flexible Rotors, ed. F.M. Dimentberg (in Russian), Moscow: Nauka 1972.

[25] REISSIG, R., SANSONE, G. & CONTI, R.: Nichtlineare Differentialgleichungen höherer Ordnung, Roma: Edizioni Cremonese 1969.

[26] SOMEYA, T.: Stabilität einer in zylindrischen Gleitlagern laufenden, unwuchtfreien Welle, Ing.-Archiv 33 (1963/64) 85-108.

[27] STAKGOLD, I.: Branching of Solutions of Nonlinear Equations, SIAM Review, 13 (1971) 289-332.

[28] STIEFEL, E.: Einführung in die numerische Mathematik, Stuttgart: Teubner 1963.

[29] SYNGE, J.L.: Classical Dynamics; in Handbuch der Physik, ed. S. Flügge, vol. III/1, Berlin/Göttingen/ Heidelberg: Springer 1960.

[30] THOMAS, H.J.: Instabile Eigenschwingungen von Turbinenläufern, angefacht durch die Spaltströmungen in Stopfbuchsen und Beschaufelungen, Bull. de L'AIM 71 (1958) 1039-1064.

[31] TOMOVIĆ, E.: Sensitivity Analysis of Dynamic Systems, New York/Toronto/London: McGraw-Hill 1963.

[32] TOMOVIC, R. & VUKOBRATOVIC, M.: General Sensitivity Theory, New York: American Elsevier 1972.

[33] TONDL, A.: Some Problems of Rotor Dynamics, London: Chapman & Hall 1965.

[34] TONDL, A.: Vibration of Rigid Rotors With Vertical Shafts Mounted in Aerostatic Bearings, Monographs and Memoranda No. 14, National Research Institute, Bechovice (CSSR) 1973.

[35] URABE, M.: Galerkin's Procedure for Nonlinear Periodic Systems, Arch. Rat. Mech. Anal. 20 (1965) 120-152.

[36] URABE, M. & REITER, A.: Numerical Computation of Nonlinear Forced Oscillations by Galerkin's Procedure, J. Math. Anal. App. 14, (1966) 107-140.

[37] URABE, M.: Nonlinear Autonomous Oscillations, New York/ London: Academic Press 1967.

[38] URABE, M.: Numerical Investigation of Subharmonic Solutions to Duffing's Equation, Publ. RIMS, Kyoto Univ. 5 (1969) 79-112.

[39] WAINBERG, M.M. & TRENOGIN, W.A.: Theorie der Lösungsverzweigung bei nichtlinearen Gleichungen, Berlin: Akademie-Verlag 1973.

[40] WARNCKE, E.: Selbsterregte Schwingung eines in Gleitlagern laufenden starren Rotors, ZAMM 53 (1973) T 60-62.

[41] WILKINSON, J.H.: The Algebraic Eigenvalue Problem, Oxford: Clarendon Press 1965.

The Vibration and Stability of Laterally Flexible Shafts Driven and Supported by Hooke's Joints

J.S. Burdess

Department of Mechanical Engineering
University of Newcastle upon Tyne

NOTATION

$\tau = \omega t$	is the non-dimensional time variable
$\hat{u} = u/\ell$	is the non-dimensional displacement of the shaft in the 0_1X direction
$\hat{v} = v/\ell$	is the non-dimensional displacement of the shaft in the 0_1Y direction
$\xi = z/\ell$	is the non-dimensional distance along the shaft axis
$\hat{T} = T\ell/EI$	is the non-dimensional torque transmitted by the shaft
ω_n	is the nth natural frequency of transverse vibration of a simply supported shaft having the same dimensions and stiffnesses as the transmission shaft
$r = \omega/\omega_1$	is the non-dimensional shaft speed
$\nu_e(\xi) = C_e(\xi)\omega_1/m$	is the external friction parameter
$\nu_j = C_j\ \ell\omega_1/EI$	is the joint friction parameter
EI	is the flexural rigidity of the shaft

1. INTRODUCTION

The introduction of Hooke's joints into a transmission which consists of a sequence of non-collinear shafts can, because of the joints' periodically varying velocity ratio, change dramatically the dynamic characteristics of the shaft system. For example if the transmission is assumed rigid the analysis presented in [1] shows that substantial speed fluctuations can occur in both input and output shafts the values of which depend upon the inertia and torque characteristics of the total system. If shaft flexibility is assumed both torsional and transverse vibrations are possible. The majority of published work relating to systems of this type has usually concentrated on the problems arising because of torsional vibration [2] although a study of the transverse motion has been undertaken in [3] and [4] .

The present paper considers the transverse behaviour of a uniform shaft driven and supported by Hooke's joints and evaluates the effect of transmitted torque and joint friction on the dynamic stability of the shaft.

2. DESCRIPTION OF SYSTEM

The transmission shown in Fig. (1) consists of a transversely flexible shaft S_2 of length l and mass per unit

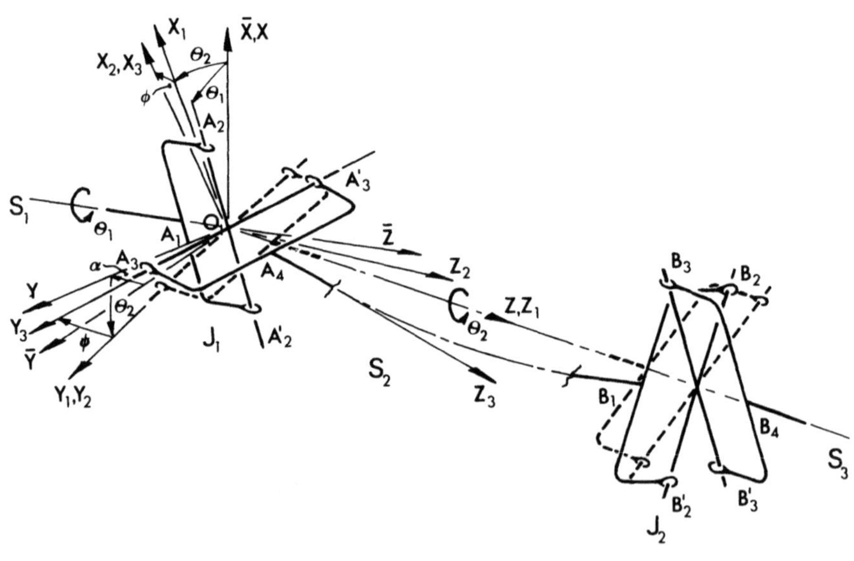

Fig. 1. The Transmission.

length m which is supported at its ends by Hooke's joints J_1 and J_2. The shaft which is assumed circular in cross-section transmits a torque T and is driven through the joint J_1 by the constant speed drive S_1. At the other end of the shaft Hooke's joint J_2, which has zero angularity connects the shaft to the output system S_3. All parts of the transmission other than the shaft are assumed rigid and viscous damping is assumed to exist at the bearings of each joint and along the length of the shaft. The damping along the shaft is not necessarily uniformly distributed.

Fig. (1) shows the transmission in its final position (full lines) and the position it would occupy if the shaft were rigid (dotted lines). Axes 0_1XYZ represent a fixed frame which corresponds to the datum position of the system with 0_1Z taken along the shaft axis. The rigid body motion of the system is derived from 0_1XYZ by rotating the shaft (0_1Z axis) through an angle Θ_2 taking 0_1XYZ to $0_1X_1Y_1Z_1$. The motion of the input system S_1 is first determined from 0_1XYZ by defining the datum position of yoke $A_1A_2A_2^1$ as axis $0_1\bar{X}$ of axes $0_1\bar{X}\bar{Y}\bar{Z}$ which are obtained from 0_1XYZ by a rotation $(-\alpha)$ about 0_1X. At any later instant the yoke is assumed to be inclined to $0_1\bar{X}$ by an amount $\Theta_1 = \omega t$.

Due to shaft flexibility the yokes $A_3A_4A_3^1$ and $B_2B_1B_2^1$ of joints J_1 and J_2 depart slightly from their rigid body position. At the input end, for example, the final position of the yoke, as measured from $0_1X_1Y_1Z_1$ may be defined by a small rotation ϕ about 0_1Y_1 taking $0_1X_1Y_1Z_1$ into $0_1X_2Y_2Z_2$ followed by a small rotation $(-\psi)$ about $0X_2$. If (U_1V_1) represents the component lateral displacements of the shaft along 0_1X_1 and 0_1Y_1, then since ϕ and ψ are both small they may be expressed in terms of the derivatives

$$(2.1) \qquad \phi = \left(\frac{\partial u}{\partial z}1\right)_{z=0} \qquad \psi = \left(\frac{\partial v}{\partial z}1\right)_{z=0}$$

With respect to the fixed axes the corresponding results are

(2.11)
$$\phi = \cos\Theta_2 \left(\frac{\partial u}{\partial z}\right)_{z=0} + \sin\Theta_2 \left(\frac{\partial v}{\partial z}\right)_{z=0}$$
$$\psi = -\sin\Theta_2 \left(\frac{\partial u}{\partial z}\right)_{z=0} + \cos\Theta_2 \left(\frac{\partial v}{\partial z}\right)_{z=0}$$

where (u,v) are the component displacements in the fixed directions.

A similar situation occurs at the input yoke of joint J_2 .

3. DYNAMICS

Equations of Motion and Boundary Conditions. The equations of motion which describe the transverse behaviour of a uniform circular shaft subject to torsion may be deduced directly from Kirchoff's equations [5] and can be written for small displacements, in the form

(3.10)
$$\frac{\partial^4 u}{\partial \xi^4} + \frac{\partial}{\partial \xi}\left(\hat{T}\,\frac{\partial^2 \hat{v}}{\partial \xi^2}\right) + n^4\pi^4 \left(\frac{\omega}{\omega_n}\right)^2 \left[\frac{\partial^2 \hat{u}}{\partial \tau^2} + \frac{\nu e(\xi)}{r}\,\frac{\partial \hat{u}}{\partial \tau}\right] = 0$$
$$\frac{\partial^4 \hat{v}}{\partial \xi^4} - \frac{\partial}{\partial \xi}\left(\hat{T}\,\frac{\partial^2 \hat{u}}{\partial \xi^2}\right) + n^2\pi^4 \left(\frac{\omega}{\omega_n}\right)^2 \left[\frac{\partial^2 \hat{v}}{\partial \tau^2} + \frac{\nu e(\xi)}{r}\,\frac{\partial \hat{v}}{\partial \tau}\right] = 0$$

Before the nature of the boundary conditions produced at the ends of the shaft by the joints J_1 and J_2 can be established it is necessary to perform a detailed analysis of the kinematics and force transmission characteristics of a Hooke's joint, when the angularity of the joint is changing due to shaft flexibility. A detailed description of this analysis is given in [6] the results of which show that time dependent bending moments are applied to the ends of the shaft in the manner described by equations (3.11). At $\xi=0$:

$$\begin{aligned}
\frac{\partial^2\hat{u}}{\partial\xi^2} = \hat{M}_u^{(0)} = {} & \frac{\hat{T}\sin\alpha\cos\alpha\sin^2\tau}{\Delta^2} + \frac{\hat{T}\cos\alpha\sin\tau}{\Delta^4}\left(\cos\tau\frac{\partial\hat{u}}{\partial\xi} + \cos\alpha\sin\tau\frac{\partial\hat{v}}{\partial\xi}\right) \\
& + \frac{\nu_j r\sin\alpha}{\Delta^4}\left(1-2\sin^2\alpha\sin^2\tau + \sin^2\alpha\sin^4\tau\right) \\
& + \frac{3\nu_j r}{\Delta^6}\sin^2\alpha\cos^2\alpha\sin^3\tau\cos\tau\,\frac{\partial\hat{u}}{\partial\xi} + \frac{\nu_j r}{\Delta^4}\left(1+\sin^2\alpha\sin^4\tau\right)\frac{\partial^2\hat{u}}{\partial\tau\partial\xi} \\
& + \frac{\nu_j r}{\Delta^6}\left(1-3\sin^2\alpha\sin^2\tau\cos^2\tau - \sin^4\alpha\sin^4\tau\right)\frac{\partial\hat{v}}{\partial\xi} \\
& - \frac{\nu_j}{\Delta^4}\sin^2\alpha\cos\alpha\sin^3\tau\cos\tau\,\frac{\partial^2\hat{v}}{\partial\tau\partial\xi}
\end{aligned}$$

$$\begin{aligned}
\frac{\partial^2\hat{v}}{\partial\xi^2} = \hat{M}_v^{(0)} = {} & -\frac{\hat{T}\sin\alpha\sin\tau\cos\tau}{\Delta^2} - \frac{\hat{T}\cos\tau}{\Delta^4}\left(\cos\tau\frac{\partial\hat{u}}{\partial\xi} + \cos\alpha\sin\tau\frac{\partial\hat{v}}{\partial\xi}\right) \\
& - \nu_j\frac{r\cos\alpha}{\Delta^6}\left(1+\sin^2\alpha\sin^2\tau-3\sin^2\alpha\sin^4\tau+\sin^4\alpha\sin^4\tau\right)\frac{\partial\hat{u}}{\partial\xi} \\
& - \nu_j r\sin^2\alpha\cos\alpha\sin^3\tau\cos\tau\,\frac{\partial^2\hat{u}}{\partial\tau\partial\xi} \\
& + \frac{\nu_j r\sin^2\alpha}{\Delta^6}\left(1-3\sin^2\tau + 2\sin^2\alpha\sin^2\tau\right)\sin\tau\cos\tau\,\frac{\partial\hat{v}}{\partial\xi} \\
& + \nu_j r\left(1-\sin^2\alpha\sin^2\tau - \sin^2\alpha\sin^4\tau + \sin^4\alpha\cdot\sin^4\tau\right)\frac{\partial^2\hat{u}}{\partial\tau\partial\xi} \\
& - \frac{\nu_j r}{\Delta^4}\sin^3\alpha\cos\alpha\sin^3\tau\cos\tau \; .
\end{aligned} \tag{3.11}$$

At $\xi = 1$:

$$\frac{\partial^2\hat{u}}{\partial\xi^2} = \hat{M}_u^{(1)} = -\hat{T}\sin\tau\left(\cos\tau\frac{\partial\hat{u}}{\partial\xi} + \sin\tau\frac{\partial\hat{v}}{\partial\xi}\right) + \nu_j r\left(\frac{\partial^2\hat{u}}{\partial\tau\partial\xi} + \frac{\partial\hat{v}}{\partial\xi}\right)$$

$$\frac{\partial^2\hat{v}}{\partial\xi^2} = \hat{M}_v^{(1)} = -\hat{T}\cos\tau\left(\cos\tau\frac{\partial\hat{u}}{\partial\xi} + \sin\tau\frac{\partial\hat{v}}{\partial\xi}\right) + \nu_j r\left(\frac{\partial^2\hat{v}}{\partial\tau\partial\xi} - \frac{\partial\hat{u}}{\partial\xi}\right)$$

where $\Delta^2 = 1-\sin^2\alpha\sin^2\tau$

The above equations together with the requirement that $\hat{u} = 0$ and $\hat{v} = 0$ at $\xi = 0$ and 1 define the boundary conditions of the problem.

4. TRANSFORMATION OF EQUATIONS OF MOTION AND MODAL SOLUTION

Before the solution of the equations of motion (3.10) satisfying boundary conditions (3.11) can be constructed it is necessary to rewrite (3.10) and (3.11) as a set of integral equations.

If (3.10) are integrated successively with respect to ξ it may be shown [7] that the displacements $\hat{u}$ and $\hat{v}$ may be expressed as follows

$$\hat{u} = -\int_0^1 K(\xi,\zeta)\cdot\{\frac{\partial}{\partial\zeta}(\frac{\hat{T}\partial^2\hat{v}}{\partial\zeta^2})+n^4\pi^4(\frac{\omega}{\omega_n})^2(\frac{\partial^2\hat{u}}{\partial\tau^2}+\frac{\nu_e(\zeta)}{r}\frac{\partial\hat{u}}{\partial\tau})\}d\zeta$$

$$(4.10)\qquad + \bar{K}(\xi,0)\,\hat{M}_u^{(0)} - \bar{K}(\xi,1)\hat{M}_u^{(1)}$$

$$\hat{v} = -\int_0^1 K(\xi,\zeta)\cdot\{-\frac{\partial}{\partial\zeta}(\frac{\hat{T}^2\hat{u}}{\partial\zeta^2})+n^4\pi^4(\frac{\omega}{\omega_n})^2(\frac{\partial^2 v}{\partial\tau^2}+\frac{\nu_e(\zeta)}{r}\frac{\partial v}{\partial\tau})\}d\zeta$$

$$+ \bar{K}(\xi,0)\,\hat{M}_v^{(0)} - \bar{K}(\xi,1)\hat{M}_v^{(1)}$$

where the functions $K(\xi,\zeta)$ and $\bar{K}(\xi,\zeta)$ are the influence functions of a simply supported shaft and represent the deflections produced at a point ξ along the shaft due to a load $\frac{EI}{\ell^2}$ and a bending moment $\frac{EI}{\ell}$ applied at the point ζ.

$$(4.11)\qquad K = \begin{cases} \frac{\zeta}{6}(1-\xi)(2\xi-\xi^2-\zeta^2) & ; \quad \xi > \zeta \\ \frac{\xi}{6}(1-\zeta)(2\zeta-\zeta^2-\xi^2) & ; \quad \zeta > \xi \end{cases}$$

$$\bar{K} = \begin{cases} \frac{\xi}{6}(6\zeta-2-3\zeta^2-\xi^2) + \frac{(\xi-\zeta)^2}{2} & ; \quad \xi > \zeta \\ \frac{\xi}{6}(6\zeta-2-3\zeta^2-\xi^2) & ; \quad \zeta > \xi \end{cases}$$

A solution to (4.10) may now be constructed by expressing $\hat{u}$ and $\hat{v}$ in the series form

(4.12)
$$\hat{u} = \sum_{n=1}^{\infty} f_n(\tau)\ \phi_n(\xi)$$
$$\hat{v} = \sum_{n=1}^{\infty} g_n(\tau)\ \phi_n(\xi)$$

The functions $\phi_n(\xi)$ describe the spatial distribution of $\hat{u}$ and $\hat{v}$ and in this case correspond to the principal modes of a pin-pinned shaft. The functions f_n and g_n are unknown functions of τ and are to be determined. Furthermore if the influence functions K and $\bar{K}$ are expressed as a series in ϕ_n ; [8] .

(4.13)
$$K = \sum_{n=1}^{\infty} \frac{2\phi_n(\xi)\phi_n(\zeta)}{n^4\pi^4}$$
$$\bar{K} = -\sum_{n=1}^{\infty} \frac{2\phi_n(\xi)\phi_n'(\zeta)}{n^4\pi^4}$$

then the equations which determine f_n and g_n follow from (4.10) and are given by

$$\frac{d^2f_n}{d\tau^2} + \left(\frac{\omega_n}{\omega}\right)^2 f_n + \sum_{k=1}^{\infty}\left\{\int_0^1 \frac{\nu e(\zeta)\phi_n(\zeta)\phi_k(\zeta)d\zeta}{r} \cdot \frac{df_k}{d\tau}\right.$$
$$+ \left(\frac{\omega_n}{\omega}\right)^2\left[\frac{\nu_j r(1+\sin^2\alpha\sin^4\tau)\phi_n'(0)\phi_k'(0)}{n^4\pi^4\Delta^4} + \frac{\nu_j r\phi_n'(1)\phi_k'(1)}{n^4\pi^4}\right]\cdot\frac{df_k}{d\tau}$$
$$+ \left(\frac{\omega_n}{\omega}\right)^2\left[\frac{\hat{T}\cos\alpha\sin\tau\cos\tau\phi_n'(0)\phi_k'(0)}{n^4\pi^4\Delta^4} - \frac{\hat{T}\sin\tau\cos\tau\phi_n'(1)\ \phi_k'(1)}{n^4\pi^4}\right.$$
$$\left.+ \frac{3\nu_j r\sin^2\alpha\cos^2\alpha\sin^3\tau\cos\tau\phi_n'(0)\phi_k'(0)}{n^4\pi^4\Delta^6}\right]\cdot f_k$$
$$- \left(\frac{\omega_n}{\omega}\right)^2 \frac{\nu_j r\sin^2\alpha\cos\alpha\sin^3\tau\cos\tau\phi_n'(0)\phi_k'(0)}{n^4\pi^4\Delta^4}\cdot\frac{dg_k}{d\tau}$$
$$+\left(\frac{\omega_n}{\omega}\right)^2\left[\int_0^1 \frac{\phi_n(\frac{\partial}{\partial\zeta}\hat{T}\phi_k'')d\zeta}{n^4\pi^4} + \frac{\hat{T}\cos^2\tau\sin^2\tau\phi_n'(0)\phi_k'(0)}{n^4\pi^4\Delta^4}\right.$$
$$+ \frac{\nu_j r(1-3\sin^2\alpha\sin^2\tau\cos^2\tau-\sin^4\alpha\sin^4\tau)\phi_n'(0)\phi_k'(0)}{n^4\pi^4\Delta^6}$$

$$- \frac{\hat{T}\sin^2\tau\phi_n'(1)\phi_k'(1)}{n^4\pi^4} + \frac{\nu_j r\phi_n'(0)\phi_k'(1)}{n^4\pi^4}\Bigg] \cdot g_k\Bigg\}$$

(4.14)

$$= -2\left(\frac{\omega_n}{\omega}\right)^2 \cdot \frac{\hat{T}\sin\alpha\cos\alpha\sin^2\tau\phi_n'(0)}{n^4\pi^4\Delta^2}$$

$$-2\left(\frac{\omega_n}{\omega}\right)^2 \cdot \frac{\nu_j r\sin\alpha(1-2\sin^2\alpha\sin^2\tau+\sin^2\alpha\sin^4\tau)\phi_n'(0)}{n^4\pi^4\Delta^4}$$

$$\frac{d^2g_n}{d\tau^2} + \left(\frac{\omega_n}{\omega}\right)^2 g_n + 2\sum_{k=1}^{\infty}\left\{\int_0^1 \frac{\nu_e(\zeta)\phi_n(\zeta)\phi_k(\zeta)\cdot d\zeta}{r} \cdot \frac{dg_k}{d\tau}\right.$$

$$+ \left(\frac{\omega_n}{\omega}\right)^2 \cdot \Bigg[-\frac{\nu_j r(1-\sin^2\alpha\sin^2\tau-\sin^2\alpha\sin^4\tau+\sin^4\alpha\sin^4\tau)\phi_r'(0)\phi_k'(0)}{n^4\pi^4}$$

$$+ \frac{\nu_j r\phi_n'(1)\phi_k'(1)}{n^4\pi^4}\Bigg] \cdot \frac{dg_k}{d\tau}$$

$$+ \left(\frac{\omega_n}{\omega}\right)^2 \Bigg[- \frac{\hat{T}\cos\alpha\sin\tau\cos\tau\phi_n'(0)\phi_k'(0)}{n^4\pi^4\Delta^4} + \frac{\hat{T}\sin\tau\cos\tau\phi_n'(1)\phi_k'(1)}{n^4\pi^4}$$

$$+ \frac{\nu_j r\sin^2\alpha(1-3\sin^2\tau+2\sin^2\alpha\sin^2\tau)\sin\tau\cos\tau\phi_n'(0)\phi_k'(0)}{n^4\pi^4\Delta^6}\Bigg] \cdot g_k$$

$$-\left(\frac{\omega_n}{\omega}\right)^2 \frac{\nu_j r\sin^2\alpha\cos\alpha\sin^3\tau\cos\tau\phi_n'(0)\phi_k'(0)}{n^4\pi^4\Delta^4} \cdot \frac{df_k}{d\tau}$$

$$+ \left(\frac{\omega_n}{\omega}\right)^2 \Bigg| - \int_0^1 \frac{\phi_n\frac{\partial}{\partial\zeta}(T\phi_k'')d\zeta}{n^4\pi^4}$$

$$- \frac{\hat{T}\cos^2\tau\,\phi_n'(0)\,\phi_k'(0)}{n^4\pi^4\Delta^4} + \frac{\hat{T}\cos^2\tau\,\phi_n'(1)\,\phi_k'(1)}{n^4\pi^4}$$

$$- \frac{\nu_j r\cos\alpha(1+\sin^2\alpha\sin^2\tau-3\sin^2\alpha\sin^4\tau+\sin^4\alpha\sin^4\tau)\,\phi_n'(0)\,\phi_k'(0)}{n^4\pi^4\Delta^6}$$

$$\left. \left. + \frac{\nu_j r\phi_n'(1)\,\phi_k'(1)}{n^4\pi^4}\right] \cdot f_k \right\}$$

$$= 2\left(\frac{\omega_n}{\omega}\right)^2 \frac{\hat{T}\sin\alpha\sin\tau\cos\tau\,\phi_n'(0)}{n^4\pi^4\Delta^2}$$

$$+ 2\left(\frac{\omega_n}{\omega}\right)^2 \frac{\nu_j r\sin^5\alpha\cos\alpha\sin^3\tau\cos\tau\,\phi_n'(0)}{n^4\pi^4\Delta^4}$$

If the quantities

$$\varepsilon=\sin^2\alpha \; , \quad \Omega_n=\omega_n/\omega \; , \quad \hat{T}=\varepsilon\hat{T}^*\sin 2\tau , \quad \beta_{nk}=2\hat{T}^* \int_0^1 \frac{\phi_n(\zeta)\,\phi_k''(\zeta)\,d\zeta}{n^4\pi^4}$$

$$\lambda_{nk}= \frac{2\{\phi_n'(1)\,\phi_k'(1)+\phi_n'(0)\,\phi_k'(0)\}}{n^4\pi^4} \qquad \sigma_{nk}= \frac{\hat{T}^*\{\phi_n'(1)\,\phi_k'(1)-\phi_n'(0)\,\phi_k'(0)\}}{n^4\pi^4}$$

$$\varepsilon\nu^*_j = \nu_j \; , \qquad \varepsilon\nu^*_e\,\mu_{nk} = 2\int_0^1 \nu_e(\zeta)\,\phi_n(\zeta)\,\phi_k(\zeta)\,d\zeta$$

are introduced then together with the assumption that the parameters associated with the joint angularity (ε) , transmitted torque $(\hat{T})$ and friction within the system (ν_e, ν_j) may be considered small, equation (4.14) may be approximated to the first order in the parameter ε as follows. It will be seen from above that the parameter associated with the external friction is determined by the quantities $\varepsilon\nu_e^*$ and μ_{nk}. The component $\varepsilon\nu_e^*$ may be considered representative of the magnitude of $\nu_e(\xi)$, whilst μ_{nk} may be taken as an index associated with the manner the friction is distributed along the length of the shaft.

$$\frac{d^2 w_n}{dt^2} + \Omega_n{}^2 w_n + \varepsilon \sum_{k=1}^{\infty} \{ (\frac{\nu_e^* \mu_{nk}}{r} + \nu_j r \Omega_n{}^2 \lambda_{nk}) \frac{dw_k}{dt} - i\nu_j r\Omega_n{}^2 \lambda_{nk} w_k$$

(4.15)

$$- \frac{\Omega_n{}^2 \sigma_{nk}}{2} (1-e^{i4\tau}) \bar{w}_k - \frac{\Omega n^2}{2} (\beta_{nk} - \sigma_{nk}) (e^{i2\tau} - e^{i2\tau}) w_k \} = 0 + 0(\varepsilon^2)$$

where

$$w_n = f_n + ig_n$$

and

$$w_n \bar{w}_n = |w_n|^2$$

To a first order of magnitude in ε equation (4.15) reveals that the harmonic nature of the system is modified by two components

(i) A stationary component associated with the frictional forces which are present within the system if the system angularity is zero and

(ii) a periodic component associated with the torque transmitted by the shaft.

In view of the natural separation of these terms it is convenient to proceed with the analysis of the system by considering the effects of friction and torque separately. Furthermore, since it is intended to examine the effect of these parameters on the stability of the transmission it is sufficient to consider the homogeneous parts of (4.15) only.

5. STABILITY

Joint Friction. If the torque transmitted by the shaft is assumed negligible then the equations which describe the free motion of the shaft have the form

$$\frac{d^2w_n}{dt^2} + \Omega_n{}^2 w_n + \varepsilon \sum_{k=1}^{\infty} \{\frac{\nu_e{}^*\mu_{nk}}{r} + \nu_j r\Omega_n{}^2 \lambda_{nk}\} \frac{dw_k}{dt} - i\nu_j^* r\Omega_n{}^2 \lambda_{nk} w_k = 0$$

(5.10)

A first order solution to the above equations is easily established using standard perturbation techniques [9] and is given by

$$w_n = P_n \exp(\varepsilon S_{1n} - i\Omega_n)\tau + Q_n \exp(\varepsilon S_{2n} + i\Omega_n)\tau + 0(\varepsilon) \tag{5.11}$$

with the exponents S_{1n} and S_{2n} given by

$$S_{1n} = -(a_{nn} + b_{nn})$$

(5.12)

$$S_{2n} = -(a_{nn} - b_{nn})$$

where

$$a_{nn} = 1/2\ (\frac{\nu_e{}^*\mu_{nn}}{r} + \nu_j{}^* r\Omega_n{}^2 \lambda_{nn})$$

and

$$b_{nn} = 1/2\ (\nu_j{}^* r\Omega_n \lambda_{nn})\ .$$

It follows from (5.11) that the motion of the nth mode takes place at a frequency Ω_n irrespective of the value of the shaft speed and is composed of (i) a component of amplitude $P_n \exp \varepsilon S_{1n} \tau$ in asynchronous precession relative to the shaft rotation and (ii) a component of amplitude $Q_n \exp \varepsilon S_{2n} \tau$ in synchronous precession relative to the shaft rotation.

It is clear from the form of (5.12) that the free motion of the shaft will be stable only if the conditions $S_{1n} < 0$ and $S_{2n} < 0$ are satisfied simultaneously. Since the quantities a_{nn} and b_{nn} are by definition positive it follows that the criterion associated with S_{1n} is always satisfied. The asynchronous component of the motion is therefore stable at all values of shaft speed and will not be considered further. However, the second condition, $S_{2n} < 0$, is not satisfied at all instances and shows, depending upon the relative magnitudes of a_{nn} and b_{nn} that the synchronous motion could be either stable or unstable. The criterion which establishes the stability of the nth mode follows from (5.12) and with the introduction of $\Omega_n = n^2/r$ and $\lambda_{nn} = 4/n^2\pi^2$ may be rewritten

$$r < n^2 + \frac{\pi^2}{4} \mu_{nn} \frac{\nu_e}{\nu_j} \quad , \quad n = 1,2,\ldots.. \tag{5.13}$$

In view of the number of independent parameters involved in inequality (5.13) the stability of the system is best discussed with reference to the following special cases.

Stability in the absence of external friction. If external friction is neglected, $\nu_e = 0$, the stability of the transmission becomes independent of the value of the joint friction and is determined only by the shaft speed

$$r < n^2 \quad \text{or} \quad \omega < \omega_n \tag{5.14}$$

It follows from (5.14) that for a shaft speed greater than ω_n the motion of the nth mode becomes unstable. However, since the frequencies ω_k satisfy the sequence $\omega_1 < \omega_2 < \ldots < \omega_k$ it follows that the shaft will become unstable as soon as its speed exceeds the first critical frequency ω_1 . The number of unstable modes will, of course, depend upon the values of ω . For example if $\omega_1 < \omega_2 < \omega < \omega_3$ then (5.14) requires that the first and second modes be unstable - and so on.

Stability with External and Joint Friction. Inequality (5.14) shows that external damping extends the range of stability by an amount which is proportional to the ratio of external friction to internal friction ν_e/ν_j and μ_{nn} the manner in which the external damping is introduced into the system. The increase in stability afforded by a fixed amount of external damping will in general depend upon the characteristics of the joint friction. For example if ν_j changes with the parameters of the system, e.g., input speed, load, etc. then the value of ν_e to ensure stability will vary according to the situation and can only be assessed from (5.13) once the functional nature of ν_j is known. However, to form an appreciation of the general effect of introducing external damping it is proposed to examine, assuming a constant value for ν_j , the situation where the damping force is represented as a concentrated load applied at some intermediate distance ξ_o along the shaft, Fig. (2). By using the results of this case it is possible to deduce the stability provided by an arbitrary distribution of external damping force.

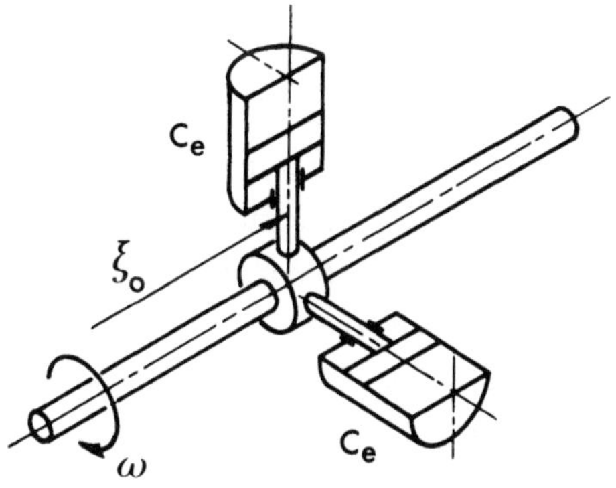

Fig. 2. Dashpot Assembly.

External Damping Applied at a Fixed Point. For this arrangement the function which corresponds to the distribution of external damping may be written as

$$(5.15) \qquad C_e(\xi) = \frac{C_o\delta(\xi,\xi_o)}{\ell}$$

and therefore

$$\nu_e = C_o/m\omega_1\ell$$

and

$$\mu_{nn} = 2\cdot\sin^2 n\pi\xi_o$$

where C_o is the coefficient of the damper and $\delta(\xi,\xi_o)$ is the Dirac Delta function.

If the transformation $\xi_o = \frac{1}{2} + \zeta_o$ is introduced so as to bring the origin of the coordinate defining the position of the damping force to the centre of the shaft then (5.13) with the aid of (5.15) may be rewritten in the form

$$(5.16) \qquad r < n^2 + \frac{\pi^2}{4}\cdot\frac{\nu_e}{\nu_j}\cdot\left[1 + (-1)^{n+1}\cos 2n\pi\zeta_o\right]$$

It is interesting to note that the stability condition (5.16) is symmetric in ζ_o which it should be from purely physical considerations. (5.16) also indicates that the stability boundaries of a particular mode, when presented in the $(r, \nu_e/\nu_j)$ plane, are straight lines the slope of which is determined by the position of the damping force. Thus for a given value of damping ratio the speed and mode at which instability is first encountered need not necessarily be associated with the shaft's first critical, as in the case when $\nu_e = 0$. For example, if $\zeta_o = 0$, then clearly the influence of the damping force would be restricted to the odd modes. Therefore, if we select a value for ν_e/ν_j which ensures that the first mode is stable in a speed range which includes ω_2 then the shaft will exhibit unstable

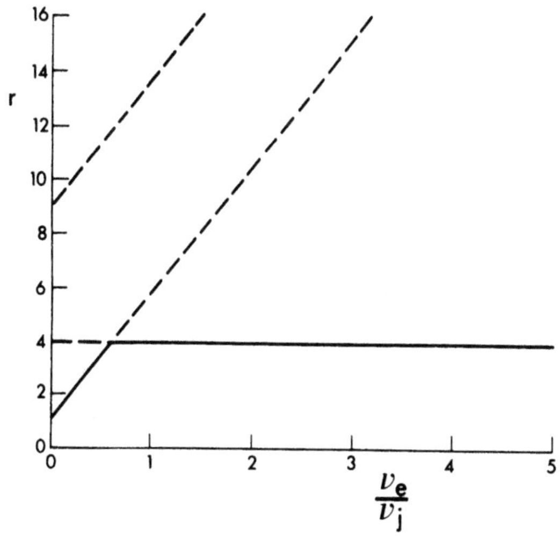

Fig. 3. Stability Chart for $\zeta_o = 0.0$.

second mode vibration as soon as its speed exceeds ω_2 . In this case instability always occurs at the second critical speed and there is no advantage to be gained by increasing the external damping beyond a value which extends the first mode stability region beyond ω_2 , Fig. (3).

However, if the damping force is situated at some general position then the speed at which instability occurs and the mode associated with the instability will change according to the value of ζ_o and $\frac{\nu_e}{\nu_j}$. Thus in order to evaluate the stability of the system as a whole it is necessary to determine the conditions relating to the stability of neighbouring modes.

If we consider the nth and k^{th} modes then it follows from (5.16) that the nth will always become unstable before the kth provided that the lines defining the stability boundaries do not intersect. This requirement is given by

$$(5.17) \qquad (-1)^n \cos 2n\pi\zeta_o > (-1)^k \cos 2k\pi\zeta_o$$

However, if (5.17) is not satisfied then the stability boundaries of the nth and k^{th} modes will intersect. If this intersection takes place at a damping ratio of $(\nu_e/\nu_j)_{nk}$ then for $\nu_e/\nu_j < (\nu_e/\nu_j)_{nk}$ instability of the nth mode will occur at a speed which is lower than that corresponding to instability of the kth and vice-versa for $\nu_e/\nu_j > (\nu_e/\nu_j)_{nk}$. The stability region of the system is therefore given by that region of the $(r,\nu_e/\nu_j)$ plane which is bounded by the segments of (5.16) which yield a minimum input speed.

Assuming a three mode approximation to describe the deflection of the shaft inequality (5.17) indicates that an overlapping between the respective stability boundaries of the first, second and third modes will take place if the damping force is located in the regions shown in Fig. (4).

With the aid of (5.16) and Fig. (4) the stability chart for the transmission for various ranges of ζ_o may be constructed and typical results are shown in Figs. (5a) to (5e).

As the interpretation of each figure is similar it is only necessary to examine the case associated with a particular value of ζ_o. The most interesting is that of Fig. (5d) where it is shown, for $\zeta_o = 0.125$, that the stability of the system is dictated, in turn, by the stability of the first, second and third modes. For (ν_e/ν_j) in the range (0,1.72) it is seen that the first mode is the primary source of instability. However, for the ranges $1.72 < \frac{\nu_e}{\nu_j} < 2.87$ and $\frac{\nu_e}{\nu_j} > 2.87$ instability is first realised in the second and then the third mode respectively. It will be noticed that the intersection which occurs between the stability boundaries of the first and third modes does not contribute towards the shape of the stability chart.

External Damping distributed along the length of the Shaft. If the damping is distributed over the length of the shaft in the manner indicated in Fig. (6) then the criterion which establishes the stability of the shaft may be deduced di-

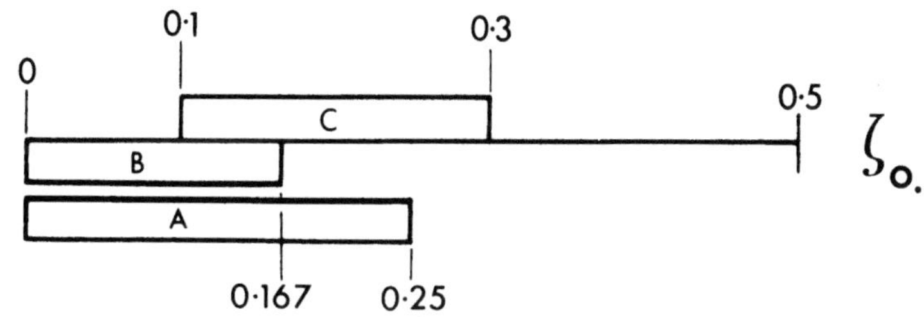

Fig. 4.
A: Intersection between 1st. and 3rd. mode boundaries
B: Intersection between 1st. and 2nd. mode boundaries
C: Intersection between 2nd. and 3rd. mode boundaries

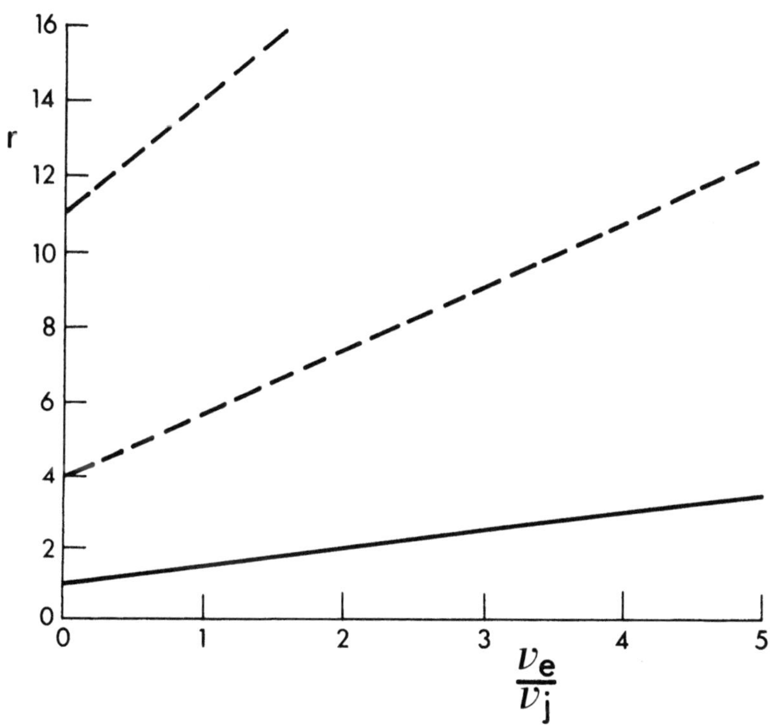

Fig. 5a. Stability Chart for $\zeta_o = 0.4$, $0.35 < \zeta_o < 0.5$.

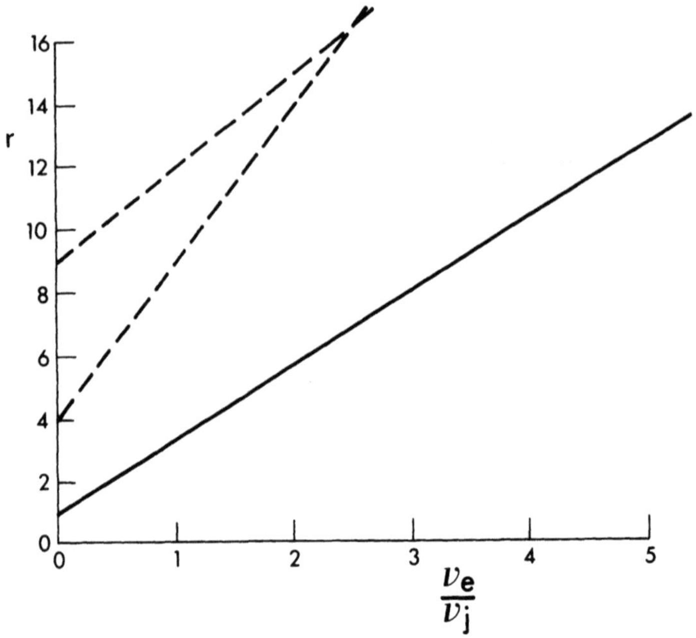

Fig. 5b. Stability Chart for $\zeta_o = 0.26$, $0.25 < \zeta_o < 0.3$.

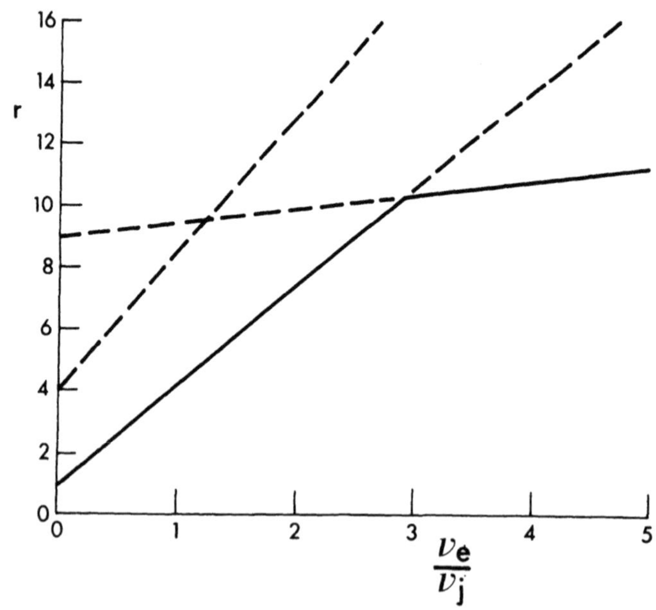

Fig. 5c. Stability Chart for $\zeta_o = 0.2$, $0.167 < \zeta_o < 0.25$.

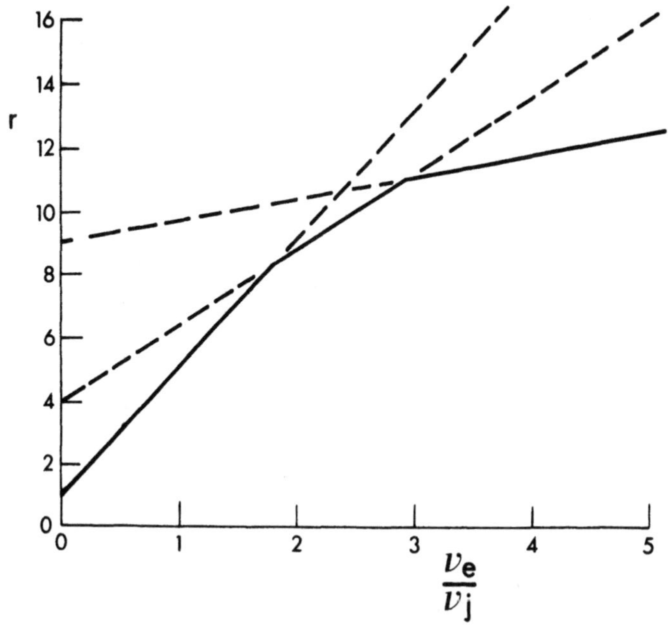

Fig. 5d. Stability Chart for $\zeta_o = 0.125$, $0.1 < \zeta_o < 0.167$.

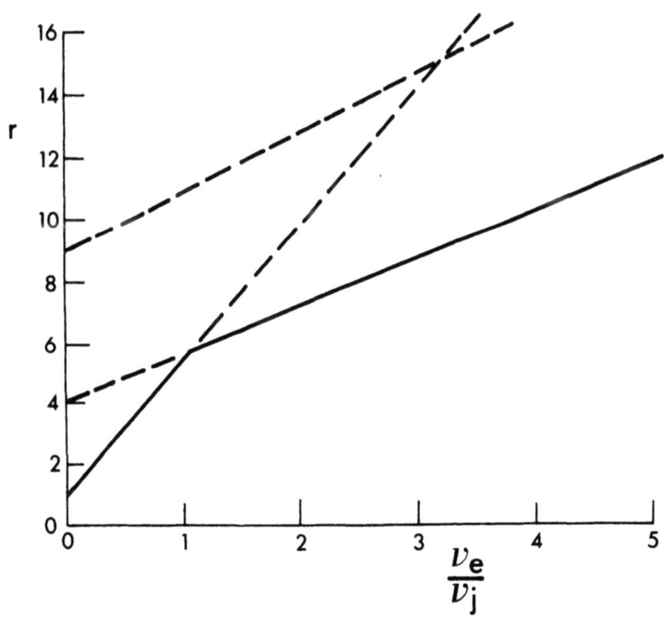

Fig. 5e. Stability Chart for $\zeta_o = 0.095$, $0 < \zeta_o < 0.1$.

rectly from (5.16) by representing the damping as a series of concentrated loads of varying magnitude $C_e(\zeta_o)d\zeta_o$ arranged along the shaft.

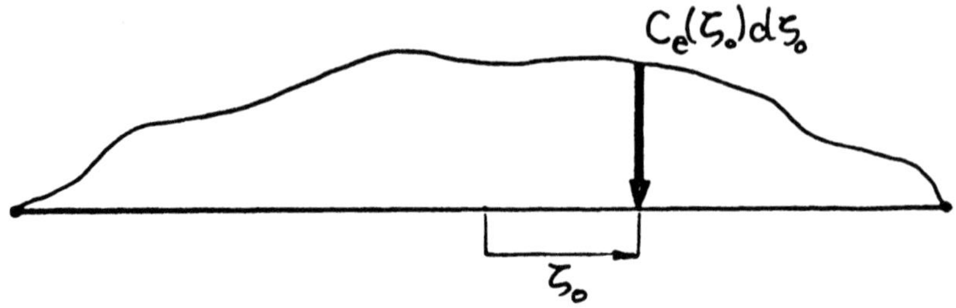

Fig. 6.

For a concentrated damping force of magnitude $C_e(\zeta_o)d\zeta_o$ situated distant ζ_o from the shaft centre the parameter ν_e and μ_{nn} follow from (5.15) and may be defined as

$$\nu_e d\zeta_o = \frac{C_e(\zeta_o)d\zeta_o}{m\omega_1}$$

(5.18)

$$\mu_{nn} = \tfrac{1}{2}[1+(-1)^{n+1}\cos 2n\pi\zeta_o]$$

Thus using (5.16) the increase in stability provided by the introduction of this damping force is given by

$$(5.19) \qquad \delta r_n = r-n^2 = \frac{\pi^2}{4}\,[1+(-1)^{n+1}\cos 2n\pi\zeta_o]\frac{\nu_e(\zeta_o)d\zeta_o}{\nu_j}$$

Therefore the equation defining the stability boundary of the nth mode follows on the integration of (5.19) over the parameter ζ_o and is given by

$$(5.20) \quad r_n = n^2 + \frac{\pi^2}{4} \int_{\zeta_o} [1+(-1)^{n+1}\cos 2n\pi\zeta_o] \frac{\nu_e}{\nu_j} (\zeta_o) d\zeta_o$$

Interpretation of (5.20) will obviously depend upon $\nu_e(\zeta_o)$ and therefore each case must be considered separately.

However, if the external damping is assumed to be uniformly distributed over the whole length of the shaft $\nu_e(\zeta_o)$ is constant and (5.20) reduces to

$$(5.21) \qquad r = n^2 + \frac{\pi^2}{4} (\nu_e/\nu_j)$$

It is clear from (5.21) that the introduction of uniformly distributed external damping increases the stability of neighbouring modes by equal amounts and as a result shows that the stability of the shaft is always dictated by the stability of the first mode.

Transmission of Periodic Torque. If the torque transmitted by the shaft is assumed to have the form

$$\hat{T} = \varepsilon\hat{T}^{*}\sin 2\tau$$

then in the absence of friction the equations which define the free motion of the shaft follow from (4.15) and are given by

$$(5.22) \quad \begin{aligned} &\frac{d^2 w_n}{dt^2} + \Omega_n{}^2 w_n - \varepsilon \sum_{k=o}^{\infty} \{\frac{\Omega_n{}^2\sigma_{nk}}{2}(1-e^{i4\tau})\bar{w}_k \\ &+ \frac{\Omega_n{}^2}{2} (\beta_{nk}-\sigma_{nk})(e^{i2\tau}-e^{-i2\tau})w_k\} = 0 \end{aligned}$$

Since these equations constitute a set of linear differential equations with periodic coefficients it is seen that the transmitted torque excites the system in a parametric manner.

The properties of equations of this type have been investigated by a number of authors and for the purpose of the foregoing analysis the work of HSU [10] and BOLOTIN [11] supply the necessary information. Generally these works indicate that for certain values of the frequency of parametric loading, systems of this class can exhibit exponential type instabilities. For example if the parametric loading has a component with a frequency of excitation $r\omega$, $r = 1.2,\ldots,$ then the theory indicates that the principal regions of instability will be located in the neighbourhoods of $\omega = \frac{2\omega_n}{r}$ and $\omega = \frac{\omega_n \pm \omega_k}{r}$, where ω_n and ω_k are natural frequencies of the unloaded system.

If a two mode approximation is assumed for the deflections $\hat{u}$ and $\hat{v}$ and the procedures associated with [10] and [11] performed for equations (5.22) it may be shown that parametric instability only occurs when the shaft speed ω is in the vicinity of regions centred on

$$\Omega_2 - \Omega_1 \doteq 2 \quad \text{and} \quad \Omega_2 + \Omega_1 \doteq 4 \tag{5.23}$$

For all shaft speeds other than those in the neighbourhood of (5.23) the transmission is stable.

The Case $\Omega_2 - \Omega_1 \doteq 2$. For the situation corresponding to a shaft speed in the vicinity of $\omega \doteq (\omega_2 - \omega_1/_2$ the criterion which establishes the stability of the first and second modes is given by

$$\mu^2 - \frac{11\Omega_1\Omega_2\hat{T}^{*2}}{36\pi^4} > 0 \tag{5.24}$$

The quantity μ is defined from the relation

$$\Omega_2 - \Omega_1 = 2 + \varepsilon\mu \tag{5.25}$$

and may be interpreted as a measure of the perturbation in $\Omega_2 - \Omega_1$ due to a small change in the shaft speed.

Thus for sufficiently small values of μ it follows from the form of inequality (5.24) that the transverse motion of the shaft can become unstable when the shaft speed is in the vicinity of $(\omega_2 - \omega_1)/_2$. The boundaries which determine the extent of this instability follow from (5.24) and (5.25) and are given by the relation

$$r = 1.5 \pm 0.056\ (\varepsilon\hat{T}^*) \tag{5.26}$$

Thus if the input speed is such that it lies in the region bounded by (5.26) then in the absence of friction the shaft will exhibit unstable first and second modes. Furthermore, since $\varepsilon\hat{T}^*$ represents the magnitude of the non-dimensional torque transmitted by the shaft expressed by $\varepsilon\hat{T}^* = \frac{T_o \ell}{EI}$, it follows from (5.26) that the unstable region is narrow for practical values of torque but will be increased if

(i) for a given shaft the magnitude of the torque is increased, and

(ii) for a given torque the length of the shaft is increased or the diameter of the shaft is reduced.

<u>The Case $\Omega_2 + \Omega_1 \doteq 4$</u>. For this situation the criterion which establishes the stability of the shaft is given by

$$\mu^2 - \frac{\Omega_1 \Omega_2 \hat{T}^{*2}}{256\pi^4} > 0 \tag{5.27}$$

where the significance placed upon the parameter μ is similar to that discussed in (5.21).

In the absence of damping the region over which this instability occurs may be shown to be contained within boundaries given by

(5.28) $$r = 1.25 \pm 0.0056\cdot(\varepsilon\hat{T}^*)$$

The interpretation of equation (5.28) is similar to that of (5.26) but it will be noticed that the width of the unstable region is much narrower than that defined by (5.26).

6. CONCLUSIONS

Attention has been focused on the dynamics of a shaft driven and supported by Hooke's joints and results have been derived which show that the stability of the transmission depends upon the frictional characteristics of the system and the torque transmitted by the shaft.

In the case of instability produced by joint friction it has been shown that the range of stable operation may be increased by introducing external damping along the length of the shaft. However, depending upon the distribution of this damping the speed and mode at which instability is first realised is shown to depend upon the ratio of joint friction to external friction.

In the absence of friction the results of a second mode approximation indicate that the transmission of periodic torque can induce parametric type instabilities when the shaft speed is in the neighbourhood of $\omega = (\omega_2 - \omega_1)/2$ or $\omega = (\omega_2 + \omega_1)/4$. The widths of these unstable regions are shown to be narrow for practical values of torque.

REFERENCES

[1] MAUNDER, L. and BURDESS, J.S.: Speed Characteristics of Hook's joint transmissions. J.M.E.S. Vol. 14, No. 4, 1972.

[2] PORTER, P.: A theoretical analysis of the torsional oscillation of a system incorporating a Hooke's joint. J.M.E.S. Vol. 3, No. 4, 1961.

[3] ROSENBERG, R.M.: On the dynamical behaviour of rotating shafts driven by Universal Couplings. J. Appl. Mech. 25, 1958.

[4] WHERLI, C.H.: Dynamisches Verhalteneiner einfach besetzten rotierenden welle mit Kordangelenken. Z. Angew. Math. Phys., 15, 1964.

[5] LOVE, A.E.H.: The Mathematical Theory of elasticity. Dover Publications.

[6] BURDESS, J.S.: The dynamics of transmission shafts driven by universal couplings. Ph.D. Thesis, University of Newcastle upon Tyne.

[7] HILDERBRAND, F.B.: Methods of Applied Mathematics. Prentice-Hall Inc., 1952.

[8] TIMOSHENKO S.P. and GERE, J.M.: Theory of Elastic Stability. McGraw Hill.

[9] COLE, J.P.: Perturbation Methods in Applied Mathematics. Blaisdell Publ. Company, 1968.

[10] HSU, C.S.: On the parametric excitation of dynamic systems having multidegrees of freedom. J. Appl. Mech. 30, 1963.

[11] BOLOTIN, V.V.: Dynamic stability of elastic systems. Holden-Day Inc., 1964.

Self-Excited Vibrations of Rotors

G. Capriz

Istituto di Elaborazione della Informazione
C.N.R. e Universitá di Pisa, Italy

ABSTRACT - Many special earlier results in the theory of oil whirl in rotors (see list of references) are shown to be consequences of appropriate general hypotheses regarding the lubricants action.

1. INTRODUCTION

The aim of the following analysis of phenomena of oil whirl in rotors is a qualitative appreciation of processes leading to unstable behaviour; therefore we consider the simplest shaft possessing the features which are essential for our description*. Precisely, as our analysis gives a special role to the proper speeds $\omega_n^{(K)}$ $(n=1,2,\ldots)$ of the shaft on elastic supports of rigidity K, varying over the whole interval $[0, +\infty]$, maximum simplicity is achieved by considering a rotor of uniform cross-section without overhangs**. The speed $\omega_1^{(\infty)}$ is of overriding importance; but it is also of interest that $\omega_1^{(o)} = \omega_2^{(o)} = 0$ and that these two nominal critical speeds correspond to the "parallel" and "conical" mode. As for the bearings, we will quote in particular either the case of full cylindrical bearings without cavitation or with cavitation; but here, however,

---

* The generality required to cover most special cases encountered in practice is pursued for instance in the linear study quoted in the last section.

** Had we accepted a massless shaft carrying a massive flywheel as a model for the rotor (as is often done for extreme simplicity), the speeds $\omega_n^{(o)}$ would have no meaning and our coordinated description of phenomena would fail.

most developments have much more general validity.

Briefly our analysis shows that:

(i) A vertical rotor (when journals and bearings are co-axial in the steady state) is, in theory at least, always unstable; but, at low speed, the transition occurs towards either a parallel or a conical half-speed whirl (a whirl which may pass unnoticed in practice because of its small amplitude or may be easily damped because of the insignificant amount of energy involved). As the speed of rotation ω is raised towards $2\omega_1^{(\infty)}$ the parallel whirl changes progressively into the lowest resonant mode corresponding to $\omega_1^{(\infty)}$; when ω grows beyond $2\omega_1^{(\infty)}$, the whirl speed stops at the value $\omega_1^{(\infty)}$, while the amplitude increases, theoretically and in absence of other damping mechanisms, beyond all limits (nor can the disturbance be easily hidden in practice, as the amount of energy involved is now considerable).

(ii) The behaviour of a horizontal heavy rotor depends on the type of bearings which support it. There is generally a range of stability against small disturbances at low speeds*, whereas transition to unstable behaviour may occur at higher speeds. If the linear response coefficients of the lubricant force (in the neighbourhood of a steady state of uniform rotation) are available, the boundary of the region of stability can be easily determined because, under critical conditions, elliptic whirls of the shaft are possible during which the response of the lubricant coincides with the response of a spring of a certain nominal rigidity K and the whirl occurs in accordance with the corresponding modal shape. This result seems at first to be in

---

* However, a heavy shaft appears to be always linearly unstable (as under point (i)) if full lubrication is assured in cylindrical full bearings (see, for instance, [1], section 6)

total variance with that described under point (i); but, on the contrary, the analytical correspondence is perfect because for a vertical shaft (and approximately for a very light shaft) the linear response coefficients are such as to lead to a vanishing value for K (and, as already recalled, $\omega_1^{(o)} = \omega_2^{(o)} = 0$).

It is much more complex to describe analytically the whirls of large amplitude of a heavy shaft (even under simple assumptions regarding the type of bearings used) and thus parallel all the results quoted under (i) for a vertical shaft. However, many numerical experiments and some still unpublished analytical results indicate that, in all instances, the linear rule of stability holds true even against large perturbations.

2. A PROPERTY OF LUBRICANT FORCES

The radial and transverse component of the lubricant force acting on a rotating journal will be called F_e , F_n with

$$F_e = Hf_e(a\, , a'\, , \psi, \psi'\, ; B) \quad , \quad F_n = Hf_n(a\, , a'\, , \psi, \psi'\, ; B) \quad ,$$

$$H = \frac{Rb^3\omega\mu}{c^2} \quad ,$$

where:

R , b , c are radius, width, radial clearance of the bearing,
ω is the rotors speed (Radians per unit time) and
μ is the viscosity of the lubricant;
f_e , f_n are non-dimensional functions of eccentricity ratio a (true eccentricity, c a), of the derivative a' of a with respect to non-dimensional time $\tau = \omega t$, of the attitude angle ψ evaluated with reference to a preferred direction (the direction of the load, for instance) and of

its derivative ψ' , and finally of a vector $\underline{B}$ of parameters characterizing the bearing (for a cylindrical full bearing $\underline{B} \equiv (c/R\ ,\ b/R)$).

If $\bar{a}$, $\bar{\psi}$ are the values of a , ψ in a steady journal position under a load W , then

$$\bar{f}_e = f_e(\bar{a}\ ,\ 0\ ,\ \bar{\psi}\ ,\ 0\ ;\ \underline{B}) = -\frac{W}{H}\cos\bar{\psi}\ ,$$

$$\bar{f}_n = f_n(\bar{a}\ ,\ 0\ ,\ \bar{\psi}\ ,\ 0\ ;\ \underline{B}) = \frac{W}{H}\sin\bar{\psi}\ .$$

It is convenient to introduce a system of reference $C = (\bar{0}\ ,\ x\ ,\ y)$ with origin in the steady-state position of the journal centre and with x-axis oriented as the attitude vector $\Omega\bar{0}$ (Ω , bearing centre) and to introduce approximate components F_x , F_y of the lubricant force in the neighbourhood of the steady state:

$$F_x \simeq H\left[\bar{f}_e - (a_{11}\xi + a_{12}\eta + b_{11}\xi' + b_{12}\eta')\right]\ ,$$

$$F_y \simeq H\left[\bar{f}_n - (a_{21}\xi + a_{22}\eta + b_{21}\xi' + b_{22}\eta')\right]\ ,$$

where $\xi = x/c$, $\eta = y/c$.

We can explore now the possibility that special eliptic trajectories exist:

$$\xi = c_1\cos\nu\tau + c_2\sin\nu\tau\ ,$$

$$\eta = c_3\cos\nu\tau + c_4\sin\nu\tau\ ,$$

(ν , ratio of whirl to running speed), such that the linearized expression of the excess force

$$(F_x - H\bar{f}_e\ ,\ F_y - H\bar{f}_n)$$

coincides, along these trajectories, with an elastic restoring force $(-Kc\xi\ ,\ -Kc\eta)$. For existence the necessary and sufficient condition is

$$\det V = 0 \quad , \tag{2.1}$$

where V is the following matrix of the type of Voigt

$$V = \begin{bmatrix} \kappa-a_{11} & -b_{11}\nu & -a_{12} & -b_{12}\nu \\ b_{11}\nu & \kappa-a_{11} & b_{12}\nu & -a_{12} \\ -a_{21} & -b_{21}\nu & \kappa-a_{22} & -b_{22}\nu \\ b_{21}\nu & -a_{21} & b_{22}\nu & \kappa-a_{22} \end{bmatrix} , \quad \kappa = \frac{cK}{H} \quad .$$

The determinant of V is the sum of two squares, so that condition (2.1) is equivalent to the two relations

$$\begin{aligned} &\nu^2(b_{12}b_{21} - b_{11}b_{22}) = a_{12}a_{21} - (\kappa-a_{11})(\kappa-a_{22}) \; , \\ &\nu[\kappa(b_{11}+b_{22}) - a_{11}b_{22} - b_{11}a_{22} + a_{12}b_{21} + b_{12}a_{21}] = 0 \; . \end{aligned} \tag{2.2}$$

There are many ways to satisfy these equations; in our context the two inequalities

$$b_{11} + b_{22} \neq 0 \quad , \quad \det b_{ij} \neq 0$$

can be presumed to apply*; hence our exclusive interest for these consequences of (2.2)**

---

* We may quote an alternative case (not belonging, however, to the class of lubricant responses), where a_{ij} is symmetric and b_{ij} is skew; in that case the second condition (2.2) is automatically satisfied, whereas κ must be a solution of the equation $(\kappa-a_{11})(\kappa-a_{22}) - a_{12}^2 = \nu^2 b_{12}^2$.

** Actually we mention later only the consequences of taking for ν a positive value.

$$\nu^2 = \frac{\det(a_{ij} - \kappa\delta_{ij})}{\det(b_{ij})} \; ,$$

(2.3)

$$\kappa = \frac{a_{11}b_{22} - a_{12}b_{21} + b_{11}a_{22} - b_{12}a_{21}}{b_{11} + b_{22}} \; .$$

The importance of these two conditions is in that they must be satisfied, as we shall see, at the boundary of the region of stability.

In some special cases of full cylindrical bearings when a_{ij}, b_{ij} are known, conditions (2.3) can be specified as follows:

(i) full lubrication, null steady load $(\bar{a} = 0)$, any ratio b/R (Muskat and Morgan)

$$\nu^2 = 1/4 \; , \; \kappa = 0 \; ;$$

(ii) full lubrication, long bearings (Sommerfeld)

$$\nu^2 = \frac{(1 - \frac{1}{2}\bar{a}^2 + \bar{a}^4)}{(2 + \bar{a}^2)^2} \; , \quad \kappa = 0 \; ;$$

notice that ν^2 falls in the interval $\left(\frac{1}{4}, \frac{1}{6}\right)$;

(iii) full lubrication, short bearings (Ocvirk)

$$\nu^2 = \frac{1}{4} \; , \; \kappa = 0 \; ;$$

(iv) full cavitation, short bearings (Holmes)

$$\nu^2 = \frac{(\pi^2/4)(1+2\bar{a}^2)(1-\bar{a}^2)(2+\bar{a}^2)^2 - 4\bar{a}^2(1+8\bar{a}^2+3\bar{a}^4)}{(1-\bar{a}^2)(2+\bar{a}^2)^2(\pi^2+2\pi^2\bar{a}^2-16\bar{a}^2)} \; ,$$

$$\kappa = \frac{\bar{a}(3+\bar{a}^2)}{(1-\bar{a}^2)^2(2+\bar{a}^2)} \; ;$$

this is the most interesting explicit case; notice that ν can be real only in an interval $(0, \bar{a}^*)$ where $\bar{a}^*$ is approximately 0.755, thus giving a first indication that transition cannot occur if $\bar{a}$ is too large; notice also that κ grows monotonically from 0 to $+\infty$ as $\bar{a}$ grows from 0 to 1.

3. SHAFT DYNAMICS

The indefinite equations of rotor whirl around a state (perhaps deflected under weight) of uniform rotation are

$$(3.1)\qquad \rho A L^4 \omega^2 \frac{\partial^2 u_i}{\partial \tau^2} + \frac{\partial^2}{\partial \zeta^2}\left(EI \frac{\partial^2 u_i}{\partial \zeta^2}\right) = 0 \quad , \quad i = 1,2 \quad ,$$

if we choose a z-axis through the journals' centres in the steady state, put $z = L\zeta$, call (cu_1, cu_2) the displacement vector, ρA the mass per unit length, EI the flexural rigidity and L the length of the shaft. If we assume that there are only two bearings and there are no overhangs, the boundary conditions are

$$(3.2)\qquad \begin{gathered} \frac{\partial^2 u_i}{\partial \zeta^2} = 0 \quad , \quad \text{for } \zeta = 0 \, , \, \zeta = 1 \; ; \\ \frac{\partial}{\partial \zeta}\left(\frac{EIc}{HL^3} \frac{\partial^2 u_i}{\partial \zeta^2}\right) = \pm \left(\sum_k a_{ik} u_k + \sum_k b_{ik} u_k'\right) \quad , \end{gathered}$$

again for $\zeta = 0$, $\zeta = 1$ but with alternate signs.

The general solution of (3.1), (3.2) is a linear combination of modes

$$(3.3)\qquad u_{s,n}(\zeta,\tau) = d_{s,n}(\zeta)\exp(i\lambda_n \tau) \quad , \quad s = 1,2 \; ; \; n = 1,2,\ldots \quad ,$$

where $d_{s,n}$ are appropriate complex functions of ζ (linear combinations themselves of exponential functions, if the shaft is uniform) and λ_n is the n-th solution of an appropriate eigenvalue equation. We will give here the details for the case of a uniform shaft rotating on two identical bearings (see [2] and [4], Part II for similar, though more special, developments); then the equation to be satisfied by λ is

$$(3.4)\quad \left[q_j-(a_{11}+i\lambda b_{11})\right]\left[q_j-(a_{22}+i\lambda b_{22})\right] = (a_{21}+i\lambda b_{21})(a_{12}+i\lambda b_{12}),$$

where q_j $(j=1,2)$ is given by

$$(3.5)\qquad q_j = Mp^3\left\{\frac{\cosh p+(-1)^j}{\sinh p} - \frac{\cos p+(-1)^j}{\sin p}\right\},$$

with

$$M = \frac{1}{\pi^4}\,\frac{c(\omega_1^{(\infty)})^2}{g}\,\frac{W}{H},$$

$$p^4 = \pi^4\left(\lambda\,\frac{\omega}{\omega_1^{(\infty)}}\right)^2,$$

and g is the acceleration due to gravity and, in our case,

$$\omega_1^{(\infty)} = \left(\frac{\pi}{L}\right)^2\left(\frac{EI}{\rho A}\right)^{1/2}, \quad W = \frac{\rho ALg}{2}.$$

If λ turns out to have a negative (positive) imaginary part, formula (3.3) describes whirls of increasing (decreasing) amplitude. Transition can be expected to occur when λ is real; then the condition to be imposed upon λ is exactly that required of ν by the first eq. (2.3); furthermore q_j must coincide with κ, as specified by the second eq. (2.3), and must satisfy, besides, the compatibility condition (3.5) on p. Transition conditions are just those which assure the existence of elliptic whirls during which the oil force acts as an elastic restoring force, as described in Section 2.

In the special cases (i), (ii), (iii) of Section 2 q_j turns out to be zero at transition (as does κ); correspondingly there is a double null root of (3.5) in p and other roots where p is such that the quantity in curly brackets in the right-hand side of (3.5) vanishes.

Once the value of p is known, so is the transition value of ω ; there is a succession of transition values, the lowest being zero in cases (i), (ii), (iii). In the special case (iv) instead κ , as already remarked, depends on $\bar{a}$ and so does K of course. The transition values of ω are given by the succession $(\omega_n^{(K)}/\nu_n)$, $n=1,2,\ldots$; these values are larger than zero for all $K > 0$. It is instructive in this special case to illustrate completely the situation with reference to the lowest (and practically most interesting) mode. The shaft-bearings system can be defined, for the purposes of the present analysis, through the two parameters

$$\gamma_1 = \frac{c(\omega_1^{(\infty)})^2}{g} \quad , \quad \sigma_1 = \frac{b^3 R\mu\omega_1^{(\infty)}}{c^2 W} = \left(\frac{H}{W}\right)_{\omega=\omega_1^{(\infty)}} \quad ,$$

and the kinematic condition by the ratio $r = \omega/\omega_1^{(\infty)}$.

In fact $\bar{a}$ is determined by the product $\sigma_1 r$ (and vice versa) and so are the coefficients a_{ij} , b_{ij} and the parameters κ and ν (i.e., q_j and λ , and hence p). Eq. (3.5) can be then considered as a compatibility condition on the choice of r (i.e. of ω) to achieve transition conditions. Fig. 1 of ref. [4], Part II shows level lines of r (called there Z_1) for all values of γ_1 and σ_1 .

4. ANALYSIS OF TRANSITION

We have found in Section 3 that transition in a shaft-bearings system occurs when ω coincides with one of the values $(\omega_n^{(k)}/\nu_n)$. We may reasonably expect that the transition occurs from stable to unstable behaviour as ω increases beyond $\omega_n^{(k)}/\nu_n$. However, as part of our analysis is general accepting almost any choice of the coefficients a_{ij} , b_{ij} , we may explore in general the forecasts from our analysis regarding the behaviour near one of the speeds

$\omega_n^{(k)}/\nu_n$. For this purpose we put in all equations

$$r = \bar{r} + \varepsilon \quad , \quad \lambda = \bar{\lambda} + \delta_1 + i\delta_2 \quad ,$$

disregard terms of higher order in ε and δ_1 , δ_2 and explore how the sign of δ_2 depends on the choice of sign for ε .

We get

$$p = \bar{p} + \frac{\pi}{2} \frac{\bar{r}(\delta_1 + i\delta_2) + \bar{\lambda}\varepsilon}{(\bar{r}\,\bar{\lambda})^{1/2}} \quad ,$$

and

$$q_j = \bar{q}_j + B_j\varepsilon + C_j(\delta_1 + i\delta_2) \quad ,$$

where

$$B_j = \frac{1}{2}\left(\frac{\bar{q}_j}{\bar{r}} - D_j\bar{\lambda}^2\right) \quad , \quad C_j = \frac{1}{2}\left(\frac{3\bar{q}_j}{\bar{\lambda}} - D_j\bar{\lambda}\,\bar{r}\right) \quad ,$$

$$D_j = (-1)^j \frac{\bar{\gamma}}{\bar{\sigma}}\left\{\frac{1}{\cosh\bar{p} + (-1)^{j+1}} + \frac{1}{\cos\bar{p} + (-1)^{j+1}}\right\} \quad ;$$

notice that $D_j < 0$, and hence $B_j > 0$, $C_j > 0$. Finally,

$$\delta_2 = \frac{B_j EF}{(G_j C_j + F)^2 + E^2 C_j^2}\,\varepsilon$$

where

$$E = \bar{\lambda}(\bar{b}_{11} + \bar{b}_{22}) \quad , \quad F = 2\bar{\lambda}(\bar{b}_{12}\bar{b}_{21} - \bar{b}_{11}\bar{b}_{22}) \quad ,$$

$$G_j = 2\bar{q}_j - \bar{a}_{11} - \bar{a}_{22} \quad ,$$

where the bars refer to the specification $\bar{a}$ for a .

Hence the final result hinges on the sign of the product EF : in all cases of interest here it turns out that this product is negative; it is thus confirmed that our scheme forecasts adequately transition from stable to unstable behaviour as ω grows beyond $\omega_n^{(k)}/\nu_n$.

5. WHIRLS OF LARGE AMPLITUDE

We have exploited thoroughly in Sections 3 and 4 the idea of examining first of all the question of the existence of paths of the journal along which the oil force has convenient properties and of checking later the compatibility of such paths with other dynamic conditions. Surprisingly this idea works also in an analysis of whirls of large amplitude. We need of course at the outset some hypothesis on f_e , f_n when a is almost 1 ; to cover all cases known explicitly, we assume

$$f_e = -\, h_e \frac{a'}{(1-a)^{\alpha}} \,[1 + o(1-a)] \quad ,$$

$$f_n = \quad h_n \frac{1-2\psi'}{(1-a)^{\alpha-1}}[1 + o(1-a)] \quad ,$$

where h_e , h_n are positive constants and α is larger than 1 .

Consider now the possibility that a and ψ evolve with τ in such a way that

$$f_e \simeq -\,\kappa a \quad , \quad f_n \simeq 0 \quad , \quad \text{for} \quad \tau \to \infty \, .$$

This possibility exists if

$$a = 1 - \left[\frac{h_e}{(\alpha-1)\kappa\tau}\right]^{\frac{1}{\alpha-1}} + o\left[\left(\frac{1}{\tau}\right)^{\frac{1}{\alpha-1}}\right] \tag{5.1}$$

and

$$\psi' = \frac{1}{2} + o\left[\left(\frac{1}{\tau}\right)\right] \quad .$$

The check of compatibility of these asymptotic formulae with the indefinite eqs. (3.1) and the boundary conditions (3.2) for the case when the bearings carry no load (vertical shaft) can be carried out as done under more special hypotheses in ref. [4], Part I; it is found that compatibility can be achieved if $\omega < 2\omega_1^{(\infty)}$. The mode of vibration at a running speed $\tilde{\omega}$ approximates, as $\tau \to \infty$, the mode of vibration of the shaft when resting on elastic supports of a rigidity K such that

$$2\omega_1^{(K)} = \tilde{\omega} \quad .$$

The amplitude of vibration along the shaft is determined by the condition that the displacement at the bearings be approximately equal to c ; hence for small ω the amplitude is approximately c everywhere (parallel whirl), whereas it grows to infinity, outside the bearings, as ω approaches $2\omega_1^{(\infty)}$.

In theory at least any of the modes associated with all $\omega_n^{(K)}$ (and not only the mode associated with $\omega_1^{(K)}$) can be brought to bear in the discussion, with obvious results; in particular one can consider the second mode, which, for small K , is an approximate picture of a conical whirl. This latter case is perhaps the only relevant one in practice as, for small K , $\omega_2^{(K)} < \omega_1^{(\infty)}$.

Consider now the case when a evolves again as specified by (5.1), whereas ψ' approximates $\omega_1^{(\infty)}/\omega$ (i.e. the whirl is "resonant"). Then f_e approximates again $-\kappa a$ whereas f_n increases indefinitely with τ . Compatibility with equations (3.1), (3.2) is now achieved only for $\omega > 2\omega_1^{(\infty)}$; the mode of vibration is approximately the fundamental mode of the shaft on rigid supports and the amplitude is proportional to τ (see [1] and [4], Part I, again for details).

The asymptotic non-linear behaviour of a horizontal heavy shaft is more difficult to follow, expecially if the case of bearings with cavitating oil film is considered; then one has to face the complexity of the expressions of the lubricant forces, complexity deriving from the need to account for the wide rocking of the lubricant film (see explicit formulae for full cylindrical bearings in ref. [2]). In a forthcoming paper [7] LARATTA attacks just this problem; of his results we quote here only the simplest one: for a heavy uniform shaft running on full cylindrical bearings with full lubrication the results indicated above for a vertical shaft maintain their validity.

6. THE CHALLENGE OF TECHNICAL FORECASTS

We have dealt so far with questions of principle with the declared intent of clarifying the issues from a qualitative point of view. The question remains of creating analytical tools for the exact forecast of behaviour of a rotor-bearings system of a particular design. A partial answer to this need are computer programs for the determination of linear transition conditions (if any); the programs can be made practical and almost as economic as those which allow the calculation of critical speeds [5]. The greater complexity is related to the need of considering elliptic (rather than circular) whirls, but this difficulty can be overcome and is met anyway even when one wants to calculate the amplitude of a synchronous vibration caused by lack of balance [3], [6].

7. REFERENCES

The following list is very parochial; its aim is not to show the Author's ignorance of the many contributions to the theory of oil whirl from Newkirk's to those of other participants in this Symposium. It is intended rather to call attention on some papers which, though rarely quoted, are strictly relevant to the analysis presented above.

[1] CAPRIZ, G.: "On the vibrations of shafts rotating on lubricated bearings". Ann. Mat. pura appl., 50 (1960), 223-248. Russian translation in Mechanika 4, 68, (1961), 47-66.

[2] CAPRIZ, G.: "Sulle vibrazioni di aste rotanti". Ann. Sc. Norm. Sup. Pisa, 17 (1963), 31-42.

[3] CAPRIZ, A.: "A numerical method for evaluating the response of shafts to excitation". Meccanica, 2 (1967), 213-217.

[4] LARATTA, A.: "Oil whirl of rotors, Parts I and II". Meccanica, 4 (1969), 1-13; 5 (1970), 126-133.

[5] CAPRIZ, G.: "Oil whirl of rotors: a numerical method of prediction". Meccanica, 5 (1970), 203-209.

[6] CAPRIZ, G., CAPOVANI, M. and MICHELINI, M.: "Automatic calculation of vibration amplitudes in rotating shafts". Quaderni Pignone, 14 (1970), 5-13.

[7] LARATTA, A.: "Large amplitude whirls of rotors". (to appear).

Helicopter Ground Resonance - A Simplified Approach

G.T.S. DONE

Department of Mechanical Engineering

University of Edinburgh, UK

ABSTRACT - Helicopter ground resonance is not a simple resonance but an instability that arises with the helicopter on the ground and the rotor running at a certain angular speed. A method of simplifying the equations of motion is suggested that allows a simple expression to be obtained for the amount of damping required to suppress the instability. A means of comparing the equations with those of more conventional non-conservative system is proposed.

1. INTRODUCTION

Helicopter ground resonance is not, as its name implies, a simple resonance deriving from some forcing excitation at an appropriate frequency, but a mechanically induced instability in the same category as fluid-dynamically induced examples such as aircraft lifting surface flutter or follower force instability. It occurs when the helicopter is on the ground and the rotor running at a certain speed, the degrees of freedom involved being the rigid body modes pitch and roll of the fuselage and some of the rotor system modes. By deliberately simplifying the mathematical model of the system, it can be shown that the instability is basically due to the fact that the rotor head is capable of substantially lateral motion and that the rotor blades can deflect in the "lag" sense (i.e. in the plane of the rotor) relative to axes of reference rotating at rotor speed. The instability can build up quickly and may result in the destruction of the helicopter unless avoiding action is taken; during a ground resonance test tethering gear is normally used to suppress rapidly the body freedoms, otherwise the pilot must fly the helicopter off the ground.

The phenomenon has been known and understood for many years and in industry sufficiently complex calculations are performed using the computer for the situation to be more or less under control. However, there has been a need too for a simplified theory of ground resonance that could be referred to when a quick but not necessarily very accurate answer was required. The facility has been available to dynamicists in the fixed wing aircraft industry in the form of the flutter calculation reduced to a binary or two degree of freedom case. This enables a certain amount of insight to be gained in a particular flutter situation and is considered valuable. Because of terms in the helicopter equations of motion that arise by virtue of the rotation present, the reduction to two degrees of freedom has previously been elusive, but an appropriate matrix transformation has been found, together with the resulting simple expressions for parameters such as the amount of damping required to suppress instability.

It has been mentioned that helicopter ground resonance is akin to other forms of instability; the system itself in fact belongs to a special class of nonconservative systems, the members of which have the same general properties, give rise the same general phenomena and are governed by more or less the same type of equations of motion. The classification of the system is discussed in Section 6.

2. SIMPLIFIED MODEL

The simplest model that is capable of ground resonance is shown in Fig. 1. The chassis mass M is considered concentrated and is constrained to move in one direction only, parallel to the plane of the rotor blades against a spring restraint. The rotor itself has three or more blades and rotates at angular velocity Ω about an axis fixed to mass M. The instability is dependent on the blades being able to de-

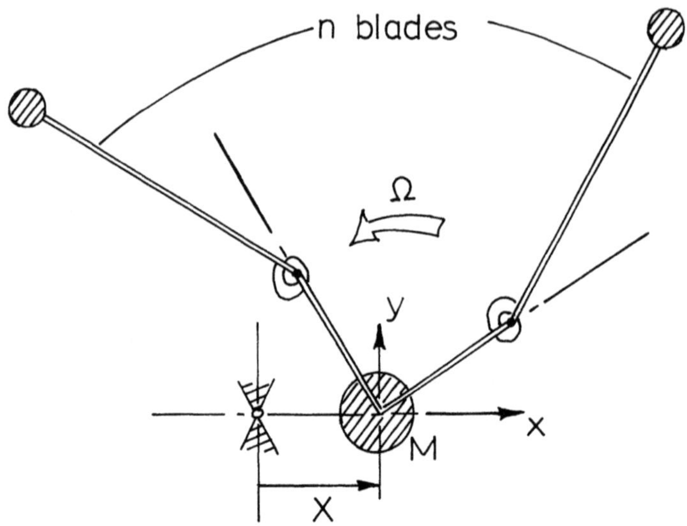

Fig. 1. Helicopter ground resonance model.

form in the plane of the rotor, and a convenient and relatively realistic representation is to consider each rotor blade as consisting of an inner arm, between each of which there is no relative movement, and an outer arm which pivots against a spring restraint and the end of the inner arm. The rotor blade mass is considered lumped at the end of this arm. Thus it will be realised that on this model an individual blade at zero chassis motion and rotor rotation has a finite "lag" natural frequency, which is in keeping with the physical properties of the modern "semi-rigid" rotor that has no lagging or flapping hinge.

The number of degrees of freedom that the model possesses is $(n+1)$ where n is the number of blades. In fact, the ground resonance equations of motion for this system involve only three degrees of freedom, and the reason for the discrepancy is as follows. Consider a transformation of the n blade lag co-ordinates into generalised co-ordinates associated with overall modes. Some of these modes involve motions of the rotor centre of mass which has only two degrees of freedom (in translation). The transformation can be made in such a way that only two rotor modes involve motions of the centre of mass; these are capable of coupling with the chassis translation mode to produce ground resonance instability. All the other (n-2) rotor modes (with no motion of the centre of mass) have no coupling effect and may be disregarded. In fact, for present purposes, the rotor could be replaced by an elastic lamina that can deform only in the plane of rotation.

The transformation mentioned is often referred to as the "Coleman transformation" after the paper by COLEMAN and FEINGOLD [1], which, although written several years ago and being mainly concerned with hinged rotors remains a standard reference on the subject of helicopter ground resonance.

3. EQUATIONS OF MOTION

The equations may be written in matrix form

$$(1)\quad \begin{bmatrix} 1 & r & 0 \\ 1 & 1 & 0 \\ 0 & 0 & 1 \end{bmatrix} \begin{bmatrix} \ddot{X} \\ \ddot{x} \\ \ddot{y} \end{bmatrix} + 2\Omega \begin{bmatrix} 0 & 0 & 0 \\ 0 & 0 & 1 \\ 0 & -1 & 0 \end{bmatrix} \begin{bmatrix} \dot{X} \\ \dot{x} \\ \dot{y} \end{bmatrix} + \begin{bmatrix} \omega_X^2 & 0 & 0 \\ 0 & c(\Omega) & 0 \\ 0 & 0 & c(\Omega) \end{bmatrix} \begin{bmatrix} X \\ x \\ y \end{bmatrix} = \begin{bmatrix} 0 \\ 0 \\ 0 \end{bmatrix}$$

where X represent non-dimensionally the chassis movement and x and y the rotor centre of mass displacement. The angular speed is Ω , r is the ratio of total blade mass to twice the overall mass, ω_X is the chassis natural circular frequency (rotor fixed) and $c(\Omega)$ is a non-dimensional stiffness term. The latter is equal to $\omega_B^2 - \Omega^2(1-d)$ where ω_B is the natural frequency of an individual blade with $\Omega = 0$ and the chassis fixed, and d is the ratio of the length of the inner arm to that of the outer in Fig. 1 ($d < 1$ for most practical configurations). The skew-symmetric terms in the matrix multiplying the velocity vector couple the rotor degrees of freedom and are of gyrodynamic origin. The terms in the mass matrix which couple the chassis motion with a rotor freedom can be made symmetric (by dividing the upper equation by r) and must be present for ground resonance to occur. Structural damping is neglected in the equations, but can be introduced by inserting terms in the appropriate matrices as in Section 5. The stiffness term $c(\Omega)$ varies linearly with Ω^2 and becomes negative above a certain value of Ω ; as the system is gyrodynamic in nature this does not imply a divergence type instability as would otherwise be the case.

Equations (1) represent the simplest mathematical expression of the system in second order form. Yet, even from these equations, it is apparent from a study of a typical solution (Fig. 2), that the instability concerns basically only two degrees of freedom in the case of the semi-rigid rotor. Three degrees of freedom are still too many to permit simple algebraic techniques to be applied, so the reduction to two as discussed in the Introduction would be most advantageous.

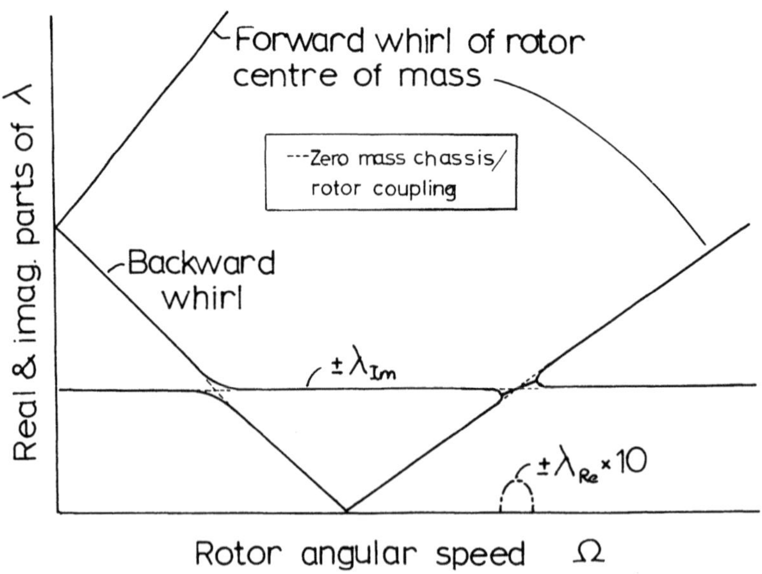

Fig. 2. Real and imaginary parts of eigenvalues for helicopter ground resonance model (Note: complete picture includes reflection below horizontal axis).

The real and imaginary parts of the eigenvalue shown in Fig. 2 follow the usual pattern for undamped nonconservative systems. The instability is concentrated at a frequency crossing of the uncoupled system, but it is interesting that the coupling when present, is apparently stabilising at the lower crossing and destablising at the upper. The eigenmodes are easily described in the absence of the mass chassis/ rotor coupling; the chassis mode appears independently whilst the rotor modes are associated each with a circular whirl of the centre of mass, the sense being either the same as that of Ω (forward whirl) or opposite (backward) as shown in the figure.

For a set of eigenvalues such as those in Fig. 2 it can be assumed that the frequency of the upper frequency (forward whirl) root becomes sufficiently high as Ω is increased for it to play little part in the instability. Thus, it would be desirable to obtain a means of ignoring this root or removing it from the mathematics. A method of doing this is described in the following section.

4. REDUCTION TO TWO DEGREES OF FREEDOM

The problem that arises in attempting to ignore the upper frequency rotor mode is that the usual methods of normal mode transformation do not work because the uncoupled (i.e. chassis/rotor coupling suppressed) and therefore conservative equations of motion do not have symmetric coefficient matrices throughout, the rotor rotation introducing skew-symmetric terms. However, it is possible to transform into "bi-normal" co-ordinates that have the same properties of orthogonality. An account of the general procedure is provided in HALFMAN [2]; the procedure as applied to the present problem is described in outline below.

The first step in the reduction of equations (1) is to re-cast them into first order form, so that velocities are thought of as co-ordinates. It is then convenient to rearrange the rows and columns so as to enable partitioning into rotor and chassis matrices. The rotor equations alone can be written

$$\underline{P}\,\dot{\underline{z}} + \underline{Q}\,\underline{z} = \underline{0} \tag{2}$$

where $\underline{z}$ is the vector of rotor co-ordinates. For this set an adjoint set is introduced

$$\underline{P}'\dot{\underline{z}}^* - \underline{Q}'\underline{z}^* = 0 \tag{3}$$

where the dash indicates a transpose and $\underline{z}^*$ is a vector of co-ordinates of the adjoint system. New co-ordinates are now defined

$$\begin{aligned} \underline{Z} &= \underline{R}\,\underline{q} \\ \underline{Z}^* &= \underline{R}^*\underline{q}^* \end{aligned} \tag{4}$$

where $\underline{R}$ and $\underline{R}^*$ are the eigenmodal matrices of the two systems. The generalised work done by the forces in the original system moving through a small displacement in the adjoint system is found, which then leads to the transformed version of equation (2)

$$\underline{R}^{*\prime}\,\underline{P}\,\underline{R}\,\dot{\underline{q}} + R^{*\prime}\,\underline{Q}\,\underline{R}\,\underline{q} = 0 \tag{5}$$

HALFMAN [2]shows that the matrices $\underline{R}^{*\prime}\,\underline{P}\,\underline{R}$ and $R^{*\prime}\,\underline{Q}\,\underline{R}$ are diagonal. The co-ordinates $\underline{q}$ are called "bi-normal".

A transformation can now be applied to the complete system (i.e. including rotor/chassis coupling) in such a way that the chassis co-ordinates are left unchanged and the rotor co-ordinates are the binormal co-ordinates found above. The upper frequency mode co-ordinates can be deleted which leaves the first order equation of motion

$$\begin{bmatrix} 0 & 1 & 0 & 0 \\ 1 & 0 & ir(\Omega-\Gamma) & -ir(\Omega-\Gamma) \\ 1 & 0 & -4i\Gamma & 0 \\ 1 & 0 & 0 & 4i\Gamma \end{bmatrix} \begin{bmatrix} \ddot{X} \\ \dot{X} \\ \dot{q}_1 \\ \dot{q}_{1c} \end{bmatrix} +$$

(6)

$$+ \begin{bmatrix} -1 & 0 & 0 & 0 \\ 0 & \omega_x^2 & 0 & 0 \\ 0 & 0 & 4\Gamma(\Gamma-\Omega) & 0 \\ 0 & 0 & 0 & 4\Gamma(\Gamma-\Omega) \end{bmatrix} \begin{bmatrix} \dot{X} \\ X \\ q_1 \\ q_{1c} \end{bmatrix} = 0$$

in which $\Gamma = \sqrt{\Omega^2 + c(\Omega)}$, q_1 is the lower frequency rotor mode co-ordinate and q_{1c} its complex conjugate. The modal frequencies for the system when the coupling between the rotor and chassis is suppressed are provided directly from equation (6) as $(\Omega-\Gamma)$ and ω_x (that of the deleted degree of freedom is $(\Omega+\Gamma)$) . The coupling terms between the chassis freedom and the lower frequency rotor mode appear in the off-diagonal partitions in the left hand coefficient matrix. Substitution of the general solution $\underline{\zeta} = \underline{\zeta}_o e^{\alpha t}$ where $\underline{\zeta} \equiv \{\dot{X}, X, q_1, q_{1c}\}$ into equation (6) leads to the determinant

$$(7) \qquad \begin{vmatrix} -1 & & 0 & 0 \\ \alpha & \omega_x^2 & ir(\Omega-\Gamma)\alpha & -ir(\Omega-\Gamma)\alpha \\ \alpha & 0 & 4\Gamma(\Gamma-\Omega-i\alpha) & 0 \\ \alpha & 0 & 0 & 4\Gamma(\Gamma-\Omega+i\alpha) \end{vmatrix} = 0$$

which gives on expansion the characteristic equation

$$(8) \qquad (\alpha^2+\omega_x^2)\{\alpha^2 + (\Omega-\Gamma)^2\} + \frac{r(\Omega-\Gamma)}{2\Gamma}\alpha^4 = 0$$

The equation can be used to obtain approximately the two lower roots only of Fig. 2. For the data actually used there was in fact a negligible discrepancy.

From study of the characteristic equation it can be shown that the presence of the term involving r is stabilising in the region of the lower frequency crossing and destabilising in that of the upper.

5. EFFECT OF STRUCTURAL DAMPING

Suppression of ground resonance is usually effected by introducing damping, so it is important to consider the system of Fig. 1 with structural damping. Velocity dependent dampers are assumed attached in parallel with the springs, so that there is a damping force acting on the chassis proportional to $\dot{X}$ and a damping torque acting on each blade proportional to $\dot{\beta}$. The modified equations of motion are, from COLEMAN [1]

$$\begin{bmatrix} 1 & r & 0 \\ 1 & 1 & 0 \\ 0 & 0 & 1 \end{bmatrix} \begin{bmatrix} \ddot{X} \\ \ddot{x} \\ \ddot{y} \end{bmatrix} + \begin{bmatrix} d_x & 0 & 0 \\ 0 & d_B & 2\Omega \\ 0 & -2\Omega & d_B \end{bmatrix} \begin{bmatrix} \dot{X} \\ \dot{x} \\ \dot{y} \end{bmatrix} +$$

$$\begin{bmatrix} \omega_x^2 & 0 & 0 \\ 0 & c(\Omega) & d_B(\Omega) \\ 0 & -d_B\Omega & c(\Omega) \end{bmatrix} \begin{bmatrix} X \\ x \\ y \end{bmatrix} = \begin{bmatrix} 0 \\ 0 \\ 0 \end{bmatrix} \qquad (9)$$

where d_x and d_B are the appropriate damping constants. The equations are re-written in first order form and transformed and reduced as outlined in Section 4. The additional elements appearing in the determinantal array in equation (9) due to the presence of d_x and d_B are,

$$(10)\qquad \begin{bmatrix} 0 & 0 & 0 & 0 \\ 0 & dx & 0 & 0 \\ 0 & 0 & 2d_B(\alpha-i\Omega) & 0 \\ 0 & 0 & 0 & 2d_B(\alpha+i\Omega) \end{bmatrix}$$

and the characteristic equation for the damped system is

$$(11)\quad (\alpha^2+d_x\alpha+\omega_x^2)\{\alpha^2+d_B\alpha+(\Omega-\Gamma)^2\}+\frac{r(\Omega-\Gamma)}{2\Gamma}\alpha^4+\frac{r(\Omega-\Gamma)}{4\Gamma^2}\Omega d_B\alpha^3 = 0$$

which may be compared with equation (8) for the undamped system.

It would be useful to know the amount of damping to just suppress the instability, i.e. make one root α purely imaginary. It can be assumed that this occurs at approximately the upper frequency crossing of the undamped and uncoupled system. The value of α under consideration is then simply $\alpha = \pm i\omega_x = \pm i(\Omega_2-\Gamma_2)$, where the suffix 2 refers to the upper frequency crossing and using this in equation (11) leads to an expression involving r, d_x, d_B. The damping terms are usually of small order and by neglecting third order small quantities the expression leads to

$$(12)\qquad d_x d_B = \omega_x^2 A(\Omega_2) = \frac{r\omega_x^3}{2(\Omega_2-\omega_x)}$$

where $A(\Omega_2) \equiv r(\Omega_2-\Gamma_2)/2\Gamma_2$. In terms of the damping $\zeta_x = d_x/2\omega_x$ and $\zeta_B = d_B/2\omega_B$ equation (12) is equivalent to

$$(13)\qquad \zeta_x\zeta_B = \frac{r\omega_x^2}{8\omega_B(\Omega-\omega_x)}$$

Thus, it is the product of the damping coefficients or ratios that must achieve a certain minimum value for the instability to be suppressed, the value itself depending directly on the mass ratio r and largely on the chassis natural frequency.

6. CLASSIFICATION OF SYSTEM

Equations (1) describe the motion of the ground resonance system but there are several reasons why it is not obvious purely from examination of the equations on a superficial basis that the system is nonconservative. First of all, the skew-symmetric terms in the matrix multiplying the velocities appear because of the gyrodynamic nature of the system and by themselves do not make the system nonconservative [3]. Secondly, if the top equation is divided by the scalar r the mass matrix is made symmetric, a property normally associated with conservative systems. Finally, the equations in second order form cannot be directly reduced from three to two whilst still retaining the instability as is the case with many known nonconservative systems. However, all the other evidence suggests that the system is indeed nonconservative; for instance, the chief parameter controlling the source of energy can be identified as Ω, the energy source as the kinetic energy of the rotor, and the instability when it occurs is an oscillatory one and is associated with a frequency coalescence.

The ground resonance equations are more amenable to manipulation when in first order form than in second, and it may be that the basis for classification of all nonconservative systems lies in expressing the equations in first order form. This is certainly the case when a simple undamped two degree of freedom system is compared with the ground resonance case. The example considered has equations

$$(14) \qquad \begin{bmatrix} 1 & a_{12} \\ a_{21} & 1 \end{bmatrix} \begin{bmatrix} \ddot{x}_1 \\ \ddot{x}_2 \end{bmatrix} + \begin{bmatrix} \omega_1^2 & c_{12} \\ c_{21} & \omega_2^2 \end{bmatrix} \begin{bmatrix} x_1 \\ x_2 \end{bmatrix} = \begin{bmatrix} 0 \\ 0 \end{bmatrix}$$

It is assumed that $a_{12}, a_{21}, c_{12}, c_{21}$ depend on the chief parameter and vanish when latter is zero. There is no loss of generality in making the diagonal mass terms unity; the diagonal stiffness terms are then equal to the squares of the natural frequencies, (which may themselves depend on the chief parameter). The equations can be written down in first order form by introducing a new co-ordinate vector $\{q_1, x_1, q_2, x_2\}$ where $q_1 = \dot{x}_1$, $q_2 = \dot{x}_2$; they are then subjected to a similarity transform into the normal co-ordinates corresponding to the uncoupled (second order) equations i.e. with $a_{12} = a_{21} = c_{12} = c_{21} = 0$. This produces

$$(15) \qquad \begin{bmatrix} \underline{G}_{11} & \underline{G}_{12} \\ \underline{G}_{21} & \underline{G}_{22} \end{bmatrix} \begin{bmatrix} \dot{\underline{p}} \end{bmatrix} + \begin{bmatrix} \underline{H}_{11} & \underline{H}_{12} \\ \underline{H}_{21} & \underline{H}_{22} \end{bmatrix} \begin{bmatrix} \underline{p} \end{bmatrix} = \begin{bmatrix} \underline{0} \end{bmatrix}$$

where $\underline{p}$ is a vector of the new co-ordinates which appear in two conjugate pairs. The sub-matrices are

$$\underline{G}_{11} = 2i\omega_1 \begin{bmatrix} 1 & 0 \\ 0 & -1 \end{bmatrix} \qquad \underline{G}_{12} = 2i\omega_2 \begin{bmatrix} 1 & 0 \\ 0 & -1 \end{bmatrix}$$

$$\underline{G}_{12} = i\omega_2 a_{12} \begin{bmatrix} 1 & -1 \\ 1 & -1 \end{bmatrix} \qquad \underline{G}_{21} = i\omega_1 a_{21} \begin{bmatrix} 1 & -1 \\ 1 & -1 \end{bmatrix}$$

$$\underline{H}_{11} = 2\omega_1^2 \begin{bmatrix} 1 & 0 \\ 0 & 1 \end{bmatrix} \qquad \underline{H}_{22} = 2\omega_2^2 \begin{bmatrix} 1 & 0 \\ 0 & 1 \end{bmatrix}$$

$$\underline{H}_{12} = c_{12} \begin{bmatrix} 1 & 1 \\ 1 & 1 \end{bmatrix} \qquad \underline{H}_{21} = c_{21} \begin{bmatrix} 1 & 1 \\ 1 & 1 \end{bmatrix}$$

From examination of the characteristic equation for the case when the coupling is absent and frequency coalescence is imminent it is easily shown that coupling introduced such that either $a_{12}\, a_{21} < 0$ or $c_{12}\, c_{21} < 0$ is destabilising. Thus, for equations (15) to represent a nonconservative system, the scalars associated with either $\underline{G}_{12}$ and $\underline{G}_{21}$ or $\underline{H}_{12}$ and $\underline{H}_{21}$ must have opposite sign.

By taking equations (6) and suitably re-arranging it is possible to achieve axactly the form shown in equations (15). In this case $\omega_1 \equiv \omega_X$, $\omega_2 \equiv (\Gamma-\Omega)$, the whirl frequency of the rotor of mass where $\Gamma = (\Omega^2+c(\Omega))^{\frac{1}{2}}$, $a_{12} \equiv r$, $a_{21} \equiv (\Gamma-\Omega)/2\Gamma$ and c_{12} and c_{21} are zero. The instability is possible when $a_{12}\, a_{21} < 0$, i.e. when $r(\Gamma-\Omega)/2\Gamma$ is negative. Examination of this term as Ω is increased from zero shows that it is positive at the lower frequency coalescence but negative at the upper, which explains why the instability occurs only at the upper frequency crossing.

7. CONCLUSIONS AND DISCUSSION

It has been shown that the ground resonance instability phenomenon of helicopters can be illuminated by considering a simple model, the number of degrees of freedom of which can be reduced to two. The resulting theory enables an immediate appreciation to be gained of some of the more important parameters controlling the instability. Much of the present work originally appeared in unpublished form in December 1971 [4] and at that time it was though that the nonconservative nature of the system rendered futile the search for a simple transformation into normal co-ordinates (i.e. as opposed to the more complicated transformation into bi-normal co-ordinates). The system, however, is unusual in that whilst maintaining its non-conservative properties its matrices can be made symmetric; the possibility of a simple transformation therefore exists and a more recent search by KING [5] for one of the similarity type has proved successful.

The minimum amount of damping needed for the suppression of ground resonance has been found to depend in a simple way on the mass ratio and the blade lag and chassis natural frequencies. The relevant expression (equation (13)) verifies the so-called "product rule" implied from COLEMAN [1], in which only the product and not each individual value of the damping ratios has to achieve a critical minimum. The simple expression has only been used several times in industry to obtain a quick but not necessarily very accurate estimate of the damping required to suppress a particular instability.

The ground resonance phenomenon follows the recognised pattern for a nonconservative system in all but the equations of motion if they are left in the more usual second order form. However, it is seen that the situation is reconciled if an appropriate transformation is made into first order co-ordinates.

REFERENCES

[1] COLEMAN, R.P. and FEINGOLD, A.M.: Theory of Self-Excited Mechanical Oscillations of Helicopter Rotors with Hinged Blades. N.A.C.A. Report 1351, 1958

[2] HALFMAN, R.L.: Dynamics Vol. II Systems. Variational Methods and Relativity. Addison-Wesley, 1962.

[3] ZIEGLER, H.: Principles of Structural Stability. Blaisdell, 1968.

[4] DONE, G.T.S.: A Simplified Approach to Helicopter Ground Resonance. British Aeronautical Research Council 33255. December, 1971 (unpublished).

[5] KING, S.: Simplified Ground Resonance Analysis. Unpublished Note, Westland Helicopters Limited, 1972.

Vibration and Its Control in Rotating Systems

R. Holmes

School of Applied Sciences

University of Sussex, Brighton, Sussex, England

ABSTRACT - The role played by sleeve bearings in determining the response and instability conditions of rigid rotors and supercritical shafting is discussed in the light of theoretical and experimental evidence. Ways of assessing methods of control are suggested.

NOTATION

a_{rr} etc.	Oil-film displacement coefficients
b	Bearing or land length
b_{rr} etc.	Oil-film velocity coefficients
c	Radial clearance
D	Bearing diameter
h	Oil-film thickness $\simeq c(1+\varepsilon\cos\theta)$
k	Shaft stiffness
m	Journal mass or rotor mass per bearing
m_ℓ	Shaft mass per bearing or land
p	Oil-film pressure
p_s	Oil-supply pressure
P	External static load
P_c	Unbalance force
P_1 , P_2	Oil-film forces
$Q_{1,2}$	Functions of ε , $\dot{\varepsilon}$, ψ , $\dot{\psi} = P_{1,2}/mc\omega^2$
Q_c	$P_c/mc\omega^2$
r	Radial co-ordinate
R	Bearing radius
R_{oxx} etc.	Oil-film receptances
R_{xx} etc.	Combined receptances
s	Transverse co-ordinates

t	Time
u	Eccentricity of rotor mass
$\bar{Y}$, $\bar{X}$	External dynamic forces in horizontal and vertical directions
y	Horizontal co-ordinate
x	Vertical co-ordinate
z	Axial co-ordinate
β	$\pi\eta Rb^3/m\omega c^3$
ε	Dynamic eccentricity ratio = dynamic journal eccentricity ÷ c
ε_o	Static eccentricity ratio
ϕ_{oxx} etc.	Oil-film complex dynamic stiffnesses
ϕ_{xx} etc.	Combined complex stiffnesses
θ	Circumferential co-ordinate measured from point of maximum film thickness in direction of journal rotation
θ_1 , θ_2	Oil-film limits
ψ	Angle between fixed vertical axis and eccentricity vector = attitude angle
η	Oil viscosity
ω	Shaft speed
ω_j	Journal speed
Ω	Frequency of incipient instability

INTRODUCTION

This paper describes work carried out by the author and his colleagues at the University of Sussex, England into the vibrational behaviour of three classes of rotating system as influenced by the bearings upon which the system is supported. These systems are

a) The rigid rotor supported on journal bearings,
b) The rigid rotor supported on squeeze-film bearings, and,
c) The supercritical shaft supported on journal bearings.

In the majority of such applications the applied forces are due to relatively small unbalance load distribution superimposed upon a static load due to rotor or shaft weight. Nevertheless the equations governing the vibration performance of rotors and shafts are highly non-linear and the effects of these non-linearities are important particularly in the first two cases. In the third case much interest is focussed on the use of linearised oil-film displacement and velocity coefficients which lend themselves to matrix manipulation in the analysis of complete rotor-bearing-pedestal systems. The effects on stability, on vibration amplitudes and on reduction of critical speeds are then of paramount importance.

THE SMALL STEAM-TURBINE ROTOR

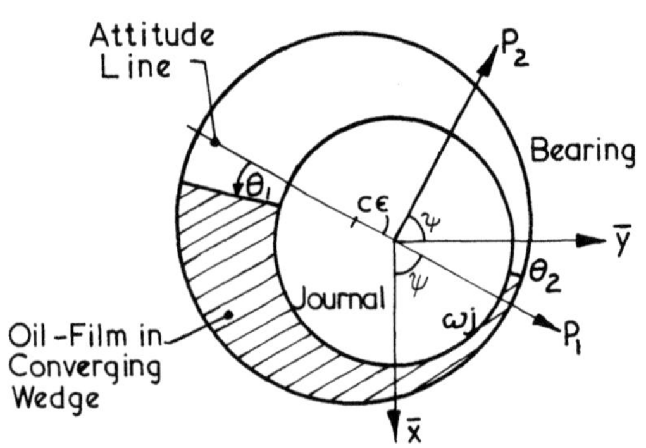

Fig. 1a. Journal-bearing system.

Consider first the case of a symmetrically-loaded rigid turbine rotor supported equally on two journal bearings, one of which is shown in fig. 1a. The equations of symmetric rotor motion may be written

$$(1) \qquad \begin{cases} P_1 + \bar{X}\cos\psi + \bar{Y}\sin\psi = mc(\ddot{\varepsilon} - \varepsilon\dot{\psi}^2) \\ P_2 + \bar{Y}\cos\psi - \bar{X}\sin\psi = mc(\varepsilon\ddot{\psi} + 2\dot{\varepsilon}\dot{\psi}) \end{cases}$$

P_1 , P_2 are the hydrodynamic forces generated in the oil film, being functions of ε_o , ε , $\dot{\varepsilon}$, ψ and $\dot{\psi}$ and m is equal to the rotor mass per bearing. The external applied forces $\bar{X}$, $\bar{Y}$ are given by

$$\bar{X} = P + P_c \cos\omega t$$

$$\bar{Y} = P_c \sin\omega t \quad ,$$

where P_c is the inertia force due to unbalance and ω is the shaft angular velocity. P is the steady load due to rotor weight which, in the absence of any other applied load, gives a steady-load position of the journal in the bearing defined by the parameter ε_o (the static eccentricity-ratio)(fig. 1b). From hydrodynamic theory applied to the short bearing it may be shown that the "load number",

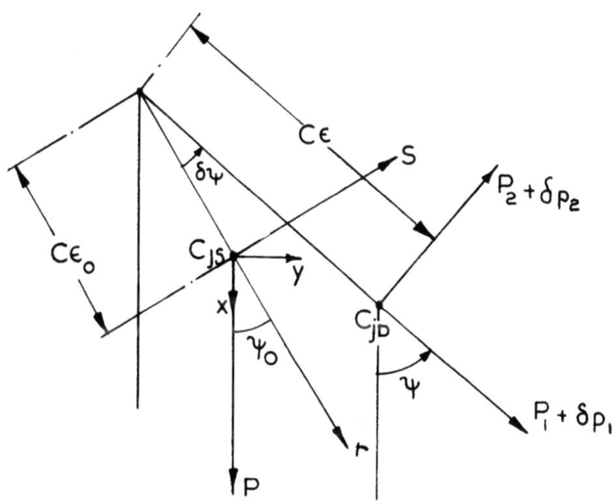

Fig. 1b. Magnified portion of fig. 1a.

(2) $$Pc^2/\eta Rb^3\omega_j = 0.25\varepsilon_o\left[\pi^2(1-\varepsilon_o^2) + 16\varepsilon_o^2\right]^{\frac{1}{2}}/(1-\varepsilon_o^2)^2$$

in which ω_j is the angular velocity of journal rotation, which in the case of the turbine bearing is equal to ω .

We may non-dimensionalize equations (1) by dividing by $mc\omega^2$. Thus,

$$
(3) \qquad \begin{cases} Q_1 + Q\cos\psi + Q_c\cos(\omega t - \psi) = \varepsilon'' - \psi'^2\varepsilon \\ Q_2 + Q\sin\psi + Q_c\sin(\omega t - \psi) = \varepsilon\psi'' + 2\varepsilon'\psi' \end{cases}
$$

where the prime (') denotes differentiation with respect to non-dimensional time, ωt and

$$
Q = P/mc\omega^2 \quad , \text{etc.}
$$

The main problem in solving these equations is to decide on the functions $Q_{1,2}(\varepsilon_o, \varepsilon, \dot{\varepsilon}, \psi, \dot{\psi})$. These have their origin in the hydrodynamics of the oil film and are deduced by integrating the oil-film pressure around the journal circumference. Hence the lubrication pressure equation must be invoked. This is known as the Reynolds equation and may be written:

$$
\frac{\partial}{\partial z}\left(h^3 \frac{\partial p}{\partial z}\right) + \frac{\partial}{\partial \theta}\left(\frac{h^3}{R} \cdot \frac{\partial p}{\partial \theta}\right) = 6\eta(\omega_j - 2\dot{\psi})\frac{\partial h}{\partial \theta} + 12\eta c\dot{\varepsilon}\cos\theta \ .
$$

Since modern turbine bearings are relatively short (length to diameter ratios of 0.5 are not uncommon), a 'short bearing' approximation is often used. This involves the assumption that in the Reynolds equation $\partial p/\partial\theta \ll \partial p/\partial z$, and hence allows the removal of the second term on the left-hand side of the Reynolds equation. This brings a great simplification in solution. There is practical evidence that this approximation is adequate for bearings having length to diameter ratios less than unity.

The oil-film forces, P_1 and P_2 are then obtained by integration as follows:

$$(4)\quad \begin{cases} P_1 = R\displaystyle\int_0^b \int_{\theta_1}^{\theta_2} p\cos\theta d\theta dz \\ P_2 = R\displaystyle\int_0^b \int_{\theta_1}^{\theta_2} p\sin\theta d\theta dz \end{cases}$$

p is found by integrating the shortened form of the Reynolds equation which gives

$$(5)\quad p = z(z-b)[12\eta c\dot{\varepsilon}\cos\theta - 6\eta(\omega_j-2\dot{\psi})c\varepsilon\sin\theta)][c(1+\varepsilon\cos\theta)]^{-3}$$

The oil-film limits, θ_1 and θ_2 which feature in equation (4) have in the past been taken to lie at the circumferential points where p becomes zero, cavitation and air entrainment being assumed to occur at all points where $p < 0$. However, experimental evidence suggests that the oil-film cannot sustain the rupture frequency implied by the above oil-film limits and a more realistic assumption for bearings having typical supply pressures appears to be,

$$\theta_1 = 0$$

$$\theta_2 = \pi \quad ,$$

that is the oil-film is regarded as oscillating to an extent defined by the attitude line. Equations (3) may be solved numerically and the results expressed in terms of independent parameters,

$$Q \; , \; Q_c \;\text{ and }\; \varepsilon_o \; .$$

Fig. 2a shows a family of whirl loci for $Q = 1.0$ and $\varepsilon_o = 0.7$, obtained by adopting in turn different values of Q_c . The increasing importance of non-linearity effects as Q_c is increased can be clearly seen. Fig. 2b shows the corresponding family of whirl loci found experimentally [1] and

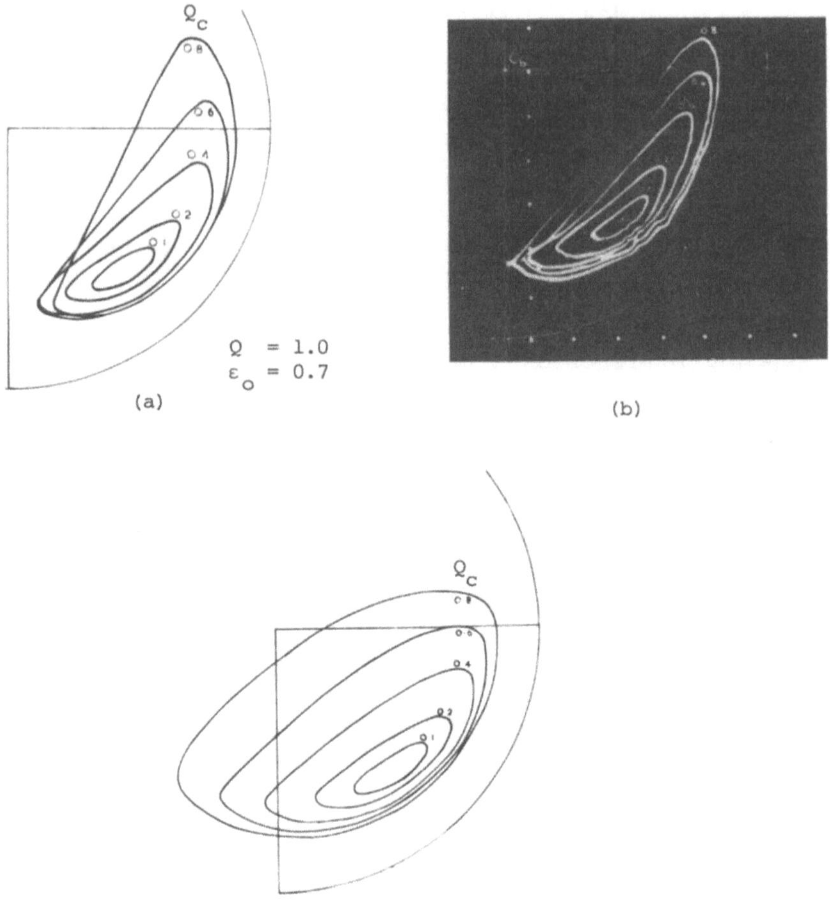

Fig. 2.

reasonable agreement is observed, particularly when it is noted that the contrasting loci of fig. 2c arise due to the adoption of the alternative oil-film limits defined by $p = 0$.

LINEARIZATION

For the load pattern imposed on a turbine bearing it is advantageous to consider the possibility of linearizing the oil-film forces P_1 , P_2 so that composite rotor-bearing-pedestal systems can be analysed in a straightforward manner using matrix techniques.

Consider fig. 1b which is a magnified portion of fig. 1a. Let c_{JS} denote the steady position of the journal due to rotor weight and let c_{JD} be a dynamic position due to the further imposition of a relatively small unbalance load. Allow P_1 and P_2 to change by small amounts δP_1 , δP_2 and consider the small changes δP_r , δP_s in the resolved forces P_r , P_s acting along axes r , s .

Since $\delta\psi$ is small, small changes in P_r and P_s are given by:

(6) $$\begin{cases} \delta P_r = \delta P_1 - P_2 \delta\psi \quad , \\ \delta P_s = P_1 \delta\psi + \delta P_2 \end{cases}$$

Now define linearised displacement and velocity coefficients of the oil-film in the following way.

$$\delta P_r = - a_{rr} r - b_{rr} \dot{r} + b_{rs} \dot{s} - a_{rs} s$$

$$\delta P_s = - a_{ss} s + a_{sr} r + b_{sr} \dot{r} - b_{ss} \dot{s} \quad .$$

From equations (4) and (5) it may be deduced that the forms of P_1 and P_2 are as follows, when oil film limits of 0 and π are applied.

(7) $$\begin{cases} P_1 = - \dot{\varepsilon} F_1(\varepsilon) - (0.5\omega_j - \dot{\psi}) h_1(\varepsilon) \\ P_2 = \dot{\varepsilon} F_2(\varepsilon) + (0.5\omega_j - \dot{\psi}) h_2(\varepsilon) \end{cases}$$

Using these expressions in equations (6) gives the linearised coefficients as

$$(8)\quad \begin{cases} a_{rr} = h_1'(\varepsilon)\omega_j/2c = h_1'(\varepsilon_o)\omega_j/2c \\ a_{rs} = h_2(\varepsilon_o)\omega_j/2c\varepsilon_o \\ a_{ss} = \omega_j h_1(\varepsilon_o)/2c\varepsilon_o \\ a_{sr} = h_2'(\varepsilon_o)\omega_j/2c \end{cases}$$

$$(9)\quad \begin{cases} b_{rr} = F_1(\varepsilon_o)/c \\ b_{rs} = h_1(\varepsilon_o)/c\varepsilon_o \\ b_{ss} = h_2(\varepsilon_o)/c\varepsilon_o \\ b_{sr} = F_2(\varepsilon_o)/c \end{cases}$$

These may be transformed to x , y coordinates as required.

Experimental work has been carried out to determine linearized oil-film coefficients for various bearing geometries [1]-[5] and their use in predicting rigid rotor response and instability has been explored.

One practical way of estimating the displacement coefficients, a on a real installation is to calculate the oil-film influence coefficients and then invert them to obtain displacement coefficients. The influence coefficients can be obtained from the locus of the journal centre relative to the bearing centre as the speed, viscosity or load is altered. Consider fig. 3a, which shows the position of a rotor journal in its bearing. For a given journal bearing geometry the displacement $c\varepsilon_o$ is governed by the parameter

$$\overline{P} = P/\eta\omega bD$$

Let $\overline{P}$ increase by $\delta\overline{P}$ and the resulting change in journal-

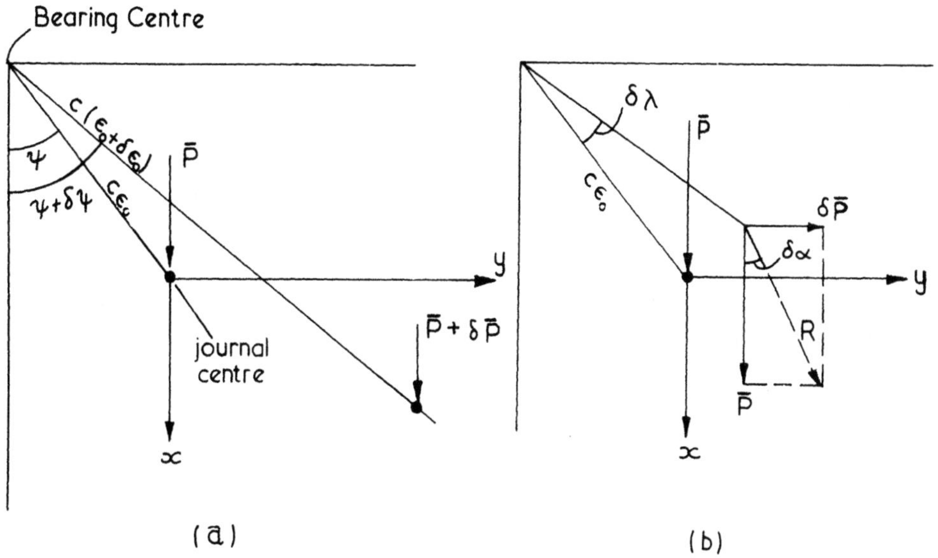

Fig. 3. Influence coefficients obtained from the journal centre locus

centre position be as shown in fig. 3a. Then two non-dimensional influence coefficients $\bar{\alpha}_{xx}$ and $\bar{\alpha}_{yx}$ may be defined as follows,

$$(10) \qquad \begin{cases} \bar{\alpha}_{xx} = \bar{x}/\delta\bar{P} \\ \bar{\alpha}_{yx} = \bar{y}/\delta\bar{P} \end{cases},$$

in which

$$\bar{x} = x/c = (\varepsilon_o+\delta\varepsilon_o)\cos(\psi+\delta\psi) - \varepsilon_o\cos\psi$$

and

$$\bar{y} = y/c = (\varepsilon_o+\delta\varepsilon_o)\sin(\psi+\delta\psi) - \varepsilon_o\sin\psi \quad ,$$

whence

$$(11) \qquad \begin{cases} \bar{\alpha}_{xx} = -\ \varepsilon_o\sin\psi\,\dfrac{d\psi}{d\bar{P}} + \dfrac{d\varepsilon_o}{d\bar{P}}\cos\psi \\ \bar{\alpha}_{yx} = \quad \varepsilon_o\cos\psi\,\dfrac{d\psi}{d\bar{P}} + \dfrac{d\varepsilon_o}{d\bar{P}}\sin\psi \end{cases}$$

These coefficients place no demands upon bearing shape.

Now consider fig. 3b, which shows the effect of applying an increment $\delta\bar{P}$ horizontally. If $\delta\bar{P}$ is sufficiently small, we may say that

$$R \simeq \bar{P} \quad .$$

Assuming the journal-centre locus may be rotated without affecting its shape, then

$$\delta\alpha = \delta\lambda$$

and non-dimensional influence coefficients $\bar{\alpha}_{yy}$ and $\bar{\alpha}_{xy}$ are given by,

$$\bar{\alpha}_{yy} = \bar{y}/\delta\bar{P} = [\varepsilon_o \sin(\psi+\delta\lambda) - \varepsilon_o \sin\psi]/\delta\bar{P}$$

$$= \varepsilon_o \cos\psi \cdot \frac{\delta\alpha}{\delta\bar{P}}$$

and

$$\bar{\alpha}_{xy} = \bar{x}/\delta\bar{P} = [-\varepsilon_o \cos\psi + \varepsilon_o \cos(\psi+d\lambda)]/\delta\bar{P}$$

$$= -\varepsilon_o \sin\psi \cdot \frac{\delta\alpha}{\delta\bar{P}} \quad .$$

Now from fig. 3b,

$$\delta\alpha \simeq \delta\bar{P}/\bar{P}$$

Hence

$$(12) \qquad \begin{cases} \bar{\alpha}_{yy} = \dfrac{\varepsilon_o}{\bar{P}} \cos\psi \\ \bar{\alpha}_{xy} = -\dfrac{\varepsilon_o}{\bar{P}} \sin\psi \quad . \end{cases}$$

The assumption that the journal-centre locus may be rotated without affecting its shape is only strictly true for a circular bearing, but many non-circular bearings are usually only loaded over a restricted arc which may be assumed circular, so errors are likely to be quite small.

THE GAS TURBINE ROTOR

Squeeze film bearings have been introduced fairly recently in the design of gas turbines, their function being to reduce rotor vibration amplitudes and the amplitudes of the forces transmitted to the engine frame. In early turbine designs, the outer races of the ball or roller bearings were supported directly by the engine frame. In some later designs, the outer races are cantilevered from the frame on a ring of flexible beams, known as a 'squirrel cage' (fig. 4). The resulting flexibility allows outer race movement

Fig. 4

without rotation. A small radial clearance between the outer race and the housing is supplied with oil and so forms the squeeze-film damping device. A particular point of interest is that, in equation (8), all the displacement coefficients are proportional to ω_j . Hence, in the absence of journal (that is outer-race) rotation, these will disappear and the velocity coefficients will remain. This means that the squeeze film bearing has no capacity to support a constant unidirectional load and it is the purpose of the squirrel cage to provide this capacity. The stiffness of the squirrel cage is usually sufficient to restrict outer-race vibration to an amplitude which allows the linear velocity coefficients to be used. For given oil-film limits these are identical with those of an ordinary journal bearing of the same geometry. However, it is debatable whether the oil-film limits can be regarded as identical with those of an ordinary journal bearing owing to the absence of journal rotation which appears to alter the mechanism of film rupture. The only application in which identical velocity coefficients can be expected is where the oil-supply pressure is sufficient to provide a full 360° film of oil.

Experimentally, these velocity coefficients have been obtained using two methods - one a drop-test [6], the other a vibratory method [7]. In the drop-tests, use was made of the manner in which a floating bearing moved relative to a fixed shaft under a gravity load. The resulting paths are known as 'squeeze paths' (fig. 5a) and their geometry and space-time relationships are sufficient to estimate the required velocity coefficients. Consider the specimen squeeze-path shown in fig. 5a. It has already been indicated in equations (9) that the velocity coefficients, b are functions of ε_o only and do not involve time-derivatives of ε or ψ . Hence despite such time derivatives existing at any point on the squeeze-path they do not influence the values of the velocity coefficients - only ε_o is important. Now for the particular case of the oil-film of 360° extent

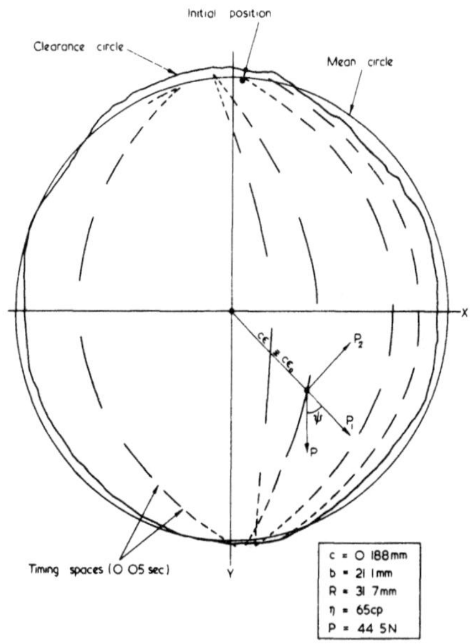

Fig. 5a. Experimental squeeze-paths.

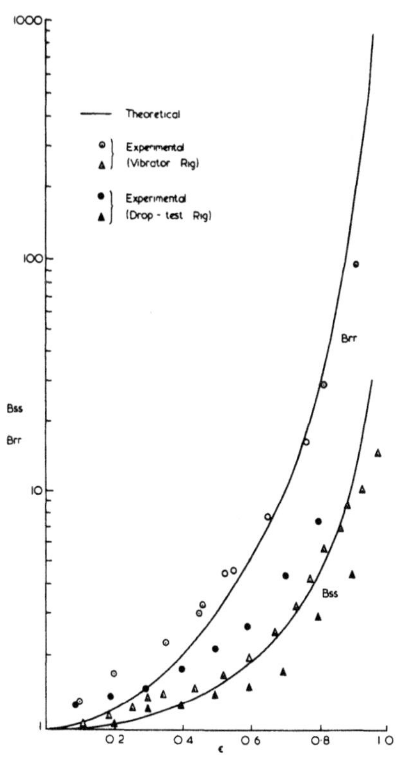

Fig. 5b. Non-dimensional velocity coefficients.

equations (7) reduce to

$$P_1 = -\dot{\varepsilon}F_1(\varepsilon)$$

$$P_2 = (0.5\omega_j - \dot{\psi})h_2(\varepsilon)$$

This may be verified by putting oil-film limits of 0 and 2π into equations (4).

Hence, from equations (9),

$$b_{rr} = F_1(\varepsilon_o)/c$$

$$b_{rs} = b_{sr} = 0$$

$$b_{ss} = h_2(\varepsilon_o)/c\varepsilon_o \quad .$$

The fact that $b_{rs} = b_{sr} = 0$ means that b_{rr} , b_{ss} are alone left to represent the damping of a 360° squeeze-film. This being the case we may also write, for a given position on the squeeze-path,

$$(13)\qquad \begin{cases} b_{rr} = \dfrac{1}{c}\left[\dfrac{\partial P_1}{\partial \dot{\varepsilon}}\right]_\varepsilon \\[2ex] b_{ss} = \dfrac{1}{c}\left[\dfrac{\partial P_2}{\partial (\varepsilon\dot{\psi})}\right]_\varepsilon \end{cases}$$

The oil-film forces P_1 and P_2 may be obtained from equations (1) in which $\overline{X} = P$, a constant gravitational force, and $\overline{Y} = 0$. If we ignore any acceleration effects in comparison with oil-film viscous effects associated with the very slow fall along the squeeze-path, then equations (1) become

$$P_1 = -P\cos\psi$$

$$P_2 = P\sin\psi \quad .$$

Hence equations (13) become

$$b_{rr} = \frac{1}{c}\left[\frac{\partial(P_1/P)}{\partial(\dot{\varepsilon}/P)}\right]_\varepsilon = -\frac{1}{c}\left[\frac{\partial(\cos\psi)}{\partial(\dot{\varepsilon}/P)}\right]_\varepsilon$$

$$b_{ss} = \frac{1}{c}\left[\frac{\partial(P_2/P)}{\partial(\varepsilon\dot{\psi}/P)}\right]_\varepsilon = \frac{1}{c}\left[\frac{\partial(\sin\psi)}{\partial(\varepsilon\dot{\psi}/P)}\right]_\varepsilon$$

If we wish we may now non-dimensionalise by writing

$$\frac{b_{rr}c^3}{\pi\eta Rb^3} = B_{rr} = -\left[\frac{\partial(\cos\psi)}{\partial(\dot{\varepsilon}/K)}\right]_\varepsilon$$

$$\frac{b_{ss}c^3}{\pi\eta Rb^3} = B_{ss} = \left[\frac{\partial(\sin\psi)}{\partial(\varepsilon\dot{\psi}/K)}\right]_\varepsilon \quad ,$$

where

$$K = Pc^2/\pi\eta Rb^3 \ .$$

If $\cos\psi$ is plotted against $\dot{\varepsilon}/K$ for different initial release positions (fig. 5a) a family of curves is obtained [6]. From this family the non-dimensional velocity coefficient B_{rr} may be obtained (for an eccentricity ratio $\varepsilon = \varepsilon_1$, say) as the slope of the line joining all the points in the family for which $\varepsilon = \varepsilon_1$. Likewise B_{ss} is obtained from the family of curves relating $\sin\psi$ and $\varepsilon\dot{\psi}/K$. $\dot{\varepsilon}$ and $\varepsilon\dot{\psi}$ were obtained experimentally by resolving the vertical and horizontal component velocities of the fall. These latter were in turn obtained by differentiating the component displacements of the fall trajectory.

In the vibratory method sinusoidal forces were applied in quadrature to a squeeze bearing supported by a squirrel-cage which could be adjusted to impose different static eccentricity ratios on the bearing. Oscillators in the horizontal and vertical planes simulated a rotating out of balance force. From the linearised equations of motion, expressions may be deduced for the velocity coefficients in terms of the applied forces, bearing displacements and phase angles between them and by experimental measurement [7] these velocity coefficients have been assessed (fig. 5b). Estimated values from the drop-tests are also included. The coefficients obtained from vibration tests were found to be reasonably valid for displacement amplitudes of about a half and a fifth of the radial clearance for static eccentricity ratios of 0.1 and 0.7 respectively.

FLEXIBLE ROTORS AND TRANSMISSION SHAFTING

A flexible rotor or shaft may be approximated as a series of concentrated masses connected by massless beam elements and, using equations (8) and (9), the oil films of the bearings may each be represented by four linear springs and four linear dampers. To exercise some vibration control one external damper mechanism may be incorporated if required and represented as a combination of vertical and horizontal linear dampers [8].

As an example of stability assessment, consider a flexible shaft-bearing system having the following design parameters,

uniform steel shaft:	span	1.52 m ,
	diameter	0.0254 m ,
	shaft mass	6 kg ,
	each journal mass	15 kg ,
	modulus of elasticity	0.21×10^{12} N/m^2 .

Each bearing has a central circumferential oil-supply-groove, giving two effective bearings at each journal of land length to diameter ratio 0.33 and radial clearance per land-load 5×10^{-7} m/N . The speed (ω) range within which the instability threshold was thought to exist was

260-300 rad/s, which for the oil used gave an eccentricity ratio of about 0.5.

A unit vertical force $1.\sin\Omega t$ was assumed at the shaft centre and the resulting central displacement x was assumed to be given by $x = X\sin(\Omega t - \theta)$. Fig. 6 shows polar plots of $Xv. \theta$ upon which values of Ω/ω have been super-imposed. It will be noted that, for a shaft speed, ω of 270 rad/s, θ increases positively with increasing Ω (fig. 6a). This exemplifies net positive damping and stable conditions [8]. For a shaft speed of 280 rad/s the response locus loops in the opposite direction (fig. 6b), indicating negative damping and unstable conditions. The instability threshold speed thus lies between 270 and 280 rad/s. Having located a borderline curve the value of forcing frequency, Ω at $\theta = \frac{-\pi}{2}$ then equals Ω_n [8]. Hence the value of Ω at $\theta = \frac{-\pi}{2}$ is an effective undamped natural frequency of the system. This is then the frequency of any incipient unstable vibration which may be initiated by any small disturbance, even if not of sinusoidal form. In the borderline cases shown in fig. 6 the appropriate values of Ω/ω are seen to be about 0.511725 and 0.4935.

Use of the above method over a wide range of conditions indicated that the effect of any external damping (in delaying the instability-onset speed) increased with increase in eccentricity-ratio, and to a slight extent with movement of the damper towards the centre of the span.

If the shaft is symmetrically supported at two effective locations a method of impedance matching may be employed. There are four mechanical impedances or complex dynamic stiffnesses ϕ_o representing a journal bearing oil film of radial clearance, c and carrying an external load, P. These are given in two mutually perpendicular coordinates x and y (fig. 1b) by

$$(14) \qquad \phi_{oxx} = a_{xx} + i\Omega b_{xx} = \frac{P}{c}\left(A_{xx} + i\frac{\Omega}{\omega_j}B_{xx}\right), \text{ etc......}$$

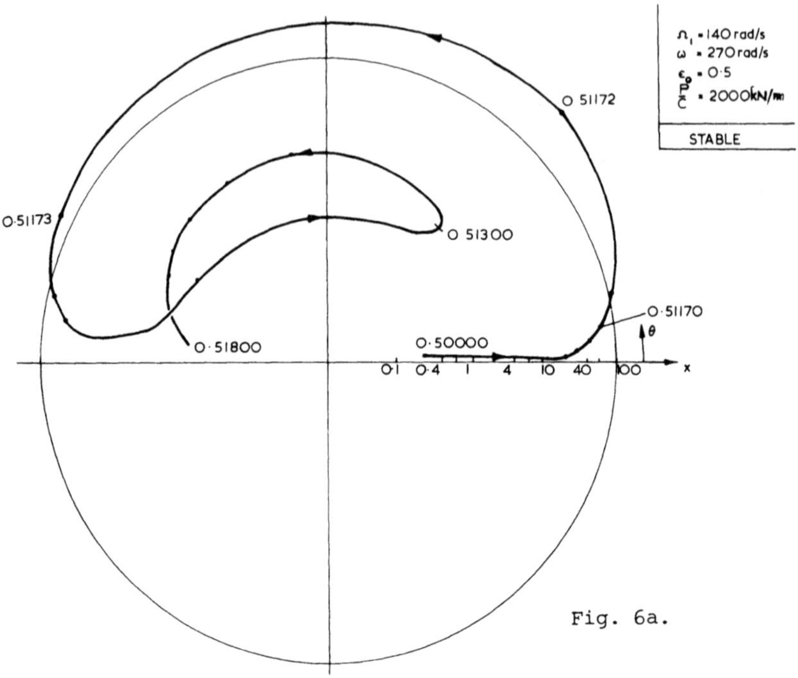
Ω₁ = 140 rad/s
ω = 270 rad/s
ε₀ = 0·5
P/C = 2000 kN/m
STABLE
0·51172
0·51173
0·51300
0·51170
0·51800
0·50000
θ
0·1 0·4 1 4 10 40 100
x

Fig. 6a.

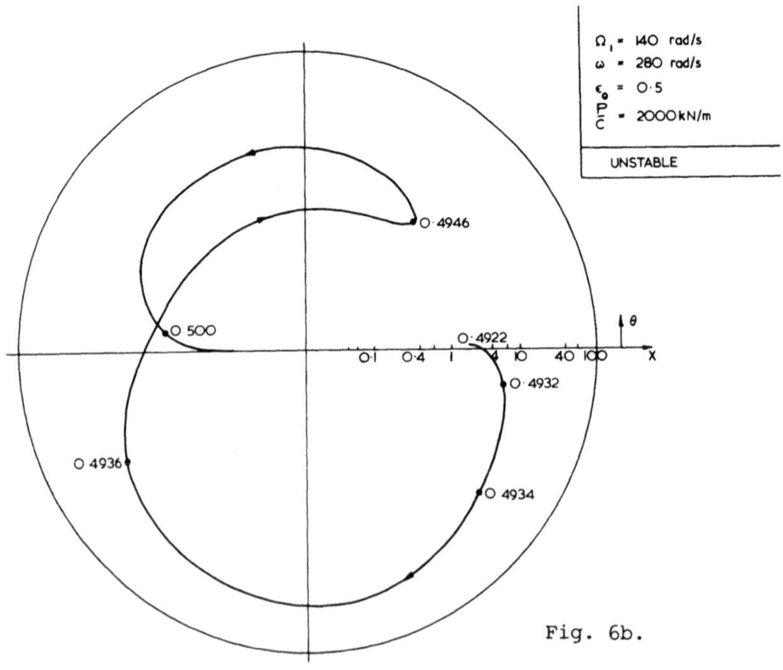
Ω₁ = 140 rad/s
ω = 280 rad/s
ε₀ = 0·5
P/C = 2000 kN/m
UNSTABLE
0·4946
0 500
0·4922
θ
0·1 0·4 1 4 10 40 100
X
0·4932
0 4936
0 4934

Fig. 6b.

in which ω_j is the journal speed, Ω the forcing frequency and A_{xx} and B_{xx} etc. depend on the eccentricity of the journal within the bearing. These are obtained by coordinate transformation using equations (8) and (9). If the oil film is in series with a flexible pedestal, as in many practical applications, the inverse of complex stiffness, namely receptance, is also a useful parameter. This is because the receptances of the oil film and the pedestal may be added directly to give a total receptance. This receptance may then be inverted to give the complex stiffnesses ϕ_{xx} , ϕ_{yy} , ϕ_{xy} and ϕ_{yx} of the combined oil film and pedestal. Oil-film receptances for unbalance forcing $(\Omega = \omega = \omega_j)$ are given in ref. [4].

MORTON [9] defined "equivalent" complex stiffnesses, ϕ , as follows

$$\phi x = \phi_{xx}x + \phi_{xy}y$$

$$\phi y = \phi_{yy}y + \phi_{yx}x$$

from which two eigenvalues may be obtained, namely

$$\phi_1 , \phi_2 = \frac{1}{2}(\phi_{xx} + \phi_{yy}) \pm \frac{1}{2}\left[(\phi_{xx} - \phi_{yy})^2 + 4\phi_{xy}\phi_{yx}\right]^{\frac{1}{2}} .$$

These are complex numbers, and if the imaginary part of either is zero, its real part represents an equivalent dynamic stiffness and may be equated to the rotor dynamic stiffness to locate an instability-onset speed.

For a given bearing load number $(Pc^2/\eta b^3 R\omega)$ (and hence, from equation (2), eccentricity ratio ε_o) the two equivalent dynamic stiffnesses $\phi_{1,2}$ may thus be calculated for a succession of forcing frequency ratios, Ω/ω . It can be seen from equations (14) that these dynamic stiffnesses will be proportional to P/c and so may be nondimensionalised by multiplying by c/P . The value of $\phi c/P$ at which the imaginary part becomes zero is recorded together with the value of Ω/ω responsible. Any motion is thus oscillatory at such a frequency ratio. The whole procedure is repeated for different values of load number. A plot is then made of

$$\phi c/P \text{ v } (\Omega/\omega)/(Pc^2/\eta b^3 R\omega) \qquad \text{(fig. 7a)}$$

On this graph appropriate values of ε_o and Ω/ω are indicated.

Any symmetrical shaft supported at each end will also have a dynamic stiffness value ϕ_s (for each forcing frequency) as seen at each bearing when vibrated at both journals simultaneously. If we now plot corresponding values of negative $\phi_s c/P$ (which again are wholly real in the absence of shaft damping) and $\Omega \,.\, \eta b^3 R/Pc^2$ on the same figure, intersections of the resulting graph and the oil film equivalent dynamic stiffness graph give the oscillatory (i.e. instability-onset) frequencies of the complete system, from which the onset rotational speeds may be deduced. This is because at the points of intersection the total dynamic stiffness,

$$\phi_T c/P = \phi_s c/P + \phi c/P = 0 \ ,$$

implying that no oscillatory force is needed to produce an oscillatory motion.

The special significance of dynamic stiffness matching is that it illustrates that the oil-films are the active determinants of whirl-frequency ratio and therefore of whirl-onset speed. The shaft dynamic stiffness merely plays a passive role, by deciding where its point of intersection with the oil film dynamic stiffness graph will exist.

There are certain generalisations which can be made with regard to the instability-onset speeds by expressing the shaft properties non-dimensionally. Consider a symmetrical shaft supported in a bearing at each end and forced at each journal with a sinusoidal excitation. Let the effective shaft mass and stiffness seen by <u>each</u> vibrator be k_ℓ and m_ℓ respectively for the lowest mode of shaft vibration. Then we may write the negative non-dimensional dynamic stiffness of the shaft at each journal as

$$(15) \qquad -\phi_s \frac{c}{P} = \bar{k}_\ell \cdot \left[\left(\frac{\bar{\Omega}_1}{\bar{\Omega}}\right)^2 - 1\right]^{-1} ,$$

where

$$\bar{k}_\ell = k_\ell c/P$$

$$\bar{\Omega} = \Omega \eta b^3 R/Pc^2$$

$$\bar{\Omega}_1 = \Omega_1 \eta b^3 R/Pc^2$$

and

$$\Omega_1 = \left(\frac{k_\ell}{m_\ell}\right)^{\frac{1}{2}} = \text{the lowest pinned-pinned critical speed of the shaft.}$$

Graphical representations of the above equation are shown in fig. 7a. These are superimposed upon the graph of $\phi \frac{c}{P}$ for the oil film. An inspection of fig. 7a shows that, the lower the value of the governing parameter $\bar{\Omega}_1$, and the lower the value of $\bar{k}_\ell$, then the more nearly will the oil-film and negative shaft dynamic stiffnesses intersect at a non-dimensional frequency, $\bar{\Omega}$ equal to the lowest non-dimensional pinned-pinned critical speed, $\bar{\Omega}_1$, thereby implying that the frequency of any unstable motion will equal the lowest critical speed of the equivalent pinned-pinned shaft. Any shaft or rotor designed to run well above its first critical speed usually falls into this category and there is a great deal of experimental evidence which supports the view that the frequency of unstable motion for such shafts is equal to the lowest critical speed. Pedestal characteristics can of course be considered by embodying any pedestal mass, stiffness or damping in an overall bearing dynamic stiffness as discussed earlier. Fig. 7b shows the dynamic stiffness graph 1 for a short cylindrical bearing

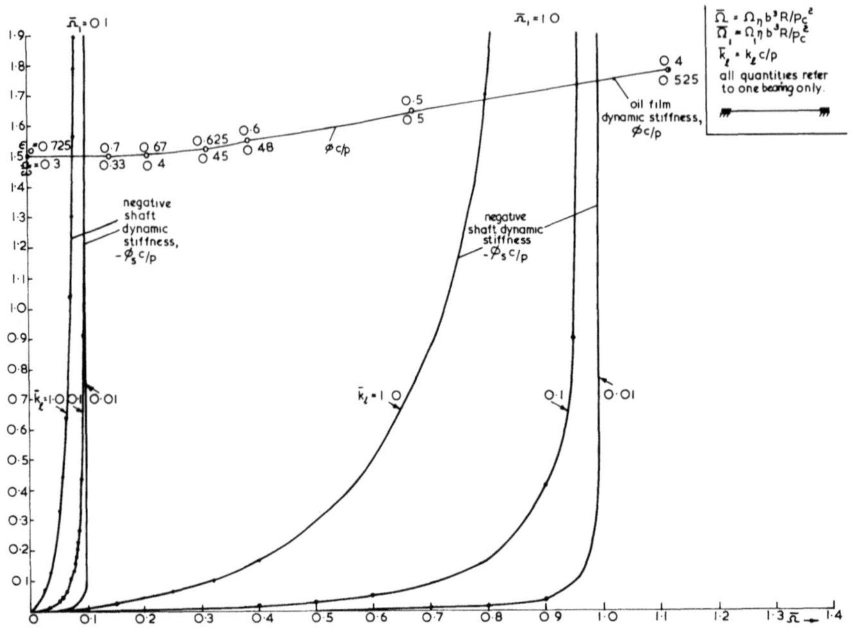

Fig. 7a. Dynamic stiffness matching.

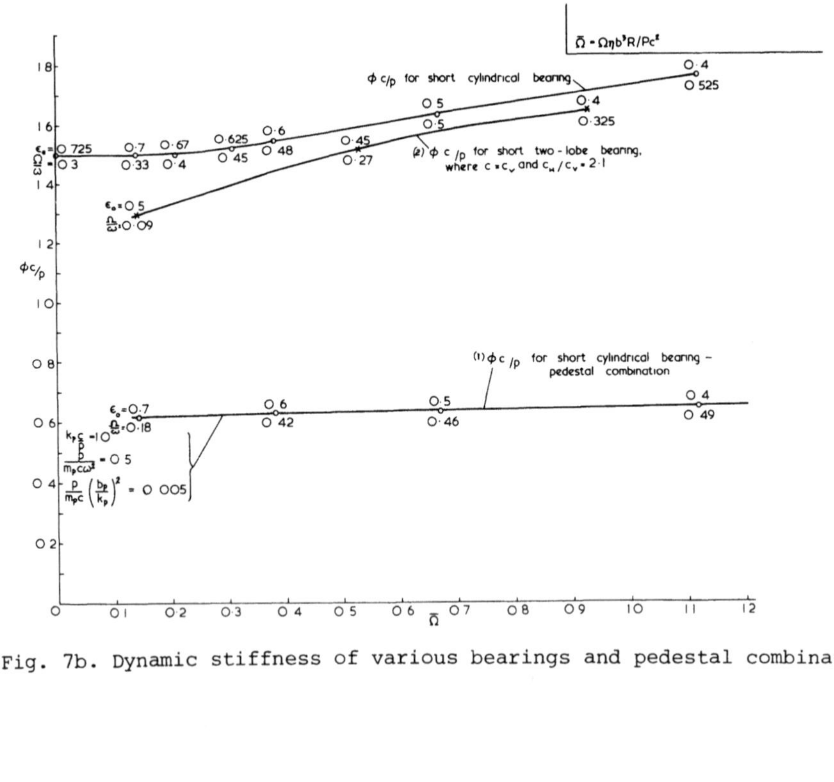

Fig. 7b. Dynamic stiffness of various bearings and pedestal combinations.

in series with a housing (or pedestal) defined by the non-dimensional parameters,

$$k_p c/P \quad , \quad \frac{P}{m_p c}\left(\frac{b_p}{k_p}\right)^2 \quad \text{and} \quad P/m_p c\omega^2 \quad ,$$

in which

$$k_p = k_{pxx} = k_{pyy}$$

$$b_p = b_{pxx} = b_{pyy} \quad .$$

For a given pedestal and bearing combination, an assessment can usually be made of the first two parameters. Let these be 1.0 and 0.005 respectively, say. If the third parameter is now assumed as, say, 0.5 then the combined dynamic stiffness graph may be drawn (fig. 7b) and a point of intersection may be found with the appropriate shaft dynamic stiffness graph. The corresponding value of ω can then be deduced from the values of $\bar{\Omega}$ and $\frac{\Omega}{\omega}$. Putting this value of ω into the parameter, $P/m_p c\omega^2$ gives the value of housing mass m_p corresponding to the stability border-line. This can be checked against the actual value of m_p used in the design to assess the likelihood of instability.

Other bearing types may be considered in the same way and also shown (curve 2) in fig. 7b is the theoretical dynamic stiffness graph for a short two-lobe bearing in which the ratio of horizontal to vertical clearances c_h/c_v is 2.1 . The specimen values of Ω/ω for given values of $\bar{\Omega}$ signify the greater stability attainable with this type of bearing.

The use of dynamic stiffness plots in predicting the performance of an experimental rotor is discussed in ref. [8].

CONCLUSIONS

The role played by circular-bearing oil-films in the vibration of rotating systems has been discussed. In particular the cases of the small rigid steam-turbine rotor, the rigid gas-turbine rotor and the flexible supercritical shaft have been considered in some detail. Comparisons of theoretical and experimental results in these areas gave rise to the following conclusions.

(1) The amplitudes and shapes of large vibratory orbits of rigid rotors is governed to a considerable extent by the oil film limits within the bearings whether they be journal bearings or squeeze-film bearings. These oil-film limits appear to be dictated by the inability of the oil-film to rupture under conditions of high frequency and amplitude of vibration, and in squeeze-film bearings those effects are particularly influencial.

(2) The amplitudes of small vibratory orbits of both rigid and flexible systems appear to be well predicted when using the oil-film coefficients corresponding to the operative bearing eccentricity-ratio.

(3) The onset of oil-whirl instability for flexible rotors is well predicted by again using the appropriate oil-film coefficients and simple prediction methods for investigating this onset have been discussed.

(4) Utilising these methods the adequacy of different methods of control (external dampers and non-circular bearings) can be readily assessed.

REFERENCES

[1] WOODCOCK, J.S.: "Dynamic characteristics of a journal-bearing oil-film", D.Phil. Thesis, University of Sussex, 1971.

[2] MITCHELL, J.R., HOLMES, R. and VAN BALLEGOOYEN, H.: "Experimental determination of a bearing oil-film stiffness", Proc. I. Mech. E 1965-6 180 (Pt3K), 90.

[3] GLIENICKE, J.: "Experimental investigation of the stiffness and damping coefficients of turbine bearings", Proc. Instn. Mech. Engrs., Vol. 181, Pt. 3B (1966-67) pp. 116-129.

[4] WOODCOCK, J.S. and HOLMES, R.: "The determination and application of the dynamic properties of a turbo-rotor bearing oil-film", Proc. Instn. Mech. Engrs., Vol. 184, Pt. 3L (1969-70) pp. 111-119.

[5] MORTON, P.G.: "Measurement of the dynamic characteristics of a large sleeve bearing", Trans. Am. Soc. Engrs., J. Lubric. Technol., Vol. 93, No. 1 (Jan. 1971) pp. 143-150.

[6] WILLIAMS, R. and HOLMES, R.: "Determination of the linear characteristics of squeeze films", Proc. I. Mech. E., Tribology Convention 1971.

[7] WILLIAMS, R.: "Squeeze-film bearing characteristics", D.Phil. Thesis, University of Sussex, 1972.

[8] DOSTAL, M., ROBERTS, J.B. and HOLMES, R.: "Stability control of flexible shafts supported on oil-film bearings", Journal of Sound & Vibration (1974) 35(3) pp. 351-377.

[9] MORTON, P.G.: "Influence of coupled asymmetric bearings on the motion of a massive flexible rotor", Proc. Instn. Mech. Engrs., Vol. 182, Pt. 1 (1967-68) pp. 255-280.

Divergence and Flutter Boundaries of Systems under Combined Conservative and Gyroscopic Forces

K. Huseyin

Department of Systems Design
University of Waterloo, Ontario, Canada

R.H. Plaut

Applied Mathematics and Engineering
Brown University, Providence, U.S.A.

1. INTRODUCTION

Vibrations and stability of conservative gyroscopic systems have been studied by several authors following THOMSON and TAIT [1]. A general discussion and a list of references can be found in ZIEGLER's book [2] and in Ref. [3]. In the latter paper, characteristic curves of free transverse vibrations of such systems are explored and flutter as well as divergence instability is discussed with the aid of these curves. An appropriate choice of frequency and loading parameters facilitates the analysis.

In this paper the vibrations and stability of conservative gyroscopic systems which are subjected to several independent conservative loads are investigated via a similar approach. The theory in section 2 and 3 applies to all conservative gyroscopic systems which contain rotating shafts as a special case. Clamped and/or simply supported pipes with internal fluid flow also fall [3] within the class of systems treated in this section. Rotating systems receive special attention in section 4 where a number of general results are obtained. Finally, several examples concerning the transverse vibrations and stability of rotating shafts with and without disks are analyzed in the light of the general discussion.

2. GENERAL GYROSCOPIC SYSTEM: TRANSVERSE VIBRATIONS

Consider a linear non-dissipative dynamical system described by n generalized coordinates. The only velocity-dependent forces present are of gyroscopic type (e.g. Coriolis force). The remaining forces are assumed to be conservative and may be represented by several independent loading parameters. At least one of these parameters is associated with the source of the gyroscopic forces (e.g., centrifugal force).

Assuming that the kinetic and potential energies are in quadratic forms as usual, and using Lagrange's equations of motion, one obtains the governing differential equation

(1) $$\hat{M}\,\ddot{q} + \hat{G}\,\dot{q} + K\,q = 0$$

where $\hat{M}$ and $\hat{K}$ are real symmetric $(n \times n)$ matrices, $\hat{M}$ is positive definite, $\hat{G}$ is real $(n \times n)$ skew-symmetric, and q is a column vector representing the generalized coordinates.

Assuming solutions in the form

(2) $$q^{(t)} = u\,e^{i\omega t}$$

one obtains

(3) $$(-\omega^2\,\hat{M} + i\,\omega\,\hat{G} + \hat{K})\;u = 0$$

In practical applications, and in particular for the analysis of rotating shafts, equation (3) takes the explicit form

(4) $$(-\omega^2\,\hat{M} + i\,\omega\,\xi\,\hat{F} + \hat{U} - \xi^2\,\hat{E}_\xi - \eta^k\,\hat{E}^k)\;u = 0$$

where ξ and ξ^2 represent the angular velocity and conservative centrifugal force respectively. The independent parameters η^k $(k = 1, 2, \ldots, m)$ describe the other conservative loading which may be acting on the system (e.g., axial loading). Summation over superscript k is implied. The matrices $\hat{U}$, $\hat{E}_\xi$, $\hat{E}^k$ are real symmetric, $\hat{G} \equiv \xi \hat{F}$ is real skew-symmetric and $\hat{U}$ is positive definite. The characteristic equation

$$(5) \quad D \equiv \det|-\omega^2 \hat{M} + i \omega \xi \hat{F} + \hat{U} - \xi^2 \hat{E}_\xi - \eta^k \hat{E}^k| = 0$$

describes (m+1) dimensional characteristic surfaces in the (m+2) dimensional $\omega^2 - \xi^2 - \eta^k$ parameter space. It is of course noted that due to the skew-symmetry of $\hat{F}$ and the symmetry of the remaining matrices, the expansion of the determinant (5) is an implicit function of ω^2, ξ^2 and η^k.

The following discussion will be facilitated by introducing the concept of "corresponding non-gyroscopic system" [3] which is defined by (5) with $\hat{F}$ identically zero. It is immediately observed that this fictitious system belongs to the class of multiple-parameter conservative systems treated in Ref. [4]; therefore, according to a theorem established in that paper, its *fundamental characteristic surface in* $\omega^2 - \xi^2 - \eta^k$ *parameter space* (i.e. closest to the origin of the parameter space) *cannot have convexity toward the origin.* It also follows that the fundamental characteristic surface of the system (5) in the subspace $\omega^2 - \eta^k$ (i.e. the intersection with $\xi^2 = 0$) has the same property (Fig. 1a) and it coincides with the fundamental characteristic surface of the $\hat{F} \equiv 0$ system. Hence, the following theorem can be stated:

Theorem 1. The fundamental characteristic surface of gyroscopic conservative systems in the $\omega^2 - \eta^k$ subspace cannot have convexity towards the origin.

On the other hand, in the $\omega^2 - \xi^2$ plane (i.e., for all $\eta^k = 0$) the two fundamental curves are distinct, and the curve belonging to system (5) is always *closer* to the origin in the stability region in which $\hat{K}$ is positive de-

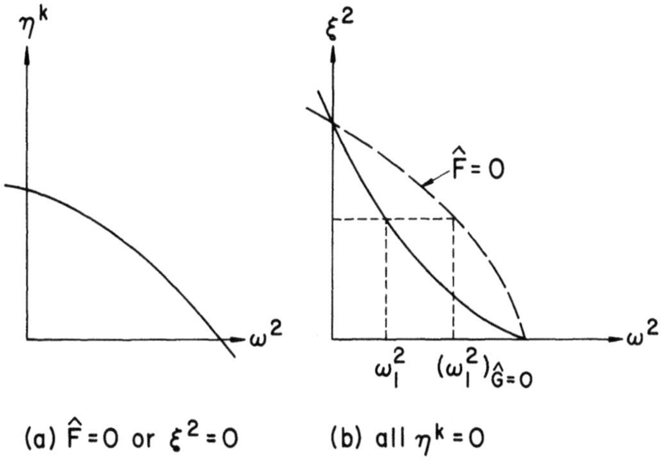

Fig. 1. Typical characteristic curves.

definite (Fig. 1b). The proof of this statement can be carried out following the approach in Ref. [3]. A simpler alternative proof is as follows:

At a specified state of loading for which the matrix $\hat{K}$ in (3) is positive definite, let the frequencies ω_j^2, which are real, be ordered as

$$0 < \omega_1^2 \leq \omega_2^2 \leq \ldots \leq \omega_n^2 \tag{6}$$

Similarly, let the roots of the corresponding non-gyroscopic system

$$(-\omega^2 \hat{M} + \hat{K})\, v = 0 \tag{7}$$

be ordered as

$$0 < (\omega_1^2)_{\hat{G}=0} \leq \ldots \leq (\omega_n^2)_{\hat{G}=0} \tag{8}$$

It is to be shown that

$$\omega_1^2 \leq (\omega_1^2)_{\hat{G}=0} \tag{9}$$

The eigenvectors corresponding to ω_1 and $(\omega_1)_{\hat{G}=0}$ are denoted by u_1 and v_1 respectively. Consider now the Hermitian matrix

$$H(\omega) = -\omega^2 \hat{M} + j \omega \hat{G} + \hat{K} ; \tag{10}$$

obviously this matrix is positive definite when $\omega = 0$, and its smallest eigenvalue λ_1 vanishes for $\omega = \omega_1$ which leads to

$$< \bar{u}_1, H(\omega_1) u_1 > = 0 \tag{11}$$

($\bar{u}_1$ is the complex conjugate of u_1)

where $< >$ denotes dot product. Using the extremum properties of the Rayleigh quotient for Hermitian matrices, then, yields

$$< v_1, H(\omega_1) v_1 > \geq 0 \tag{12}$$

where v_1 is assumed to be normalized with respect to $\hat{M}$ such that

$$< v_1, \hat{M} v_1 > = 1 . \tag{13}$$

It then follows from (12) that

$$-\omega_1^2 + (\omega_1^2)_{\hat{G}=0} \geq 0 \tag{14}$$

since $< v_1, \hat{G} v_1 > = 0$ due to v_1 being real, and $(\omega_1^2)_{\hat{G}=0} = < v_1, \hat{K} v_1 >$ when (13) is satisfied. Assertion (9) is therefore proved. This result can be stated in the form of a theorem:

<u>Theorem 2</u>. The fundamental characteristic curve of the corresponding non-gyroscopic system constitutes an upper bound for that of the gyroscopic system in the $\omega^2 - \xi^2$ subspace in the region where $\hat{K}$ is positive definite.

3. GENERAL GYROSCOPIC SYSTEM: STABILITY

Interest now focusses on the stability properties of the system (4) as reflected in the $\xi^2 - \eta^k$ loading space. If all the loading parameters are zero one has real and positive frequencies, ω_1^2 , and the system is stable. As the values of ξ^2 and/or $|\eta^k|$ are raised, instability may occur by divergence or flutter. Initial instability, however, will always be of divergence type characterized by $\omega_j^2 = 0$ for some j , due to the fact that the system is conservative [3]. The locus of all divergence points in $\xi^2 - \eta^k$ space defines the "divergence boundary" [5], and the following theorem can be stated:

Theorem 3. The initial divergence boundary cannot have convexity toward the primary region of stability.

This result is contained as a special case in Theorem 1 concerning the convexity of the characteristic surface mentioned earlier, and the reader is referred to Ref. [4] for the proof.

Depending on the values of η^k , the gyroscopic forces may or may not stabilize the system which appears to be unstable in the absence of such forces. Fig. 2 shows the sta-

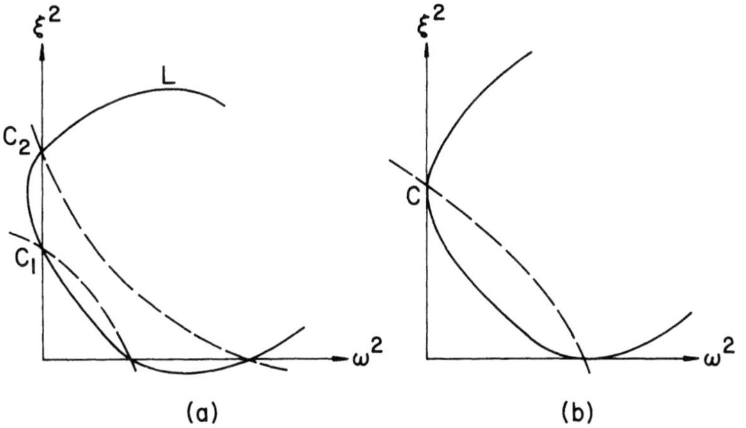

Fig. 2. Characteristic curves for gyroscopic and corresponding nongyroscopic systems.

bilizing effect of the gyroscopic forces (dashed lines show the characteristic curves of the $\hat{F} \equiv 0$ system). As Fig. 2(b) illustrates, it may happen that the characteristic curve in the $\omega^2 - \xi^2$ plane is tangential to the ξ^2 -axis at the critical point C, the system being stable for both $\xi^2 > \xi_C$ and $\xi^2 < \xi_C$ at least in the vicinity of C. Part of the divergence boundary may, then, lie between two stable regions while the remaining part associated with η^k demarcates stable and unstable regions. This distinction arises close to the η^k axes where the characteristic surface tends to change curvature and approach that of the corresponding non-gyroscopic system.

As Fig. 2(a) indicates, for some combinations of the loading parameters the characteristic curves (5) may have extrema, like L, resulting in coalescence of two frequencies ω_j^2. At such a point the system loses its stability by flutter, and the locus of all such points in the $\xi^2 - \eta^k$ loading space defines the "flutter boundary". Analytically, the flutter boundary can also be defined by (5) and

$$\frac{\partial D}{\partial \omega^2} = 0 \tag{15}$$

It will be seen later that the flutter boundary may have convexity properties in contrast with the divergence boundary.

The flutter instability gains significance, particularly when it occurs after the gyroscopic forces have stabilized the system.

4. ROTATING SYSTEMS

In this section, more specific results will be obtained with regard to rotating systems which fall within the scope of the general gyroscopic systems discussed in the preceding sections.

Assuming that the rotating system is analyzed in a reference frame attached to the system and thus rotating with the same angular velocity ξ , and considering only one additional parameter η , the linearized equations of motion can be written as

$$
\begin{aligned}
&M\ddot{p} - 2\xi M\dot{q} + (U - \xi^2 M - \eta E)\, p = 0 \\
(16)\qquad & \\
&M\ddot{q} + 2\xi M\dot{p} + (V - \xi^2 M - \eta E)\, q = 0
\end{aligned}
$$

where M, U, V, E are real symmetric (N x N) matrices, and M, U, V are positive definite. The column vectors p and q represent deflections (generalized coordinates) in two orthogonal directions in the rotating frame. The system (16) is a special case of the system (4) such that

$$
\hat{M} = \begin{bmatrix} M & 0 \\ 0 & M \end{bmatrix}, \quad \hat{F} = 2\begin{bmatrix} 0 & -M \\ M & 0 \end{bmatrix}, \quad \hat{U} = \begin{bmatrix} U & 0 \\ 0 & V \end{bmatrix}
$$

(17)

$$
\hat{E}_\xi = \begin{bmatrix} M & 0 \\ 0 & M \end{bmatrix}, \quad \hat{E}^1 = \begin{bmatrix} E & 0 \\ 0 & E \end{bmatrix}
$$

with $m = 1$, $\eta^1 = \eta$, $N = \frac{n}{2}$. Note that $\hat{F}$ corresponds to the Coriolis force and $\hat{E}_\xi$ to the centrifugal force. The characteristic equation (3) takes the form

$$
(18)\qquad \det \begin{vmatrix} U - \eta E - (\xi^2 + \omega^2)\, M & -2i\omega\xi M \\ 2i\omega\xi M & V - \eta E - (\xi^2 + \omega^2)\, M \end{vmatrix} = 0
$$

The divergence boundary in the $\eta - \xi^2$ loading plane is now described by

$$
(19a)\qquad \det\, |U - \eta E - \xi^2 M| = 0
$$

(19b) $$\det |V - \eta E - \xi^2 M| = 0$$

If $U = V$, indicating equal rigidity in the two principal directions, the $2N$ branches of (19) coalesce into N branches.

Among all branches of (19), the initial divergence boundary is the one closest to the origin in the $\eta - \xi^2$ plane and, as stated earlier, this branch cannot have convexity toward the origin. In the special case when the matrices U, E and M can simultaneously be reduced to diagonal form, the divergence curves described by (19a) consist of straight lines (e.g., simply supported rotating shafts). A similar result can be stated for (19b); besides, in most practical problems V happens to be simply a multiple of U and simultaneous reduction of (19a) implies the same for (19b), resulting in divergence curves all in the form of straight lines.

Now consider the flutter boundary in the $\eta - \xi^2$ plane, associated with double roots ω_j^2 of (5). The following properties of the flutter boundary can be established:

I. If $U = V$, then the flutter boundary consist of straight lines, $\eta = \eta_j$, parallel to the ξ^2 axis, where η_j are the critical divergence points on the η axis, i.e.

(20) $$\det |U - \eta_j E| = 0$$

(e.g., see Fig. 3).

II. If M, E, U and V can be similtaneously reduced to diagonal matrices, then the flutter boundary consists of branches of hyperbolas with the η axis as a horizontal asymptote and with vertical asymptotes midway between pairs of critical divergence points on the η axis (e.g., see Fig. 5).

In order to prove statement I, consider the determinant (18). If one adds i times the last N rows to the first N rows, respectively, and then substracts i times

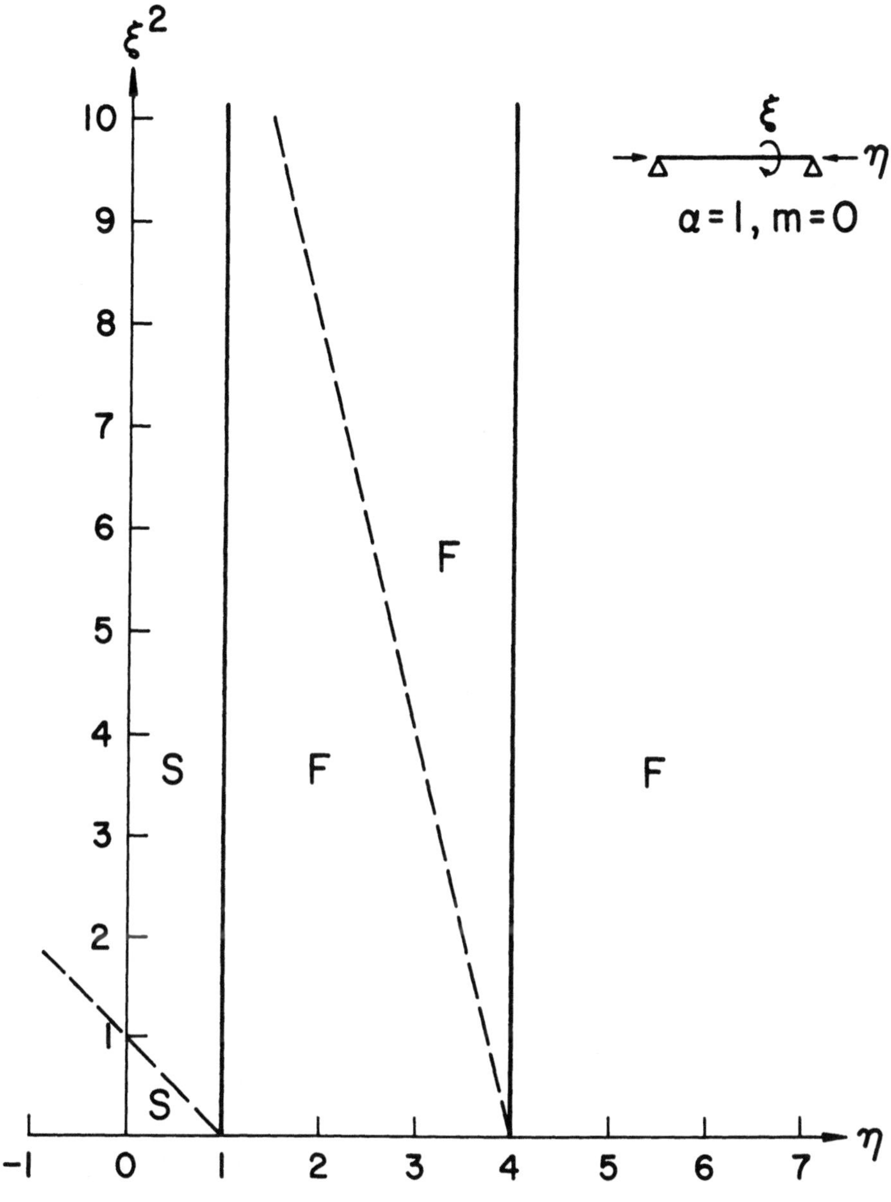

Fig. 3. Simply supported shaft, equal rigidities.

the first N columns from the last N columns, respectively, where $i = \sqrt{-1}$, one obtains

$$\text{(21)} \qquad \det \begin{vmatrix} U - \eta E - (\xi+\omega)^2 M & i\,(U-V) \\ 2i\omega\xi M & V - \eta E - (\xi-\omega)^2 M \end{vmatrix} = 0$$

If $U = V$, this leads to the conditions

$$(22)\quad \det|U-\eta E-(\xi+\omega)^2 M| = 0 \quad \text{or} \quad \det|U-\eta E-(\xi-\omega)^2 M = 0$$

Since the matrices in (22) are symmetric and M is positive definite, the roots $(\xi+\omega)^2$ and $(\xi-\omega)^2$ are real. Considering, for instance, the matrix equation corresponding to the first determinant in (22)

$$[U - \eta E - (\xi + \omega)^2 M] z = 0$$

one has

$$< z , [U - \eta E - (\xi + \omega)^2 M] z > = 0$$

Differentiating with respect to ω yields

$$(23)\qquad < z , [- \frac{\partial\eta}{\partial\omega} - 2(\xi+\omega)(\frac{\partial\xi}{\partial\omega} + 1) M] z > = 0$$

which at flutter point results in

$$(\xi + \omega) < z , M z > = 0$$

and

$$\xi + \omega = 0 .$$

Similar arguments hold for the second equation in (22), and (20) follows.

For statement II, assume that M , E , U and V are in diagonal form with diagonal elements m_j , e_j , u_j and v_j respectively. Equation (18) can be separated into the product of 2×2 determinants of the form

$$(24)\qquad \det \begin{vmatrix} u_j-\eta e_j-(\xi^2+\omega^2) m_j & -2i\omega\xi m_j \\ 2i\omega\xi m_j & v_j-\eta e_j-(\xi^2+\omega^2) m_j \end{vmatrix} = 0$$

The condition for a double root ω^2 of (24) is then found to be

$$\xi^2 = \frac{(u_j - v_j)^2}{8m_j\,(2\eta e_j - u_j - v_j)} \quad , \quad j = 1, \ldots , N \tag{25}$$

These curves are branches of hyperbolas with horizontal asymptote $\xi^2 = 0$ and vertical asymptote

$$\eta = \frac{u_j + v_j}{2e_j} \tag{26}$$

if $e_j \neq 0$. They are tangential to the divergence boundary at

$$\xi^2 = \left|\frac{u_j - v_j}{4m_j}\right| \tag{27}$$

where $\omega^2 = 0$ is a double root.

5. EXAMPLES

Consider an elastic shaft rotating with a constant angular velocity $\bar{\xi}$ and loaded with a constant axial force $\bar{\eta}$ (positive in compression). Let z denote the axial coordinate, $w_j\,(z,t)$ the transverse displacements in the principal directions $(j = 1,2)$, EI_j the flexural rigidities in the principal directions, ℓ the length, and μ the mass per unit length of the shaft. A disk of mass $\bar{m}$ may be located at $z = \ell_o$, where $0 < \ell_o < \ell$, and its displacements are $W_j(t) = w_j(\ell_o,t)$. Let Δ represent changes of quantities at the disk, e.g.

$$\Delta W_j(t) = w_j(\ell_o+,t) - w_j(\ell_o-,t) \ , \tag{28}$$

and it is assumed that $\Delta(\partial W_j/\partial z) = 0$.

Neglecting the effects of rotatory inertia and gyroscopic moments caused by the disk, the equations of motion of the shaft are assumed to be [6]

(29)
$$\mu \frac{\partial^2 w_1}{\partial t^2} - 2\bar{\xi}\mu \frac{\partial w_2}{\partial t} + EI_1 \frac{\partial^4 w_1}{\partial z^4} + \bar{\eta} \frac{\partial^2 w_1}{\partial z^2} - \bar{\xi}^2 \mu w_1 = 0$$

$$\mu \frac{\partial^2 w_2}{\partial t^2} + 2\bar{\xi}\mu \frac{\partial w_1}{\partial t} + EI_2 \frac{\partial^4 w_2}{\partial z^4} + \bar{\eta} \frac{\partial^2 w_2}{\partial z^2} - \bar{\xi}^2 \mu w_2 = 0$$

and the transition conditions at the disk are given by

(30)
$$\bar{m} \frac{\partial^2 W_1}{\partial t^2} - 2\bar{\xi}\bar{m} \frac{\partial W_2}{\partial t} + EI_1\Delta \frac{\partial^3 W_2}{\partial z^3} - \bar{\xi}^2 \bar{m} W_1 = 0$$

$$\bar{m} \frac{\partial 2 W_2}{\partial t^2} + 2\bar{\xi}\bar{m} \frac{\partial W_1}{\partial t} + EI_2\Delta \frac{\partial^3 W_2}{\partial z^3} - \bar{\xi}^2 m W_2 = 0$$

$$\Delta \frac{\partial^2 W_k}{\partial z^2} = 0 \ , \qquad k = 1,2 \ .$$

The ends of the shaft are either simply supported or clamped, so that the boundary condition at each end is either

(31) $$w_j = 0 \ , \quad \frac{\partial^2 w_j}{\partial z^2} = 0 \quad \text{or} \quad w_j = \frac{\partial w_j}{\partial z} = 0 \quad j = 1,2 \ .$$

Assuming motion of the form

(32) $$w_1 (z,t) = y_1 (z) e^{i\bar{\omega}t} \ , \quad w_2 (z,t) = y_2(z) e^{i\bar{\omega}t}$$

equations (29) become

(33)
$$-\bar{\omega}^2 \mu y_1 - 2i\bar{\omega}\bar{\xi}\mu y_2 + EI_1 y_1'''' + \bar{\eta} y_1'' - \bar{\xi}^2 \mu y_1 = 0$$

$$-\bar{\omega}^2 \mu y_2 + 2i\bar{\omega}\bar{\xi}\mu y_1 + EI_2 y_2'''' + \bar{\eta} y_2'' - \bar{\xi}^2 \mu y_2 = 0$$

where primes denote differentiation with respect to z. Now consider the approximations

$$(34) \qquad y_1(z) = \sum_{k=1}^{N} a_k\phi_k(z) \quad , \quad y_2(z) = \sum_{k=1}^{N} b_k\phi_k(z)$$

where the functions $\phi_k(z)$ satisfy the boundary conditions.

Following the Galerkin procedure, (32) and (34) are applied to the equations of motion, transition conditions and boundary conditions. Multiplying the resulting equations of motion by ϕ_j and integrating, one gets

$$(35) \qquad \begin{aligned} &\sum_{k=1}^{N} \{[EI_1\int\phi_j\phi_k''''dz+\bar{\eta}\int\phi_j\phi_k''dz-(\bar{\xi}^2+\bar{\omega}^2)\mu\int\phi_j\phi_k dz]a_k - \\ &\qquad\qquad 2i\bar{\omega}\bar{\xi}\mu\int\phi_j\phi_k dzb_k\} = 0 \\ &\sum_{k=1}^{N} \{2i\bar{\omega}\bar{\xi}\mu\int\phi_j\phi_k dza_k + \\ &\qquad [EI_2\int\phi_j\phi_k''''dz+\bar{\eta}\int\phi_j\phi_k''dz-(\bar{\xi}^2+\bar{\omega}^2)\mu\int\phi_j\phi_k dz]b_k\} = 0 \end{aligned}$$

Now integrating by parts and using boundary and transition conditions, one obtains (18) with $U = (u_{jk})$, $V = (v_{jk})$, $M = (m_{jk})$, $E = (e_{jk})$ where

$$u_{jk} = EI_1\int_0^{\ell}\phi_j''\phi_k''dz \ , \quad v_{jk} = EI_2\int_0^{\ell}\phi_j''\phi_k''dz \ ,$$

$$(36) \qquad m_{jk} = \mu\int_0^{\ell}\phi_j\phi_k dz + \bar{m}\phi_j(\ell_o)\phi_k(\ell_o)$$

$$e_{jk} = \int_0^{\ell}\phi_j'\phi_k'dz$$

and η, ξ, ω are replaced by $\bar{\eta}$, $\bar{\xi}$, $\bar{\omega}$. Define the non-dimensional quantities

$$x = z/\ell\ ,\quad \psi_j(x) = \phi_j\ (\ell x)/\ell$$

(37)

$$\xi = \frac{\bar{\xi}\ell^2}{\pi^2}\sqrt{\frac{\mu}{EI_1}}\ ,\quad \eta = \bar{\eta}\ell^2/\pi^2 EI_1\ ,\quad \omega = \frac{\bar{\omega}\ell^2}{\pi^2}\sqrt{\frac{\mu}{EI_1}}$$

$$x_o = \ell_o/\ell\ ,\quad m = \bar{m}/\mu\ell\ ,\quad \alpha = EI_2/EI_1\ .$$

If primes now denote differentiation with respect to x, one then finds

$$u_{jk} = \frac{1}{\pi^4}\int_o^1 \psi_j''\psi_k''dx\ ,\quad v_{jk} = \frac{\alpha}{\pi^4}\int_o^1 \psi_j''\psi_k''dx = \alpha u_{jk}$$

(38)

$$m_{jk} = \int_o^1 \psi_j\psi_k dx + m\psi_j\ (x_o)\ \psi_k(x_o)\ ,$$

$$e_{jk} = \frac{1}{\pi^2}\int_o^1 \psi_j'\psi_k'dx\ .$$

More generally, with disks of mass $m_1, \ldots, m_r$ at $x = x_1 \ldots, x_r$,

(39) $$m_{jk} = \int_o^1 \psi_j\psi_k dx + \sum_{\beta=1}^{r} m_\beta\psi_j\ (x_\beta)\ \psi_k\ (x_\beta)\ .$$

Computation were carried out using the first two vibration modes for the cases of a shaft simply-supported at both ends (A) and a shaft having one end clamped and the other simply-supported (B), with $\alpha = 1$ and $\alpha = 0.5$, and with $m = 0$ and $m = 1$ at $x_o = 0.5$ (i.e., a disk at the center of the shaft with mass equal to the shaft mass). The resulting stability and instability regions are depicted in Figs. 3 - 10, as follows:

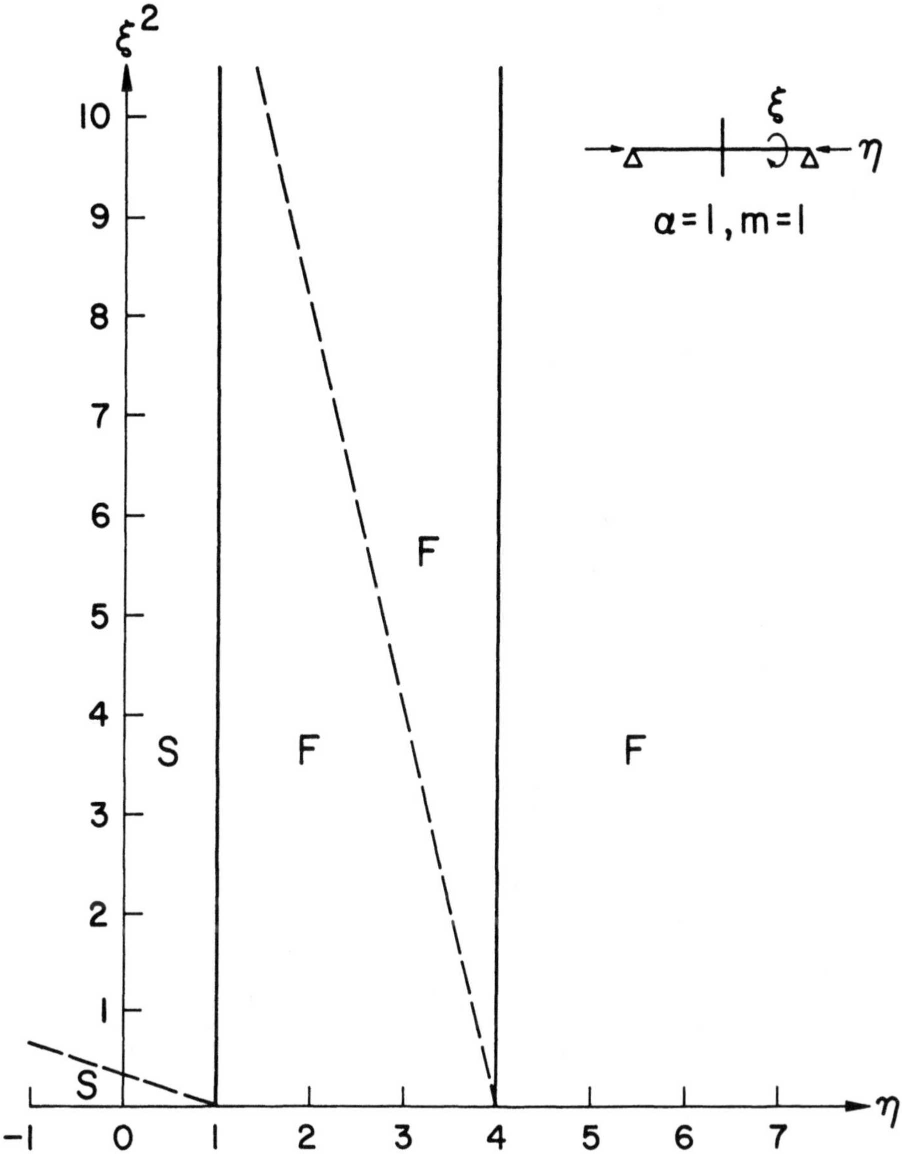

Fig. 4. Simply supported shaft with disk, equal rigidities.

Fig.	Ends	α	m
3	A	1	0
4	A	1	1
5	A	0.5	0
6	A	0.5	1
7	B	1	0

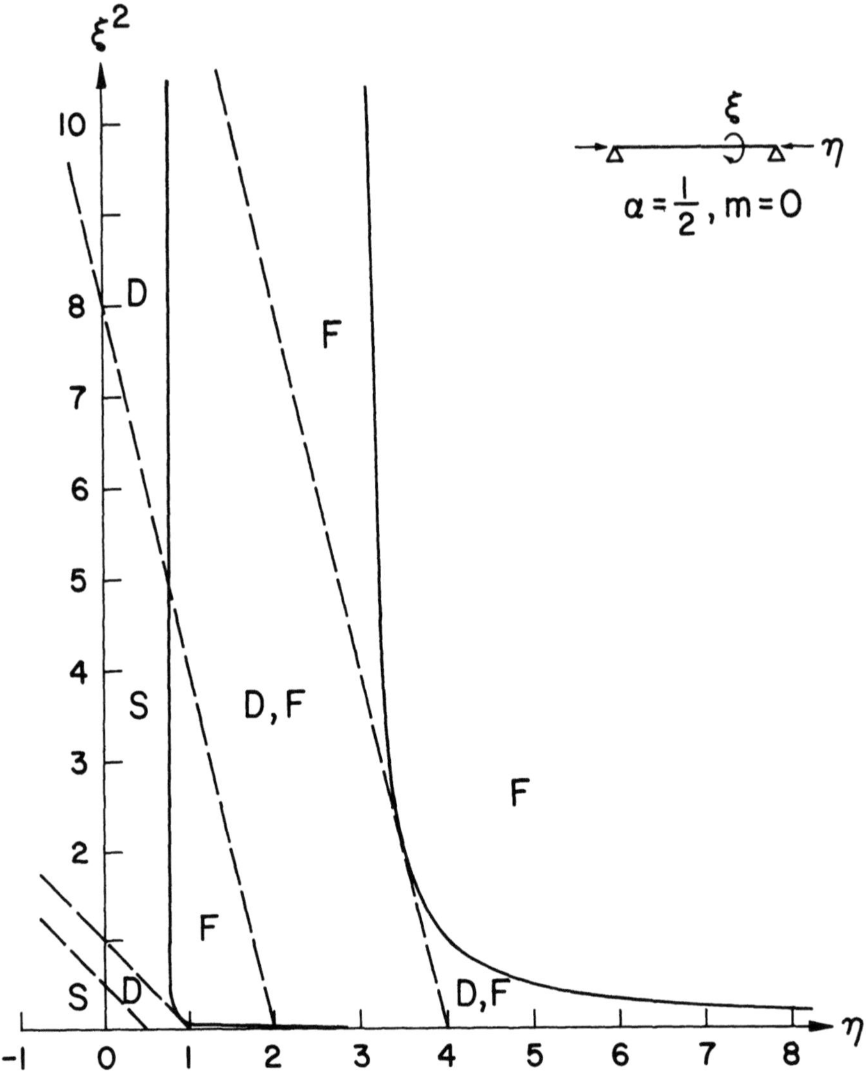

Fig. 5. Simply supported shaft, unequal rigidities.

Fig.	Ends	α	m
8	B	1	1
9	B	0.5	0
10	B	0.5	1

S denotes a region of stability, D denotes a region in which divergence instability occurs, and F designates a region of flutter instability (vibration with increasing

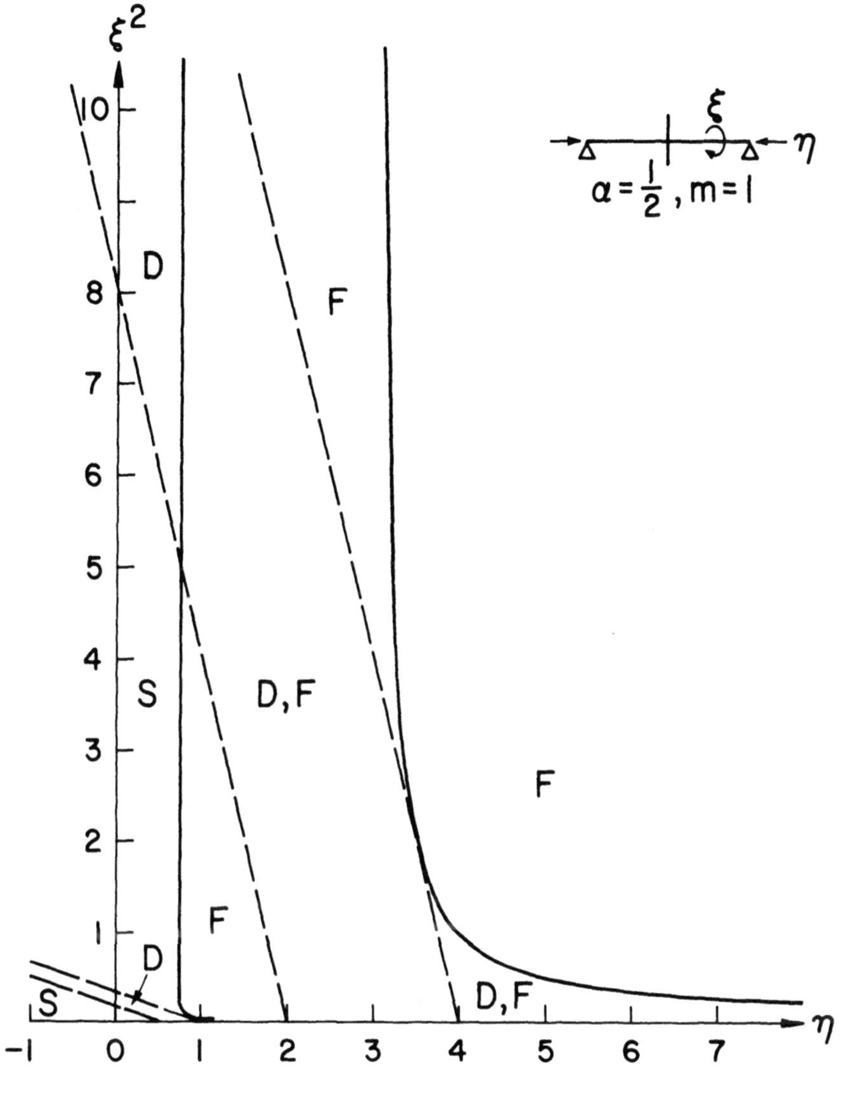

Fig. 6. Simply supported shaft with disk, unequal rigidities.

amplitude). The divergence boundary is shown with dashed lines and the flutter boundary with solid lines.

For Figs. 3, 4, 7 and 8, $\alpha = 1$ $(U = V)$ and hence the branches of the flutter boundary are vertical straight lines. Flutter instability may occur everywhere to the right of the first vertical line (e.g., for $\eta \geq 1$ and any $\xi^2 > 0$ in Figs. 3 and 4). The divergence boundary consists

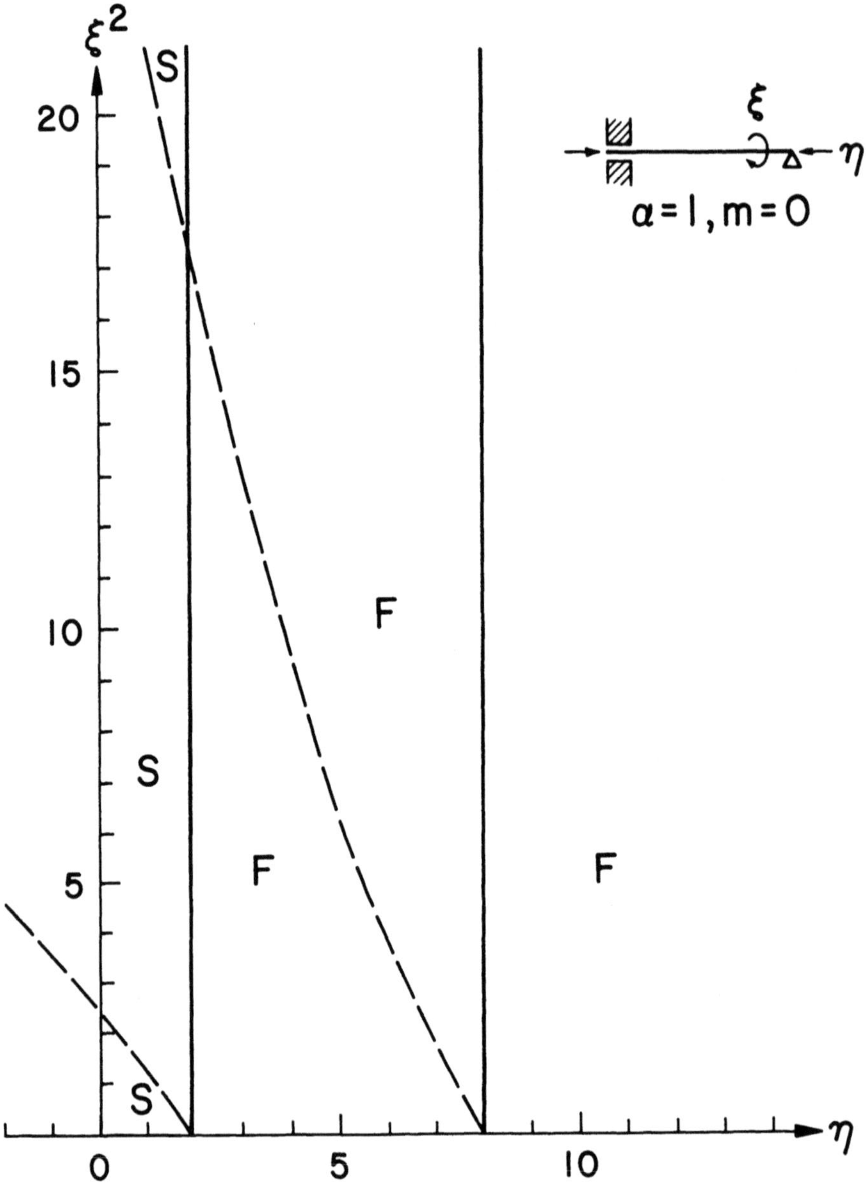

Fig. 7. Clamped - simply supported shaft, equal rigidities.

of two brances; in Figs. 3 and 4 these brances are straight since M , E , U and V are all diagonal matrices, while in Figs. 7 and 8 the initial divergence boundary is concave toward the origin and the final one is convex toward the origin. There are no regions of divergence instability in

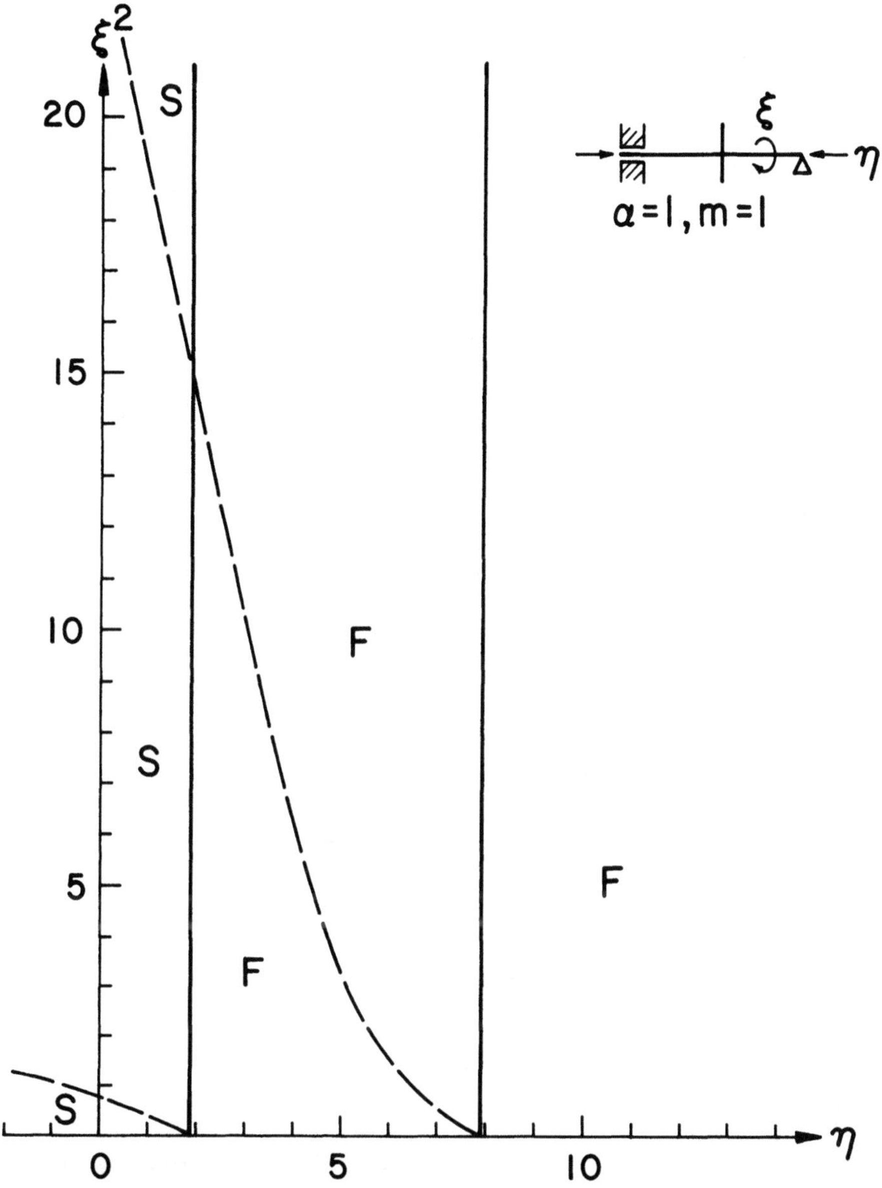

Fig. 8. Clamped - simply supported shaft with disk, equal rigidities.

these cases, i.e., divergence only occurs on these branches. The presence of the disk has no effect on the flutter boundary in these cases, but it tends to bring the divergence boundary closer to the origin (except in special cases such as the second divergence branch in Fig. 4).

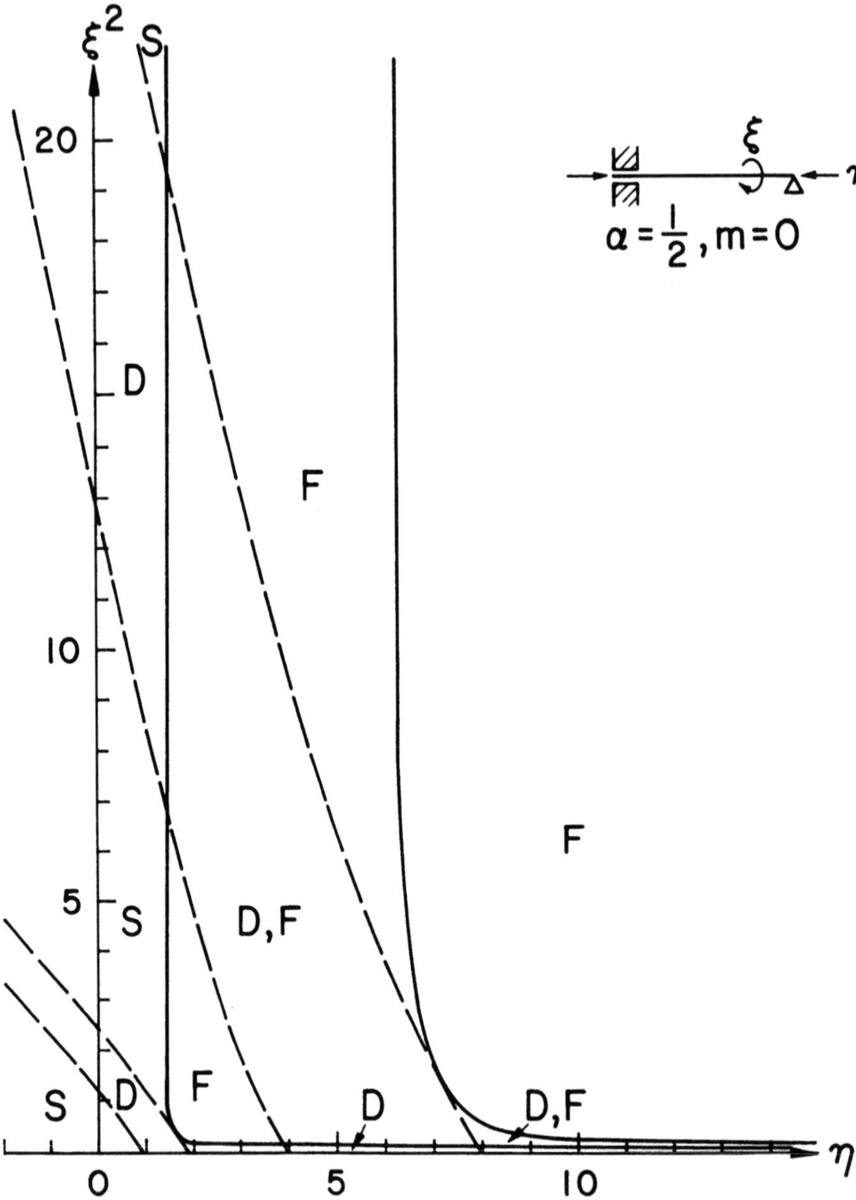

Fig. 9. Clamped - simply supported shaft, unequal rigidities.

In the remaining figures the shaft has unequal rigidities in its two principal directions $(U \neq V)$ and the divergence boundary consists of two pairs of branches. A divergence region exists between each of these pairs, and a branch of the flutter boundary is tengential to each of

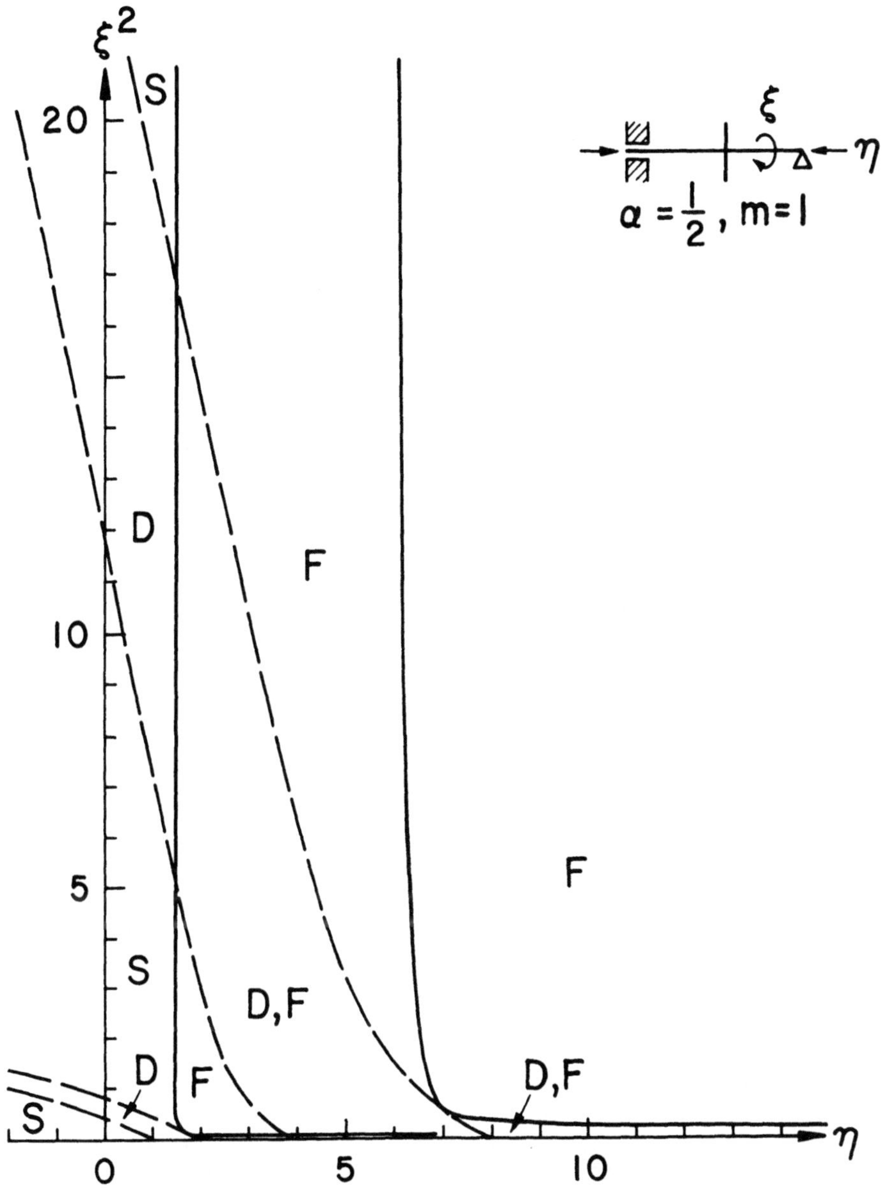

Fig. 10. Clamped - simply supported shaft with disk, unequal rigidities.

these regions. Divergence also occurs in the regions which are under the flutter branches and to the right of the divergence branches.

The vertical asymptotes of the flutter boundary as

$\xi^2 \to \infty$ occur midway between pairs of divergence points on the η axis. In Figs. 5 and 6, for example, the asymptotes are $\eta = 0.75$ and $\eta = 3$. In these two cases the flutter boundary consists of brances of hyperbolas (since M , E , U and V are diagonal) and no flutter is possible for $\eta < 0.75$. The leftmost vertical asymptote in Figs. 9 and 10 also provides a bound on flutter; however, this property is not a general one for all systems of the form (16). That is, a branch of the flutter boundary might have a portion lying to the left of its vertical asymptote in some cases. The convexity properties of the flutter boundaries in Figs. 5, 6, 9 and 10 also may not exist in all cases; that is, it cannot be shown that the branches of the flutter boundary are always convex toward the origin.

ACKNOWLEDGEMENT. This research was supported in part (K. Huseyin) by the National Research Council of Canada, and in part (R.H. Plaut) by the U.S. Army Research Office, Durham, under Grant DA-ARO-D-31-124-73-G130 and the National Science Foundation under Grant GK-41223 .

REFERENCES

[1] THOMSON W. & TAIT, G.: A Treatise on Natural Philosophy, Vol. 1, Part I, Cambridge University Press, Cambridge, England, 1879.

[2] ZIEGLER, H.: Principles of Structural Stability, Blaisdell, Waltham, Mass., 1968.

[3] HUSEYIN, K. & PLAUT, R.H.: Transverse vibrations and stability of gyroscopic systems, J. Struc. Mech. 3(2), 163-177 (1974). Also 7th U.S. National Congress of Applied Mechanics, Boulder, Colorado (1974).

[4] HUSEYIN, K. & ROORDA, J.: The loading-frequency relationship in multiple eigenvalue problems, J. Appl. Mech., 38 : 1007-1011 (1971).

[5] HUSEYIN, K. & PLAUT, R.H.: The elastic stability of two-parameter nonconservative systems, J. Appl. Mech. 40 : 175-180 (1973).

[6] SHIEH, R.C.: Energy and variational principles for generalized (gyroscopic) conservative problems, Int. J. Non-linear Mech. 5 : 495 - 509 (1971).

Limits to Modal Balancing of Flexible Rotors

W. Kellenberger

BBC BROWN, BOVERI & Company Limited
Baden, Switzerland

INTRODUCTION

Modal balancing of flexible rotors is based on the well-known fact that the deflection of the rotor due to unbalance forces, can be represented by the sum of various modes. With a precisely defined set of balancing weights it is thus possible to eliminate the first, the second and each successive mode up to the n-th. It is a characteristic of the set of balancing weights for the n-th mode that those modes which have already been balanced (and others, if desired) are no longer disturbed, or affected. The relative magnitudes of the sets of weights can be calculated in advance, i.e. before balancing is carried out, before anything is known about the unbalance distribution. Their absolute magnitude and phase relationship must, however, be obtained by measurement (during the balancing process). When those modes which are disturbing in the operating range have finally been reduced to a tolerable level, so are the undesired rotating bearing forces. In other words, the rotor is balanced.

Among other things, the procedure assumes bearing conditions such that the modes are "orthogonal" with respect to one another. This procedure is known for "round" and "oval" shafts *without* bearing damping or "coupling" of the bearing rigidities and therefore has its theoretical and practical limitations, like any technical process [3], [7], [9], [18], [19], [20]. In the present paper an attempt is made to determine these limits and inside them to formulate the systems of equations with the aid of which the sets of balancing weights can be computed. It has been found that this method of "modal balancing", though with rather more computing work, can also be extended to rotors *with* bearing damping. However, it breaks down when "coupled" bearing conditions (e.g. oil film) are involved. Other results are given in the conclusions.

Modal balancing has for years been the established method of balancing the flexible rotors used in machines of very high output (turbo-alternators and turbosets). Since it is still commonly used, although the requirements laid down for smooth running have become more exacting, it may be assumed that the theory of inaccessible influence is of minor significance.

1. THE ISOTROPIC (ROUND) SHAFT

<u>1.1 Differential equation</u>. The following points are assumed, as they are largely fulfilled by slender shaft assemblies, such as those of turbosets.

- The flexural rigidity and mass distribution down the length of the rotor are variable and given,
- Any number of intermediate bearings can be taken into account,
- The bearings exert only transverse forces on the shaft, no bending moments,
- Gyratory effects and shear deformation are ignored,
- The actual weight of the rotor is ignored because it is expressed by a static sag of the shaft, constant with respect to time,
- External and internal friction forces in the bearings are assumed to be proportional to the mass distribution down the length of the rotor,
- External friction is proportional to the absolute velocity $\dot{r}$ *, the internal friction being proportional to the

---

* Dots (•) denote derivatives with respect to t
Strokes (') denote derivatives with respect to z

relative velocity $\dot{\rho}$.

We can therefore write

$$\frac{\text{friction force}}{\text{mass}} = \begin{cases} \dfrac{b_a(z)\dot{r}(z)}{m(z)} = \mu\dot{r}(z) & ; \ \mu = \dfrac{b_a(z)}{m(z)} = \text{konstant} \\ \dfrac{b_i(z)\dot{\rho}(z)}{m(z)} = \nu\dot{\rho}(z) & ; \ \nu = \dfrac{b_i(z)}{m(z)} = \text{konstant} \end{cases}$$

The constants μ and ν are thus representatives of the external and internal damping of the shaft. A stationary, rectangular system of coordinates (x, y, z) and a rotating system of coordinates (u, v, z) revolving with the angular velocity Ω, are employed (see Fig. 1). The transformation

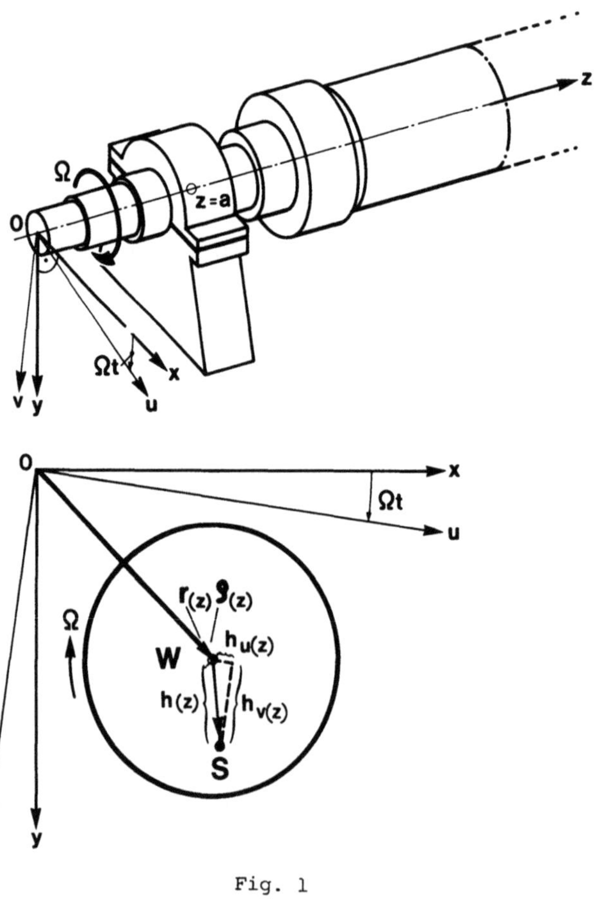

Fig. 1

formulas when calculating with complex numbers are as follows

(1a)
$$\underline{r = \rho e^{i\Omega t}} \quad \begin{cases} r = x + iy \\ \rho = u + iv \end{cases}$$

(1b)
$$\begin{vmatrix} x \\ y \end{vmatrix} = \underline{\underline{M_r}} \begin{vmatrix} u \\ v \end{vmatrix} \quad ; \quad \begin{vmatrix} u \\ v \end{vmatrix} = \underline{\underline{M_a}} \begin{vmatrix} x \\ y \end{vmatrix}$$

$$\underline{\underline{M_r}} = \begin{vmatrix} \cos\Omega t \, , & -\sin\Omega t \\ \sin\Omega t \, , & \cos\Omega t \end{vmatrix} \qquad \underline{\underline{M_a}} = \begin{vmatrix} \cos\Omega t \, , & \sin\Omega t \\ -\sin\Omega t \, , & \cos\Omega t \end{vmatrix}$$

Deviation of the centre of gravity from the geometrical centre of the shaft - the eccentricity of the centre of gravity - is denoted by $h(z) = h_u(z) + ih_v(z)$, where $h_u(z)$ and $h_v(z)$ are regarded as components which are constant with respect to time in the rotating system, and thus given.

The differential equation for the deflection $r(x, y, z, t)$ of the shaft in the stationary system, under the above conditions, is

(2a)
$$\frac{1}{m(z)}\Big(EI(z)r''\Big)'' + \ddot{r} + (\mu+\nu)\dot{r} - i\nu\Omega r = h(z)\Omega^2 e^{i\Omega t}$$

Substituting $r = x + iy$, this "vectorial equation" can be broken down into two component equations by separating the real and imaginary terms

(2b)
$$\boxed{\begin{aligned} &\frac{1}{m(z)}\Big(EI(z)x''\Big)'' + \ddot{x} + (\mu+\nu)\dot{x} + \nu\Omega y = \mathcal{Re}\Big(h(z)\Omega^2 e^{i\Omega t}\Big) \\ &\frac{1}{m(z)}\Big(EI(z)y''\Big)'' + \ddot{y} + (\mu+\nu)\dot{y} - \nu\Omega x = \mathcal{Jm}\Big(h(z)\Omega^2 e^{i\Omega t}\Big) \end{aligned}}$$

In order to define the boundary value problem, the end- and

bearing-conditions still have to be stated for differential equation (2). It should be noted that equation (2) is independent of the kind of bearing.

<u>1.2 End and bearing conditions</u>. The shaft assembly (see Fig. 2) is assumed as not being subjected to forces at its ends $(z = 0, \ell)$, i.e. the bending moments EIx'' , EIy'' and the shear forces $(EIx'')'$, $(EIy'')'$ vanish, so that the end conditions are

$$x''(0) = 0 \ , \ (EIx''(0))' = 0 \ ; \ x''(\ell) = 0 \ , \ (EIx''(\ell))' = 0$$

(3)

$$y''(0) = 0 \ , \ (EIy''(0))' = 0 \ ; \ y''(\ell) = 0 \ , \ (EIy''(\ell))' = 0$$

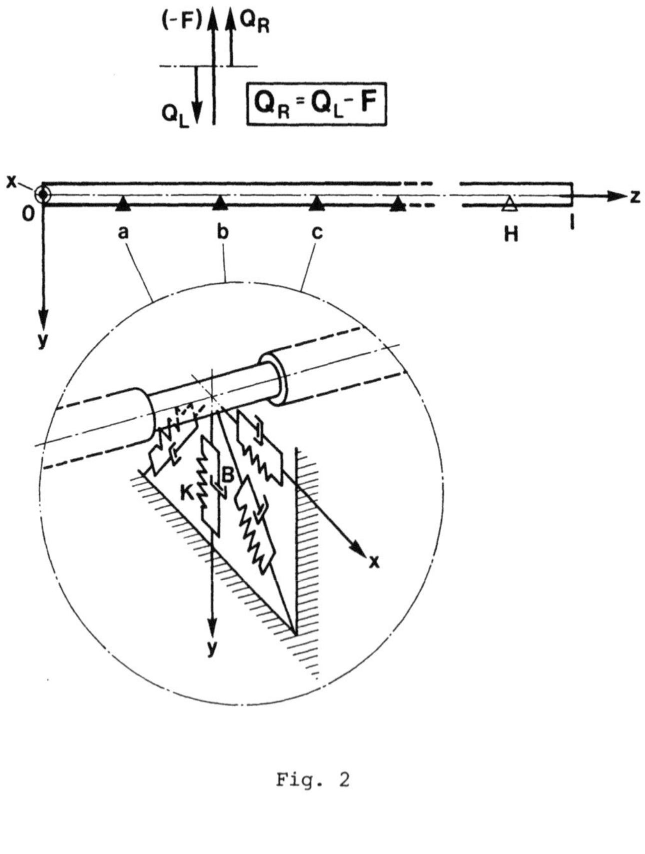

Fig. 2

Let us assume that the bearings supporting the shaft are located at points $z = a, b, c, \dots H$. Since the bearings only exert transverse forces on the shaft, the bending moments, angles of flexion and deflections to the left (L) and right (R) of the bearings must be equal, i.e.

$$\text{(4a)}\qquad \begin{matrix} x''_R = x''_L & x'_R = x'_L & x_R = x_L \\ y''_R = y''_L & y'_R = y'_L & y_R = y_L \end{matrix} \qquad (z = a, b, \dots H)$$

The bearing forces $F(a)$, $F(b)$, ... $F(H)$ are assumed to be proportional to the deflection and the rate of deflection of the bearings, i.e. the following relationship exists for each bearing force $F(z)$

$$\text{(4b)}\quad F(z) = \begin{vmatrix} F_x(z) \\ F_y(z) \end{vmatrix} = \begin{vmatrix} K_{xx}x + K_{xy}y + B_{xx}\dot{x} + B_{xy}\dot{y} \\ K_{yx}x + K_{yy}y + B_{yx}\dot{x} + B_{yy}\dot{y} \end{vmatrix} \quad (z = a, b, \dots H)$$

$$= \begin{vmatrix} K_{xx}, & K_{xy} \\ K_{yx}, & K_{yy} \end{vmatrix} \begin{vmatrix} x \\ y \end{vmatrix} + \begin{vmatrix} B_{xx}, & B_{xy} \\ B_{yx}, & B_{yy} \end{vmatrix} \begin{vmatrix} \dot{x} \\ \dot{y} \end{vmatrix} = \underline{\underline{K}} \begin{vmatrix} x \\ y \end{vmatrix} + \underline{\underline{B}} \begin{vmatrix} \dot{x} \\ \dot{y} \end{vmatrix}$$

Thus the bearings are defined by the stiffness matrix $\underline{\underline{K}}$ having four spring constants and the damping matrix $\underline{\underline{B}}$ having four damping constants. Basically, the constants may depend on the angular velocity Ω and may be different for each bearing.

If x and y vary exponentially with respect to time t, e.g. $x \sim e^{i\omega t}$, $y \sim e^{i\omega t}$ we may write

$$\text{(4c)}\quad F(z) = \underline{\underline{K}} \begin{vmatrix} x \\ y \end{vmatrix} + \underline{\underline{B}} \begin{vmatrix} i\omega x \\ i\omega y \end{vmatrix} = (\underline{\underline{K}} + i\omega\underline{\underline{B}}) \begin{vmatrix} x \\ y \end{vmatrix} = \begin{vmatrix} Z_{xx}, & Z_{xy} \\ Z_{yx}, & Z_{yy} \end{vmatrix} \begin{vmatrix} x \\ y \end{vmatrix} = \underline{\underline{Z}} \begin{vmatrix} x \\ y \end{vmatrix}$$

$$\begin{matrix} Z_{xx} = K_{xx} + i\omega B_{xx}\ ; & Z_{xy} = K_{xy} + i\omega B_{xy} \\ Z_{yx} = K_{yx} + i\omega B_{yx}\ ; & Z_{yy} = K_{yy} + i\omega B_{yy} \end{matrix} \qquad (z = a, b, \dots H)$$

because of $\dot{x} = i\omega x$ and $\dot{y} = i\omega y$.

The terms Z_{xx}, Z_{xy}, Z_{yx}, Z_{yy} may be regarded as dynamic spring constants (dynamic modulus).

If $K_{xy} = K_{yx} = B_{xy} = B_{yx} = 0$ the bearing is referred to as being *"uncoupled"*. We must distinguish between four cases of uncoupled bearings (see Table 1).

Type	Bearing-condition		$\underline{\underline{K}}$	$\underline{\underline{B}}$	$\left.\begin{matrix}\dot{x} = i\omega x\\ \dot{y} = i\omega y\end{matrix}\right\}\underline{\underline{Z}} = \begin{vmatrix} Z_{xx} & 0 \\ 0 & Z_{yy}\end{vmatrix}$
1,1	anisotropic	damped	$\begin{vmatrix} K_{xx} & 0 \\ 0 & K_{yy}\end{vmatrix}$	$\begin{vmatrix} B_{xx} & 0 \\ 0 & B_{yy}\end{vmatrix}$	$Z_{xx} = K_{xx} + i\omega B_{xx}$ $Z_{yy} = K_{yy} + i\omega B_{yy}$
1,2	anisotropic	undamped	$\begin{vmatrix} K_{xx} & 0 \\ 0 & K_{yy}\end{vmatrix}$	$\begin{vmatrix} 0 & 0 \\ 0 & 0\end{vmatrix}$	$Z_{xx} = K_{xx}$ $Z_{yy} = K_{yy}$
2,1	isotropic	damped	$\begin{vmatrix} K & 0 \\ 0 & K\end{vmatrix}$	$\begin{vmatrix} B & 0 \\ 0 & B\end{vmatrix}$	$Z_{xx} = Z_{yy} = K + i\omega B$
2,2	isotropic	undamped	$\begin{vmatrix} K & 0 \\ 0 & K\end{vmatrix}$	$\begin{vmatrix} 0 & 0 \\ 0 & 0\end{vmatrix}$	$Z_{xx} = Z_{yy} = K$

Table 1

With uncoupled bearings it is, in principle, possible to take complex bearing conditions into account, e.g. any arrangement of parallel and series spring, damping and mass elements (Fig. 3).

At the bearings $z = a, b, \ldots H$, the following transfer conditions apply to the shear forces $Q_x = (EIx'')'$ and $Q_y = (EIy'')'$, (see Fig. 2).

(4d)
$$Q_{xR} = Q_{xL} - F_x$$
$$Q_{yR} = Q_{yL} - F_y$$

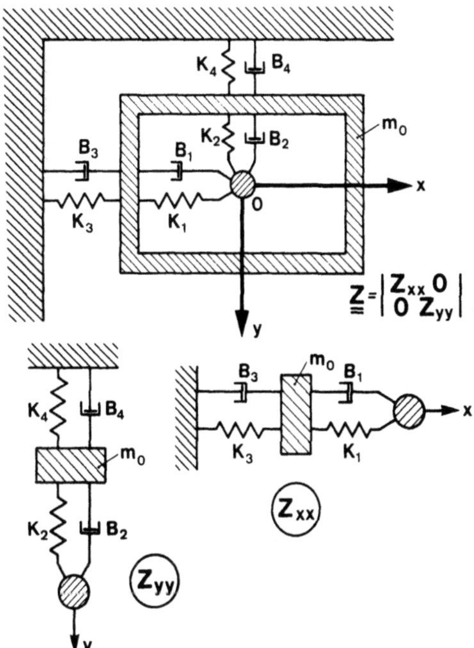

$$\dot{x} = i\omega x, \quad \dot{y} = i\omega y \qquad Z_{xx} = \frac{(K_1 + i\omega B_1)(K_3 + i\omega B_3 - m_0\omega^2)}{(K_1 + i\omega B_1) + (K_3 + i\omega B_3 - m_0\omega^2)}$$

$$Z_{yy} = \frac{(K_2 + i\omega B_2)(K_4 + i\omega B_4 - m_0\omega^2)}{(K_2 + i\omega B_2) + (K_4 + i\omega B_4 - m_0\omega^2)}$$

Fig. 3

With (4c) we then obtain the following transfer conditions for the shear forces at the bearings:

$$\text{(4e)} \qquad \begin{aligned} (EIx'')'_R &= (EIx'')'_L - Z_{xx}x - Z_{xy}y \\ (EIy'')'_R &= (EIy'')'_L - Z_{yx}x - Z_{yy}y \end{aligned} \qquad (z = a, b, \ldots H)$$

<u>1.3 Solution of the homogeneous boundary value problem (free oscillation)</u>. Our aim is to balance the rotor, i.e. to eliminate deflections caused by unbalance. These deflections are found analytically as particular solutions of the inhomogeneous differential equation with the end and bearing conditions. However, we require the solution in the form of a sum of modal functions originating from the homogeneous

equation. It is therefore essential for the homogeneous limit problem to be solved first. It is convenient to utilize the modal functions of the undamped shaft in damped bearings. For the undamped shaft $\mu = \nu = 0$ the homogeneous equation (2b) is

$$\begin{aligned} &\frac{1}{m(z)}(EI(z)x'')'' + \ddot{x} = 0 \\ &\frac{1}{m(z)}(EI(z)y'')'' + \ddot{y} = 0 \end{aligned} \tag{5}$$

The solution in generalized coordinates is then

$$\begin{aligned} x(z,t) &= \sum_n X_n(z)w_n(t) \\ y(z,t) &= \sum_n Y_n(z)w_n(t) \end{aligned} \tag{6}$$

If the n-th term (6) is inserted in (5) and the first equation divided by $w_n X_n$, the second by $w_n Y_n$, we obtain

$$\frac{(EIX_n'')''}{mX_n} + \frac{\ddot{w}_n}{w_n} = 0$$

$$\frac{(EIY_n'')''}{mY_n} + \frac{\ddot{w}_n}{w_n} = 0$$

These two equations can be separated, using an initially unknown constant ω_n^2. Then

$$\boxed{\begin{aligned} &(EIX_n'')'' - m\omega_n^2 X_n = 0 \\ &(EIY_n'')'' - m\omega_n^2 Y_n = 0 \end{aligned}} \tag{7a}$$

$$\boxed{\ddot{w}_n + \omega_n^2 w_n = 0} \tag{7b}$$

The general solution of the homogeneous equation (7b) is

$$(8) \qquad w_n(t) = w_{on} e^{i(\omega_n t + \varepsilon_n)} = w_{on} e^{i\varepsilon_n} e^{i\omega_n t}$$

With w_{on} and ε_n as integration constants, ω_n obviously has the significance of an angular frequency of the free oscillation. ω_n may be real or complex. If $\omega_n = \alpha_n + i\beta_n$ is complex, the solution is

$$w_n(t) = w_{on} e^{i\varepsilon_n} e^{i(\alpha_n + i\beta_n)t} = w_{on} e^{i\varepsilon_n} e^{-\beta_n t} e^{i\alpha_n t}$$

With (6) and (8) the solution for the n-th term becomes

$$x_n = X_n w_n = X_n w_{on} e^{i\varepsilon_n} e^{i\omega_n t} = \sim e^{i\omega_n t}$$

$$y_n = Y_n w_n = Y_n w_{on} e^{i\varepsilon_n} e^{i\omega_n t} = \sim e^{i\omega_n t}$$

and thus varies exponentially with respect to time. Hence the bearing conditions (4e) apply.

The end conditions (3) and the bearing conditions (4a) and (4e) are rendered independent of time by substituting (6)

$$(9a) \qquad \begin{matrix} X_n'' = 0 & (EIX_n'')' = 0 \\ Y_n'' = 0 & (EIY_n'')' = 0 \end{matrix} \qquad (z = 0\ ,\ \ell)$$

$$(9b) \qquad \begin{cases} X_{nR}'' = X_{nL}'' \quad X_{nR}' = X_{nL}' \quad X_{nR} = X_{nL} \\ Y_{nR}'' = Y_{nL}'' \quad Y_{nR}' = Y_{nL}' \quad Y_{nR} = Y_{nL} \\ (EIX_n'')_R' = (EIX_n'')_L' - Z_{xx} X_n - Z_{xy} Y_n \\ (EIY_n'')_R' = (EIY_n'')_L' - Z_{yx} X_n - Z_{yy} Y_n \end{cases} \qquad (z = a\ ,\ b\ ,\ \ldots H\ )$$

With the end- and bearing conditions (9) the homogeneous system of equations (7a) can be solved. The boundary value problem has become a problem of eigenvalues with the eigenvalue ω_n^2 . The most convenient method of solution is a matrix transfer method, such as that of Myklestad - Prohl. It should be noted that, although equations (7a) are not coupled, a coupling nevertheless exists due to the bearing conditions (9b).

With coupled bearing conditions $(Z_{xy} \neq 0\ ,\ Z_{yx} \neq 0)$ or uncoupled bearing conditions $(Z_{xy} = Z_{yy} = 0)$ but with bearing damping (Table 1), the eigenvalue is complex and consequently so are the modes. Hence the latter are no longer situated in planes passing through the z axis, but are "twisted".

<u>1.4 Orthogonality of the modal functions with different bearing conditions</u>. In the next chapter 1.5 the movement of the shaft due to unbalance will be derived. Therein use will be made of the orthogonality properties of the modal functions $X_n(z)$ and $Y_n(z)$, i.e. it is assumed that

$$
(10) \qquad \left.\begin{aligned} \int_0^\ell mX_nX_m dz &= 0 \\ \int_0^\ell mY_nY_m dz &= 0 \end{aligned}\right\} \begin{aligned} &\text{Indices:} \\ &m \neq n \end{aligned} \qquad \begin{aligned} \int_0^\ell mX_n^2 dz &= N_{nx} \\ \int_0^\ell mY_n^2 dz &= N_{ny} \\ \int_0^\ell mX_nY_n dz &= N_{nxy} \end{aligned}
$$

What we have to investigate is to what extent the bearing conditions (9b) satisfy this requirement.

The equation (7a) for $n = n$ and $n = m$ are

$$
(EIX_n'')'' - m\omega_n^2 X_n = 0 \qquad (EIX_m'')'' - m\omega_m^2 X_m = 0
$$

If we multiply the first equation by X_m and the second by X_n subtract the second from the first and integrate over the length of the rotor, we obtain

$$
\int_0^\ell \left[X_m (EIX_n'')'' - X_n (EIX_m'')''\right] dz = \int_0^\ell mX_nX_m(\omega_n^2 - \omega_m^2) dz
$$

Integration by parts supplies

$$A_x - B_x = (\omega_n^2 - \omega_m^2) \int_0^{\ell} m X_n X_m dz \ , \qquad A_x = \Big[X_m (EIX_n'')' - X_n (EIX_m'')' \Big]_0^{\ell} \qquad B_x = \Big[EI(X_m' X_n'' - X_n' X_m'') \Big]_0^{\ell}$$

An analogous equation is obtained from the second equation (7a)

$$A_y - B_y = (\omega_n^2 - \omega_m^2) \int_0^{\ell} m Y_n Y_m dz \ , \qquad A_y = \Big[Y_m (EIY_n'')' - Y_n (EIY_m'')' \Big]_0^{\ell} \qquad B_y = \Big[EI(Y_m' Y_n'' - Y_n' Y_m'') \Big]_0^{\ell}$$

If the orthogonality conditions (10) are to apply, the terms A_x , B_x and A_y , B_y must vanish. Below we will examine whether this is in fact the case for the general bearing condition (9b).

a. Terms B_x, B_y. For B_x we have

$$B_x = \Big[EI(X_m' X_n'' - X_n' X_m'') \Big]_0^{\ell} = \Big[f_x(z) \Big]_0^{\ell} = \Big[f_x \Big]_0^{a_L} + \Big[f_x \Big]_{a_R}^{b_L} + \ldots + \Big[f_x \Big]_{H_R}^{\ell} =$$

$$= - f_x(0) + f_x(a_L) - f_x(a_R) + f_x(b_L) - f_x(b_R) + \ldots + f_x(\ell)$$

Owing to (9a) and (9b)

$$f_x(0) = 0 \ , \ f_x(\ell) = 0$$

$$f_x(a_L) - f_x(a_R) = 0 \ , \ f_x(b_L) - f_x(b_R) = 0 \ , \ \ldots f_x(H_L) - f_x(H_R) = 0$$

so that $B_x = 0$ and similarly $B_y = 0$.

b. Terms A_x, A_y. For A_x we have

$$A_x = \left[X_m(EIX_n'')' - X_n(EIX_m'')'\right]_0^{\ell} = \left[g_x(z)\right]_0^{\ell} =$$

$$= \left[g_x\right]_0^{a_L} + \left[g_x\right]_{a_R}^{b_L} + \ldots + \left[g_x\right]_{H_R}^{\ell} =$$

$$= -g_x(0) + g_x(a_L) - g_x(a_R) + g_x(b_L) - g_x(b_R) + \ldots + g_x(\ell)$$

According to (9a), the boundaries are given by

$$(11) \qquad \begin{matrix} g_x(0) = 0 \\ g_x(\ell) = 0 \end{matrix} \qquad \text{analog} \qquad \begin{matrix} g_y(0) = 0 \\ g_y(\ell) = 0 \end{matrix}$$

At the bearings, e.g. when $z = a$, we obtain from (9b) that

$$\left[EIX_n''(a_R)\right]' = \left[EIX_n''(a_L)\right]' - Z_{xx}X_n(a) - Z_{xy}Y_n(a)$$

$$\left[EIY_n''(a_R)\right]' = \left[EIY_n''(a_L)\right]' - Z_{yx}X_n(a) - Z_{yy}Y_n(a)$$

If we multiply the first equation by $X_m(a)$ and the second by $X_n(a)$ afterwards exchanging the indices m and n, we obtain four equations. If we now subtract the equations with exchanged indices from the original equations, we arrive at

$$X_m(a)\left[EIX_n''(a_R)\right]' - X_n(a)\left[EIX_m''(a_R)\right]' = X_m(a)\left[EIX_n''(a_L)\right]' -$$
$$X_n(a)\left[EIX_m''(a_L)\right]' - Z_{xx}\cdot 0 - Z_{xy}\left[Y_n(a)X_m(a) - Y_m(a)X_n(a)\right]$$

$$Y_m(a)\left[EIY_n''(a_R)\right]' - Y_n(a)\left[EIY_m''(a_R)\right]' = Y_m(a)\left[EIY_n''(a_L)\right]' -$$
$$Y_n(a)\left[EIY_m''(a_L)\right]' - Z_{yx}\left[X_n(a)Y_m(a) - X_m(a)Y_n(a)\right] - Z_{yy}\cdot 0$$

Taking into account that, according to (9b)

$$X_n(a) = X_n(a_L) = X_n(a_R) \quad ; \quad Y_n(a) = Y_n(a_L) = Y_n(a_R)$$

the above expressions may be reduced to

$$(12) \qquad \begin{aligned} g_x(a_R) &= g_x(a_L) - Z_{xy}\Big[Y_n(a)X_m(a) - Y_m(a)X_n(a)\Big] \\ g_y(a_R) &= g_y(a_L) = Z_{yx}\Big[X_n(a)Y_m(a) - X_m(a)Y_n(a)\Big] \end{aligned}$$

Analogous equations apply to all other bearings ($z = b, \ldots H$).

With (11) and (12) the expressions we are looking for become

$$A_x = \sum_{z-a,\ldots}^{H} Z_{xy}\Big[Y_n(z)X_m(z) - Y_m(z)X_n(z)\Big] \; ;$$

$$A_y = \sum_{z=a,\ldots}^{H} Z_{yx}\Big[X_n(z)Y_m(z) - X_m(z)Y_n(z)\Big]$$

It is evident that the expressions only vanish when $Z_{xy} = Z_{yx} = 0$ i.e. when the bearings are uncoupled.

Hence, the *orthogonality conditions* (10) is *only satisfied* for *uncoupled bearings* as per Table 1.

<u>1.5 Forced movement due to unbalance of the round shaft in uncoupled bearings</u>. With the generalized coordinates

$$(13) \qquad \begin{aligned} x(z, t) &= \sum_n X_n(z)q_{nx}(t) \\ y(z, t) &= \sum_n Y_n(z)q_{ny}(t) \end{aligned}$$

the following form for the n-th term can be derived from the differential equation (2b)

$$\frac{(EIX_n'')''}{mX_n} + \frac{\ddot{q}_{nx}}{q_{nx}} + (\mu+\nu)\frac{\dot{q}_{nx}}{q_{nx}} + \nu\Omega\frac{Y_n q_{ny}}{X_n q_{nx}} = \mathcal{R}e\left(\frac{h\Omega^2 e^{i\Omega t}}{X_n q_{nx}}\right)$$

(14)

$$\frac{(EIY_n'')''}{mY_n} + \frac{\ddot{q}_{ny}}{q_{ny}} + (\mu+\nu)\frac{\dot{q}_{ny}}{q_{ny}} - \nu\Omega\frac{X_n q_{nx}}{Y_n q_{ny}} = \mathcal{J}m\left(\frac{h\Omega^2 e^{i\Omega t}}{Y_n q_{ny}}\right)$$

If the symbols $\mathcal{R}e$ and $\mathcal{J}m$ are omitted, in order to simplify the calculation, it should be noted that only the real part of x and the imaginary part of y represent physically usable solutions. The positional vector of the curve followed is then given by

(13a) $$r_p = \mathcal{R}e(x) + i\mathcal{J}m(y) = x_p + iy_p \qquad \begin{cases} x_p = \mathcal{R}e(x) \\ y_p = \mathcal{J}m(y) \end{cases}$$

with the projections x_p and y_p .

In the differential equation (14) the quantity $h(z) = h_u(z) + ih_v(z)$ occurs. If we form the product with the mass $m(z)$

$$m(z)h(z) = m(z)[h_u(z) + ih_v(z)] = p(z)$$

the continuous unbalance $p(z)$ of the shaft is obtained. In view of the fact that we are trying to compensate this existing unbalance by placing a finite number of discrete unbalance "forces" at definite places $z = z_k$ $(k = 1,2,\ldots K)$ down the shaft, we may regard the unbalance as the sum of continuous unbalance "forces" $p(z)$ and the sum of K individual unbalance "forces" $P_k = P_{ku} + iP_{kv}$. We may then write

(15) $$m(z)h(z) = p(z) + \sum_k^K P_k\delta(z - z_k)$$

with δ as "delta function". Let us represent the continuous unbalance by the sum of modal functions X_n and Y_n i.e. we can write

(16) $p(z) = \sum_n m(z) a_n X_n(z) = \sum_n m(z) b_n Y_n(z) \qquad (n=1,2,\ldots N)^*$

The constants a_n and b_n can be calculated formally from the relationship

$$a_n = \frac{1}{N_{nx}} \int_0^{\ell} p(z) X_n(z)\,dz \quad , \quad b_n = \frac{1}{N_{ny}} \int_0^{\ell} p(z) Y_n(z)\,dz$$

when the conditions for orthogonality (10) are satisfied.

Reverting to differential equation (14) we will equate the first term in each equation to a constant ω_n^2. Then

(17a)
$$\boxed{\begin{aligned} (EIX_n'')'' - m\omega_n^2 X_n &= 0 \\ (EIY_n'')'' - m\omega_n^2 Y_n &= 0 \end{aligned}}$$

and automatically for the other terms

$$\ddot{q}_{nx} + (\mu+\nu)\dot{q}_{nx} + \nu\Omega \frac{Y_n}{X_n} q_{ny} + \omega_n^2 q_{nx} = \frac{h\Omega^2 e^{i\Omega t}}{X_n}$$

$$\ddot{q}_{ny} + (\mu+\nu)\dot{q}_{ny} - \nu\Omega \frac{X_n}{Y_n} q_{nx} + \omega_n^2 q_{ny} = \frac{h\Omega^2 e^{i\Omega t}}{Y_n}$$

If we multiply both sides of these equations by mX_n^2 or mY_n^2 and substitute (16) for the relationship in (15), integrating both sides of the equations over the length of the rotor, we obtain the following equations from which z has been eliminated.

---

* The number N depends on the unbalance distribution. N can, but need not, be infinitely large.

$$(17b)\quad \boxed{\begin{aligned} &\ddot{q}_{nx}+(\mu+\nu)\dot{q}_{nx}+\nu\Omega\frac{N_{nxy}}{N_{nx}}q_{ny}+\omega_n^2 q_{nx} = \frac{\Omega^2}{N_{nx}}\left(a_n N_{nx}+\sum_k^K P_k X_{nk}\right)e^{i\Omega t}\\ &\ddot{q}_{ny}+(\mu+\nu)\dot{q}_{ny}-\nu\Omega\frac{N_{nxy}}{N_{ny}}q_{nx}+\omega_n^2 q_{ny} = \frac{\Omega^2}{N_{ny}}\left(b_n N_{ny}+\sum_k^K P_k Y_{nk}\right)e^{i\Omega t}\end{aligned}}$$

Therein use was made of the condition for orthogonality (10) and the abbreviations

$$X_{nk} = X_n(z_k) \qquad Y_{nk} = Y_n(z_k)$$

Comparing (17a) with (17b) we find that ω_n^2 represents the real or complex eigenvalues and X_n , Y_n the real or complex modes of the undamped shaft, damped (or not as the case may be) in the bearings. Eigenvalues and modes can thus be calculated in accordance with 1.3 and regarded as *given*. They do not depend on the internal and external damping of the shaft, but on the existing bearing condition. A particular integral for the n-th term of the system (17b) is

$$(18)\qquad q_{nx} = A_n e^{i\Omega t} \quad , \quad q_{ny} = B_n e^{i\Omega t}$$

Substituting in (17b) we arrive at the following system of equations to calculate the unknowns A_n and B_n

$$(18a)\qquad \begin{vmatrix} \omega_n^2-\Omega^2+i(\mu+\nu)\Omega , & \nu\Omega N_{nxy}/N_{nx} \\ -\nu\Omega N_{nxy}/N_{ny} & , \omega_n^2-\Omega^2+i(\mu+\nu)\Omega \end{vmatrix} \begin{vmatrix} A_n \\ B_n \end{vmatrix} = \begin{vmatrix} \frac{\Omega^2}{N_{nx}}\left(a_n N_{nx}+\sum_k^K P_k X_{nk}\right) \\ \frac{\Omega^2}{N_{ny}}\left(b_n N_{ny}+\sum_k^K P_k Y_{nk}\right) \end{vmatrix}$$

with the determinant

$$(18b)\qquad \text{Det} = \left[\omega_n^2 - \Omega^2 + i(\mu+\nu)\Omega\right]^2 + \nu^2\Omega^2\,\frac{N_{nxy}^2}{N_{nx}N_{ny}} \neq 0$$

and the solutions

$$A_n = \frac{1}{\mathrm{Det}}\left\{\frac{\Omega^2}{N_{nx}}\left[\omega_n^2 - \Omega^2 + i(\mu+\nu)\Omega\right]\left(a_n N_{nx} + \sum_k^K P_k X_{nk}\right) - \right.$$

$$\left. - \frac{\nu\Omega^3 N_{nxy}}{N_{nx}N_{ny}}\left(b_n N_{ny} + \sum_k^K P_k Y_{nk}\right)\right\}$$

(18c)

$$B_n = \frac{1}{\mathrm{Det}}\left\{\frac{\Omega^2}{N_{ny}}\left[\omega_n^2 - \Omega^2 + i(\mu+\nu)\Omega\right]\left(b_n N_{ny} + \sum_k^K P_k Y_{nk}\right) + \right.$$

$$\left. + \frac{\nu\Omega^3 N_{nxy}}{N_{nx}N_{ny}}\left(a_n N_{nx} + \sum_k^K P_k X_{nk}\right)\right\}$$

The solution (13a) for the n-th term in (13) is obtained with the aid of (18), and works out at

$$x_{np} = \mathcal{R}e[x_n(z\,,\,t)] = \mathcal{R}e\left[A_n X_n(z) e^{i\Omega t}\right]$$

(19)

$$y_{np} = \mathcal{J}m[y_n(z\,,\,t)] = \mathcal{J}m\left[B_n Y_n(z) e^{i\Omega t}\right]$$

With complex values $A_n X_n = C_r + iC_i$ or $B_n Y_n = D_r + iD_i$ (19) becomes

$$x_p(z\,,\,t) = \sum_n x_{np}(z\,,\,t) \quad , \quad x_{np} = C_r \cos\Omega t - C_i \sin\Omega t$$

(19a)

$$y_p(z\,,\,t) = \sum_n y_{np}(z\,,\,t) \quad , \quad y_{np} = D_r \sin\Omega t + D_i \cos\Omega t$$

i.e. when z is constant, the parametric representation of an *ellipse*.

1.6 Balancing the round shaft in uncoupled bearings. The shaft is balanced when the deflections x_p and y_p vanish, with the result that there are no longer any rotating forces transferred to the bearings. The conditions for x_p and y_p vanishing are obtained from (19a), (19) and (18c) as follows

$$a_n N_{nx} + \sum_k^K P_k X_{nk} = 0$$

(20) $\qquad (n=1,2,\ldots N)$

$$b_n N_{ny} + \sum_k^K P_k Y_{nk} = 0$$

and may thus be regarded as the decisive equations for the flexible balancing of flexible round shafts in uncoupled bearings. In matrix form resolved into the individual terms, (20) may be written as

(21a)
$$\begin{array}{c} \\ n \\ \downarrow \\ N \end{array} \overset{\xrightarrow{\quad k \quad} \quad K}{\begin{vmatrix} X_{11} & X_{12} & X_{13} & \cdots \\ X_{21} & X_{22} & X_{23} & \\ X_{31} & X_{32} & X_{33} & \\ \vdots & & & \end{vmatrix}} \begin{Vmatrix} P_1 \\ P_2 \\ P_3 \\ \vdots \end{Vmatrix} = - \begin{vmatrix} a_1 N_{1x} \\ a_2 N_{2x} \\ a_3 N_{3x} \\ \vdots \end{vmatrix} ;$$

$$\left(a_n N_{nx} = \int_0^{\ell} p(z) X_n(z) dz \right)$$

(21b)
$$\begin{array}{c} \\ n \\ \downarrow \\ N \end{array} \overset{\xrightarrow{\quad k \quad} \quad K}{\begin{vmatrix} Y_{11} & Y_{12} & Y_{13} & \cdots \\ Y_{21} & Y_{22} & Y_{23} & \\ Y_{31} & Y_{32} & Y_{33} & \\ \vdots & & & \end{vmatrix}} \begin{Vmatrix} P_1 \\ P_2 \\ P_3 \\ \vdots \end{Vmatrix} = - \begin{vmatrix} b_1 N_{1y} \\ b_2 N_{2y} \\ b_3 N_{3y} \\ \vdots \end{vmatrix} ;$$

$$\left(b_n N_{ny} = \int_0^{\ell} p(z) Y_n(z) dz \right)$$

We thus obtain two systems of linear equations containing N equations for K unknown sets of balancing weights P_k. For a unique solution the same number of unknowns are required as there are equations, or in short

$$K = N$$

Thus, for N modes N balancing weights are needed and therefore inevitably N balancing planes. The set of weights for the n-th mode $(n=1,2,\ldots N)$ is obtained by equating all terms on the right-hand side of (21), except for $a_n N_{nx}$ or $b_n N_{ny}$, to zero. The final state of balance is obtained by adding the N sets of weights together.

Hence for $N = 2$, substituting in (21a), we obtain

$$\begin{vmatrix} X_{11} & X_{12} \\ X_{21} & X_{22} \end{vmatrix} \begin{vmatrix} P_1 \\ P_2 \end{vmatrix} = - \begin{vmatrix} a_1 N_{1x} \\ a_2 N_{2x} \end{vmatrix} = \underset{\to\, n=1}{\begin{vmatrix} -a_1 N_{1x} \\ 0 \end{vmatrix}} + \underset{\to\, n=2}{\begin{vmatrix} 0 \\ -a_2 N_{2x} \end{vmatrix}}$$

<u>n = 1</u>

$$P_1 = \frac{- a_1 N_{1x} X_{22}}{X_{11} X_{22} - X_{12} X_{21}} \quad , \; P_1/P_1 = 1$$

$$P_2 = \frac{a_1 N_{1x} X_{21}}{X_{11} X_{22} - X_{12} X_{21}} \quad , \; P_2/P_1 = - \frac{X_{21}}{X_{22}}$$

<u>n = 2</u>

$$P_1 = \frac{a_2 N_{2x} X_{12}}{X_{11} X_{22} - X_{12} X_{21}} \quad , \; P_1/P_1 = 1$$

$$P_2 = \frac{- a_2 N_{2x} X_{11}}{X_{11} X_{22} - X_{12} X_{21}} \quad , \; P_2/P_1 = - \frac{X_{11}}{X_{12}}$$

and similarly for the system in (21b).

When balancing in practice it should be borne in mind that the right-hand side of (21) depends on the existing continuous unbalance $p(z)$ which is unknown to begin with. Consequently only the relative values P_1/P_1 , P_2/P_1 , ... or $P_1 : P_2 : P_3 : \ldots$ i.e. the *relative* weights can be calculated in advance.

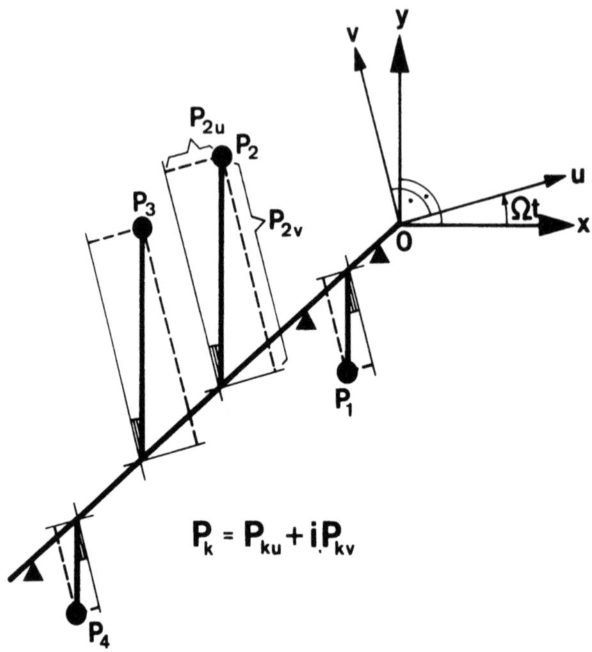
y
v
u
x
O
Ωt
P2u
P2
P2v
P3
P1
P4
Pk = Pku + iPkv

Fig. 4

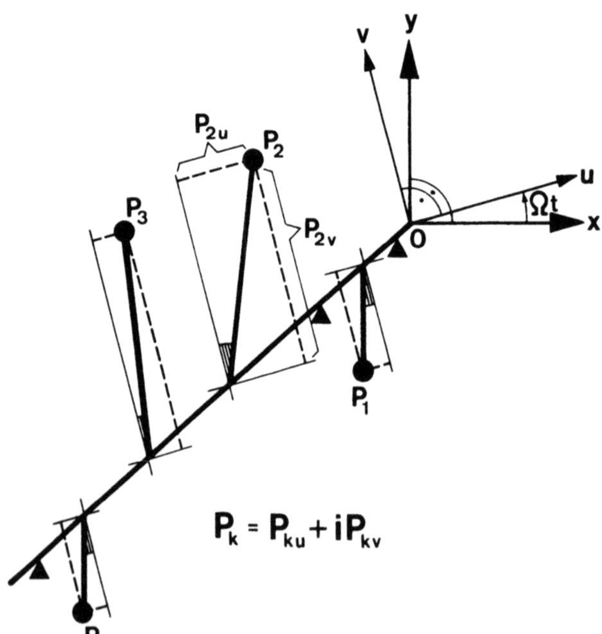
y
v
u
x
O
Ωt
P2u
P2
P2v
P3
P1
P4
Pk = Pku + iPkv

Fig. 5

The latter do not depend on the balance and are pure functions of the modes. The absolute magnitude and angular relationship of the balancing weights have to be determined by measurement. This should be carried out as near as possible to the critical speed at which the n-th mode is most pronounced.

<u>1.6.1 Isotropic bearings</u> (see Table 1, 2.1 and 2.2). In this case $Y_{nk} = X_{nk}$ and $b_n N_{ny} = a_n N_{nx}$ so that the two systems in (21) coincide, forming one system, e.g. (21a). If X_n is real (case 2.2), it is a well-known fact that all the balancing weights of a set are situated in one plane though the axis (Fig. 4).

If, on the other hand, X_n is complex (case 2.1) the balancing weights P_k are no longer situated in one plane through the axis, but are twisted (Fig. 5).

The degree of twist is smaller, the less the bearings are damped.

<u>1.6.2 Anisotropic bearings</u> (Table 1, 1.1 and 1.2). If X_n and Y_n are real (case 1.2), all balancing weights for X_n and Y_n are situated on one plane through the axis (Fig. 4) because $P_{ku}/P_{kv} = \text{constant}$. If, on the other hand, X_n and Y_n are complex (case 1.1) the balancing weights are twisted. (Fig. 5). Due to the anisotropy it is known that the N critical speeds Ω_n and the modes (of the round shaft in isotropic bearings) are split up into $2N$ critical speeds Ω_{nx} and Ω_{ny} and $2N$ modes. This means that the n-th mode of system (21a) has to be balanced near the critical speed Ω_{nx} and the n-th mode of system (21b) near the critical speed Ω_{ny} . Whether Ω_{nx} is larger or smaller than Ω_{ny} depends on the stiffnesses of the bearings in the two directions.

2. THE ANISOTROPIC (OVAL) SHAFT

2.1 Differential equation. In contrast to the round shaft, the oval shaft has two different principal moments of inertia I_u and I_v, the direction of the principal axes coinciding with the coordinates u and v of the rotating system (Fig. 6). The ratio of I_u to I_v is assumed to be

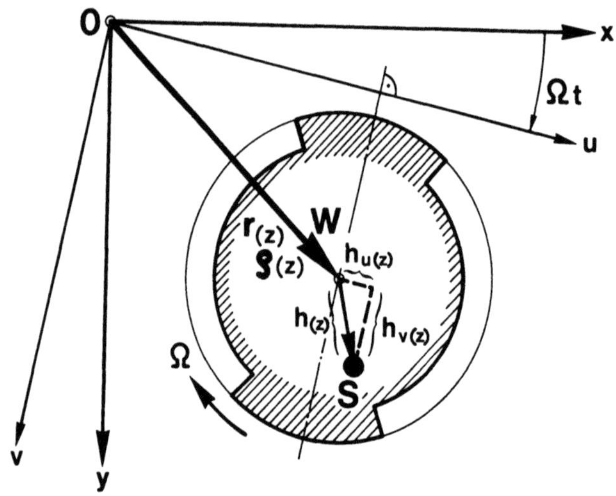

Fig. 6

constant over the length of the rotor. In addition, the same conditions are assumed to apply as for the round shaft. Using the abbreviations

$$B(z) = \frac{1}{2}\left[EI_u(z) + EI_v(z)\right]$$

(22)

$$\theta(z) = \frac{1}{2}\left[EI_u(z) - EI_v(z)\right]$$

the differential equation of the rotor revolving at constant angular velocity Ω in stationary coordinates $x = x(z , t)$, $y = y(z , t)$

$$\frac{1}{m(z)}\Big[B(z)x''\Big]'' + \frac{1}{m(z)}\Big[\theta(z)x''\Big]''\cos 2\Omega t + \frac{1}{m(z)}\Big[\theta(z)y''\Big]''\sin 2\Omega t$$

$$+\ddot{x} + (\mu+\nu)\dot{x} + \nu\Omega y = \mathcal{Re}\left(h(z)\Omega^2 e^{i\Omega t}\right)$$

(23)

$$\frac{1}{m(z)}\Big[B(z)y''\Big]'' - \frac{1}{m(z)}\Big[\theta(z)y''\Big]''\cos 2\Omega t + \frac{1}{m(z)}\Big[\theta(z)x''\Big]''\sin 2\Omega t$$

$$+\ddot{y} + (\mu+\nu)\dot{y} - \nu\Omega x = \mathcal{Im}\left(h(z)\Omega^2 e^{i\Omega t}\right)$$

In the special case of the round shaft $(I_u = I_v = I\ ,\ B = EI\ ,\ \theta = 0)$. Equation (23) coincides with equation (2b) for the round shaft. It is convenient to transform differential equation (23) to rotating coordinates. Using the transformation formula (1a) equation (23) becomes

(24)

$$\frac{1}{m(z)}\Big[EI_u(z)u''\Big]'' + \ddot{u} + (\mu+\nu)\dot{u} - 2\Omega\dot{v} - \Omega^2 u - \mu\Omega v = \Omega^2 h_u(z)$$

$$\frac{1}{m(z)}\Big[EI_v(z)v''\Big]'' + \ddot{v} + (\mu+\nu)\dot{v} + 2\Omega\dot{u} - \Omega^2 v + \mu\Omega u = \Omega^2 h_v(z)$$

2.2 End and bearing conditions. The end and bearing conditions have to be formulated in the rotating system of coordinates. The end and the first bearing condition are similar to (3) and (4a) if x and y are changed by u and v. The second bearing condition - after transforming the bearing forces into the rotating system - are in analogy to (4e)

(25)

$$(EI_u u'')'_R = (EI_u u'')'_L - F_u = (EI_u u'')'_L - M_{uu}u - M_{uv}v$$

$$(EI_v v'')'_R = (EI_v v'')'_L - F_v = (EI_v v'')'_L - M_{vu}u - M_{vv}v \qquad (z=a,b,\ldots H)$$

$$M_{uu} = Z_{xx}\cos^2\Omega t + (Z_{yx} + Z_{xy})\sin\Omega t\cos\Omega t + Z_{yy}\sin^2\Omega t$$

$$M_{uv} = (Z_{yy} - Z_{xx})\sin\Omega t\cos\Omega t + Z_{xy}\cos^2\Omega t - Z_{yx}\sin^2\Omega t$$

$$M_{vu} = (Z_{yy} - Z_{xx})\sin\Omega t\cos\Omega t - Z_{xy}\sin^2\Omega t + Z_{yx}\cos^2\Omega t$$

$$M_{vv} = Z_{xx}\sin^2\Omega t - (Z_{xy} + Z_{yx})\sin\Omega t\cos\Omega t + Z_{yy}\cos^2\Omega t$$

It can be seen from (25) that the transfer conditions become dependent on time. Such bearing conditions can no longer be solved with the formulas given in the following section 2.3 and 2.4.

The only possible means of support that does not vary with time depends on the condition that

$$\left.\begin{aligned} Z_{xy} = Z_{yx} = 0 \\ Z_{xx} = Z_{yy} = Z \end{aligned}\right\} \quad \underline{\underline{Z}} = \begin{vmatrix} Z & 0 \\ 0 & Z \end{vmatrix} \quad , \quad Z = K + i\omega B \tag{26}$$

This implies an *uncoupled, isotropic*, damped or undamped set of bearings as per Table 1 (2.2 and 2.1).

From (25) two uncoupled sets of bearing conditions, independent of one another are given by (26)

$$\begin{aligned} (EI_u u'')'_R &= (EI_u u'')'_L - Zu \\ (EI_v v'')'_R &= (EI_v v'')'_L - Zv \end{aligned} \qquad (z=a,b,\ldots H) \tag{25a}$$

<u>2.3 Solution of the homogeneous boundary value problem (free oscillation)</u>. For the undamped $(\mu = \nu = 0)$ and stationary $(\Omega = 0)$ shaft, equation (24) can be separated with the generalized coordinates

$$\begin{aligned} u(z\,,\,t) &= \sum_n U_n(z)w_{nu}(t) \\ v(z\,,\,t) &= \sum_n V_n(z)w_{nv}(t) \end{aligned} \tag{27}$$

and with the aid of two constants ω_{nu}^2 and ω_{nv}^2 which are unknown to begin with. This leads to the following systems

$$\text{(28a)} \quad \boxed{\begin{aligned} (EI_u U_n'')'' - m\omega_{nu}^2 U_n &= 0 \\ (EI_v V_n'')'' - m\omega_{nv}^2 V_n &= 0 \end{aligned}} \qquad \text{(28b)} \quad \boxed{\begin{aligned} \ddot{w}_{nu} + \omega_{nu}^2 w_{nu} &= 0 \\ \ddot{w}_{nv} + \omega_{nv}^2 w_{nv} &= 0 \end{aligned}}$$

The general solutions of (28b) are known to be

$$\text{(29)} \quad \begin{aligned} w_{nu} &= w_{nuo} e^{i(\omega_{nu} t + \varepsilon_{nu})} = w_{nuo} e^{i\varepsilon_{nu}} e^{i\omega_{nu} t} \\ w_{nv} &= w_{nvo} e^{i(\omega_{nv} t + \varepsilon_{nv})} = w_{nvo} e^{i\varepsilon_{nv}} e^{i\omega_{nu} t} \end{aligned}$$

ω_{nu} and ω_{nv} are obviously the real or complex angular frequencies of the flexural vibrations in the (u, z) and the (v, z) planes, i.e. in the "soft" and "hard" principal axes of the shaft cross-sections. Since the deflections (29) vary exponentially with time, the bearing conditions are independent of time.

Since neither the differential equations (28a) nor the end and bearing conditions are coupled in U_n and V_n, equations (28a) can be calculated independently. The solution in this case is also most conveniently obtained by using a matrix transfer method. The angular frequencies ω_{nu} and ω_{nv} as well as the modes U_n and V_n are assumed to be *known* in what follows. With damped bearings the angular frequencies and modes become complex. With uncoupled bearing conditions the modes U_n and V_n each become separately orthogonal. Therefore the condition for orthogonality (10) applies in the appropriate sense, (x and y changed by u and v).

<u>2.4 Forced movement due to unbalance of the oval shaft in uncoupled isotropic bearings</u>. In the terms on the right-hand side of equation (24) are the expressions $h_u(z)$ and $h_v(z)$. The product $m(z)h(z) = m(z)h_u(z) + im(z)h_v(z)$ is

the unbalance of the shaft, constant with respect to time in the rotating system. We regard it as the sum of a continuous unbalance $p(z) = p_u(z) + ip_v(z)$ plus the sum of K discrete unbalance "forces" $P_k = P_{ku} + iP_{kv}$ at the points $z = z_k$, $(k=1,2,...K)$. The individual unbalance "forces" are used as balancing weights to compensate for the existing continuous unbalance of the shaft. Then

$$(30)\qquad \begin{aligned} m(z)h_u(z) &= p_u(z) + \sum_k^K P_{ku}\delta(z - z_k) \\ m(z)h_v(z) &= p_v(z) + \sum_k^K P_{kv}\delta(z - z_k) \end{aligned}$$

with δ as "delta function". The continuous unbalances $p_u(z)$ and $p_v(z)$ are assumed to be expressed by the sum of modal functions $U_n(z)$ and $V_n(z)$, in other words we may write

$$(31)\qquad \begin{aligned} p_u(z) &= \sum_n m(z)a_nU_n(z) \\ p_v(z) &= \sum_n m(z)b_nV_n(z) \end{aligned} \qquad (n=1,2,...N)$$

The number N depends on the unbalance distribution.

Equation (24) can be separated with the generalized coordinates in the rotating system

$$(32)\qquad \begin{aligned} u(z, t) &= \sum_n U_n(z)q_{nu}(t) \\ v(z, t) &= \sum_n V_n(z)q_{nv}(t) \end{aligned}$$

and with the aid of the two constants ω_{nu}^2 and ω_{nv}^2, at present unknown. This gives

$$(33a)\qquad \boxed{\begin{aligned} (EI_uU_n'')'' - m\omega_{nu}^2U_n &= 0 \\ (EI_vV_n'')'' - m\omega_{nv}^2V_n &= 0 \end{aligned}}$$

$$(33b)\quad \begin{array}{l} \ddot{q}_{nu} + (\mu+\nu)\dot{q}_{nu} - 2\Omega\dfrac{N_{nuv}}{N_{nu}}\dot{q}_{nv} + (\omega_{nu}^2 - \Omega^2)q_{nu} - \mu\Omega\dfrac{N_{nuv}}{N_{nu}}q_{nv} = \\ \dfrac{\Omega^2}{N_{nu}}\left(a_n N_{nu} + \sum\limits_k^K P_{ku}U_{nk}\right) \\ \\ \ddot{q}_{nv} + (\mu+\nu)\dot{q}_{nv} + 2\Omega\dfrac{N_{nuv}}{N_{nv}}\dot{q}_{nu} + (\omega_{nv}^2 - \Omega^2)q_{nv} + \mu\Omega\dfrac{N_{nuv}}{N_{nv}}q_{nu} = \\ \dfrac{\Omega^2}{N_{nv}}\left(b_n N_{nv} + \sum\limits_k^K P_{kv}V_{nk}\right) \end{array}$$

Therein use was made of the orthogonality of the modes U_n and V_n .

A particular integral for the n-th term in system (33b) can be written as follows

$$(34)\quad \begin{array}{l} q_{nu}(t) = A_n = \dfrac{1}{Det}\left[\dfrac{\Omega^2}{N_{nu}}(\omega_{nv}^2 - \Omega^2)\left(a_n N_{nu} + \sum\limits_k^K P_{ku}U_{nk}\right) + \right. \\ \left. \dfrac{\mu\Omega^3 N_{nuv}}{N_{nu}N_{nv}}\left(b_n N_{nv} + \sum\limits_k^K P_{kv}V_{nk}\right)\right] = q_{nu} \\ \\ q_{nv}(t) = B_n = \dfrac{1}{Det}\left[\dfrac{\Omega^2}{N_{nv}}(\omega_{nu}^2 - \Omega^2)\left(b_n N_{nv} + \sum\limits_k^K P_{kv}V_{nk}\right) - \right. \\ \left. \dfrac{\mu\Omega^3 N_{nuv}}{N_{nu}N_{nv}}\left(a_n N_{nu} + \sum\limits_k^K P_{ku}U_{nk}\right)\right] = q_{nv} \end{array}$$

with the determinant

$$Det = (\omega_{nu}^2 - \Omega^2)(\omega_{nv}^2 - \Omega^2) + \mu^2\Omega^2\frac{N_{nuv}^2}{N_{nu}N_{nv}} \neq 0$$

and the abbreviation

$$U_{nk} = U_n(z_k) \quad ; \quad V_{nk} = V_n(z_k)$$

Comparing (33a) with (28a) we find that ω_{nu}^2 and ω_{nv}^2 represent the real or complex eigenvalues and $U_n(z)$ and $V_n(z)$ the real or complex modes of the flexural vibrations of the shaft, damped only in the bearings (or not at all, as the case may be) in the (u, z) and (v, z) plane. Thus eigenvalues and modes can be calculated according to 2.3 and regarded as *given*. They are independent of internal and external shaft damping, but they depend on any bearing damping that may be present.

With the aid of (34), solution (32) for the n-th term in the rotating system becomes

$$(35)\qquad \begin{aligned} u_n(z, t) &= u_n(z) = A_n U_n(z) \\ v_n(z, t) &= v_n(z) = B_n V_n(z) \end{aligned}$$

With damped bearings U_n, V_n and A_n, B_n become complex. Then for u_n only the real part and for v_n only the imaginary part applies

$$\begin{aligned} u_{np}(z) &= Re\left[A_n U_n(z)\right] \\ v_{np}(z) &= Jm\left[B_n V_n(z)\right] \end{aligned}$$

and in the stationary system of coordinates using the transformation formula (1a)

$$(36)\qquad \begin{aligned} x_p(z,t) &= \sum_n x_{np}(z,t)\ , \quad x_{np}(z,t) = u_{np}(z)\cos\Omega t - v_{np}(z)\sin\Omega t \\ y_p(z,t) &= \sum_n y_{np}(z,t)\ , \quad y_{np}(z,t) = u_{np}(z)\sin\Omega t + v_{np}(z)\cos\Omega t \end{aligned}$$

which, when z is constant, is the parametric representation of a *circle*.

2.5 Balancing the oval shaft in isotropic bearings. The shaft is balanced when the deflections $x(z, t)$ and $y(z, t)$ vanish, with the result that there are no longer any rotating forces transferred to the bearings. According to (36) and (34) the condition for this is similar to that for the round shaft

$$(37) \qquad a_n N_{nu} + \sum_k^K P_{ku} U_{nk} = 0$$

$$b_n N_{nv} + \sum_k^K P_{kv} V_{nk} = 0$$

In matrix form and resolved into the individual terms, the condition for balancing is

$$(38a) \qquad \begin{array}{c} \\ | \\ n \\ \downarrow \\ N \end{array} \begin{vmatrix} U_{11} & U_{12} & U_{13} & \cdots \\ U_{21} & U_{22} & U_{23} & \\ U_{31} & U_{32} & U_{33} & \\ \vdots & & & \end{vmatrix}^{\;\overset{k \;\longrightarrow\; K}{}} \begin{vmatrix} P_{1u} \\ P_{2u} \\ P_{3u} \\ \vdots \end{vmatrix} = - \begin{vmatrix} a_1 N_{1u} \\ a_2 N_{2u} \\ a_3 N_{3u} \\ \vdots \end{vmatrix}$$

$$\left(a_n N_{nu} = \int_0^{\ell} P_u(z) U_n(z) dz \right)$$

$$(38b) \qquad \begin{array}{c} \\ | \\ n \\ \downarrow \\ N \end{array} \begin{vmatrix} V_{11} & V_{12} & V_{13} & \cdots \\ V_{21} & V_{22} & V_{23} & \\ V_{31} & V_{32} & V_{33} & \\ \vdots & & & \end{vmatrix}^{\;\overset{k \;\longrightarrow\; K}{}} \begin{vmatrix} P_{1v} \\ P_{2v} \\ P_{3v} \\ \vdots \end{vmatrix} = - \begin{vmatrix} b_1 N_{1v} \\ b_2 N_{2v} \\ b_3 N_{3v} \\ \vdots \end{vmatrix}$$

$$\left(b_n N_{nv} = \int_0^{\ell} P_v(z) V_n(z) dz \right)$$

Again we obtain two linear systems of equations with N equations for K unknown weights P_{ku} and P_{kv}. For a unique solution it is again necessary that $K = N$, i.e. the number of weights and planes must be the same as the number of modes to be balanced. The set of weights for the n-th mode is obtained by equating all terms on the right-hand side of (38), except $a_n N_{nu}$ or $b_n N_{nv}$, to zero.

Only the relative weights can be calculated $P_{1u} : P_{2u} : \ldots$, $P_{1v} : P_{2v} : \ldots$, these being independent of the absolute magnitude of the unbalance and pure functions of the modes. The absolute magnitude and angular relationship of the weights have to be determined by measurement, this being carried out as near as possible to the critical speed at which the n-th mode is most pronounced.

Owing to the shaft being oval, it is known that the critical speeds Ω_n (of the round shaft in isotropic bearings) are split up into two sets of critical speeds Ω_{nu} and Ω_{nv} of the oval shaft, where $\Omega_{nu} < \Omega_{nv}$ (the u-direction being "softer" than the v-direction, see Fig. 7). Equation (38a) applies for balancing the modes at Ω_{nu} for the "softer" direction and equation (38b) for balancing the modes at Ω_{nv} .

If the difference between the flexural rigidities EI_u and EI_v is not too large, twice as many modes have to be balanced in the same operating range as for a round shaft.

With damping in the bearings, the eigenfunctions $U_n(z)$ and $V_n(z)$ become complex arising out of the complex calculation of flexural vibrations in the u- and v-directions. Despite the complex coefficients in the two systems of equations (38), the balancing weights P_{ku} , and similarly P_{kv} , are co-planar even with damping in the bearings. This is in contrast to a round shaft.

Also noteworthy is that the angular relationship of the balancing weights is determined by the geometry of the rotor. With P_{ku} is in the u-direction balancing being performed at Ω_{nu} only for $p_u(z)$; if it is in the v-direction, only $p_v(z)$ at Ω_{nv} is balanced, see Fig. 7. This is a well-known state of affairs for rotors with only one degree of freedom and supported in "ideal" bearings.

In the rotors of turbo-generators the flexural rigidities are largely evened out by constructional precautions, also the bearings of the balancing pit are largely

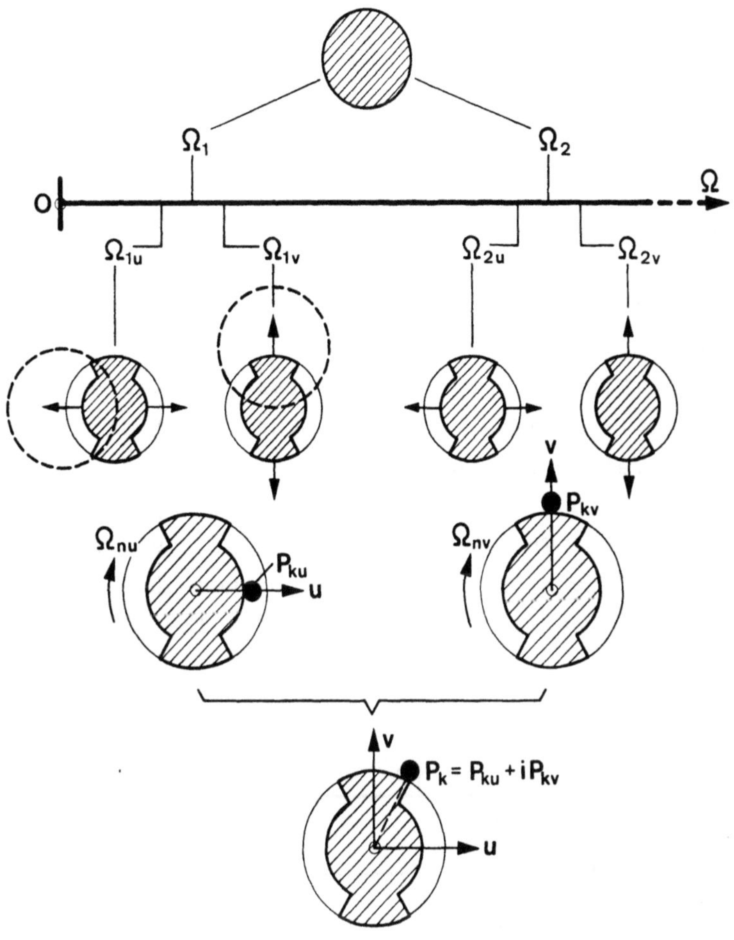

Fig. 7

isotropic. The two values of the critical speeds Ω_{nu} and Ω_{nv} are then very close together and there is hardly any difference between the coefficients U_{nk} and V_{nk} and the matrix for systems (38a) and (38b), so that may be combined to form a single system. In this case a round shaft is balanced.

3. SOME REMARKS ON BALANCING PRACTICE

To begin with, one will consider how many modes (critical speeds) have to be balanced in the operating range. In some cases it may be advantageous also to balance the next mode (critical speed) following the maximum service speed. Higher modes than this can practically be neglected. The number N of the modes is then fixed.

With N modes, $K = N$ balancing planes have to be prepared (see Fig. 8a, $K = 4$) . A start is made with balancing the first mode $(n = 1)$. System (21) or (38) provides the relative weights (four in all) such that the higher-order modes $(n = 2,3,\ldots N)$ are not excited by these balancing weights. The measurement nearest the first critical speed Ω_1 , gives the absolute magnitude and angular relationship of the first set of weights. The first mode

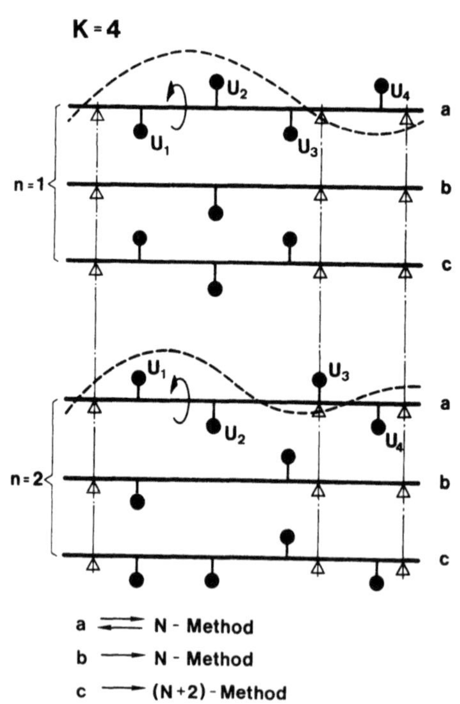

Fig. 8

has been balanced. Then the speed is raised until it is near the critical speed Ω_2 for the second mode the procedure adopted being similar for this and subsequent modes. Of course, it is assumed that the critical speeds at which balancing is performed are not in the unstable region. This applies particularly to oval shafts.

This procedure can be simplified by adopting the method of *"balancing forwards"*: This means that N balancing planes are made available, a start being made with the first mode. Deliberately no attention is paid to higher-order modes, so that only the first equation of (21) or (38) is compulsory for the first set of weights. A single weight is inserted, preferably at the point when the maximum deflection of the first mode occurs (Fig. 8b). Thus the first mode is balanced. The same procedure is adopted for the second mode, no attention being paid to the modes $n > 2$, but the already balanced first mode must not be excited. Thus for the second set of weights the first two equations in (21) or (38) become decisive. Two planes are chosen, preferably where the deflection of the second mode is large, the rotor being balanced with two weights (Fig. 8b). This procedure is then continued until the N-th mode has been balanced.

However, both methods break down when the first mode is excited to such an extent that it is impossible to reach the first critical speed. Then one is forced to balance the rotor at a lower speed. But at such speeds the rotor hardly bends at all and is to all intents and purposes a "rigid" rotor. The conditions for balancing a rigid rotor (equilibrium of forces and moments) are known to be

$$\sum_k^K P_k = \int_0^\ell p(z)dz$$

(39)

$$\sum_k^K \frac{z_k}{\ell} P_k = \frac{1}{\ell}\int_0^\ell zp(z)dz$$

These are two additional linear equations for the unknown balancing weights P_k. If they are included in system (21) or (38), equation (21a) then reads (and likewise 21b, 38a, 38b):

$$
\text{(21c)}\qquad
\begin{array}{c}
\\ \\ n+2 \\ \downarrow \\ N+2 \\ \\
\end{array}
\overset{\longleftarrow\; k+2 \;\longrightarrow \qquad K+2}{\begin{Vmatrix}
1 & 1 & 1 & \cdots \\
z_1/\ell & z_2/\ell & z_3/\ell & \\
X_{11} & X_{12} & X_{13} & \\
X_{21} & X_{22} & X_{23} & \\
X_{31} & X_{32} & X_{33} & \\
\vdots & & &
\end{Vmatrix}}
\begin{vmatrix}
P_1 \\ P_2 \\ P_3 \\ P_4 \\ P_5 \\ \vdots
\end{vmatrix}
= -
\begin{vmatrix}
\int_0^\ell p(z)\,dz \\
\frac{1}{\ell}\int_0^\ell z\,p(z)\,dz \\
a_1 N_{1x} \\
a_2 N_{2x} \\
a_3 N_{3x} \\
\vdots
\end{vmatrix}
$$

Then, obviously, $(N+2)$ balancing planes and weights are needed, i.e. two planes more than before. At least three weights are required to balance the first mode, in order not to affect the "rigid balancing" (Fig. 8c). For the second mode four planes are needed - this being the well-known $(N+2)$ method [9], [10], [11], [14].

It can be shown [14] that the condition (39) is satisfied by (21a), provided all modes excited by the unbalance are balanced. The number of excited modes is, however, generally much larger than the number of modes (N) which have to be balanced in the operating range, so that condition (39) is only partly fulfilled. From this it follows that the rather laborious system (21c) for many unbalance distributions ensures more accurate balancing than the simpler system (21a) including those occasions when the first mode is attained at the first attempt.

4. CONCLUSIONS

The process of "modal balancing" can also be extended to rotors in damped bearings if the mode shapes are referred to the undamped shaft in damped bearings. Damped bearings result in complex modal functions, which are no longer situated in a single plane through the axis of the rotor, but are slightly twisted. The amount of twist is smaller the less the bearings are damped. Due to use of complex algebra the computation of mode shapes and sets of weights is cumbersome.

One may ask whether this additional effort is worth while, in view of the accuracy attainable by balancing, or whether it is technically inevitable. This question will have to be answered by practical experience. Also worth bearing in mind is the fact that coupled bearing conditions, such as accompany the oil film, cannot be dealt with by this method.

It would thus appear that this is a limit to a method which is systematical, clear and easy to use. A method which, paying no attention to bearing damping and coupling, has provided its worth in practice.

REFERENCES

[1] BISHOP, R.E.D. and GLADWELL, G.M.L.: "The vibration and balancing of an unbalanced flexible rotor", J. Mech. Engineering Science, Vol. 1 (1959).

[2] CHURCH, A.H. and PLUNKETT, R.: "Balancing flexible rotors", Trans. ASME, J. of Engineering for Industry, (1961), pp. 383-389.

[3] PARKINSON, A.G. and BISHOP, R.E.D.: "Vibration and balancing of rotating continuous shafts", J. Mech. Engr. Science, Vol. 3, No. 3, (1961).

[4] BISHOP, R.E.D. and PARKINSON, A.G.: "On the isolation of modes in the balancing of flexible shafts", Proc. Instn. Mech. Engrs., Vol. 177, No. 16, (1963).

[5] KELLENBERGER, W.: "Die stabilität rotierender wellen infolge innerer und äusserer dämpfung", Ing. Archiv 32, (1963) 5.

[6] LINDLEY, A.L.G. and BISHOP, R.E.D.: "Some recent research on the balancing of large flexible rotors", Proc. Instn. Mech. Engrs., Vol. 177, No. 30, (1963).

[7] PARKINSON, A.G.: "The vibration and balancing of shafts rotating in asymmetric bearings", J. Sound Vib., (1965) 2 (4), pp. 477-501.

[8] HUNDAL, M.S. and HARKER, R.J.: "Balancing of flexible rotors having arbitrary mass and stiffness distribution", Trans. ASME, J. of Engineering for Industry, (1966), pp. 217-223.

[9] KELLENBERGER, W.: "Balancing flexible rotors on two generally flexible bearings", Brown Boveri Review, (1967).

[10] MIWA, S.: "Theoretical consideration on general condition of balancing a flexible rotor", Japan Society of Mech. Engineers, Meeting Sept. 1968.

[11] MIWA, S.: "balancing of a flexible rotor by means of mode separation", Japan Society of Mech. Engineers, Meeting Nov. 1969.

[12] KELLENBERGER, W.: "Balancing flexible turbo-rotors with the aid of the theory of orthogonal functions", CIGRE 1970, Report 11-10.

[13] RIEGER, N.F.: "Unbalance response of an elastic rotor in damped flexible bearings at supercritical speeds", Trans. ASME, J. of Engineering for Power, April 1971, pp. 265-278.

[14] KELLENBERGER, W.: "Should a flexible rotor be balanced in N or (N+2) planes?", J. of Engineering for Industry, Trans. ASME, Serie B 94, 2, pp. 548-560, May 1972.

[15] LUND, J.W. and TONNESEN, J.: "Analysis and experiments on multi-plane balancing of a flexible rotor", Trans. ASME, J. of Engineering for Industry, 1972, pp. 233-242.

[16] OMORI, T., IIDA, S. and IWATSUBO, T.: "Balancing of flexible rotors", Bulletin of the JSME, Vol. 15, No. 87, (1972).

[17] LUND, J.W.: "Modal response of a flexible rotor in fluid-film bearings", J. of Engineering for Industry, Trans. ASME, paper No. 73-DET-98, 1973.

[18] PARKINSON, A.G.: "Balancing of flexible shafts rotating in massive flexible bearings", J. Mech. Engr. Science, Vol. 15, No. 6, (1973).

[19] KELLENBERGER, W.: "Biegeschwingungen einer unrunden, rotierenden welle in horizontaler lage", Ing. Archiv 26 (1958) 4.

[20] BISHOP, R.E.D. and PARKINSON, A.G.: "Second order vibration of flexible shafts", Phil. Trans. Roy. Soc. London, 259 (1965), No. 1095.

[21] IWATSUBO, T. and MAKAMURA, M.: "Balancing of flexible rotors with asymmetric shaft stiffness", Memoirs of the faculty of Engineering, Kobe University, Japan, No. 15 (1968).

[22] BUCHSER, B.: "Die rotierende unrunde welle als stabilitätsproblem eines linearen periodischen systems", ZAMP, Vo. 23 (1972).

[23] IWATSUBO, T., TOMITA, A. and KAWAI, R.: "Vibrations of asymmetric rotors supported by asymmetric bearings", Ing. Archiv, 42 (1973), pp. 416-432.

The Effect of Support Flexibility on the Stability of Rotors Mounted in Plain Cylindrical Journal Bearings

R.G. Kirk

Pratt & Whitney Aircraft, U.S.A.

P. De Choudhury

Elliott Corporation, U.S.A.

E.J. Gunter

University of Virginia, U.S.A.

ABSTRACT - This analysis treats the effect of support damping and flexibility on the response characteristics of a symmetric rotor mounted in journal bearings. The successful operation of turbomachinery utilizing fluid-film bearings is shown to be very dependent upon the bearing support structure for high speed applications. A stability analysis is discussed based upon the application of both the Routh criteria and the calculation of damped eigenvalues. Results are plotted for typical rotor support properties. The analysis also discusses the influence of rotor imbalance on the stability of rotor systems, and time transient orbits of shaft and support motions clearly indicate their dynamic behavior for both the stable and unstable operating speeds. The special case of a vertical or zero-g rotor system is discussed and transient response orbits indicate that it is possible to achieve successful operation only for the case of an unbalanced rotor. The analysis of damped eigenvalues indicates three modes of instability when rotor flexibility is included in the equations of motion. In addition to the classical half-frequency whirl, both support resonant whip and resonant shaft whip are identified in an example case study.

NOMENCLATURE

Symbol	Definition	Dimension
a	Growth factor	(T^{-1})
c	Journal bearing radial clearance	(L)
C_{xx}, C_{yy}	Bearing direct damping	(FTL^{-1})
C_{xy}, C_{yx}	Bearing cross coupled damping	(FTL^{-1})
C_{1xx}, C_{1yy}	Support direct damping	(FTL^{-1})
C_{1xy}, C_{1yx}	Support cross coupled damping	(FTL^{-1})
$\bar{C}_B$	Support damping	(DIM.)
D	Journal bearing diameter	(L)

e_μ	Imbalance of rotor/m	(L)
E_μ	Imbalance ratio, $= e_\mu/c$	(DIM.)
F_x, F_y	Journal fluid-film force	(F)
g	Gravitational constant	(LT^{-2})
h	Film thickness	(L)
H	Film thickness ratio, = h/c	(DIM.)
$\bar{K}_B$	Support stiffness	(DIM.)
K_S	Shaft stiffness	(FL^{-1})
K_{xx}, K_{yy}	Bearing direct stiffness	(FL^{-1})
K_{xy}, K_{yx}	Bearing cross coupled stiffness	(FL^{-1})
K_{1xx}, K_{1yy}	Support direct stiffness	(FL^{-1})
K_{1xy}, K_{1yx}	Support cross coupled stiffness	(FL^{-1})
L	Journal bearing length	(L)
m_2	Rotor mass	(FT^2L^{-1})
m_1	Support mass	(FT^2L^{-1})
m	One half equivalent rotor mass, $m_j+m_2/2$	(FT^2L^{-1})
m_j	Journal bearing mass	(FT^2L^{-1})
δ_m	Mass ratio, = ml/m	(DIM.)
N	Rotor spin, $\omega/2\pi$	(REV/T)
R	Journal radius	(L)
S	Sommerfeld number	(REV.)
S_s	Ocvirk number, $= S(L/D)^2$	(REV.)
w	Parameter $= \dfrac{\mu R L^3 \omega}{2c^2}$	(FL^{-2})
W	Bearing loading	(F)
$\bar{W}$	Load number $\equiv \dfrac{2\pi W}{\mu L D}\left(\dfrac{2c}{L}\right)^2\sqrt{\dfrac{mc}{W}}$	$(REV.^{-1})$
x_1, y_1	Support displacement	(L)
x_2, y_2	Rotor absolute displacement	(L)
x_j, y_j	Journal relative motion	(L)
X, Y	Displacement ratio, = x/c, y/c	(DIM.)
X_o, Y_o	Journal equilibrium position	(DIM.)
ε	Journal eccentricity ratio	(DIM.)
ε_o	Journal equilibrium eccentricity	(DIM.)
λ	Complex eigenvalue	(T^{-1})

μ	Viscosity of lubricant	(FTL^{-2})
ϕ	Attitude angle	(Rad.)
ω_b	Bearing spin rate	(T^{-1})
ω_j	Rotor spin rate	(T^{-1})
ω	Sum of $\omega_j + \omega_b$	(T^{-1})
ω_s	Stability parameter, $= \omega\sqrt{\frac{mc}{W}}$	(DIM.)
Ω	Whirl rate ratio, $= \dot{\phi}/\omega_j$	(DIM.)
CL	Journal clearance	(L)
CBX, CBY	Support damping	(FTL^{-1})
EMU	Journal eccentricity ratio, $\equiv E_\mu$	(DIM.)
ES	Journal equilibrium eccentricity ratio, $\equiv \varepsilon_o$	(DIM.)
ESU	ε_o calculated for LOAD = FU	(DIM.)
FO	Rotating load	(F)
FMAX	Maximum fluid-film force	(F)
FMAXB	Maximum force to bearing	(F)
FMAXS	Maximum force to support	(F)
FU	Imbalance load	(F)
FURATIO	Imbalance ratio, FU/W	(DIM.)
MU@5	Viscosity of lubricant $(x10^5)$	(FTL^{-2})
SS	Ocvirk number	(Rev.)
TRDMAX	Maximum dynamic load/me μ ω_j^2	(DIM.)
TRSMAX	Maximum film load/W	(DIM.)
W	Bearing load from rotor	(F)
WB	Bearing weight	(F)
WJ	Journal weight	(F)
WS	Stability parameter, ω_s	(DIM.)
WT	(Total load on bearing) / W	(DIM.)

1. INTRODUCTION

The dynamics of rotor bearing systems has become of extreme importance due to the increased speed and power level requirements of advanced design rotating equipment. The aircraft jet engine industry has as their ultimate goal a lighter, more powerful, more efficient, and more dependable power plant. Compressor and industrial gas turbine manufacturers are concerned with reduced development costs, high reliability, and low maintenance costs over extended useful service life. These factors are responsible for the interest in and growth of the study stability and forced response of rotor bearing systems. Power plants for ground installation and out space are relying on fluid-film bearing to provide long, maintenance free service. To avoid the classical instability associated with plain cylindrical bearings the designers have used multi-lobe, tilting pad, pressure dam, and other designs such as herringbone groove to increase the stable speed range of their equipment.

Jet engine manufacturers are designing advanced engines with anti-friction bearings in series with fluid-film dampers to reduce engine sensitivity to imbalance. In qualitative terms, the same system parameters which promote optimum forced response characteristics are also conducive to an increased stable operating speed range. While this is true in general, it is extremely important for the dynamist to fully understand the cause and nature of instabilities associated with fluid-film bearings, internal friction damping, and aerodynamic cross-coupling. The present discussion will be concerned with only those instabilities associated with fluid-film bearings. The instability associated with the plain cylindrical journal bearing has received considerable attention in the literature [1, 2, 3, 4, 5]. PINKUS [6] reported that flexible mountings gave greater stability to rotor bearing systems while PORITSKY [7] and HAGG and WARNER [8] found that support flexibility lowered the stability threshold speed. GUNTER [9] explained the discrepancy when it was shown that a symmetrical undamped support lowered the threshold speed whereas damped and/or asymmetric supports can greatly increase the stability of a rotor system.

HORI [10] presented an analysis of a long bearing with negative fluid film pressures neglected (to model cavitation) which indicated that the journal was not inherently unstable at all speeds as ROBERTSON [3] has shown. Hori's analysis was in agreement with practical experience and all future investigations have included the cavitation effects unless supply pressures are in excess of the peak film pressures in which case the film does not cavitate. HORI concluded that the vertical balanced journal would be unstable at all speeds. This fact will be demonstrated in the following analysis and the results of an investigation of the influence of imbalance on the stability of vertical rotors will be discussed.

An investigation by BADGLEY [11] reported on a time transient analysis method of obtaining the stability threshold speed of short, long and finite bearings. The orbits of a balanced horizontal rotor on rigid supports were examined for various methods of perturbation to determine the influence on the stability boundary. He indicated that the threshold speed at high eccentricities is reduced by large initial velocities. (The results obtained for the short bearings are shown in Fig. (6) of this report). LUND [12] examined the stability of a flexible rotor with damped supports and concluded that damped flexible supports can considerably increase the threshold speed. Stability curves for specific rotor systems were presented for rotors supported in cylindrical gas bearings. GUNTER [13] presented a linear analysis of the influence of damped supports on the response characteristics of a rigid rotor including gyroscopic effects. The analysis of a typical rotor system indicated significant reductions of forces transmitted for damped bearing supports.

Recent investigations of rotor stability [14, 15] have extended the rotor models to general multi-mass flexible rotors with damped bearings. These analyses are generally intended to study complex rotor systems and will add enormous capability to the rotor dynamist in future studies of stability.

The purpose of the following discussion is to present the results of an extensive study of the influence of damped flexible supports on the stability threshold speed of a rotor in nonlinear fluid film bearings [16, 17]. Stability maps are presented in dimensionless form for a range of support stiffness, damping, and mass ratios. Numerous time transient response orbits will indicate the nature of stable and unstable response orbits for both balanced and unbalanced systems. Examination of the complex eigenvalues for a specific rotor system will give further insight to explain why rotors may pass through an unstable speed range and restabilize, only to go unstable at some higher speed in the classical "half-frequency whirl".

When shaft flexibility is included in the analysis another mode of instability is introduced which corresponds to the condition of "resonant whip" as reported by NEWKIRK [18]. This phenomena can be shown analytically by examination of damped eigenvalues. The examination of the real part of the eigenvalue gives not only an indication of whether the system is stable or unstable, but also an accurate indication of the rate of growth of small perturbation. This additional information has proven to be extremely useful in giving added insight to the cause and nature of rotor-bearing instability.

2. DESCRIPTION OF ROTOR BEARING SYSTEM

2.1 Rotor and Support System. The following discussion will present the analysis of a symmetric rotor system such as the one shown in Fig. 1. The rotor is considered to be supported upon nonlinear fluid-film bearings which have a length to diameter ratio of less than one (i.e., $L/D < 1.0$). The bearings are therefore modeled by the short bearing approximation (19, 22). A derivation of the fluid-film forces expressed in fixed cartesian coordinates is presented in the following section. For the stability analysis the rotor is assumed to be perfectly balanced and supported on flexible, damped pedestals as indicated in Fig. 1. The transient re-

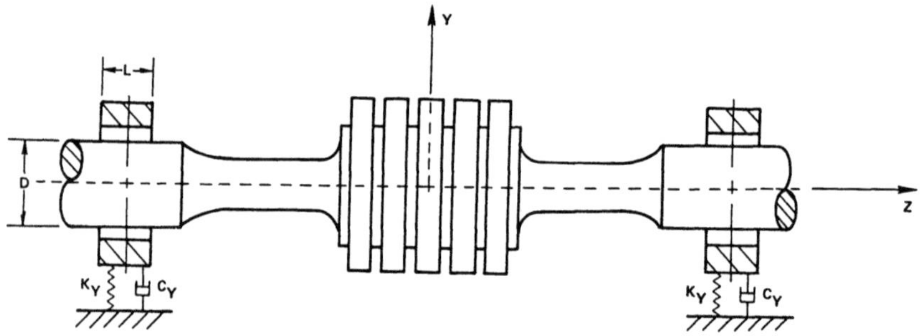

Fig. 1. Typical Dynamics Model for a Symmetric Rotor with Flexible Supports.

sponse analysis assumes the rotor to have an effective imbalance located at midspan and hence the rotor and support can only be excited in the cylindrical modes. The conical mode is not considered in the following analysis. Considering these restrictions, the rotor system may be fully described for dynamic simulation by the coordinates as shown in Fig. 2. The support motion is given by (x_1, y_1), the bearing journal relative motion is denoted by (x_j, y_j), and the rotor absolute motion is described in the fixed coordinate system by (x_2, y_2). The equations of motion of the system may therefore be written as six coupled second order differential equations with the fluid-film bearing forces ex-

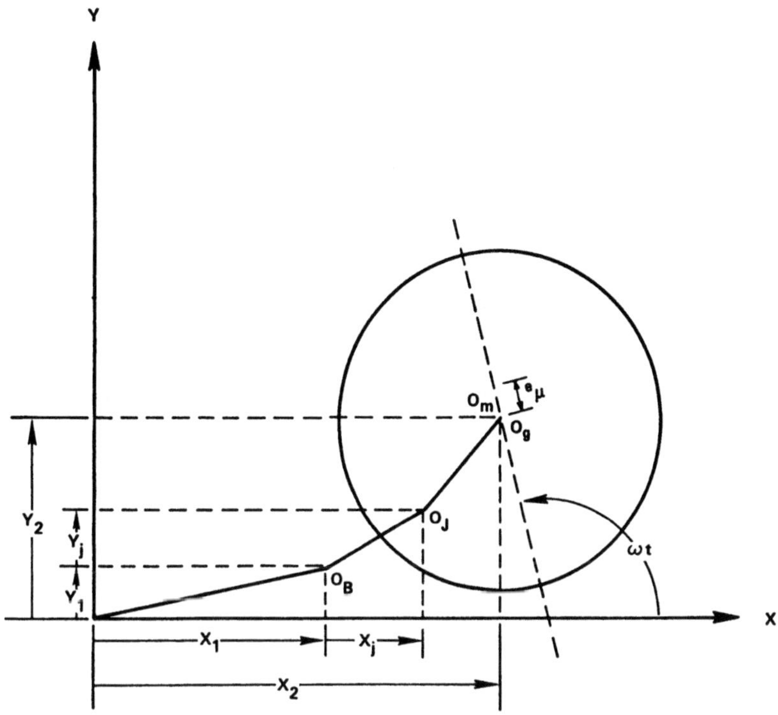

Fig. 2. Cross-Section of Deflected Rotor at Time ωt Showing Displacement Nomenclature.

pressed as a nonlinear function of the journal relative displacement and velocity. If the static deflection of the rotor midspan is much less than the journal clearance $(\delta/c<1.0)$, the rotor may be considered to be rigid and hence $(x_2, y_2)= (x_1 + x_j, y_1 + y_j)$. This reduces the analysis to the system shown in Fig. 3. The fluid-film will be considered to cavitate when the fluid pressure drops below ambient. The support system is considered to be linear but the stability analysis may be applicable to rotors with nonlinear squeeze film damper supports which have preloaded hairsprings in parallel with the damper. The damping characteristics of such a support system may be found in the literature (13,17).

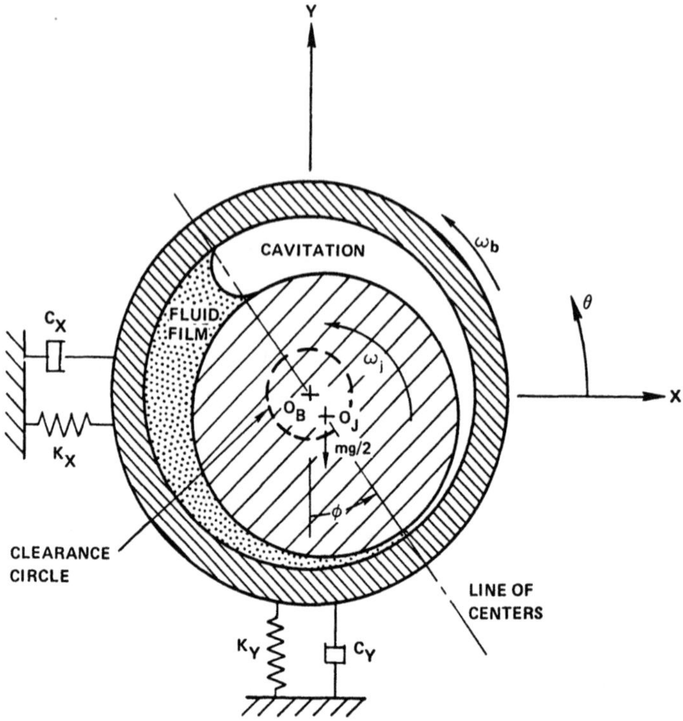

Fig. 3. Cross-Section of Rigid Rotor System Showing Fluid-Film and Region of Cavitation.

2.2. Fluid Film Bearing Characteristics. The nonlinear fluid-film forces generated by the journal bearings may be derived by considering the solution to the general Reynolds equation. Expressed in terms of a fixed cartesian coordinate system, the Reynolds equation may be written as follows:

$$(1)\qquad \frac{\partial}{\partial z}\left[\frac{h^3}{6\mu}\frac{\partial p}{\partial z}\right] + \frac{1}{R^2}\frac{\partial}{\partial \theta}\left[\frac{h^3}{6\mu}\frac{\partial p}{\partial \theta}\right] = (\omega_b + \omega_j)\frac{\partial h}{\partial \theta} + 2\frac{\partial h}{\partial t}$$

where

$h = c - x_j \cos(\theta) - y_j \sin(\theta)$

θ = angle measured from positive X-axis in direction of rotation

ω_j = journal spin rate

ω_b = bearing spin rate

The standard solution technique for bearings having $L/D < 1$ is to neglect the term $\frac{\partial p}{\partial \theta}$ in Eq. (1). This permits an approximate solution to be written for the pressure profile. Considering ambient pressure at the bearing axial boundaries, the pressure profile is given as:

$$p(\theta,z)= \frac{3\mu z(z-L)}{h^3}\left[(\omega_b+\omega_j)(x_j\sin(\theta)-y_j\cos(\theta)) - 2(\dot{x}_j\cos(\theta)+\dot{y}_j\sin(\theta))\right] \quad (2)$$

The components of force on the journal in the x-y coordinate system may be found by integrating the pressure profile over the bearing surface. This is expressed as:

$$(3a) \quad \begin{cases} F_x = \int_{z=0}^{L}\int_{\theta=0}^{2\pi} p(\theta,z)R\cos(\theta)d\theta dz \\ F_y \int_{z=0}^{L}\int_{\theta=0}^{2\pi} p(\theta,z)R\sin(\theta)d\theta dz \end{cases}$$

Integrating the axial component, the nonlinear forces generated by the journal bearing may be expressed by the following integral:

$$(3b) \quad \begin{Bmatrix} F_x \\ \hline F_y \end{Bmatrix} = \frac{\mu R L^3}{2} \cdot \int_0^{2\pi} \frac{(\omega_b+\omega_j)(x_j\sin(\theta)-y_j\cos(\theta))-2(\dot{x}_j\cos(\theta)+\dot{y}_j\sin(\theta))}{(c-x_j\cos(\theta)-y_j\sin(\theta))^3}\left\{\frac{\cos(\theta)}{\sin(\theta)}\right\}d\theta$$

It may be noted that the term $\omega_b + \omega_j$ can be replaced by ω, where it is to be understood that the solution is valid for bearings having both surfaces spinning. However, caution is due when defining the dimensionless velocities for transient response calculations. If the velocity terms are to be made dimensionless using the reference spin rate of the journal (i.e., ω_j), the variables should be defined such that:

$$X_j = \frac{x_j}{c} \; ; \; Y_j = \frac{y_j}{c} \; ; \; \dot{X}_j = \frac{\dot{x}_j}{c\omega_j} = \frac{dX_j}{dT} \; ; \; \dot{Y}_j = \frac{\dot{y}_j}{c\omega_j} = \frac{dY_j}{dT} \; ; \; T = \omega_j t$$

The equation for the fluid-film forces may then be expressed as:

$$(4) \qquad \left\{\frac{F_x}{F_y}\right\} = \frac{\mu R L^3 \omega}{2c^2}$$

$$\cdot \int_0^{2\pi} \frac{X_j \sin(\theta) - Y_j \cos(\theta) - \frac{\omega_j}{\omega}(\dot{X}_j \cos(\theta) + \dot{Y}_j \sin(\theta))}{(1 - X_j \cos(\theta) - Y_j \sin(\theta))^3} \left\{\frac{\cos(\theta)}{\sin(\theta)}\right\} d\theta$$

Fluid-film cavitation is accounted for in the analysis by neglecting the contribution of negative pressures when evaluating the integral given by Eq. (4).

The incremental fluid-film forces may be expressed by linear stiffness and damping coefficients in the small region near the steady state equilibrium position (x_o, y_o). The incremental forces are thus expressed by

$$(5a) \qquad \delta F_x = -[K_{xx}\, x + C_{xx}\, \dot{x} + K_{xy}\, y + C_{xy}\, \dot{y}]$$

$$(5b) \qquad \delta F_y = -[K_{yy}\, \dot{y} + C_{yy}\, y + K_{yx}\, x + C_{yx}\, \dot{x}]$$

where

$$x = x_j - x_o \; ; \; y = y_j - y_o$$

with

$$F_x = F_{xo} + \delta F_x$$

$$F_y = F_{yo} + \delta F_y$$

and

(6a) $$K_{xx} = -\left.\frac{\partial F_x}{\partial x_j}\right|_{\substack{x_j = x_{oj} \\ y_j = y_{oj}}}$$

$$= \frac{-\mu RL^3\omega}{2c^3}\int_0^{2\pi} \frac{\sin(\theta)\cos(\theta)H_o+3\cos^2(\theta)(X_{oj}\sin(\theta)-Y_{oj}\cos(\theta))}{H_o^{\ 4}}\, d\theta$$

(6b) $$K_{yy} = -\left.\frac{\partial F_y}{\partial y_j}\right|_{\substack{x_j = x_{oj} \\ y_j = y_{oj}}}$$

$$= \frac{-\mu RL^3\omega}{2c^3}\int_0^{2\pi} \frac{3\sin^2(\theta)(X_{oj}\sin(\theta)-Y_{oj}\cos(\theta))-\sin(\theta)\cos(\theta)H_o}{H_o^{\ 4}}\, d\theta$$

(6c) $$C_{xx} = \left.\frac{-\partial F_x}{\partial \dot{x}}\right|_{\substack{x_j = x_{oj} \\ y_j = y_{oj}}} = \frac{\mu RL^3}{2c^3}\int_0^{2\pi} \frac{2\cos^2(\theta)}{H_o^{\ 3}}\, d\theta$$

(6d) $$C_{yy} = \left.\frac{-\partial F_y}{\partial \dot{y}}\right|_{\substack{x_j = x_{oj} \\ y_j = y_{oj}}} = \frac{\mu RL^3}{2c^3}\int_0^{2\pi} \frac{2\sin^2(\theta)}{H_o^{\ 3}}\, d\theta$$

(6e) $$K_{xy} = \left.\frac{-\partial F_x}{\partial y}\right|_{\substack{x_j = x_{oj} \\ y_j = y_{oj}}}$$

$$= \frac{\mu RL^3\omega}{2c^3}\int_0^{2\pi} \frac{3\sin(\theta)\cos(\theta)(X_{oj}\sin(\theta)-Y_{oj}\cos(\theta))-H_o\cos^2(\theta)}{H_o^{\ 4}}\, d\theta$$

(6f) $$K_{yx} = \left. \frac{-\partial F_y}{\partial x} \right|_{\substack{x_j = x_{oj} \\ y_j = y_{oj}}}$$

$$= \frac{-\mu R L^3 \omega}{2c^3} \int_0^{2\pi} \frac{3\sin(\theta)\cos(\theta)(X_{oj}\sin(\theta) - Y_{oj}\cos(\theta)) + H_o \sin^2(\theta)}{H_o^4} \, d\theta$$

(6g) $$C_{xy} = \left. \frac{-\partial F_x}{\partial \dot{y}} \right|_{\substack{x_j = x_{oj} \\ y_j = y_{oj}}} = \frac{\mu R L^3}{2c^3} \int_0^{2\pi} \frac{2\sin(\theta)\cos(\theta)}{H_o^3} \, d\theta$$

(6h) $$C_{yx} = \left. \frac{-\partial F_y}{\partial \dot{x}} \right|_{\substack{x_j = x_{oj} \\ y_j = y_{oj}}} = \frac{\mu R L^2}{2c^3} \int_0^{2\pi} \frac{2\sin(\theta)\cos(\theta)}{H_o^3} \, d\theta$$

These integrals are to be calculated over the area of positive pressure which is present under steady state conditions.

The steady state equilibrium position ε_o, may be found for any load by an iterative routine using the OCVIRK number [19] which is expressed as:

(7) $$S_s = \frac{\mu N}{P} \left(\frac{R}{c}\right)^2 \left(\frac{L}{D}\right)^2 = \frac{(1-\varepsilon^2)^2}{\pi\varepsilon[\pi^2(1-\varepsilon^2)+16\varepsilon^2]^{\frac{1}{2}}}$$

where

$$P = \frac{\text{bearing load}}{L \times D}$$

N = bearing effective speed (rev./sec.)

Hence, for a given bearing geometry, load, and speed the equilibrium position is expressed as:

$$X_{oj} = \varepsilon_o \sin(\phi_o)$$

(8)

$$Y_{oj} = -\varepsilon_o \cos(\phi_o)$$

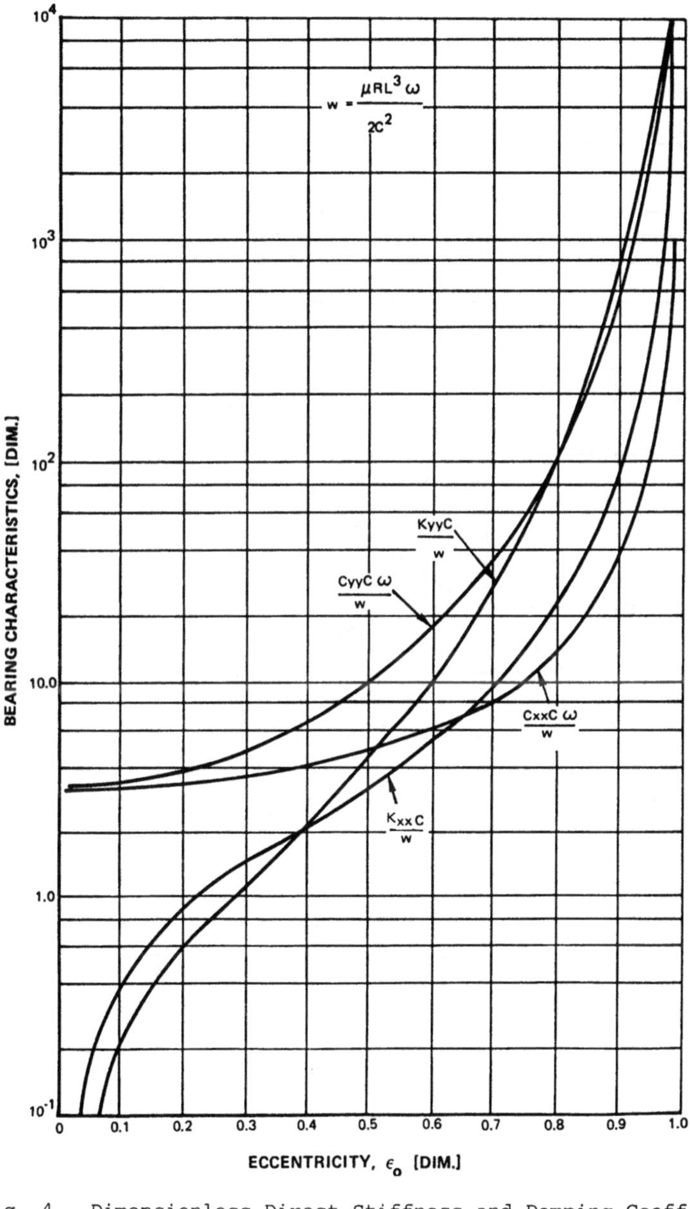

Fig. 4. Dimensionless Direct Stiffness and Damping Coefficients for the Short Journal Bearing.

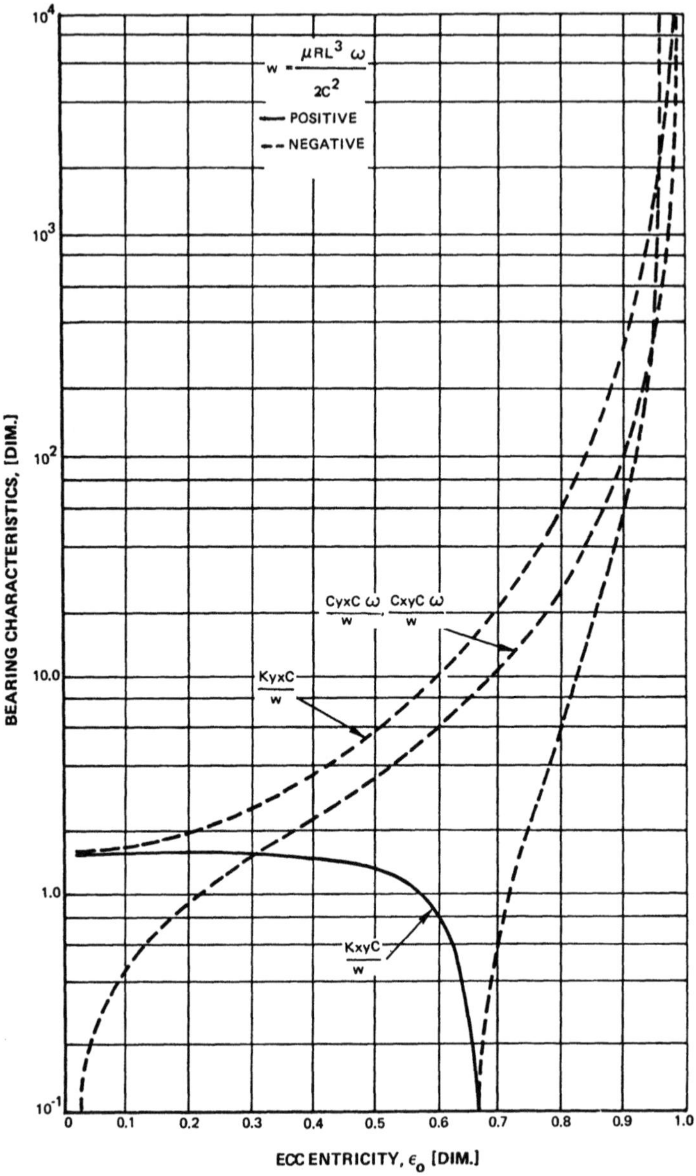

Fig. 5. Cross-Coupled Stiffness and Damping Coefficients for the Short Journal Bearing.

where:

(9) $$\phi_o = \text{attitude angle} = \tan^{-1}\left[\frac{\pi}{4}\ \frac{\sqrt{1-\varepsilon_o{}^2}}{\varepsilon_o}\right]$$

and ε_o is found by using Eq. 7.

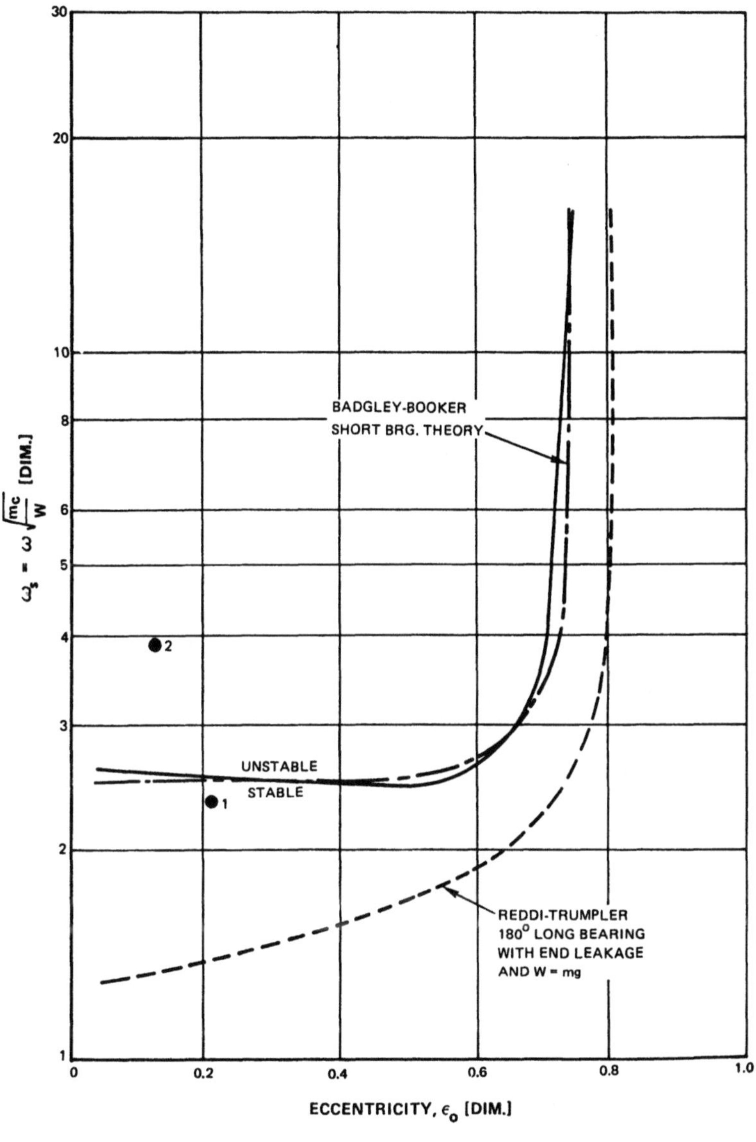

Fig. 6. Stability Map for the Short Journal Bearing with Rigid Supports.

The bearing characteristics may therefore be expressed as dimensionless values as a function of eccentricity and the resulting values are shown in Figs. 4-5. These bearing characteristics have been used in generating the stability boundaries in the following section of this report. In Fig. 6 the results of a simple stability analysis of an isolated journal bearing using the above characteristics are compared with results generated by BADGLEY-BOOKER using the transient response method of analysis [20]. Their results are shown to be in good agreement with the stability boundary generated using the above characteristics. The stability boundary generated by REDDI-TREMPLER [21] for the 180° bearing with end leakage indicates a lower threshold speed for all eccentricities. The results of the present analysis are found to first decrease with increasing steady state eccentricity, then increase being completely stable beyond $\varepsilon_o = 0.7$, which agrees with the results stated by HORI [10].

The typical characteristics of a rotor operating below the stability threshold speed is shown in Fig. 7a. The transient path of the journal is plotted in the reference clearance circle. The rotor is operating at 6500 rpm and at time $t = 0$, the rotor is dropped from the center (X=Y=0) . The resulting path of the journal is indicated to be stable and spiraling into the equilibrium eccentricity (ε_o = ES = 0.211) . The small circles on the orbit indicate one spin revolution of the rotor. The asterisk on the orbit is the point of maximum load transmitted and for this case it is indicated to be 64.4 lb. (286.45 N.)

If the system properties are used to calculate the position on the stability map of Fig. 6 the point labeled 1 is obtained. The system is thus seen to be operating just below the threshold speed. An increase of rotor speed to 10,500 rpm gives the position on the map labeled 2 which is indicated to be unstable. This is shown clearly in Fig. 7b where the journal is spiraling outward at a whirl rate very close to one-half operating speed. A continuation of the response as shown in Fig. 7c indicates that the system is nearly in the limit cycle with the bearing dynamic loading indicated to be 3.96 times the static load of the rotor.

These transient response plots clearly indicate the nature of response for a rotor system operating above the threshold speed. The stability and transient response of a more complex rotor-support system will be presented following

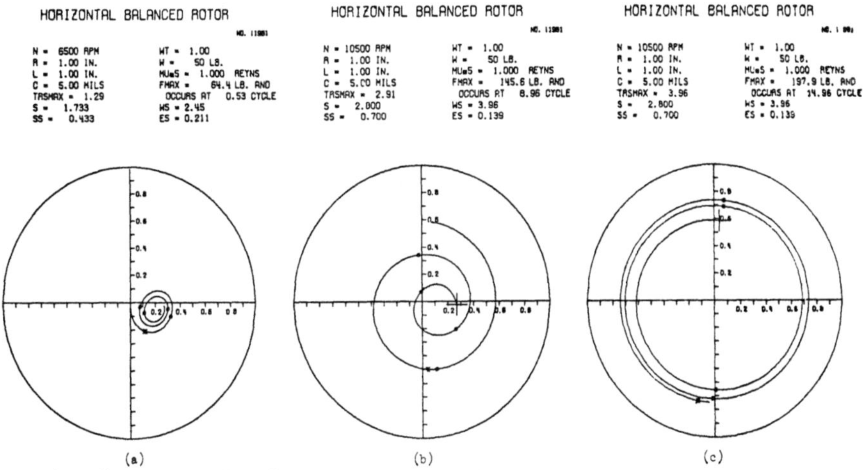

Fig. 7a. Journal Orbit of a Balanced Horizontal Rotor on Rigid Supports (N = 6500 RPM , W = 50 lb. (22.68 Kg) , C = 0.005 mils (0.0127 cm) , L/D = 1/2) .

7b. Journal Orbit of a Balanced Rotor for Cycles 6-10 (N = 10500 RPM)

7c. Journal Orbit of a Balanced Rotor for Cycles 11-15 Showing Motion Near Limit Cycle (N = 10500 RPM)

the discussion of the equations of motion.

2.3. Equations of Motion. The equations of motion for the symmetric rotor system as shown in Fig. 1 may be expressed as follows with the rotor characterized by a shaft stiffness, K_s , and absolute damping, C_s .

Rotor:

$$m_2\ddot{x}_2+K_s(x_2-x_j-x_1) + C_s\dot{x}_2 = m_2\, e\mu_2\, \omega_j^2 \cos(\omega_j t)+f_x(t) \tag{10a}$$

$$m_2\ddot{y}_2+K_s(y_2-y_j-y_1) + C_s\dot{y}_2 = m_2 e\mu_2\omega_j^2\sin(\omega_j t)-m_2 g+f_y(t) \tag{10b}$$

Journal:

$$m_j(\ddot{x}_j+\ddot{x}_1)-F_x(x,y,\dot{x},\dot{y})_j - \frac{K_s}{2}(x_2-x_j-x_1) = 0 \tag{11a}$$

$$m_j(\ddot{y}_j+\ddot{y}_1)-F_y(x,y,\dot{x},\dot{y})_j - \frac{K_s}{2}(y_2-y_j-y_1) = -m_j g \tag{11b}$$

Support:

(12a) $m_1\ddot{x}_1+F_x(x,y,\dot{x},\dot{y})_j+K_{1xx}x_1+K_{1xy}y_1+C_{1xx}\dot{x}_1+C_{1xy}\dot{y}_1 = F_{px}$

(12b) $m_1\ddot{y}_1+F_y(x,y,\dot{x},\dot{y})_j+K_{1yy}y_1+K_{1yx}x_1+C_{1yy}\dot{y}_1+C_{1yx}\dot{x}_1=-m_1g+F_{py}$

where

C_s = absolute damping on rotor midspan

$\bar{f}(t)$ = any time dependent or steady loading acting on the rotor (including effects of aerodynamic loading or internal friction).

$\bar{F}_p$ = preload force on the bearing support system.

$\bar{F}(x,y,\dot{x},\dot{y})_j$ = fluid-film forces expressed by Eq. (3).

If the rotor shaft stiffness may be considered to the rigid ($\frac{m_2 g}{K_s}$ << c) then the equation may be reduced still further to the following system of equations.

(13a)

$$m(\ddot{x}_1+\ddot{x}_j)-F_x(x,y,\dot{x},\dot{y})_j = m\varepsilon_\mu\omega_j^2\cos(\omega_j t)+1/2f_x(t)-1/2C_s(\dot{x}_1+\dot{x}_j)$$

(13b)

$$m(\ddot{y}_1+\ddot{y}_j)-F_y(x,y,\dot{x},\dot{y})_j=m\varepsilon_\mu\omega_j^2\sin(\omega_j t)+1/2f_y(t)-mg-1/2C_s(\dot{y}_1+\dot{y}_j)$$

(14a) $m_1\ddot{x}_1+F_x(x,y,\dot{x},\dot{y})_j+K_{1xx}x_1+K_{1xy}y_1+C_{1xx}\dot{x}_1+C_{1xy}\dot{y}_1 = F_{px}$

(14b) $m_1\ddot{y}_1+F_y(x,y,\dot{x},\dot{y})_j+K_{1yy}y_1+K_{1yx}x_1+C_{1yy}\dot{y}_1+C_{1yx}\dot{x}_1=F_{py}-m_1g$

These are the equations required for time transient response calculations. For stability consideration, the homogeneous equation written about the steady state equilibrium position may be expressed in dimensionless form considering the bearing characteristics as given by Eq. 5. The cross-coupling terms for the flexible support are neglected to reduce the number of variables in the analysis. The resulting equations are expressed as follows:

(15a) $$(1+\delta m)X_1'' + \bar{C}_x X_1' + \bar{K}_x X_1 + X_2'' = 0$$

(15b) $$(1+\delta m)Y_1'' + \bar{C}_y Y_1' + \bar{K}_y Y_1 + Y_2'' = 0$$

(15c) $$X_1'' + X_j'' + \bar{K}_{xx} x_j + \bar{C}_{xx} X_j' + \bar{K}_{xy} Y_j + \bar{C}_{xy} Y_j' = 0$$

(15d) $$Y_1'' + Y_j'' + \bar{K}_{yy} Y_j + \bar{C}_{yy} Y_j' + \bar{K}_{yx} X_j + \bar{C}_{yx} X_j' = 0$$

where

$$\delta m = m_1/m \ ; \quad m = m_j + \frac{m_2}{2}$$

$$\bar{C}_x = C_{1xx}/(m\omega) \ ; \ \bar{C}_y = C_{1yy}/(m\omega)$$

$$\bar{K}_x = K_{1xx}/(m\omega^2) \ ; \bar{K}_y = K_{1yy}/(m\omega^2)$$

$$\bar{K}_{ij} = \alpha_k K_{ij}; \quad \bar{C}_{ij} = \alpha_c C_{ij}, \quad \text{where} \quad i,j = 1,2(1=x;\ 2=y)$$

and

$$\alpha_k = \frac{c}{w}\alpha; \quad \alpha_c = \frac{c\omega}{w}\alpha$$

$$\alpha = \frac{\mu N}{P}\left(\frac{R}{c}\right)^2\left(\frac{L}{D}\right)^2/[\omega\sqrt{\frac{mc}{2\pi W}}]^2 = S_s/\Omega_c^2$$

$$w = \frac{\mu R L^3 \omega}{2c^2}$$

It should be noted that the above equations have been written such that the differentiation with respect to time, denoted by the primes, are defined such that

$$X' = \frac{1}{\omega}\frac{dX}{dt} \ ; \quad X'' = \frac{1}{\omega^2}\frac{d^2X}{dt^2}$$

The support characteristics are assumed to be symmetric to further reduce the number of dimensionless parameters in the study. Thus the extent of the stable region will be the lowest obtainable since bearing support asymmetry will in general increase the stable operating speed range. The following dimensionless support characteristics variables are defined.

$$\bar{K}_B = K_B/m\omega_g^2 \; ; \quad K_B = K_{1xx} = K_{1yy}$$
$$\bar{C}_B = C_B/m\omega_g \; ; \quad C_B = C_{1xx} = C_{1yy} \tag{16}$$

where,

$$\omega_g = \sqrt{\frac{W}{mc}}$$

The stability of the rotor system may now be calculated in terms of the dimensionless parameters given in the preceding discussion. From the above dimensionless parameters, it may be shown that the stability threshold speed may be expressed as

$$\omega\Big|_{\substack{\text{Threshold}\\ \text{Speed}}} = f\left(\sqrt{\frac{W}{mc}}\,,\ \delta m,\ \varepsilon,\ \bar{K}_B,\ \bar{C}_B\right) \tag{17}$$

These parameters will be used to plot the results of the stability analysis presented in the following discussion.

3. STABILITY ANALYSIS

The equations of motion as presented in Eq. (15) may be analyzed for stability by examination of the system characteristic equation. Assuming solutions of the form $X = \bar{X}e^{\lambda t}$, the following characteristic determinant may be written.

$$(18)\qquad \begin{vmatrix} (1+\delta m)\lambda^2 + \bar{C}_x\lambda + \bar{K}_x & 0 & \lambda^2 & 0 \\ 0 & (1+\delta m)\lambda^2 + \bar{C}_y\lambda + \bar{K}_y & 0 & \lambda^2 \\ \lambda^2 & 0 & \lambda^2+\bar{C}_{xx}\lambda + \bar{K}_{xx} & \bar{C}_{xy}\lambda+\bar{K}_{xy} \\ 0 & \lambda^2 & \bar{C}_{yx}\lambda+\bar{K}_{yx} & \lambda^2+\bar{C}_{yy}\lambda + \bar{K}_{yy} \end{vmatrix} = 0$$

The resulting 8th order polynomial in λ may be expressed as (16)

$$(19)\qquad A_o\lambda^8 + A_1\lambda^7 + A_2\lambda^6 + \ldots + A_7\lambda + A_8$$

where

$$(20)\qquad A_i = \alpha_i + \beta_i \text{ ; } i = 0, 1, 2, \ldots, 7, 8$$

and

$$(21)\qquad \alpha_i = (1+\delta m)a_{i+2} + \bar{C}_x a_{i+1} + \bar{K}_x a_{i+2}; \; i = 0,1,2,\ldots 7,8$$

with

$$a_o = a_1 = 0$$

$$a_2 = \delta m$$

$$a_3 = \bar{C}_y + \bar{C}_{yy} + \delta m(\bar{C}_{xx} + \bar{C}_{yy})$$

$$a_4 = \bar{K}_y + \bar{K}_{yy} + \bar{C}_{xx}\bar{C}_{yy} - \bar{C}_{xy}\bar{C}_{yx}$$
$$+ \delta m(\bar{K}_{xx} + \bar{K}_{yy} + \bar{C}_{xx}\bar{C}_{yy} - \bar{C}_{xy}\bar{C}_{yx}) + C_y(\bar{C}_{xx} + \bar{C}_{yy})$$

$$a_5 = (1 + \delta m)[\bar{C}_{yy}\bar{K}_{xx} + \bar{C}_{xx}\bar{K}_{yy} - \bar{C}_{yx}\bar{K}_{xy} - \bar{C}_{xy}\bar{K}_{yx}]$$
$$+ \bar{C}_y\ [\bar{K}_{xx}+\bar{K}_{yy}+\bar{C}_{xx}\bar{C}_{yy} - \bar{C}_{xy}\bar{C}_{yx}] + \bar{K}_y[\bar{C}_{xx} + \bar{C}_{yy}]$$

$$a_6 = (1 +\delta m)[\bar{K}_{xx}\bar{K}_{yy} - \bar{K}_{xy}\bar{K}_{yx}]$$
$$+ \bar{C}_y[\bar{C}_{yy}\bar{K}_{xx} + \bar{C}_{xx}\bar{K}_{yy} - \bar{C}_{yx}\bar{K}_{xy} - \bar{C}_{xy}\bar{K}_{yx}]$$
$$+ \bar{K}_y[\bar{K}_{xx} + \bar{K}_{yy} + \bar{C}_{xx}\bar{C}_{yy} - \bar{C}_{xy}\bar{C}_{yx}]$$

$$a_7 = \bar{C}_y[\bar{K}_{xx}\bar{K}_{yy} - \bar{K}_{xy}\bar{K}_{yx}] + \bar{K}_y[\bar{C}_{yy}\bar{K}_{xx} + \bar{C}_{xx}\bar{K}_{yy}$$
$$- \bar{C}_{yx}\bar{K}_{xy} - \bar{C}_{xy}\bar{K}_{yx}]$$

$$a_8 = \bar{K}_y[\bar{K}_{xx}\bar{K}_{yy} - \bar{K}_{xy}\bar{K}_{yx}]$$

Also,

$$\beta_o = -\delta m$$
$$\beta_1 = -\bar{C}_y - \bar{C}_{yy}(1 +\delta m)$$
$$\beta_2 = -\bar{K}_y - \bar{C}_y\bar{C}_{yy} - (1 + \delta m)\bar{K}_{yy}$$
$$\beta_3 = -\bar{K}_y\bar{C}_{yy} - \bar{C}_y\bar{K}_{yy}$$
$$\beta_4 = -\bar{K}_y\bar{K}_{yy}$$
$$\beta_i = 0, i = 5,6,7,8$$

The characteristic equation given by Eq. (19) may be examined by the Routh criteria (16) to determine the onset of instability. The equations were programmed on a digital computer and an extensive study of the lowest stability threshold speed was determined for various values of dimensionless support parameters. The results of this investigation are plotted as a dimensionless stability parameter, $\omega_s = \omega\sqrt{\frac{mc}{W}}$, versus the dimensionless steady state equilibrium position of the journal which is denoted as ε_o. For the horizontal rotor system under 1-g loading the stability parameter reduces to $\omega_s = \omega\sqrt{\frac{c}{g}}$. This parameter has been widely used to express the stability of short journal bearings (11,14,20,21). The equilibrium position of the journal may be expressed in terms of the Ocvirk number, S_s. In addition, it is easily shown that

$$\omega\sqrt{\frac{mc}{W}} = \overline{W}\,S_s \tag{22}$$

where,

$$\overline{W} = \frac{2\pi W}{\mu LD}\left(\frac{2c}{L}\right)^2\sqrt{\frac{mc}{W}} \tag{23}$$

This parameter is defined as the load number of the bearing and may be used on the stability map to determine the operating eccentricity for a given rotor bearing configuration.

The stability map shown as Fig. 8 was plotted for a specified value of support stiffness, $\overline{K}_B$, and mass ratio, $\delta m = m_1/m$. Stability boundaries for numerous of support damping are shown on the figure to illustrate the effect of damping on the threshold speed. For example, Fig. (8) indicates that (for $\overline{K}_B = 0.1$, $\delta m = 0.1$) the threshold speed may be raised significantly above the rigid support threshold speed (shown as dash-dot line) for a damping value of $\overline{C}_B \sim 0.5$. In addition, it is easily observed that a lightly damped support system $(C_B \sim 0.001)$ reduces the onset of instability for bearings having $\varepsilon_o < 0.7$. This effect will be examined in greater detail later in the discussion.

The load lines indicated by dashed lines trace the journal eccentricity of a rotor as the running speed is increased or decreased. For heavily loaded journals $(\overline{W} > 100)$ it is evident that the stability is greatly increased beyond the corresponding onset speed for a lightly loaded journal $(\overline{W} < 50)$. The stability map also indicates an optimum damping value for maximum increase of the stable operating speed range. An overdamped support system $(C_B > 10.0)$, reduces the threshold speed to the rigid support onset speed.

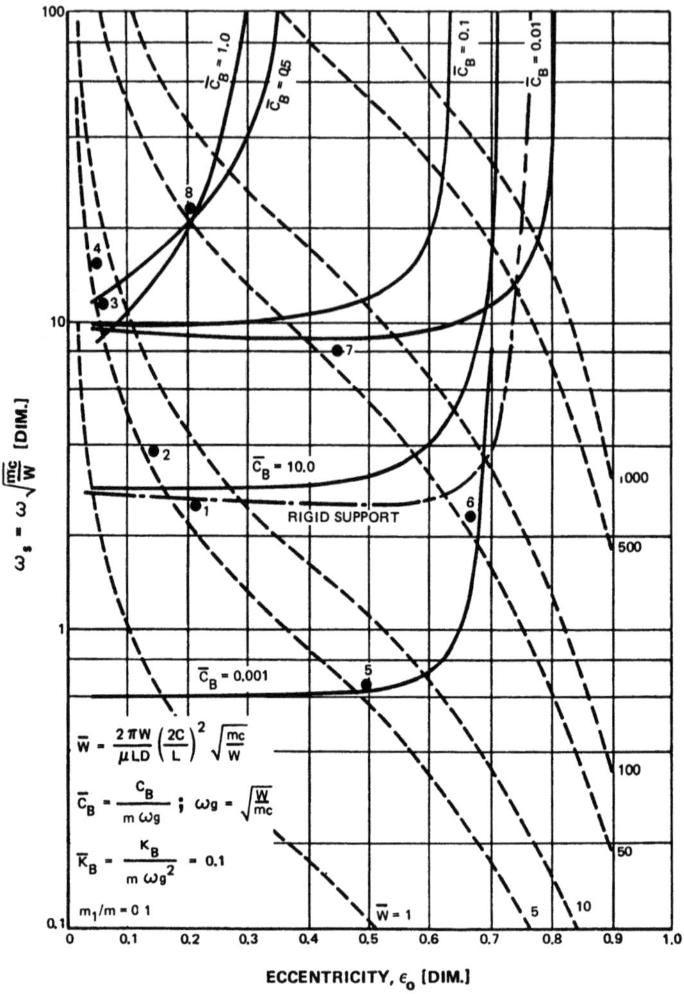

Fig. 8. Stability Map for the Damped, Flexible Support Rotor on Short Journal Bearings ($K_B = 0.1$, $m_1/m = 0.1$)

Consider the following rotor system which will be used to illustrate the effects of support properties on stability and transient response.

Example 1:

Rotor weight - 50 lb. (22.68 Kg)

Journal weight (each) - 25 lb. (11.34 Kg)

Journal radius - 1 inch (2.54 cm)

Journal clearance - .005 inch (.0127 cm)

Support weight (each) - 5.0 lb. (2.268 Kg)

Support stiffness - 1000.0 lb./in.
(1751.26 N/cm)

Support damping - 20 lb.-sec./in.
(35.03 N-sec./cm

Viscosity - 1.0 x 10^{-5} reyns

This rotor system produces the following dimensionless values:

$$\overline{K}_B = \frac{1000.0}{\frac{50}{386}\left(\frac{50\ (386)}{50(.005)}\right)} = 0.1$$

$$\overline{C}_B = \frac{20}{\frac{50}{386}\sqrt{\frac{50\ (386)}{50(.005)}}} = 0.556$$

$$\overline{W} = \frac{2\pi(50)}{1\text{x}10^{-5}(1)(2)} \left(\frac{.010}{1}\right)^2 \sqrt{\frac{50(.005)}{386(50)}} = 5.65$$

If the rotor speed is 10,500 rpm the previous discussion of rigid support bearings indicated that the rotor would be unstable. The stability parameter is calculated to be:

$$\omega_s = 10500\ \left(\frac{2\pi}{60}\right)/\ \sqrt{\frac{50\ (386)}{50(-005)}} = 3.96$$

The equilibrium eccentricity is calculated to be 0.139 and the point labeled 2 on Fig. 8 indicates the system with the damped support giving $\overline{C}_B = 0.556$ should be very stable. The transient response of the balanced rotor is shown in Fig. 9a. The rotor is indicated to be very stable as shown by the rapid decay of the transient of the steady state eccentricity. An increase of rotor speed to 31,500 rpm produces the point 3 on Fig. 8 and is very near the threshold speed. This is clearly indicated by the transient response shown as Fig. 9b where the journal continues to orbit at approximately half-frequency indicating near zero damping at this speed. An increase of speed to 45,000 rpm places the journal above the threshold as shown by point 4 of Fig. 8.

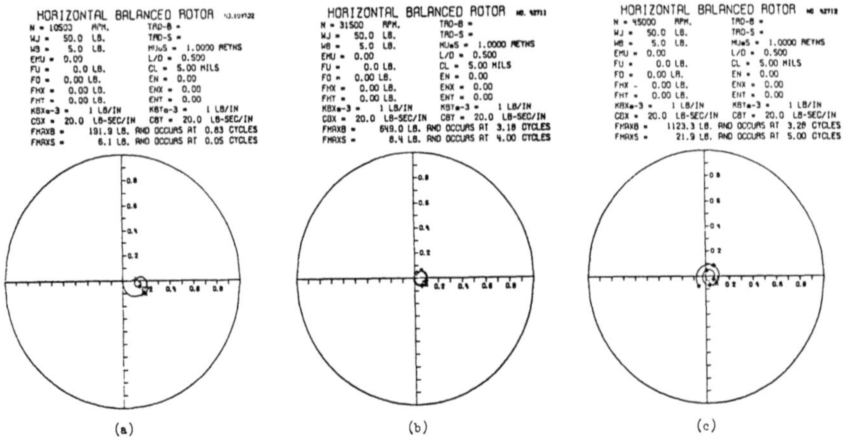

Fig. 9a. Rotor Relative Motion Showing Stabilized Response (N = 10500 RPM, $\omega_s = 3.96$).

9b. Rotor Relative Motion at the Threshold of Stability (N = 31500 RPM, $\omega_s = 11.9$).

9c. Rotor Relative Motion for Five Cycles Showing Instability (N = 45000 RPM, $\omega_s = 17$).

The response for this condition is given in Fig. 9c and the motion is growing exponentially and would eventually grow into a large limit cycle response. The support system has increased the stability threshold speed by a factor of 4.75 beyond the rigid support threshold speed.

The effect of decreasing the support stiffness, $\bar{K}_B = 0.01$ is shown in Fig. 10a and it is indicated that for values of damping $\bar{C}_B < 0.01$ the threshold speed has been increased significantly beyond that for $\bar{K}_B = 0.1$. By increasing the support stiffness to $\bar{K}_B = 1.0$ with $\delta m = 0.1$, the optimum damping values remain at approximately $\bar{C}_B \sim 0.5\text{-}1.0$ and the lightly damped support threshold speed is slightly better than in Fig. (8), but still below the rigid support onset speed (see Fig. 10b). The stability maps shown as Figs. 10c and 10d clearly indicate that an increase of support mass greatly reduces the maximum onset speed whereas a lighter support system can be designed to increase the threshold speed to extremely high values for support

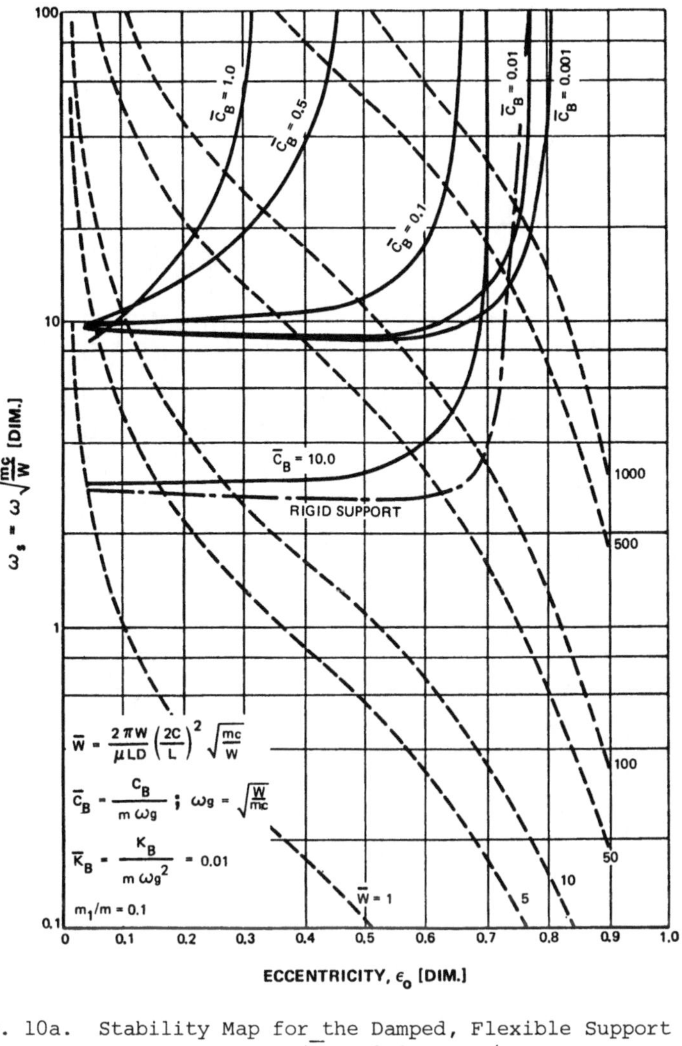

Fig. 10a. Stability Map for the Damped, Flexible Support Rotor on Short Bearings ($\bar{K}_B = 0.01$, $m_1/m = 0.1$).

damping values between $\bar{C}_B = 0.01$ and 1.0 . The overdamped condition ($\bar{C}_B > 10$) remains at about the same threshold speed for all the variations considered in this series of stability maps.

The importance of proper design of support structures is further illustrated in Fig. 11a where, for the specific condition of $\bar{K}_B = 1.0$ and $\delta m = 0.1$, the optimum support damping is indicated to be $\bar{C}_B \sim 1.0$. This value of damping

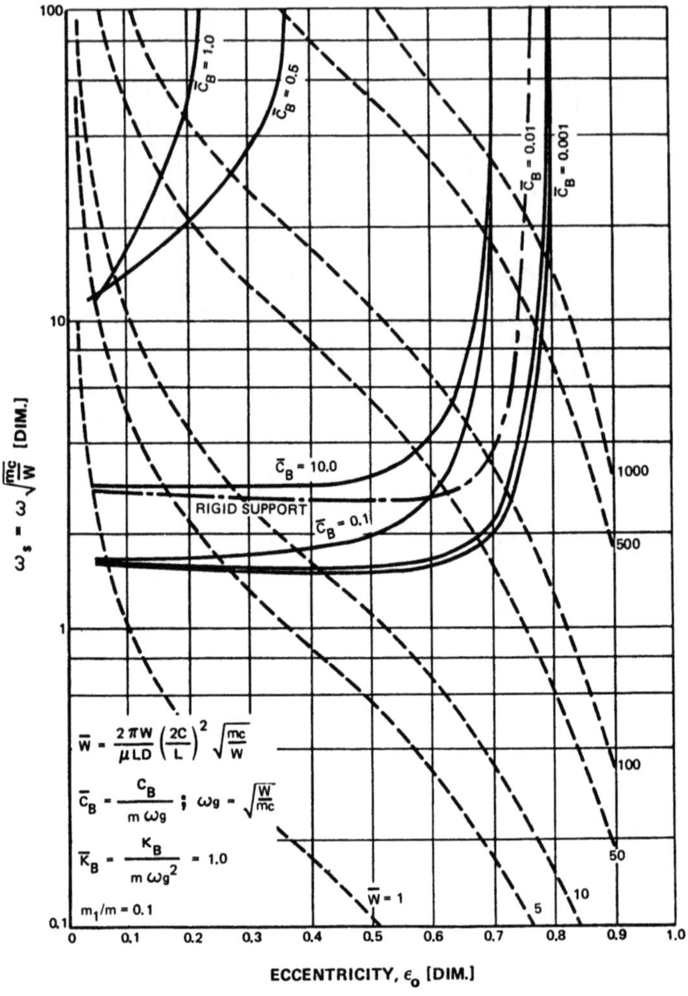

Fig. 10b. Stability Map for the Damped, Flexible Support Rotor on Short Journal Bearings ($\bar{K}_B = 1.0$, $m_1/m = 0.1$).

would be ideal for both the lightly loaded and heavily loaded condition. Fig. 11b indicates that by reducing $\bar{K}_B$ to 0.1 with $\delta = 0.01$ the range of acceptable support damping is greatly increased. However, for the heavily loaded system ($\bar{W} > 100$) , the optimum damping must be properly chosen to ensure the greatest stable operating speed range.

These stability maps indicate that the plain journal bearing can be designed with a flexible damped support such that the stable operating speed range is greatly extended beyond the rigid support threshold speed. Equally important, the analysis indicates that improper support design could possibly lower the threshold speed below that which would be obtained for rigid supports.

4. TRANSIENT RESPONSE ANALYSIS

Stability maps such as the ones presented in the previous

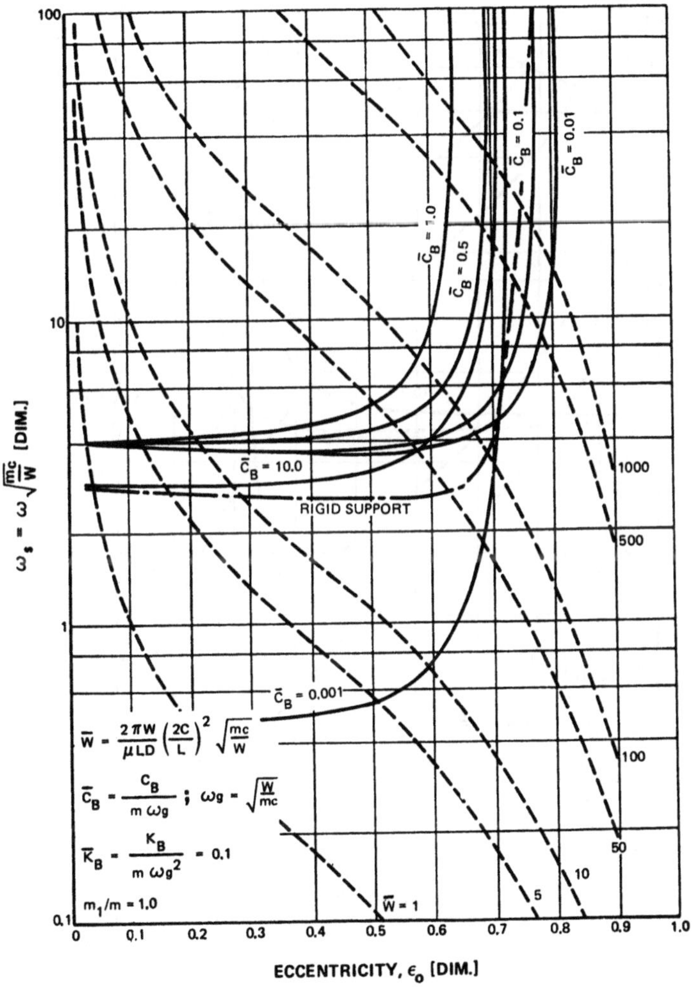

Fig. 10c. Stability Map for the Damped, Flexible Support Rotor on Short Journal Bearings ($\bar{K}_B = 0.1$, $m_1/m = 1.0$).

section of this paper are of great value due to the amount of useful information contained on a single plot. The analysis of time-transient response of rotor systems cannot be presented in such a compact form. However, the amount of design information obtained from an actual time-transient response analysis for a specific rotor system easily justifies

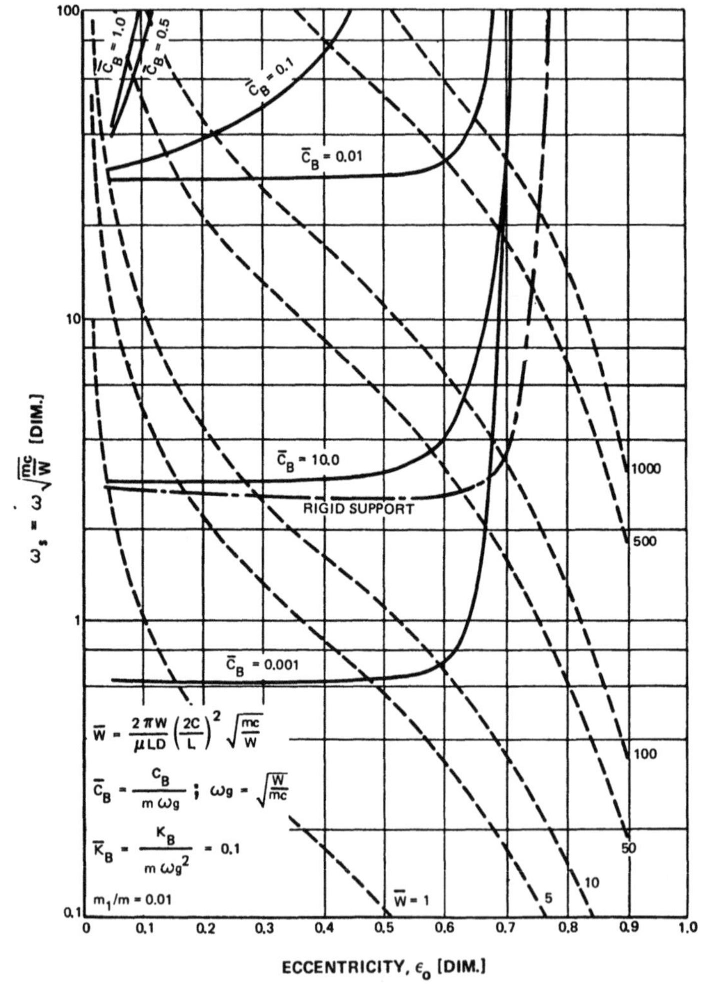

Fig. 10d. Stability Map for the Damped, Flexible Support Rotor on Short Journal Bearings ($\bar{K}_B$ = 0,1, m_1/m = 0.01).

the effort required to produce and interpret the results.

The technique for calculation of the transient response has been discussed in detail in several references (11,22). The initial value problem requires only that the displacements and velocities of the rotor system be given at a

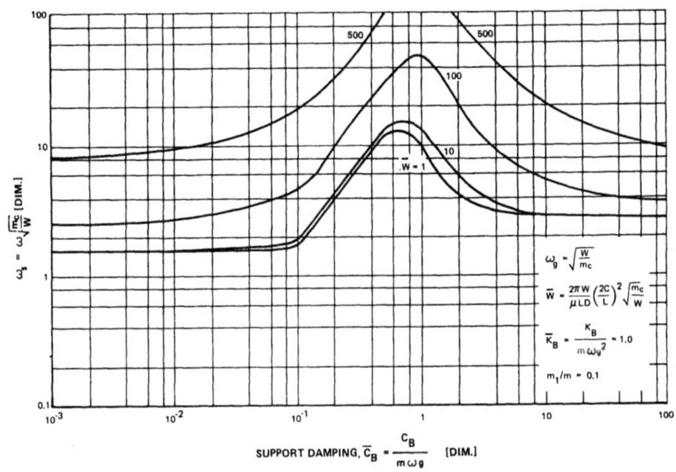

Fig. 11a. Stability Map Showing the Threshold Speed Versus Support Damping ($\bar{K}_B = 1.0$, $m_1/m = 0.1$).

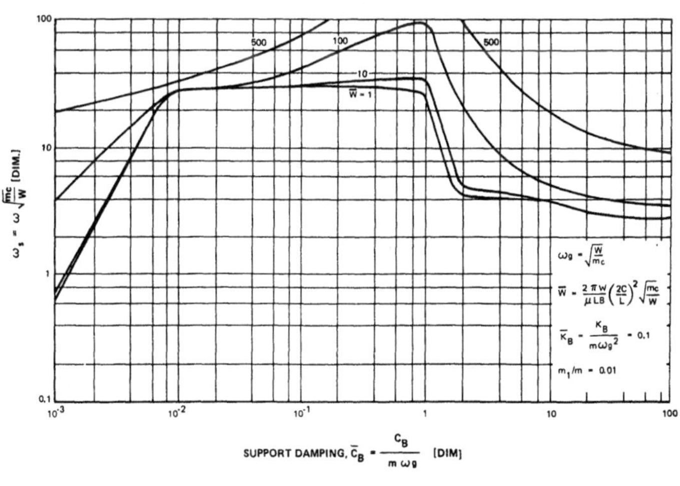

Fig. 11b. Stability Map for Reduced Support Stiffness and Damping Indicating a Larger Region of Stable Operation ($\bar{K}_B = 0.1$, $m_1/m = 0.01$).

specific initial starting time. From this information the nonlinear fluid-film forces may be calculated (as shown in Eq. 3) and any other forces specified as a function of time, displacement, and velocity may also be included in the simulation.

The equations of motion in dimensionless form are expressed as follows:

(24a) $\ddot{X}_1 + \ddot{X}_j =$

$$\frac{F_x(X,Y,\dot{X},\dot{Y})_j}{mc\omega_j^2} + E_\mu \cos(T) + \frac{1}{2mc\omega_j^2} f_x(t) - \frac{C_s}{2m\omega_j}(\dot{X}_1+\dot{X}_j)$$

(24b) $\ddot{Y}_1 + \ddot{Y}_j =$

$$\frac{F_y(X,Y,\dot{X},\dot{Y})_j}{mc\omega_j^2} + E_\mu \sin(T) + \frac{1}{2mc\omega_j^2} f_y(t) - \frac{g}{c\omega_j^2} - \frac{C_s}{2m\omega_j}(\dot{Y}_1+\dot{Y}_j)$$

(24c) $$\delta m\ddot{X}_1 = \frac{1}{mc\omega_j^2}(F_{Px} - F_x(X,Y,\dot{X},\dot{Y})_j) -$$

$$\left[\frac{K_{1xx}}{m\omega_j^2} X_1 + \frac{K_{1xy}}{m\omega_j^2} Y_1 + \frac{C_{1xx}}{m\omega_j} \dot{X}_1 + \frac{C_{1xy}}{m\omega_j} \dot{Y}_1\right]$$

(24d) $$\delta m\ddot{Y}_1 = \frac{1}{mc\omega_j^2}(F_{py} - F_y(X,Y,\dot{X},\dot{Y})_j) -$$

$$\left[\frac{K_{1yy}}{m\omega_j^2} Y_1 + \frac{K_{1yx}}{m\omega_j^2} X_1 + \frac{C_{1yy}}{m\omega_j} \dot{Y}_1 + \frac{C_{1yx}}{m\omega_j} \dot{X}_1\right] - \delta m \frac{g}{m\omega_j^2}$$

where the equation for $\overline{F}(x,y,\dot{x},\dot{y})$ is given by Eq. 4.

The response of fluid-film bearings are typically plotted as the motion of the geometric center of the journal in the clearance circle. This is equivalent to the orbit trace that would be observed on an oscilloscope showing the output of displacement probes mounted on the bearing support and monitoring the journal motion. The physical interpretation of an instability was demonstrated in Fig. 7b where the rotor speed was 10,500 rpm (point 2 on Fig. 6). The rotor was spiraling outward at a very high growth rate. In five revolutions of the shaft the response had doubled. Thus the average growth rate, a , for this system is calculated to be:

$$R = R_o e^{at} \quad ; \quad t = 0.005714 \text{ sec./rev.}$$

therefore,

$$\frac{R}{R_o} = 2.0 = e^{a(.02857)}$$

and hence

$$a = 24.26$$

The continuation of the motion for another five cycles was shown in Fig. 7c. The growth rate was greatly reduced ($a = 8.53$) and a limit cycle was forming which is as expected for the nonlinear system. More details of the significance of the growth factor will be discussed in the next section. The stability analysis which uses linearized equations cannot predict the limit cycle condition as was shown in the transient orbit.

The analysis of imbalance response in a nonlinear system may be obtained by transient response techniques. The mass of the rotor is considered to be displaced from the geometric center by an amount $E_\mu = EMU = e_{\mu/c}$ where c is the radial clearance of the bearing. Consider initially a rigid support system and an imbalance given as $EMU = 0.20$. For a rotor speed of 6500 rpm the orbit is shown in Fig. 12a for the case of a suddenly applied imbalance and release of the journal from $x = y = 0$ the time $t = 0$. The response has both synchronous and non-synchronous components but at the end of only five cycles of motion the system is reducing to predominantly a synchronous response. An increase

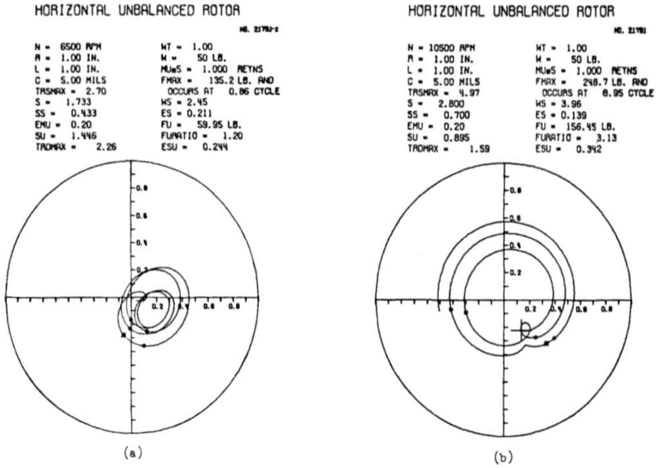

Fig. 12a. Journal Orbit of an Unbalanced Rotor on Rigid Supports for Five Cycles (N = 6500 RPM, W = 50 lb. (22.68 K_g), C = 0.005 mils (0.0127 cm.), L/D = 1/2, E_μ = 0.2).[9]

12b. Journal Orbit of an Unbalanced Rotor Above the Stability Threshold Speed (N = 10500 RPM, ω_s = 3.96).

of rotor speed to 10,500 rpm (above the threshold speed) produces the unstable orbit shown in Fig. 12b.

The addition of a flexible, damped support system has

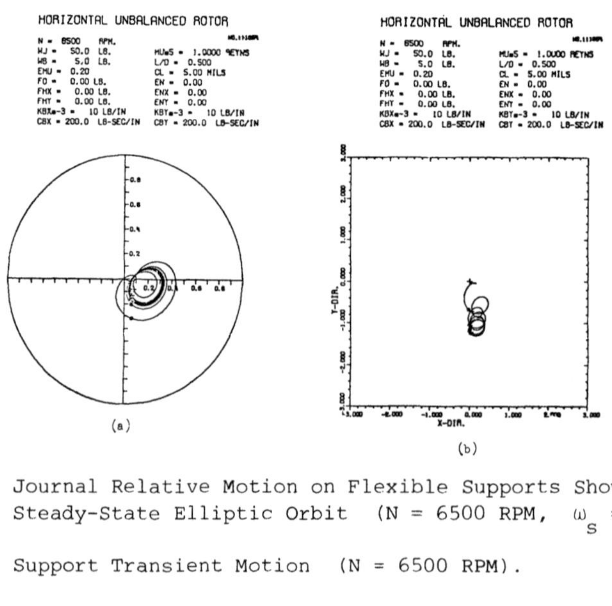

Fig. 13a. Journal Relative Motion on Flexible Supports Showing Steady-State Elliptic Orbit (N = 6500 RPM, ω_s = 2.45).

13b. Support Transient Motion (N = 6500 RPM).

been shown in Example 1 to greatly increase the threshold speed if the support is properly designed. There are instances where the support can be poorly designed and produce no harmful results as long as the rotor speed remains below the stability threshold. However, the bearing and support loading will in most all cases be higher than if the support system were properly designed. Consider the system where the support stiffness and damping are relatively large ($\bar{K}_B = 1.0$, $\bar{C}_B = 5.56$). The motion at 6500 rpm shows a very stable elliptic orbit with very little support motion. An increase of speed to 10,500 rpm results in the large amplitude response shown in Fig. 14. By reducing the support stiffness

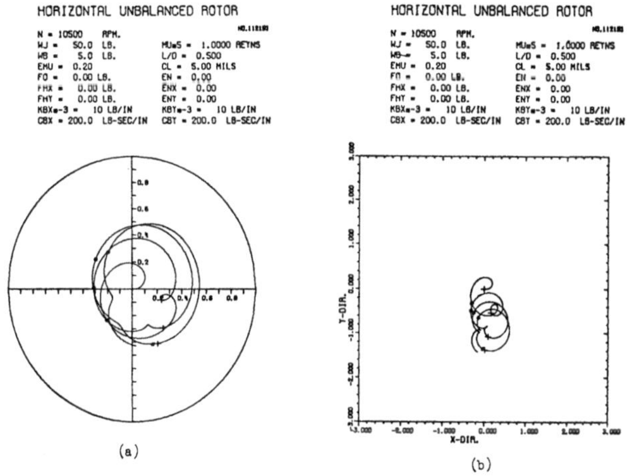

Fig. 14a. Journal Transient Response for an Overdamped Support ($N = 10500$ RPM, $E\mu = 0.2$).

14b. Support Response ($\bar{K}_B = 1.0$, $\bar{C}_B = 5.56$).

and damping to the values given in Example 1 (i.e., $\bar{K}_B = 0.1$, $\bar{C}_B = 0.556$, $\delta m = 0.1$) the response at 10,500 rpm can be reduced to that shown in Fig. 15. The maximum support loading is only 33.5 lb. (149.0N.) above the static loading condition although the rotor is being driven by an imbalance load of 313.2 lb. (1393.2 N.) or 156.6 lb. (696.59 N.)

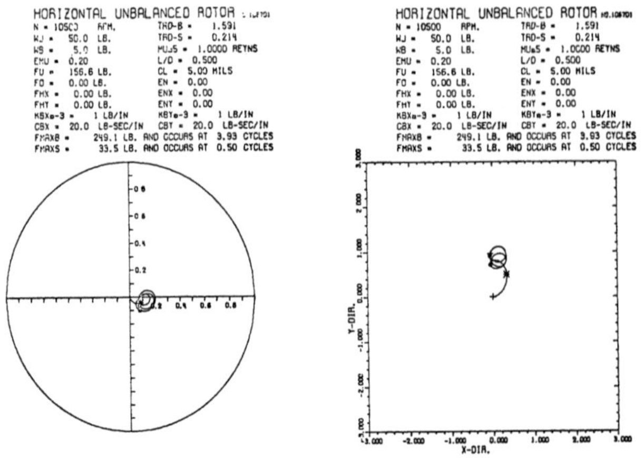

(a) (b)

Fig. 15a. Journal Response for Improved Support Characteristics (N = 10500 RPM, Eμ = 0.2).

15b. Support System Response for Five Cycles ($\bar{K}_B$ = 0.1, $\bar{C}_B$ = 0.556).

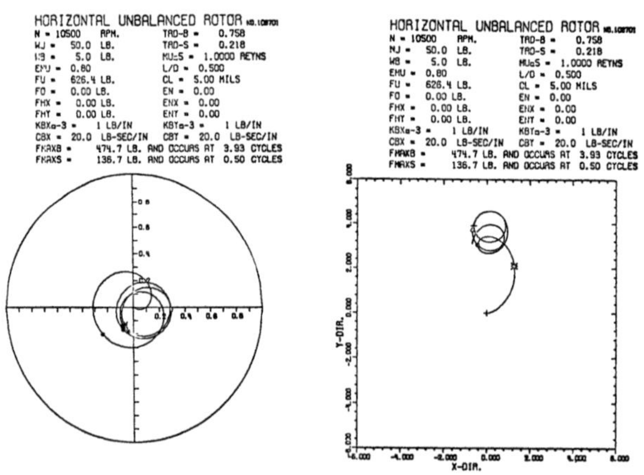

(a) (b)

Fig. 16a. Journal Response for Increased Imbalance (N = 10500 RPM, Eμ = 0.8).

16b. Support System Response ($\bar{K}_B$ = 0.1, $\bar{C}_B$ = 0.556)

on each bearing. Even if the imbalance level is increased by a factor of four (see Fig. 16) the resulting journal orbit is less than the rigid support system with one fourth the imbalance and running at only 6500 rpm. The maximum force transmitted to the support system is indicated to be only 136.7 lb. (608.07 N.) per support while the rotor is being driven by a rotating load of 1252.8 lb. (5572.73 N.).

These transient response orbits clearly indicate the advantages in optimum design consideration for both stability and forced response.

The instability of the plain cylindrical journal bearing for vertically mounted rotors or rotors operating in reduced or zero-g environment has been reported in the literature. From the classical theory of linearized stability this is the only conclusion possible.

With the time transient response analysis investigation some interesting results have been observed which indicate that the vertical plain journal bearing can be made to run with a small synchronous limit cycle. The rotor must have an imbalance level sufficiently high to produce synchronous response orbits yet the imbalance must not produce dynamic loads which are excessive.

Consider the same rotor as in the previous example mounted vertically and with a small imbalance, EMU = 0.01 to perturb the system at time $t = 0$. As the response of Fig. 17 indicates the system is very unstable and is orbiting outward a limit cycle response condition. The same response would occur at all rotor speeds, the only difference being the growth rate. If this system is now mounted in the optimum support system which was discussed for the horizontal rotor of Example 1 the growth rate is substantially reduced as indicated by the response shown in Fig. 18. However, the system remains unstable for this balanced rotor configuration.

The stability parameter for a short journal bearing is given as

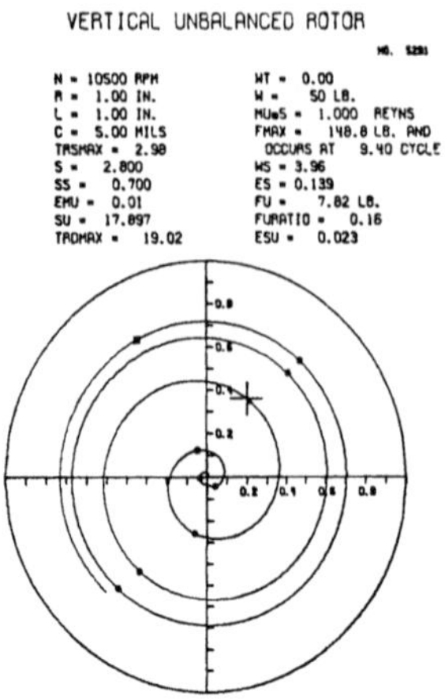

Fig. 17. Journal Orbit of a Vertical Rotor for Ten Cycles Showing Half-Frequency Whirl (N = 10500 RPM, Eμ = 0.01).

$$(25)\quad \omega_s = \omega\sqrt{\frac{mc}{W}}$$

If the load on the system is thought of as that arising from the imbalance, then it is possible to write

$$(26)\quad \Omega_s = \omega\sqrt{\frac{mc}{me_\mu\omega^2}}$$

or

$$\sqrt{\frac{c}{e_\mu}} > \Omega_s$$

for instability.

This can also be expressed as

$$\sqrt{\overline{EMU}} = \sqrt{E_\mu} > \frac{1}{\Omega_s}$$

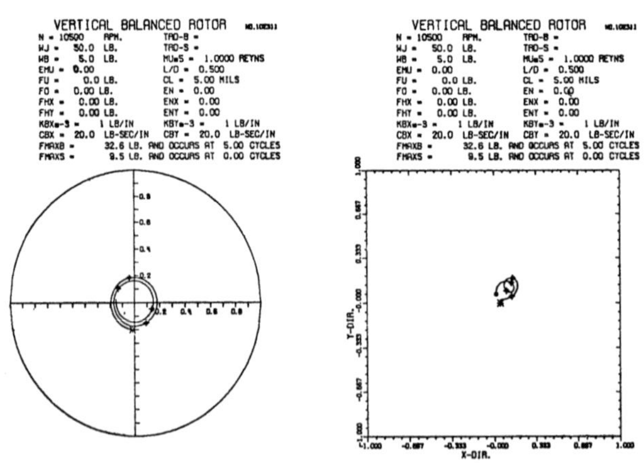

(a) (b)

Fig. 18a. Journal Relative Motion of a Balanced Vertical Rotor with Flexible Supports (N = 10500 RPM).

18b. Support System Response ($\bar{K}_B = 0.1$, $\bar{C}_B = 0.556$).

Now $\Omega_s \sim 2.5$ for the rigid support journal bearing, hence

(27) EMU > .16 (for stability)

This is a necessary condition for stability (but not sufficient). Consider the same rotor as in Fig. 17 with the imbalance increased to a value of EMU = 0.20 . The response for ten cycles is shown in Fig. 19a where the motion is observed to have both synchronous and half-synchronous components as indicated by the one inner loop and the occurrence of the marks. The response is going into a small orbit as indicated in Fig. 19b. The response for a higher speed is

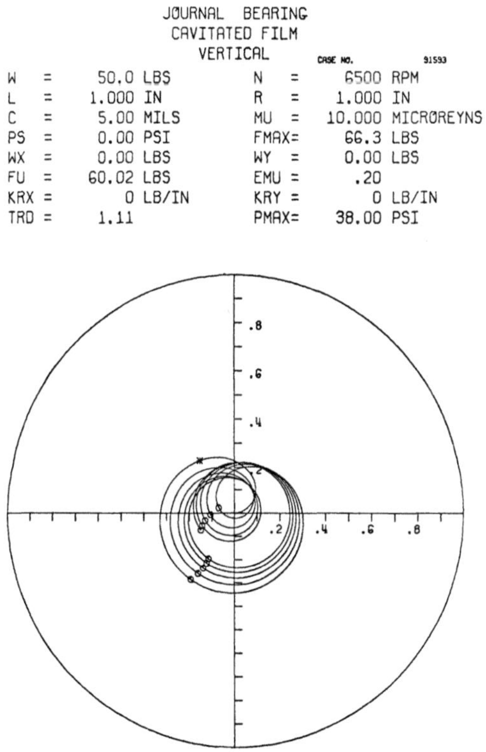

Fig. 19a. Journal Orbit of an Unbalanced Vertical Rotor Showing Motion Tending to Synchronous Response.

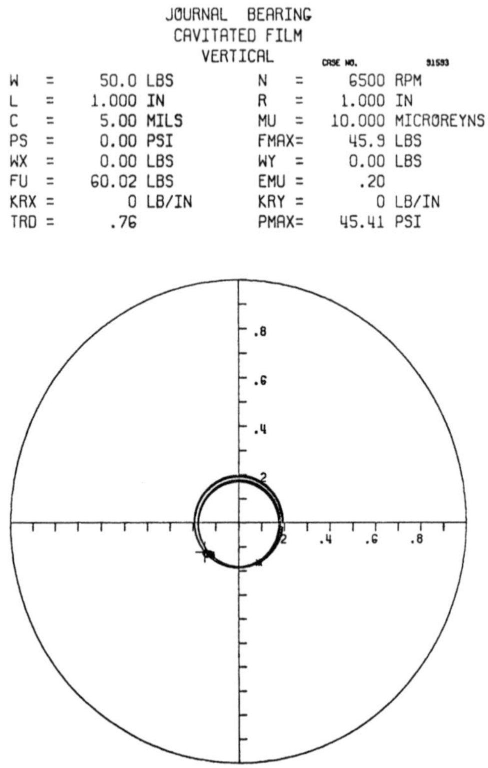

19b. Journal Orbit Motion Very Near Stable Synchronous Response.

shown in Fig. 20 where the limit cycle is composed of a much larger component of half-frequency whirl. The dynamic transmissibility is noted to be 1.28 for Fig. 20 whereas the synchronous response orbit of Fig. 19b was noted to have a ratio of only 0.76 . This means that the maximum load transmitted to the bearing was 76% of the imbalance load. This condition is very desirable as compared to the value of 1.28 in Fig. 20.

Extensive analysis of many time transient orbits for the vertical rotor mounted on rigid supports have allowed the following conclusions to be drawn (22).

1. The balanced, unloaded vertical rotor is always unstable.

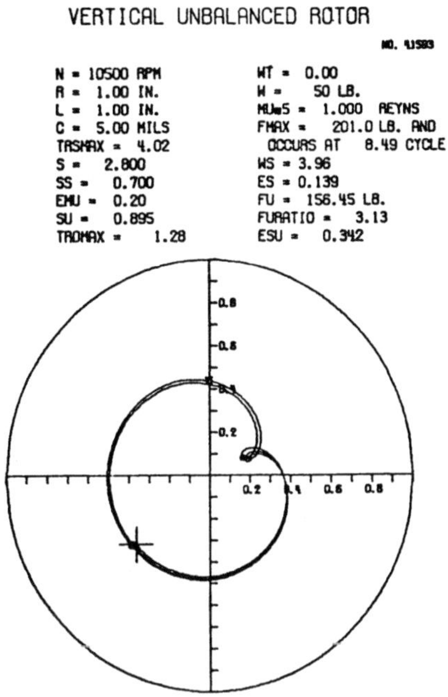

Fig. 20. Journal Limit Cycle for Increased Speed Showing Predominant Half-Frequency Component.

2. Unidirectional loading can stabilize a rotor by creating an effective g-loading on the bearing.

3. For very small values of imbalance (EMU < 0.16) the rotor will exhibit non-synchronous motion at all speeds.

4. The necessary and sufficient condition for stable synchronous small orbit response is that the imbalance be such that the value of EMU is slightly larger than 0.16 and the speed parameter, $\omega \sqrt{\frac{c}{g}}$ remains below a value of 2.5 .

The design of an optimum support is equally important for vertical rotors as is indicated in Fig. 21 where the large orbit response at 10,500 rpm in Fig. 20 has been substantially reduced. The maximum force transmitted to the support is only 20% of the load generated by the imbalance.

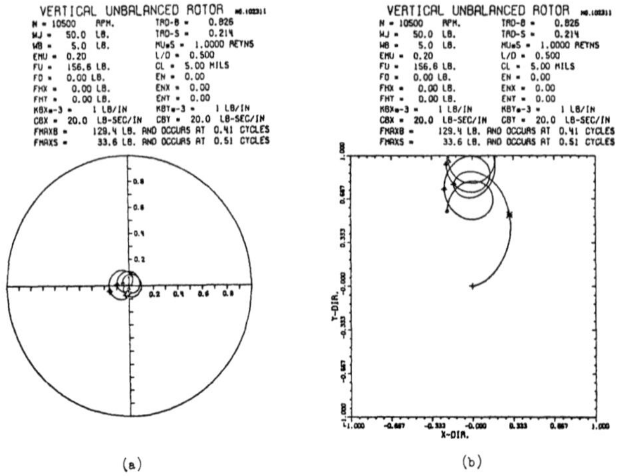

Fig. 21a. Journal Orbit for Flexible Support Vertical Unbalanced Rotor Showing Small Synchronous Response

21b. Support Transient Motion Showing Predominant Synchronous Response.

While the results are not all inclusive, there is enough evidence to indicate that the vertical journal can be stabilized by proper design of support characteristics and the addition of a small rotor imbalance. More extensive analysis of this subject area would be of great benefit to many manufacturers who use this type bearing in the design of water pumps and other similar equipment.

5. EXAMINATION OF ROTOR SYSTEM EIGENVALUES

The discussion of stability in the earlier section treated only the determination of the lowest threshold speed of the rotor system. The method of analysis (i.e., Routh criteria) cannot give any further information concerning the region of instability which is useful to the analyst. For this reason it is instructive and desirable to examine the complex eigenvalues of the characteristic equation. This is in fact the calculation of the damped natural frequency of the rotor-bearing system which can be expressed as:

(28) $$\lambda_k = a_k + i\omega_k$$

where

λ_k = kth eigenvalue

a_k = real part of eigenvalue and is denoted as the growth factor

ω_k = imaginary part of the eigenvalue and is the damped natural frequency corresponding to a specific natural mode of vibration.

The condition for stability is that the real part, ak , must have a negative value. The threshold speed is defined to be the speed at which the growth factor is zero. The larger the value of a_k the more stable or unstable is the corresponding mode depending on whether the real part is negative or positive respectively.

To take advantage of the additional information that is contained in the eigenvalue it is necessary to be able to solve for the roots of a polynomial with real coefficients. There are standard iterative schemes available in the literature for solving such problems. The particular method chosen for calculation of results in the following discussion was a modified BAIRSTOW's method [23]. In order to obtain the maximum information the characteristic determinant was set up for the rotor system of Fig. 1 including the flexibility of the rotor shaft. The resulting determinant is a 6 x 6 matrix with elements quadratic in λ and is given in Eq. 29. The procedure for expanding the determinant and solving for the roots of the resulting polynomial was programmed on a digital computer including the automatic calculation of stiffness and damping of the fluid-film journal bearing (24). Hence, the stability maps presented in the previous discussion can be checked approximately by introducing a very high stiffness in the rotor shaft with the proper weight distributed along the rotor. The program may be made to calculate only the first threshold speed by applying the Routh criteria to the polynomial or it may calculate all roots for a given speed range to allow a plot to be made of the change of the growth factor as the rotor speed is increased. The former method was applied to verify the stability boundaries generated by the Routh criteria as discussed in the previous section of this paper. Of particular interest was the stability boundary for the value of $\bar{C}_B = 0.001$ on Fig. 8. The rotor considered in the earlier discussion and described in Example 1 will be examined in more detail in the following discussion. The shaft stiffness was considered to be 1×10^6 lb./in. (1.75×10^6 N./cm.). If the support damping is reduced to a value of 0.4 lb.-sec./in. (.07N.-sec./cm), the value of $\bar{C}_B$ is calculated to be 0.001 . The stability threshold was found to be at the point indicated as 5 on Fig. 8.

(29)

$$
\begin{array}{r}
X_2: \\ Y_2: \\ X_1+X_j: \\ Y_1+Y_j: \\ X_1: \\ Y_1:
\end{array}
\left[\begin{array}{c|c|c|c|c|c}
m_2\lambda^2+C_s\lambda+K_s & 0 & -K_s & 0 & 0 & 0 \\
\hline
0 & m_2\lambda^2+C_s\lambda+K_s & 0 & -K_s & 0 & 0 \\
\hline
-K_s/2 & 0 & m_j\lambda^2+C_{xx}\lambda+K_s/2+K_{xx} & C_{xy}\lambda+K_{xy} & -C_{xx}\lambda-K_{xx} & -C_{xy}\lambda-K_{xy} \\
\hline
0 & -K_s/2 & C_{yx}\lambda+K_{yx} & m_j\lambda^2+C_{yy}+K_s/2+K_{yy} & -C_{yx}\lambda-K_{yx} & -C_{yy}\lambda-K_{yy} \\
\hline
0 & 0 & -C_{xx}\lambda-K_{xx} & -C_{xy}\lambda-K_{xy} & m_b\lambda^2+(C_{xx}+C_{1xx})\lambda+(K_{xx}+K_{1xx}) & (C_{xy}+C_{1xy})\lambda\;(K_{xy}+K_{1xy}) \\
\hline
0 & 0 & -C_{yx}\lambda-K_{yx} & -C_{yy}\lambda-K_{yy} & +(C_{yx}+C_{1yx})\lambda+(K_{yx}+K_{1yx}) & m_b\lambda^2+(C_{yy}+C_{1yy})\lambda+(K_{yy}+K_{1yy})
\end{array}\right]
\left[\begin{array}{c} A_1 \\ A_2 \\ A_3 \\ A_4 \\ A_5 \\ A_6 \end{array}\right]
=
\left[\begin{array}{c} 0 \\ 0 \\ 0 \\ 0 \\ 0 \\ 0 \end{array}\right]
$$

By increasing the rotor system properties by a factor of ten the load line shifted to follow the points indicated as 6, 7, 8 on Fig. 8. Point 6 was for the value of $\bar{C}_B = 0.001$, while point 8 was found for the value of $\bar{C}_B = 0.01$. This point is just below the rigid rotor threshold speed. For an increase of support damping to give $\bar{C}_B = 0.0556$ the threshold was calculated to lie at point 8 on Fig. 8. The points calculated verify the validity of the stability maps generated by the analysis of the characteristic equation given by Eqs. 19-21).

The nature of the instability given by the lightly damped support will now be examined by plotting the growth factors of the eigenvalues versus rotor speed.

<u>Example 2</u>:

Case (a) - very rigid rotor

The same rotor as considered for Example 1 with the following modifications.

K_s - shaft stiffness = 10×10^6 lb./in. (17.5×10^6 N./cm.)

C_{1xx} = 0.04, 0.08, 0.2, 0.4, 4.0, 20.0, 400.0 lb.-sec./in.

or (= 0.07, 0.14, .35, .70, 7.0, 35.0, 700.0 N.-sec./cm.)

$\bar{C}_B$ = 0.001, 0.002, 0.005, 0.01, .556, 11.12

The value of shaft deflection δ divided by the journal clearance is 0.001 for this rotor. As shown by RUHL [14], when $\delta/c < 0.1$, the shaft flexibility does not change the stability threshold speed (on rigid supports). Hence, this rotor may be considered as rigid. The plot of the growth factors for the various values of support damping are shown in Fig. 22a. For the value of $\bar{C}_B = 0.001$ the instability as predicted in Fig. 8 is indicated to initiate at a whirl frequency ratio of $\Omega = 0.5$. The value of the speed parameter $(\omega = 0.62)$ is in full agreement with Fig. 8. As the rotor speed is increased, it is noted that the whirl rate ratio decreases, indicating that the instability is exciting

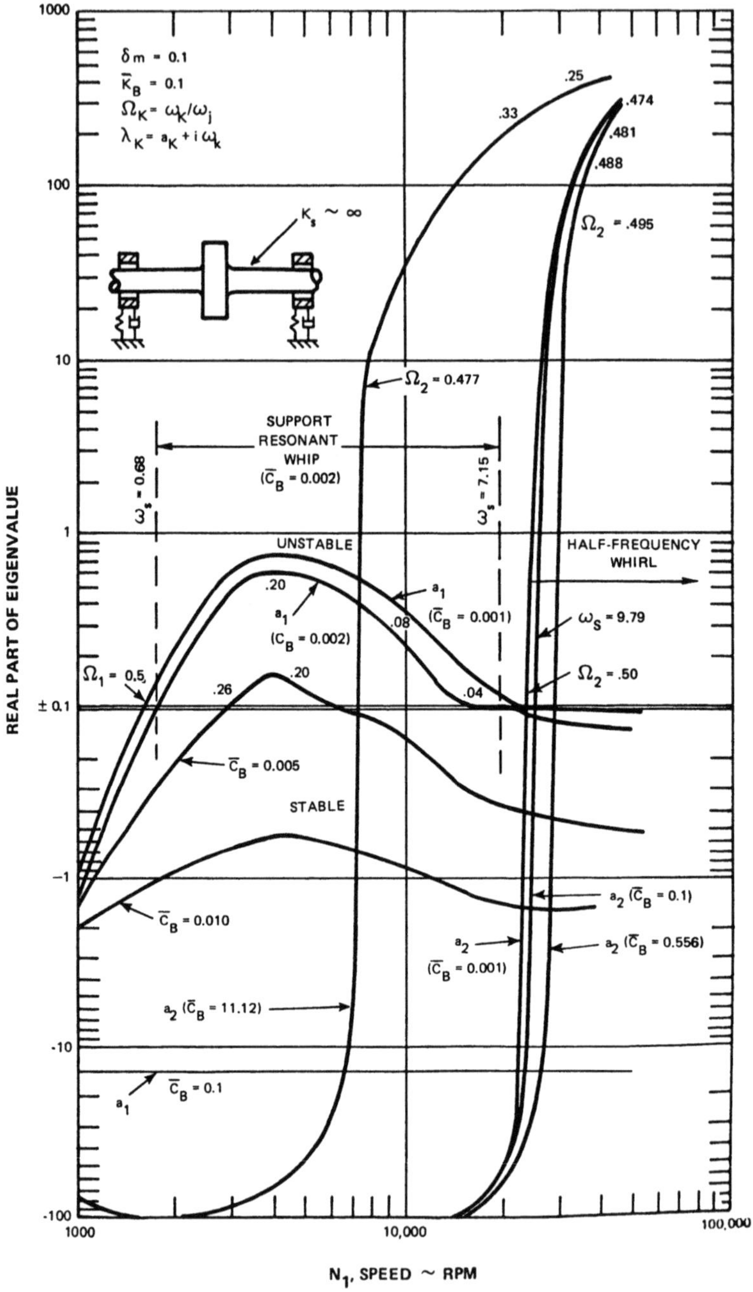

Fig. 22a. Stability Plot for a Rigid Rotor with Flexible Supports for Various Values of Support Damping.

the lowest natural frequency in the system and can be classified as a support resonant whip. This region of instability extends beyond the onset speed of a much more severe instability (as indicated by the large growth rate). The whirl rate for this region remains at just below one-half rotor speed and is the classical half-frequency whirl instability normally associated with journal bearing instability. As the support damping is increased the extent and severity of the support whip is reduced until it is completely suppressed as is indicated for the value of $\bar{C}_B = 0.01$. The onset speed of the half-frequency whirl is not greatly influenced by the lower values of support damping considered for this case study. The speed parameter at the threshold for $\bar{C}_B = 0.1$ is shown to be at $\omega_s = 9.41$ which is in good agreement with the curve of Fig. 8.

For the optimum damping of $\bar{C}_B = 0.556$ the instability threshold speed is increased to 30,000 rpm ($\omega_s = 11.2$) as was indicated in the stability map (see Fig. 8). For the overdamped support condition ($\bar{C}_B = 11.12$) the stability threshold speed is reduced to 7000 rpm ($\omega_s = 2.63$), which is very near the rigid support value of $\omega_s = 2.5$. The influence of rotor shaft flexibility on stability will be demonstrated by considering the above rotor system with reduced rotor shaft stiffness.

<u>Example 2</u>:

Case (b)

Consider the same basic system as in case (a) with

$K_s = 1.0 \times 10^6$ lb./in. (1.75×10^6 N./cm.)
$C_{1xx} = 0.08$ lb.-sec./in. ($\bar{C}_B = 0.002$)
$\delta/C = 0.01$

The growth factor is plotted in Fig. 22b where it is observed that the region of support resonant whip is very nearly the same as shown in Fig. 22a. The growth factor for the flexible shaft mode are plotted as a_3, a_4. The large negative value for these factors indicate a very stable condition.

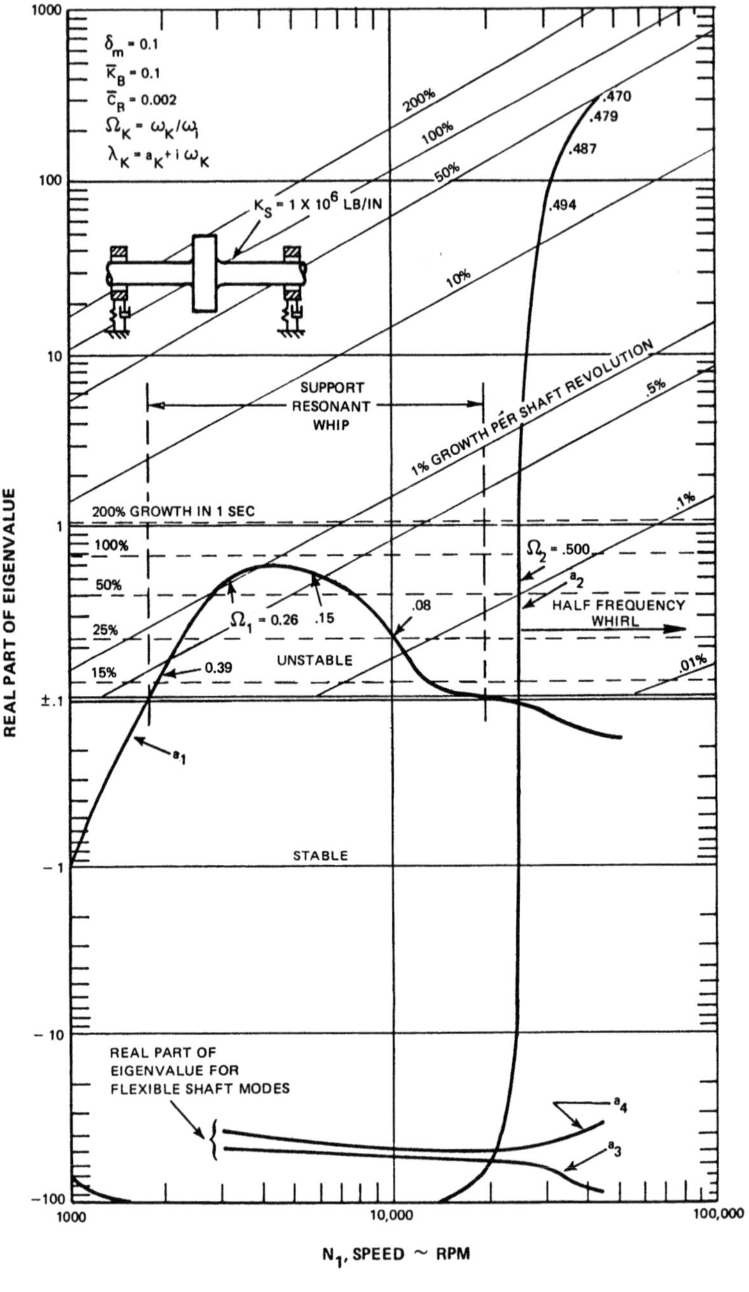

Fig. 22b. Stability Plot for a Rotor with High Shaft Stiffness.

The growth factor related to the half-frequency whirl is very much the same as that shown in the previous plot.

An indication of the severity of the instability can be shown by considering the percentage growth of a perturbation per revolution of the rotor. This is calculated as

$$(30) \qquad a = \frac{N}{60} \ell n \left(\frac{\%\ \text{Growth per revolution}}{100} + 1\right)$$

The diagonal lines in Fig. 22b are plotted for various percentage increase of perturbation per revolution. It is noted that the resonant whip has less than 1.0% growth per shaft revolution whereas the half-frequency whirl has a growth of 10% just beyond the threshold speed. The system being studied is nonlinear and hence the growth rate changes as the limit cycle is approached. The initial rate of growth, however, should agree with those calculated from linear theory. This growth factor interpretation gives a good basis for selection of time duration for transient response simulation. Several rotor revolutions would be required to produce substantial growth for the speeds near the threshold condition of the support resonant whip.

An alternative approach to the interpretation of the growth rate would be to calculate the percentage growth of a perturbation per second of real time. The horizontal lines on Fig. 22b give this information. This interpretation shows clearly the severity of the half-frequency whirl instability.

<u>Example 2:</u>

Case (c)

Consider now a rotor with increased shaft flexibility.

$K_s = 1.0 \times 10^5$ lb./in. (1.75×10^5N./cm.)

$C_{1xx} = 0.08$ lb.-sec./in. ($\overline{C}_B = 0.002$)

$\delta/C = 0.1$

The lower shaft stiffness produces a very interesting result as indicated in Fig. 22c. The region of support resonant whip remains unchanged but an instability associated

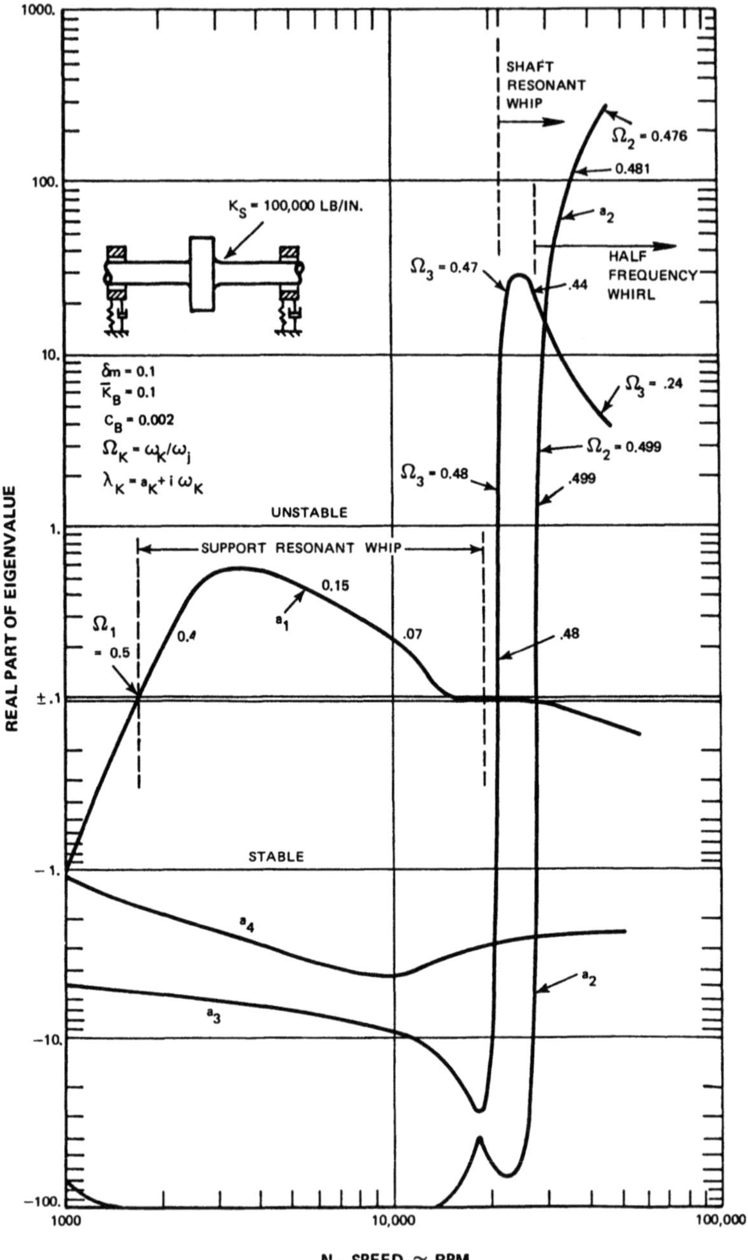

Fig. 22c. Stability Plot for a Flexible Rotor Showing Three Modes of Instability.

with the flexible shaft is excited just before the half-frequency whirl. This region of stability may be referred to as shaft resonant whip. Newkirk reported the occurrence of this type instability in a flexible shaft experimental rig. As indicated by the whirl rates the instability is exciting a lower system critical which does not increase substantially with increased speed. The third type of instability shown on the plot is the half-frequency whirl which is indicated to be the most severe of the instabilities. Considering the previous discussion of growth per second it is very unlikely that this system could be driven through the shaft resonant whip instability. These results give a physical meaning to the stability maps and a better understanding of the phenomena of rotor restabilization that has been observed in experimental test rigs.

6. CONCLUSIONS

The following observations and conclutions may be stated considering the results that have been presented in this discussion.

1. The stable operating speed range of a rotor in fluid-film bearing may be greatly extended by the proper design of the support flexibility, inertia, and damping characteristics. Stability maps have been presented for various dimensionless support characteristics.
2. Imbalance will not extend the region of stability for a horizontal rotor excited by the half-frequency whirl instability.
3. The addition of a small imbalance may allow a vertical or o-g rotor to operate with a small synchronous response orbit.
4. Excessive imbalance will force a rotor to execute synchronous motion above the stability threshold. This mode

of operation is very undesirable due to the excessive bearing loading produced by the imbalance.

5. A lightly damped support structure can develop a subsynchronous instability (support resonant whip) that could in some cases be driven through to reach a stable operating speed. The journal bearing instability will be excited at a rotor speed higher than that which would exist were the rig on rigid supports.

6. Excessive support damping ($\overline{C}_B > 10.0$) reduces the flexible support stability threshold speed to that which would be obtained were the system on rigid bearing supports.

7. Rotor shaft flexibility may introduce a condition of shaft resonant whip which could occur before or after the support instability depending on rig properties.

8. A computer program has been developed to rapidly calculate the stability boundaries of a symmetric flexible rotor mounted on fluid-film bearings and having damped support structure. The calculation of the nonlinear floating bush viscous damper characteristics has been included in the computer program.

REFERENCES

[1] NEWKIRK, B.L.: Shaft Whipping, Gen. Elec. Rev., Vol. 27, 1924, p. 169.

[2] NEWKIRK, B.L. and TAYLOR, H.D.: Shaft Whipping Due to Oil Action in Journal Bearings, Gen. Elec. Rev., Vol. 28, 1925, pp. 985-988.

[3] ROBERTSON, D.: Whirling of a Journal in a Sleeve Bearing, Phil. Mag., Series 7, Vol. 15, No. 66, January 1933.

[4] HAGG, A.C.: The Influence of Oil Film Journal Bearings on the Stability of Rotating Machines, Journal of Applied Mechanics, Vol. 68, 1946, p. 211.

[5] ORBECK, F.: Theory of Oil Whip for Vertical Rotors Supported by Plain Journal Bearings, Trans. ASME, Oct. 1958, pp. 1497-1502.

[6] PINKUS, O.: Experimental Investigation of Resonant Whip, Trans. ASME, Vol. 78, 1956, pp. 975-983.

[7] PORITSKY, H.: Contribution to the Theory of Oil Whip, Trans. ASME, Aug. 1953, pp. 1153-1161.

[8] HAGG, A.C. and WARNER, P.C.: Oil Whip of Flexible Rotors", Trans. ASME, Vol. 75, October, 1953, pp. 1339-1344.

[9] GUNTER, E.J.: Dynamic Stability of Rotor-Bearing Systems, NASA SP-113, 1966.

[10] HORI, Y.: Theory of Oil Whip, Journal of Applied Mechanics, Vol. 26, Trans. ASME, Series E, Vol. 81, 1959, p. 189.

[11] BADGLEY, R.H.: Turborotor Instability - Dynamic Unbalance, Gyroscopic, and Variable-Speed Effects with Finite-Length, Cavitated, Fluid-Film Bearings, PH.D. Dissertation, Cornell University, June 1967.

[12] LUND, J.W.: The Stability of an Elastic Rotor in Journal Bearings with Flexible, Damped Supports, Journal of Applied Mechanics, Trans. ASME, Series E, Vol. 87, No. 4, December 1965, pp. 911-920.

[13] GUNTER, E.J.: The Influence of Flexibility Mounted Rolling Element Bearings on Rotor Response, Part 1: Linear Analysis, Journal of Lubrication Technology, Trans. ASME, pp. 59-75, January 1970.

[14] RUHL, R.L.: Dynamics of Distributed Parameter Rotor Systems: Transfer Matrix and Finite, Ph.D. Dissertation, Cornell University, January, 1970.

[15] LUND, J.W.: Stability and Damped Critical Speeds of a Flexible Rotor in Fluid-Film Bearings, Trans. ASME, Journal of Engineering for Industry, Vol. 96, Series B, No. 2, May, 1974, pp. 509-517.

[16] CHOUDHURY, P.De.: Dynamic Stability of Flexible Rotor-Bearing Systems, Ph.D. Dissertation, University of Virginia, June 1971.

[17] KIRK, R.G. and GUNTER, E.J.: Nonlinear Transient Analysis of Multi-Mass Flexible Rotors-Theory and Applications, NASA CR-2300, September, 1973.

[18] NEWKIRK, B.L.: Varieties of Shaft Disturbances Due to Fluid Films in Journal Bearings, Trans. ASME, Vol. 78, 1956, p. 985.

[19] OCVIRK, F.W.: Short-Bearing Approximation for Full Journal Bearings, NACA TN 2808, 1952.

[20] BADGLEY, R.H. and BOOKER, J.F.: Turborotor Instability: Effect of Initial Transient on Plane Motion, Journal of Lub. Tech., Trans. ASME, October, 1969, pp. 625-633.

[21] REDDI, M.M. and TRUMPLER, P.R.: Stability of the High-Speed Journal Bearing Under Steady Load - 1. The Incompressible Film, Journ. of Engineering for Insustry, Trans. ASME, Series B, Vol. 84, 1962, pp. 351-358.

[22] KIRK, R.G. and GUNTER, E.J.: Transient Journal Bearing Analysis, NASA CR-1549, Washington, D.C., June, 1970.

[23] BECKETT, R. and HURT, J.: Numerical Calculations and Algorithms, McGraw Hill Book Co., 1967, p. 70.

[24] KIRK, R.G.: Nonlinear Transient Analysis of Mult-Mass Flexible Rotors, Ph.D. Dissertation, University of Virginia, June, 1972.

Rotating Blade Analysis by the Finite Element Method

M. Lalanne, R. Henry and P. Trompette

Institut National des Sciences Appliquees
Laboratoire de Mecanique des Structures, Villeurbanne, France

INTRODUCTION

Blades are important and expensive part of turbomachinery; they can be excited by harmonic external forces. At the frequencies of resonances, the level of stresses may be very high due to the low loss factor of the material. In order to avoid these resonances, it is important to predict accurately the dynamical behaviour of the blade that is the behaviour of the assembly disc-blades and shroud.

In a recent interesting survey [1] J.S. RAO presents many papers on natural frequencies of turbine blading with a special emphasis on steam turbines and various solutions of several types of beams. Very recently, papers on calculation of blades using finite element method and sometimes, with application to existing blades have been published.
M.A. DOKAINISH and S. RAWTANI [2] present the calculations of rotating thin plates and deduce modified SOUTHWELL coefficients. In [3], the calculation of thin rotating blades is presented by M.A. DOKAINISH and D.M. GOSSAIN. H.J. BARTEN, J.A. SCHEURENBRAND and D.D. SCHEER [4] determine stresses, frequencies and mode shapes of an existing non rotating thin blade. R. AHMAD, R.G. ANDERSON and O.C. ZIENKIEWICZ [5] calculate frequencies and mode shapes of an existing thick blade. M.A.J. BOSSAK and O.C. ZIENKIEWICZ [6] give the extension of paper [5] to thick rotating blades and to buckling phenomena. Precisions on the calculation of existing non rotating blades are given by L.D. HOFFMEISTER and D.A. EVENSEN [7]. J. THOMAS and C.A. MOTTA SOAREZ [8] give the frequencies and mode shapes of thick rotating blades

with application to rotating plates as in paper [2]. P.O. ARALDSEN apply the superelement method to an existing thick non rotating blade [9]. In [10] and [11], the calculation of frequencies and mode shapes of existing thin rotating blade is presented, and a design procedure using the results of paper [2] and of a previous paper [12] is suggested. In [13] and [14], the frequencies and mode shapes of an existing thick rotating blade are given. The influence of the temperature and of the root condition is analysed.

In this paper, we present how to obtain easily using Lagrange's equations and finite element method a mathematical model of rotating blades but the various matrices are not presented as they are given in [10], [11], [13], [14]. The method is then applied to existing systems: a gas turbine and an aircraft engine. At last, some suggestions are presented on future work.

MATHEMATICAL MODELISATION OF THE STRUCTURE

The expressions of the kinetic and potential strain energies are respectively:

(1) $$T = \frac{1}{2} \int_V \rho V^2 dv$$

(2) $$U = \frac{1}{2} \int_V \varepsilon^t \sigma dv$$

with:

- V , velocity
- ρ , mass per unit volume
- t , matrix transposition symbol
- ε , strains
- σ , stresses

Between stresses and strains the relation is:

(3) $$\sigma = D \cdot \varepsilon$$

with:

D , matrix function of the material characteristics.

For large strains, the expressions of ε must include non linear terms.

Then,

(4) $$U = I_1 + I_2$$

I_1 , known expression of the potential strain energy

I_2 , expression of the potential large strain energy obtained by linearisation of the initial stresses.

A finite element modelisation is now used. In the case of thin blades, the structure is modeled by triangular plane elements with three nodes and six degrees of freedom by node (fig. 1). In the case of thick blades, the structure is modeled by isoparametric elements with 24 nodes and 3 degrees of freedom by node (fig. 2).

Then, for the whole structure:

(5) $$\frac{\partial U}{\partial \delta} = [K_E + K_G(\sigma_o)]\delta$$

with:

δ , nodal displacement vector

σ_o , initial stresses

K_E , classical stiffness matrix

K_G , geometric stiffness matrix

and

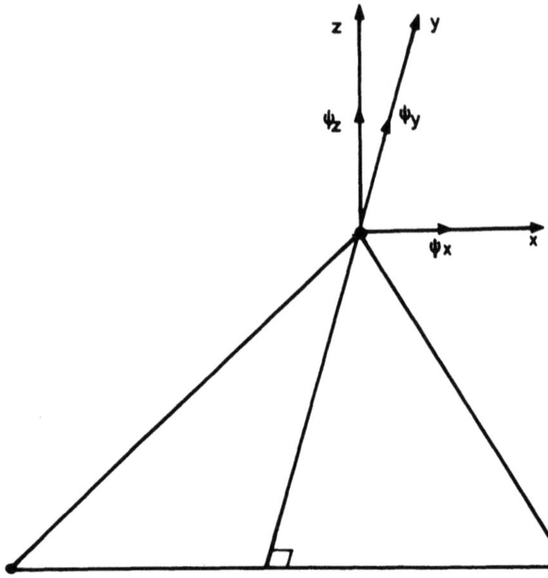

Fig. 1.

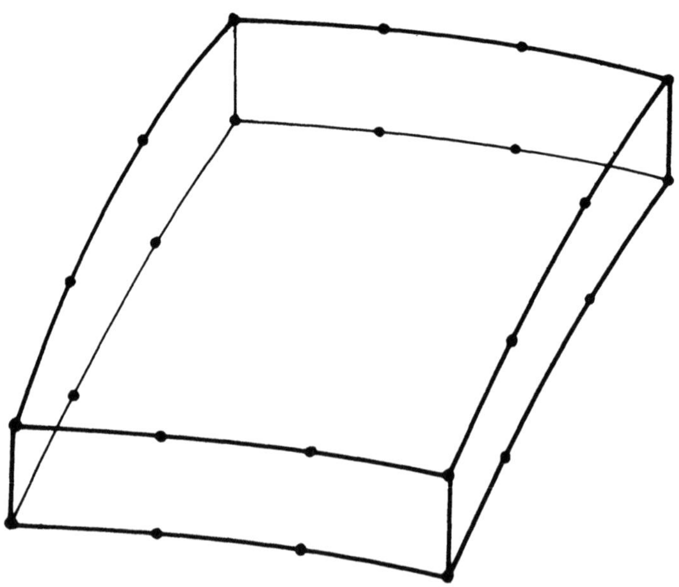

Fig. 2.

(6) $$\frac{d}{dt}\left[\frac{\partial T}{\partial \dot{\delta}}\right] - \frac{\partial T}{\partial \delta} = M_E \delta^{oo} + C\delta^{o} - \Omega^2 M_G \delta - F(\Omega^2)$$

with:

t	,	time
C	,	coriolis matrix
Ω	,	speed of rotation
$\Omega^2 M_G$	,	supplementary stiffness matrix
$F(\Omega^2)$	,	centrifugal forces

The matrix equation of the whole structure according to Lagrange equation is:

(7) $$M_E \delta^{oo} + C\delta^{o} + [K_E + K_G - \Omega^2 M_G]\delta = F(\Omega^2)$$

The influence of C on blade behaviour is small [6] and may be neglected, then (7) becomes:

(8) $$M_E \delta^{oo} + [K_E + K_G(\sigma_o) - \Omega^2 M_G]\delta = F(\Omega^2)$$

The stresses due to rotation are deduced from:

(9) $$[K_E + K_G(\sigma_o) - \Omega^2 M_G]\delta = F(\Omega^2)$$

through an iterative process. Initially, the stresses are taken equal to zero, it appears that for practical applications the procedure can be stopped after the first iteration.

The lowest frequencies and mode shapes are obtained from:

(10) $$[K_E + K_G(\sigma_o) - \Omega^2 M_G]\delta = \omega^2 M_E \delta$$

where here δ is the amplitude of the motion about the equilibrium position.

As shown in [6], the differential equation (8) may also be used for buckling phenomena.

Frequencies and mode shapes are obtained using a simultaneous iterative method [15] without reduction of the total number of degrees of freedom. The numerical integration of the isoparametric element is performed with 32 $(4 \times 4 \times 2)$ Gauss points. For the plane triangular thin shell element, the various integrals are calculated exactly.

In the computer program, stiffness and mass matrices of the structure are partitionned in square blocks, dimensions of which is the bandwith. For systems which are not too large, this partitionning allows us to minimize the total computer time. All the calculations have been performed on a CDC 6600 Computer.

APPLICATION

The following examples which illustrate the general method described above are chosen in two important fields of the present day mechanical engineering. They concern aircraft and gas turbine turbomachinery. In these two applications, triangular plane and isoparametric elements are used respectively to model thin and thick blades.

<u>I GAS TURBINE</u>. Figure 3 shows the different parts of the gas turbine which have been studied:

Inducer blade ⓐ
Exducer blade ⓓ
Turbine blade ⓔ

<u>Inducer Blade</u>. (fig. 4). Mechanical characteristics of the constituting material AU2GN are:

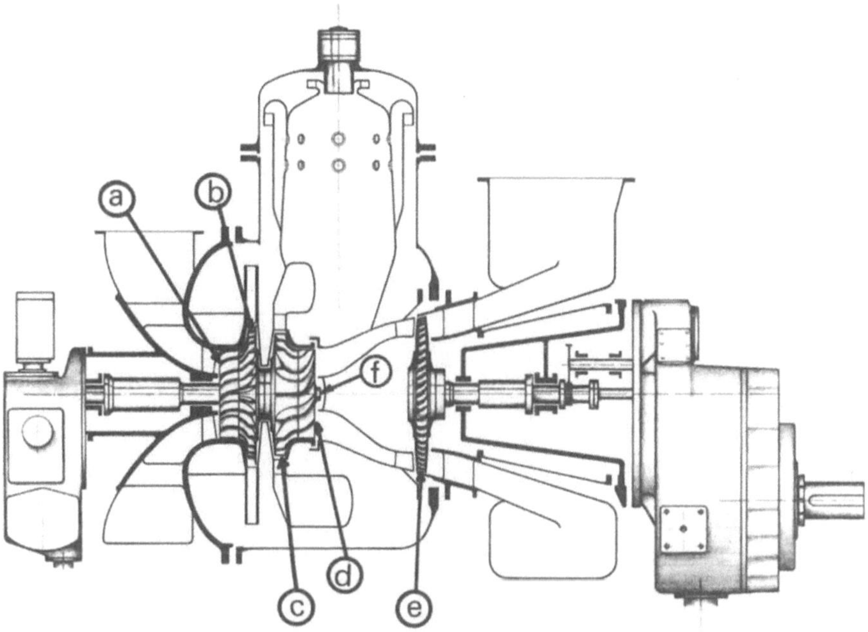

Fig. 3.

$$E = 0.736\ 10^{11}\ N/m^2$$
$$\nu = 0.34$$
$$\rho = 2.75\ 10^3\ kg/m^3$$

Boundary conditions between inducer (a) and compressor impeller (b) are given by the following conditions for the 2 nodes a , b (fig. 6).

$u = w = 0$ (displacements along 0x and 0z)

$\psi x = \psi z = 0$ (rotations around 0x and 0z)

Fig. 6 shows that the agreement between theoretical and experimental values of both frequencies and mode shape is good. At the operating speed, the frequencies are slighty higher and could be calculated approximately by the formulas of paper [2]. The maximum intrados and extrados principal stress σ_1 obtained by the finite element method is presented in fig. 7.

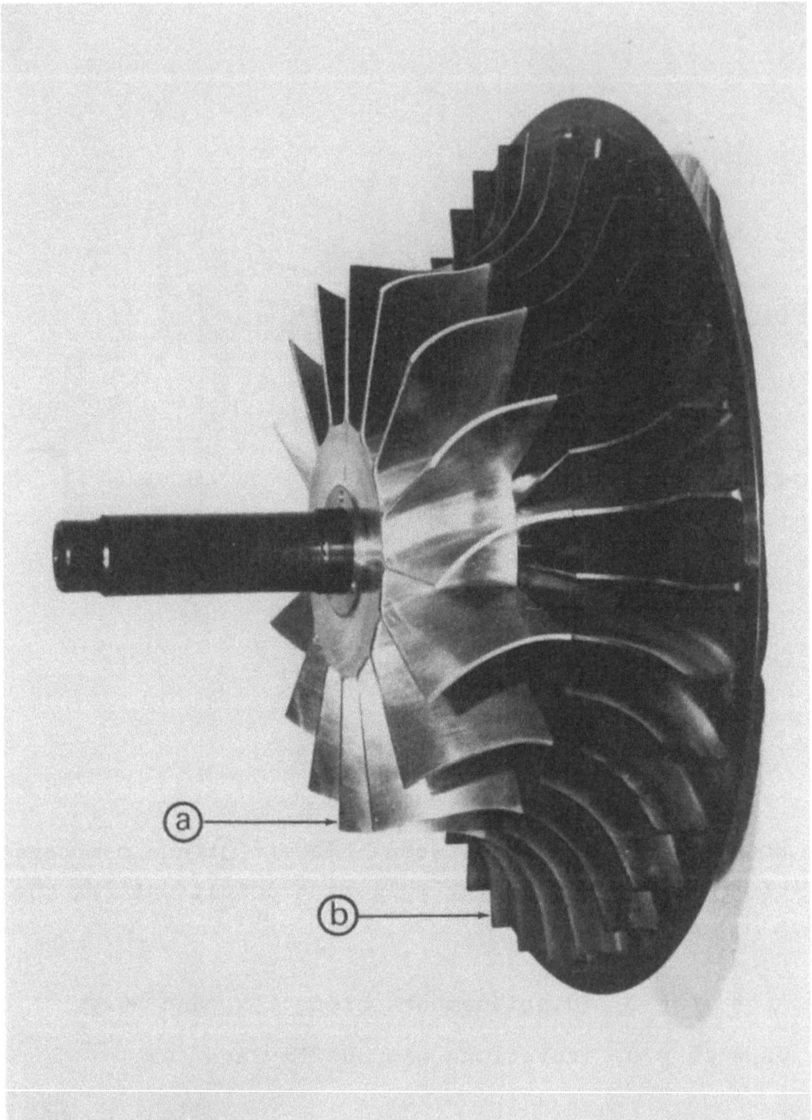

Fig. 4

<u>Exducer Blade</u>. (fig. 5). Mechanical characteristics of the constituting material Inconal 713 C are:

$$E = 2.06\ 10^{11}\ N/m^2$$
$$\nu = 0.3$$
$$\rho = 7.91\ 10^{3}\ kg/m^3$$

At the operating speed, the temperature is about 700^{0} C and E equal to $1.70\ 10^{11}\ N/m^2$.

The exducer blades ⓓ are tighten on the turbine impeller blades ⓒ through the bolt ⓕ .

In this case, the finite element boundary conditions at nodes a , b , c (fig. 8) are approximated:

Fig. 5

INDUCER BLADE

Frequencies and mode-shapes

f_{Hz}	Experim. (- - - -)	Fin. Elem. (——)	
	N = 0. rpm	N=0	N=29 700
1	1 330	1 280	1 495
2	3 450	3 613	3 755
3	4 920	5 337	5 440
4	6 460	6 795	6 940
5	-	8 280	8 350

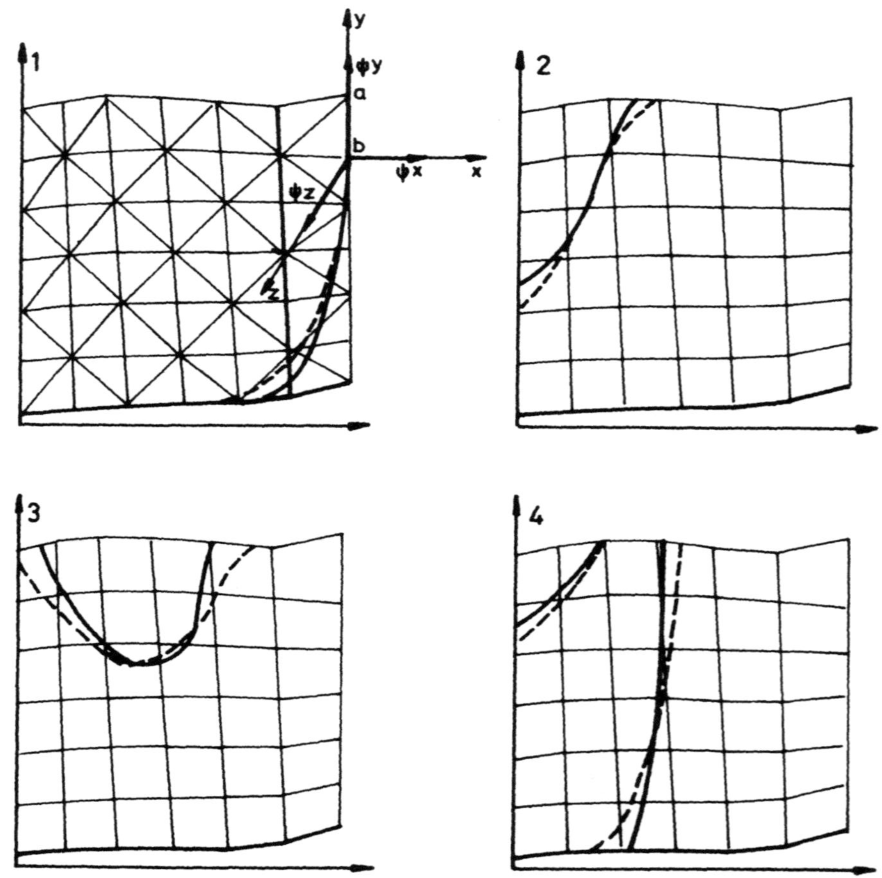

Fig. 6.

INDUCER BLADE STRESSES AT N = 29700. rpm

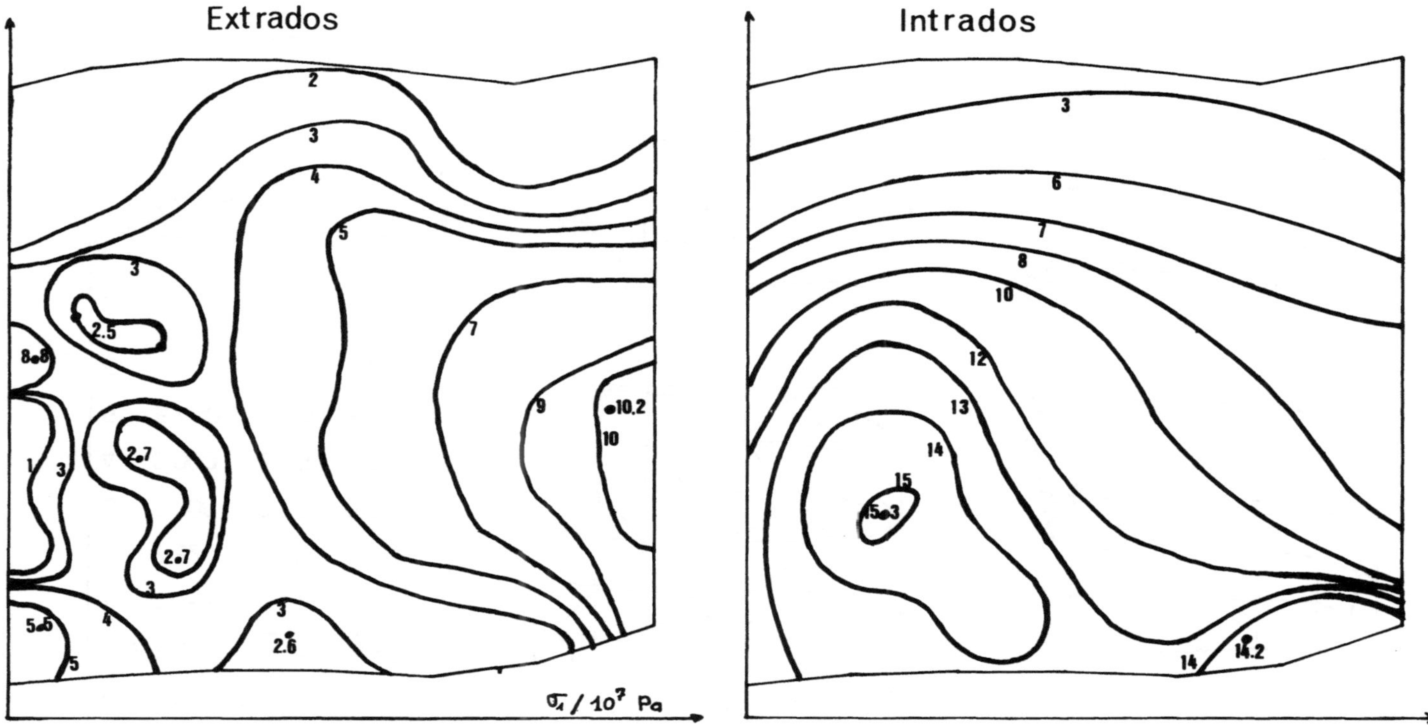

Fig. 7.

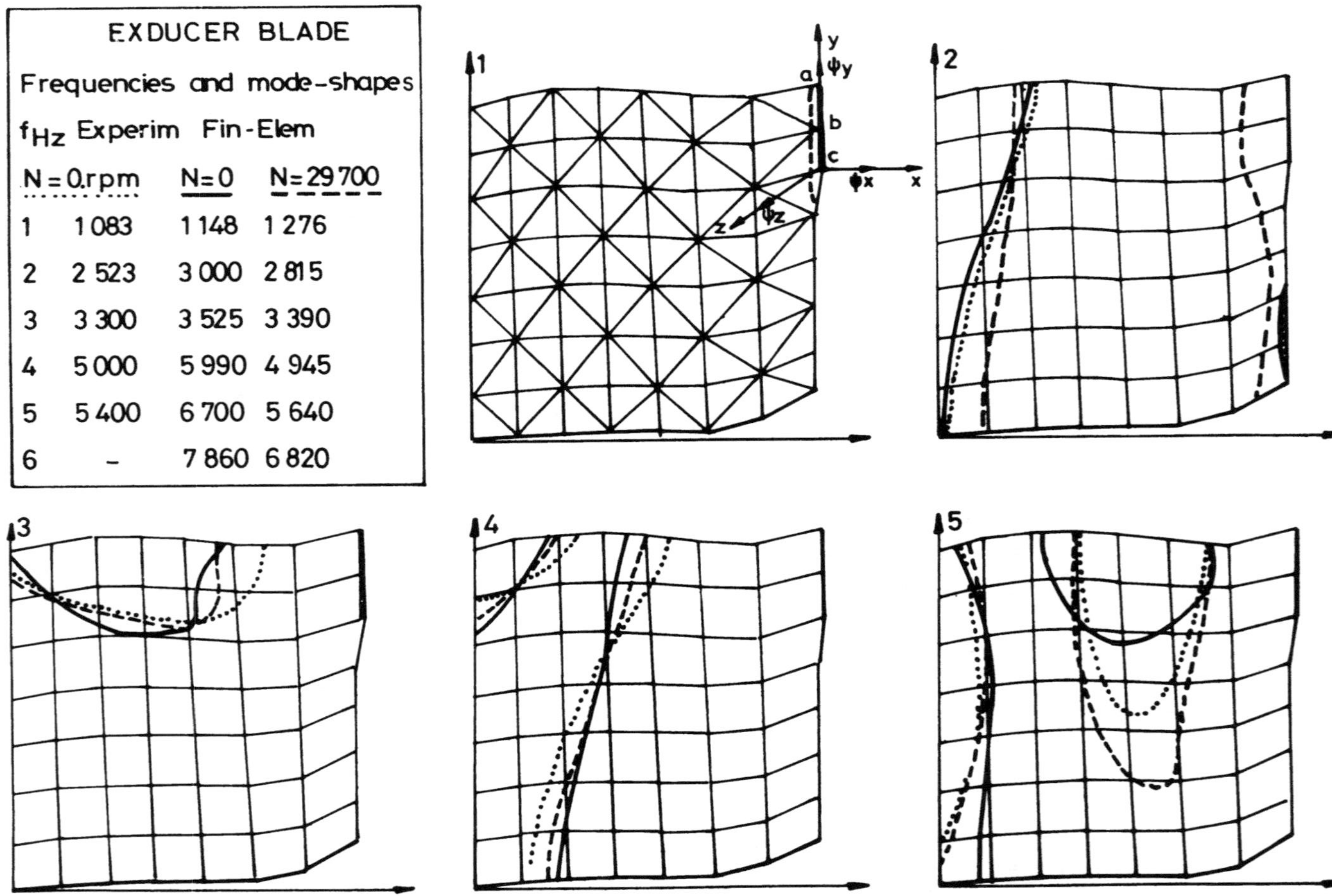

EXDUCER BLADE

Frequencies and mode-shapes

f_{Hz}	Experim	Fin-Elem	
	N = 0.rpm	N=0	N=29 700
1	1 083	1 148	1 276
2	2 523	3 000	2 815
3	3 300	3 525	3 390
4	5 000	5 990	4 945
5	5 400	6 700	5 640
6	-	7 860	6 820

Fig. 8.

- At rest, they are chosen as follows:

$$u = v = w = 0$$
$$\Psi x = \Psi z = 0$$

- At operating speed, it has been verified that rotation unloses slightly (d) from (c) so the boundary conditions have to be modified and are:

$$u = 0$$

$$\Psi z = 0$$

Fig. 8 shows the experimental and theoretical frequencies at rest and at operating speed. The difficulty to express the right boundary conditions is the prime cause of the relative inaccuracy on frequencies, nevertheless the agreement on mode shapes is rather satisfactory.

The maximum intrados and extrados principal stress σ_1 obtained by the finite element method is presented in fig. 9.

Turbine Blade. (fig. 3. (e)). It is made with NIMONIC 90 mechanical characteristics of which are:

$$E = 1{,}97\ 10^{11}\ N/m^2$$
$$\nu = 0.3$$
$$\rho = 8.27\ 10^3\ kg/m^3$$

Fig. 10 illustrates the agreement between experimental and theoretical results, which is good for the three first frequencies with only four elements.

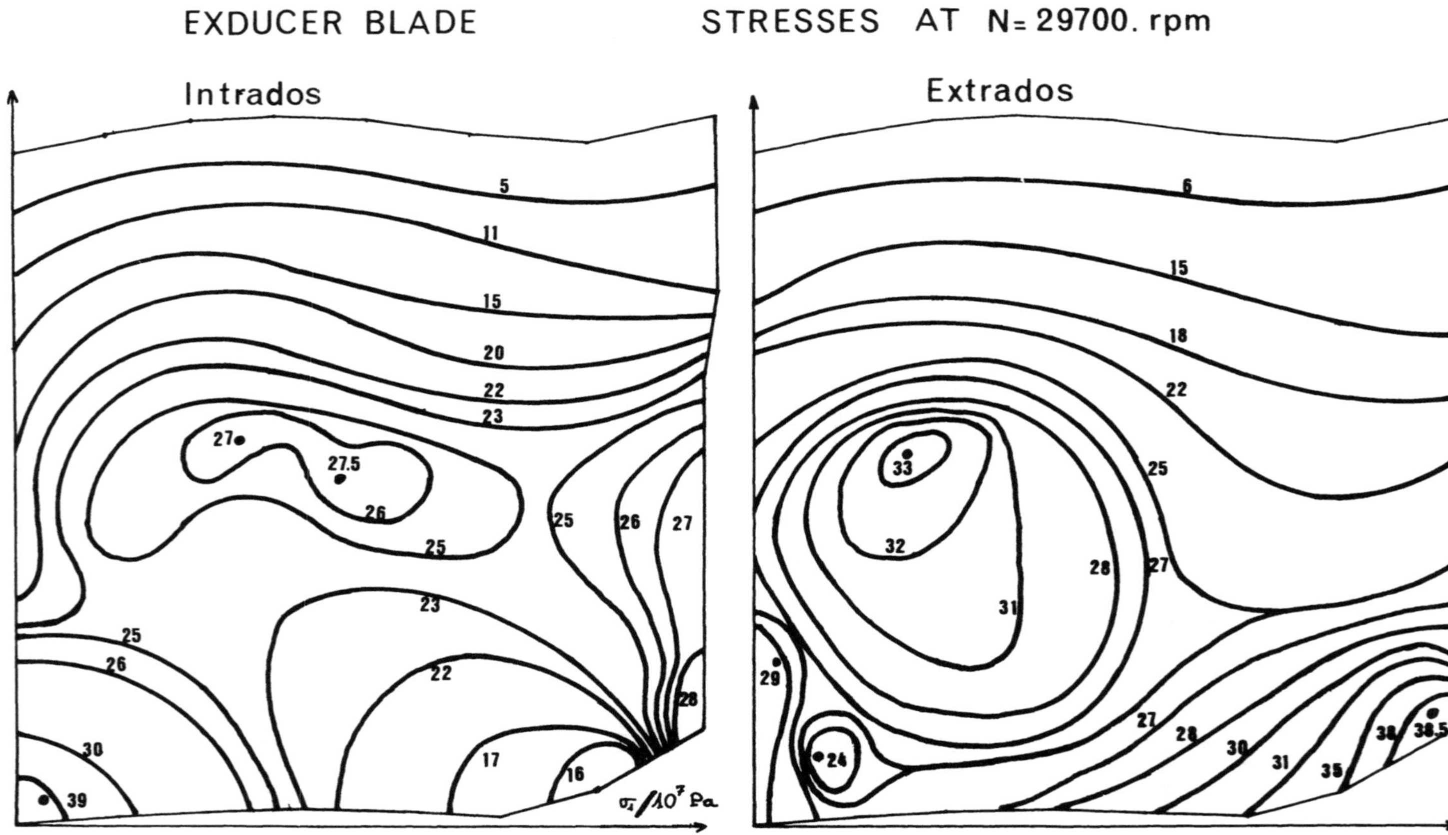
EXDUCER BLADE
STRESSES AT N= 29700. rpm
Intrados
Extrados
5
11
15
20
22
23
27
27.5
26
25
25
26
27
23
25
26
22
28
30
17
16
39
$\sigma_I/10^7$ Pa
6
15
18
22
25
33
32
28
27
31
29
24
27
28
30
31
35
38
38.5

Fig. 9.

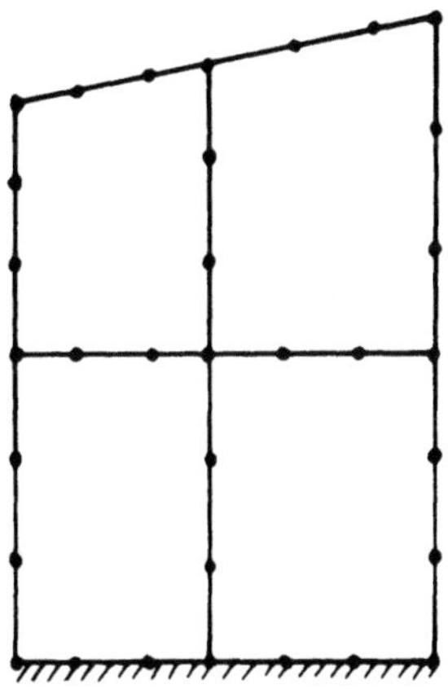

Freq. Hz	1 EL	2 EL	4 EL	Exp.	%
N_1	3 405	2 541	2 308	2 282	1
N_2	7 444	5 158	4 619	4 150	11
N_3	9 598	7 850	7 187	7 660	7

Fig. 10.

II AIRCRAFT ENGINE.

Turbofan Blade. In this part, the most important aim was to know the principal stresses near the root due to centrifugal effects. Fig. 11 presents the finite element mesh, fig. 12 the different results.

The experiments were very difficult to carry out; several models in araldite were built and tested. Due to the difficulties of these experiments, the results may be considered as good. Let us mention also that the influence of large strains, matrix $K_G\ (\sigma_o)$, is very important.

Turbine Blade. The purpose is to obtain frequencies and modes shapes of a blade clamped on to the platform and

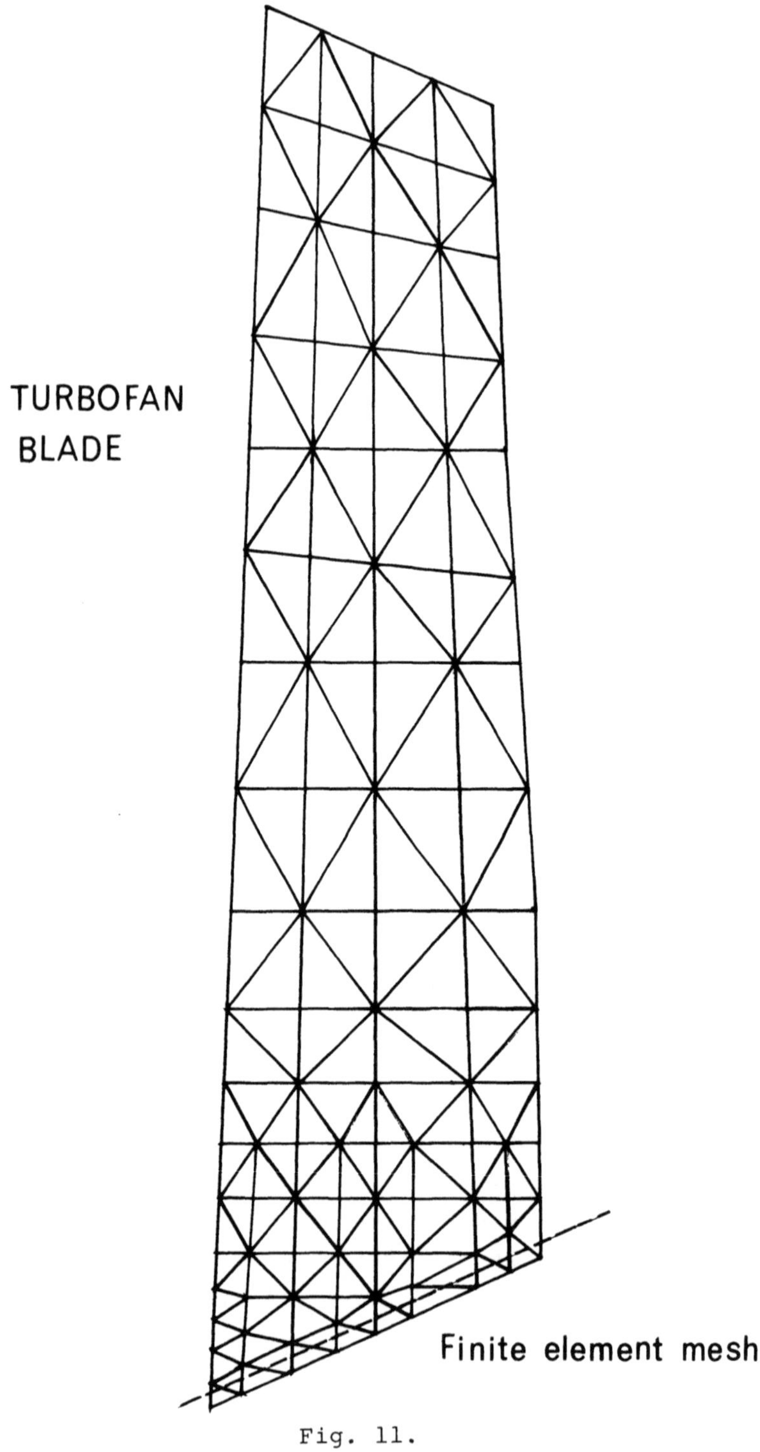

Fig. 11.

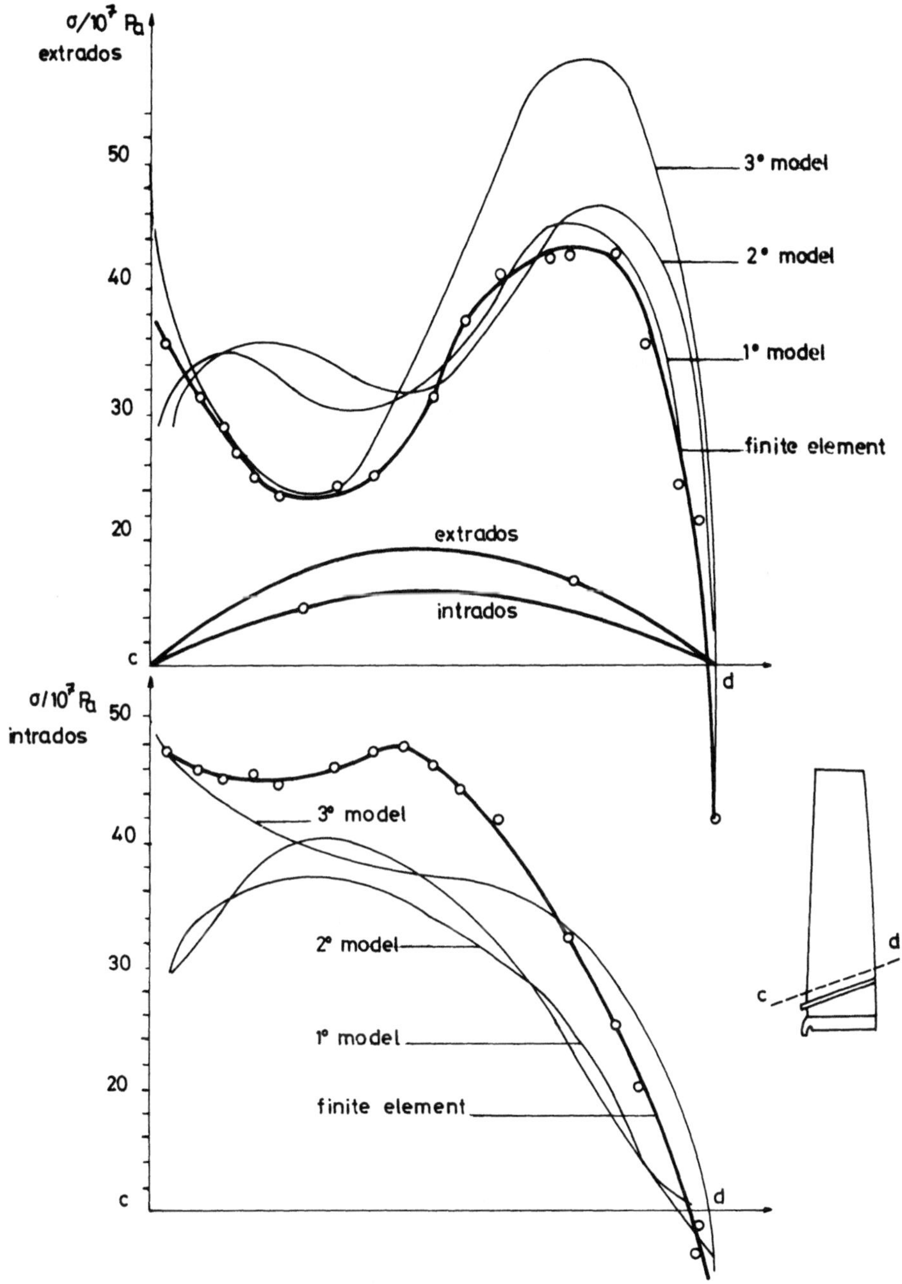

σ/10⁷ Pa
extrados
50
40
30
20
c
3° model
2° model
1° model
finite element
extrados
intrados
d
σ/10⁷ Pa
intrados
50
40
30
20
c
3° model
2° model
1° model
finite element
d
c
d

Fig. 12.

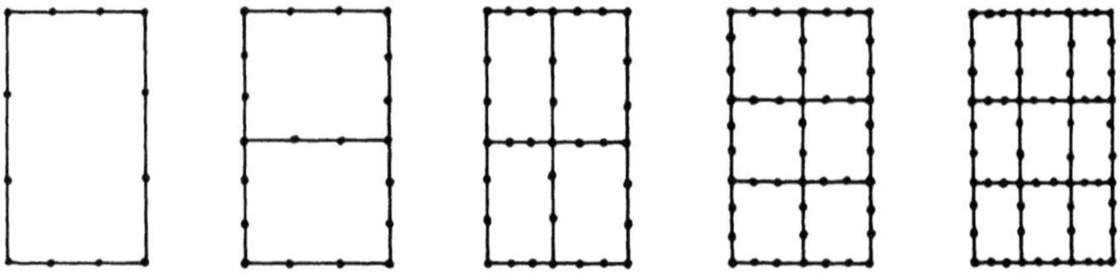

Freq. Hz	1 EL	2 EL	4 EL	6 EL	9 EL	Exp.	%
N_1	2 143	1 898	1 713	1 678	1 588	1 498	6
N_2	5 655	4 978	4 444	3 980	3 852	3 464	11
N_3	10 905	7 147	6 128	5 635	5 248	4 975	6

Fig. 13.

modelled with isoparametric elements. In fig. 13 the table shows the results obtained with different numbers of elements. With nine elements the agreement is good for practical purposes.

DISCUSSION AND CONCLUSION

It is obvious that the finite element method gives good results in blade calculations, to obtain stresses as well as frequencies and mode shapes. Further the influence of rotation can relatively easily be included in these calculations. One of the most important question arises when boundary conditions are difficult to specify, for example in turbine part of aircraft engines. Another important limitation about the blade calculation is due to the possible influence of the disc and the shroud. In this field interesting papers [16], [17], [18] have been published, but we think that a lot more work must be done.

On the other hand much work is done to day on damping and recently a paper was presented which studies a hollow blade damped by a vibrating beam [19]. We think it is a new research area.

ACKNOWLEDGEMENTS - The major part of this work was supported by the Direction des Recherches et Moyens d'Essais under contract 70.688. The authors are indepted to DRME for permission to publish this paper. The authors are also indepted to CIT Alcatel - SACM Mulhouse for permission to publish the results on gas turbine and to SNECMA for permission to publish the results on aircraft engines.

REFERENCES

[1] RAO, J.S.: "Natural frequencies of turbine blading, a survey", The Shock and Vibration Digest, Oct. 1973.

[2] RAWTANI, S. and DOKAINISH, M.A.: "Vibration analysis of rotating cantilever plates", International Journal for Numerical Methods in Engineering, Vol. 3, 1971.

[3] DOKAINISH, M.A. and GOSSAIN, D.M.: "Deflection and vibration analysis of a flexible fan turbine", Journal of Mechanical Engineering Science, Vol. 15, No. 6, 1973.

[4] BARTEN, H.G., SHEURENBRAND, J.A. and SCHEER, D.D.: "Stresses and vibration analysis of inducer baldes using finite element techniques", AIAA paper 70-630.

[5] AHMAD, R., ANDERSON, R.G. and ZIENKIEWICZ, O.C.: "Vibration of thick curved shells with particular reference to turbine blades", Journal of Strain Analysis, No. 5, 1970.

[6] BOSSAK, M.A.J. and ZIENKIEWICZ, O.C.: "Free vibrations of initially stressed solids with particular reference to centrifugal force effects in rotating machinery", Journal of Strain Analysis, No. 4, 1973.

[7] HOFMEISTER, L.D. and EVENSEN, D.A.: "Vibration problems using isoparametric shell elements", International Journal for Numerical Methods in Engineering, No. 1, 1972.

[8] THOMAS, J. and MOTA SOAREZ, C.A.: "Finite elements analysis of rotating shells", ASME paper 73 DET 94.

[9] "The application of the superelement method in analysis and design of ship structures and machinery component", National Symposium on Computerized Structural Analysis and Design, Washington D.C., 1972.

[10] HENRY, R.: "Calcul des coques minces en grandes déformations par éléments finis. Application aux aubes de compresseurs en rotation", Thése de Docteur Ingénieur, Lyon, 1973.

[11] HENRY, R. and LALANNE, M.: "Vibration analysis of rotating compressor blades", ASME paper 73 DET 140. The Journal of Engineering for Industry, August 1974, pp 1028-1035.

[12] PETRICONE, R. and SISTO, F.: "Vibration characteristics of low aspect ratio compressor baldes", Journal of Engineering for Power, Jan. 1970.

[13] LALANNE, M., TROMPETTE, P. and CHAMBAUD, M.T.: "Calcul de structures tridimensionnelles. Application aux aubes de turbines en rotation", Rapport DRME, Mai 1973.

[14] LALANNE, M. and TROMPETTE, P.: "Vibration analysis of rotating turbine blades", ASME paper 74-WA/DE-23.

[15] JENNINGS, A. and CLINT, M.: "The evaluation of eigenvalues and eigenvectors of real symetric matrices by simultaneous iteration", The Computer Journal, No. 1, 1970.

[16] ARMSTRONG, E.K., CHRISTIE, P.I. and HAGUE, W.M.: "Natural frequencies of bladed discs", Proceedings Institution of Mechanical Engineers, London, 1965-1966.

[17] EWINS, D.J.: "Vibration characteristics of bladed disc assemblies", Journal of Mechanical Engineering Science, 1973.

[18] COTTNEY, D.J. and EWINS, D.J.: "Toward the efficient vibration analysis of shrouded bladed disc assemblies", ASME paper 73 DET 144. Journal of Engineering for Industry, Aug. 1974, pp 1054-1059.

[19] JONES, D.I.G., NASHIF, A.D. and STARGARDTER, J.: "Vibration beam dampers for reducing vibration in gas turbine blades", ASME paper 74 GT 95. To be published in the Journal of Engineering for Power.

Response Characteristics of a Rotor with Flexible, Damped Supports

J.W. Lund

Department of Machine Design
The Technical University of Denmark, Lyngby, Denmark

ABSTRACT - A study is performed of the response behaviour of a flexible rotor with resilient supports. To simulate a variety of support conditions such as oil lubricated bearings, gas bearings, pivoted shoe bearings and flexible pedestals with complex impedance, the linear support characteristics are permitted to have anisotropic, non-conservative, frequency dependent properties. The analysis is based on a normal coordinate resolution of the response, and considers first the response to excitation from the foundation, including shock excitations and random vibration. Secondly, the response to mass unbalance is analyzed, either as the probable response to a random distribution of the unbalance, or as the modal response caused by a specified unbalance. In the latter case, a procedure for modal balancing is discussed. The practical application of the analysis is illustrated by sample numerical calculations for a compressor rotor supported in oil lubricated bearings.

INTRODUCTION

In rotor dynamics analysis an important topic is the response of the rotor. Of most practical concern is the response to mass unbalance, but occasionally other forms of excitation may also be of interest. Among these the present analysis considers the amplitude response caused by excitation from the foundation as it may occur in machinery onboard various types of vehicles or, for stationary machinery, in case of ground excitation. Foundation excitations are frequently of transient nature, either as shock pulses with a specified time profile or as random vibrations with a given spectral distribution. Both cases are analyzed and illustrated by numerical examples.

Existing methods for calculating mass unbalance response prove quite satisfactory in predicting the amplitude response to a specified unbalance. To complement such methods, this analysis considers two problems of a more special kind. First, as the details of the mass unbalance distribution often may not be known in practise, an analysis is performed for the case of a random unbalance in the rotor. The resulting amplitude is expressed in terms of its root-mean-square value, defining the amplitudes probability distribution.

Secondly, the basis for using the characteristic functions of the rotor-support-system to generate the unbalance response is explored for the purpose of rotor balancing. The approach is similar to the one utilized in modal balancing techniques with the difference that modal separation is not possible in the present non-conservative, non-selfadjoint system.

The rotor system model, employed in the analysis, treats the shaft as a non-uniform beam, incorporating shear deformation, rotary inertia, gyroscopic moments and internal hysteresis damping. The supports are assumed to have linear characteristics which can be represented by impedances. As these are allowed to be non-symmetric and frequency dependent, a wide variety of support conditions may be accomodated, including oil lubricated bearings, gas bearings, pivoted shoe bearings, and resilient pedestals with inertia.

ROTOR SYSTEM EQUATIONS

The lateral rotor displacements in orthogonal directions are x and y with corresponding angular displacements θ and ϕ. They are functions of the coordinate z along the rotor axis.

With bending moments M_x and M_y and shear forces V_x and V_y, the equations for the shaft deformations, including the contributions from shear deformation and internal hysteresis damping, are:

$$\frac{\partial\theta}{\partial z} = \frac{1}{EI}(M_x\cos\gamma - M_y\sin\gamma)$$

(1)

$$\frac{\partial\phi}{\partial z} = \frac{1}{EI}(M_x\sin\gamma + M_y\cos\gamma)$$

$$\theta = \frac{\partial x}{\partial z} + \frac{1}{\kappa GA}(V_x\cos\gamma - V_y\sin\gamma)$$

(2)

$$\phi = \frac{\partial y}{\partial z} + \frac{1}{\kappa GA}(V_x\sin\gamma + V_y\cos\gamma)$$

where E is Young's modulus, G is the shear modulus, A and I are the cross-sectional area and transverse moment of inertia, respectively, κ is a cross-section coefficient to account for the shear stress distribution, and γ is a material property, defining the internal hysteresis damping such that the dissipated energy per cycle equals $\pi E\varepsilon^2\sin\gamma$ (ε is the strain amplitude).

At a given angular speed of rotation, Ω, the differential equations of motion are:

$$\mu\frac{\partial^2 x}{\partial t^2} + \frac{\partial V_x}{\partial z} + \sum_k \delta(z-z_k)(R_x)_k = F_x$$

$$\mu\frac{\partial^2 y}{\partial t^2} + \frac{\partial V_y}{\partial z} + \sum_k \delta(z-z_k)(R_y)_k = F_y$$

(3)

$$j_T\frac{\partial^2\theta}{\partial t^2} + \Omega j_P\frac{\partial\phi}{\partial t} + V_x - \frac{\partial M_x}{\partial z} + \sum_k \delta(z-z_k)(R_\theta)_k = F_\theta$$

$$j_T\frac{\partial^2\phi}{\partial t^2} - \Omega j_P\frac{\partial\theta}{\partial t} + V_y - \frac{\partial M_y}{\partial z} + \sum_k \delta(z-z_k)(R_\phi)_k = F_\phi$$

where μ is the mass per unit length and j_T and j_P are the accompanying transverse and polar mass moments of inertia. F_x, F_y, F_θ and F_ϕ are the external forces and moments per unit length.

The rotor supports are located at $z = z_k$ ($\delta(z)$ is the unit delta function) and, assuming linear characteristics, the reaction forces and moments may be expressed as:

$$R_x = K_{xx}x + K_{xy}y + K_{x\theta}\theta + K_{x\phi}\phi + B_{xx}\frac{dx}{dt} + B_{xy}\frac{dy}{dt} + B_{x\theta}\frac{d\theta}{dt} + B_{x\phi}\frac{d\phi}{dt} \tag{4}$$

and analogously for R_y, R_θ and R_ϕ. For oillubricated bearings, the coefficients are functions of the rotational speed (they are functions of the bearing Sommerfeld number). For gas lubricated bearings, they are not only functions of speed but, because of compressibility effects, they also depend on the time history of the journal motion. Thus, for harmonic motion the coefficients become functions of frequency and must be treated as impedances. Hence, with the understanding that the motion can be described by an exponential solution $(x = X \cdot e^{st})$, eq. (4) may be written in a more general form as:

$$R_x = Z_{xx}X + Z_{xy}Y + Z_{x\theta}\Theta + Z_{x\phi}\Phi \tag{5}$$

where the impedances are:

$$Z_{xx} = K_{xx} + sB_{xx} \tag{6}$$

and similarly for the remaining terms. As discussed above, the coefficients themselves (K_{xx}, B_{xx}, etc.) may also depend on s. Eq. (5), therefore can accomodate impedances of a more general form that given by eq. (6) and may incorporate inertial effects when desired.

To arrive at a solution of the rotor system equations by means of modal expansion, a solution to the homogeneous equations may be written as:

$$
\begin{aligned}
x &= X_n e^{s_n t} \\
\theta &= \Theta_n e^{s_n t} \\
M_x &= M_{xn} e^{s_n t} \\
V_x &= V_{xn} e^{s_n t}
\end{aligned} \tag{7}
$$

and similarly for y , ϕ , M_y and V_y where s_n is the complex eigenvalue:

$$
s_n = \lambda_n + i\omega_n \tag{8}
$$

With these substitutions eqs. (3) become:

$$
\begin{Bmatrix}
s_n^2 \mu X_n \\
s_n^2 \mu Y_n \\
s_n^2 j_T \Theta_n + s_n \Omega j_P \Phi_n \\
s_n^2 j_T \Phi_n - s_n \Omega j_P \Theta_n
\end{Bmatrix}
+ \sum_k \delta(z-z_k) (\underline{\underline{Z}}_n)_k
\begin{Bmatrix}
X_n \\ Y_n \\ \Theta_n \\ \Phi_n
\end{Bmatrix}_k
= -
\begin{Bmatrix}
\dfrac{dV_{xn}}{dz} \\
\dfrac{dV_{yn}}{dz} \\
V_{xn} - \dfrac{dM_{xn}}{dz} \\
V_{yn} - \dfrac{dM_{yn}}{dz}
\end{Bmatrix} \tag{9}
$$

where $\underline{\underline{Z}}_n$ is a 4 by 4 impedance matrix whose elements Z_{xx} , Z_{xy} , etc. are evaluated at $s = s_n$. Hence, it is in general necessary to employ an iterative procedure to solve the indicated eigenvalue problem. An example of a numerical method may be found in ref. 6.

If, in eqs. (9), $(\underline{\underline{Z}}_n)_k$ is replaced by its transposed matrix while the sign of Ω is reversed and, also, the sign of γ in eqs. (1)-(2) is changed, the modified equations will have the same eigenvalues as the original equations but yield a new set of characteristic functions: X_n^* , Θ_n^* , M_{xn}^* , V_{xn}^* , etc., which are the adjoint characteristic functions. By multiplying the four eqs. (9) by X_m^* , Y_m^* , Θ_m^* and Φ_m^* , respectively, adding and integrating over the rotor length ℓ , and by subtracting from this result the similar result derived on the basis of the adjoint equations with $s = s_m$, the following orthogonality relationships are obtained (see ref. 7):

$$(10)\int_0^{\ell}[(s_n+s_m)(\mu(X_m^*X_n+Y_m^*Y_n)+j_T(\Theta_m^*\Theta_n+\Phi_m^*\Phi_n))+\Omega j_P(\Theta_m^*\Phi_n-\Phi_m^*\Theta_n)]dz$$

$$+\sum_k\{X_m^*,Y_m^*,\Theta_m^*,\Phi_m^*\}_k\frac{1}{s_n-s_m}(\underline{\underline{Z}}_n-\underline{\underline{Z}}_m)_k\begin{Bmatrix}X_n\\Y_n\\\Theta_n\\\Phi_n\end{Bmatrix}_k=\begin{cases}0 & \text{if } n\neq m\\ N_n & \text{if } n=m\end{cases}$$

$$(11)\int_0^{\ell}\Big\{\frac{1}{EI}((M_{xm}^*M_{xn}+M_{ym}^*M_{yn})\cos\gamma-(M_{xm}^*M_{yn}-M_{ym}^*M_{xn})\sin\gamma)$$

$$+\frac{1}{\kappa GA}((V_{xm}^*V_{xn}+V_{ym}^*V_{yn})\cos\gamma-(V_{xm}^*V_{yn}-V_{ym}^*V_{xn})\sin\gamma)$$

$$-s_ns_m(\mu(X_m^*X_n+Y_m^*Y_n)+j_T(\Theta_m^*\Theta_n+\Phi_m^*\Phi_n))\Big\}dz$$

$$+\sum_k\{X_m^*,Y_m^*,\Theta_m^*,\Phi_m^*\}_k\frac{1}{s_n-s_m}(s_n\underline{\underline{Z}}_m-s_m\underline{\underline{Z}}_n)_k\begin{Bmatrix}X_n\\Y_n\\\Theta_n\\\Phi_n\end{Bmatrix}_k=\begin{cases}0 & \text{if } n\neq m\\ -s_nN_n & \text{if } n=m\end{cases}$$

The modal norm N_n is defined by eq. (10) and may also, following the method of ref. 2, be expressed as:

$$(12)\qquad N_n=\left[\Theta_n^*\frac{dM_{xn}}{ds}+\Phi_n^*\frac{dM_{yn}}{ds}-X_n^*\frac{dV_{xn}}{ds}-Y_n^*\frac{dV_{yn}}{ds}\right]_{z=0}^{z=\ell}$$

Eqs. (11) and (12) assume that the shaft ends are free ($M_x = M_y = V_x = V_y = 0$ at $z = 0, \ell$) .

To evaluate eqs. (10) and (11) for $n = m$, the following relationships apply:

$$\left(\frac{1}{s_n-s_m}\,(\underline{\underline{Z}}_n-\underline{\underline{Z}}_m)\right)_{n=m} = \left(\frac{d}{ds}\,\underline{\underline{Z}}\right)_n$$

(13)

$$\left(\frac{1}{s_n-s_m}\,(s_n\underline{\underline{Z}}_m-s_m\underline{\underline{Z}}_n)\right)_{n=m} = \underline{\underline{Z}}_n - s_n\left(\frac{d}{ds}\,\underline{\underline{Z}}\right)_n$$

On this basis eqs. (1) to (3) may be transformed into decoupled, ordinary differential equations by introducing normal coordinates q_n :

$$x = \sum_n X_n q_n$$

(14)

$$\frac{dx}{dt} = \sum_n s_n X_n q_n = \sum_n X_n \frac{dq_n}{dt}$$

and similarly for y , θ , ϕ , M_x , M_y , V_x and V_y .

By these substitutions and the orthogonality relationships a set of first order equations are obtained:

$$(15)\quad \frac{dq_n}{dt} - s_n q_n = \frac{1}{N_n}\int_0^{\ell} (X_n^* F_x + Y_n^* F_y + \Theta_n^* F_\theta + \Phi_n^* F_\phi)\,dz = F_n(t)$$

with the general solution:

$$(16)\qquad q_n = q_n(t{=}0)\cdot e^{s_n t} + \int_0^t e^{s_n(t-\tau)}\,F_n(\tau)\,d\tau$$

The initial value of q_n is determined from the initial conditions for x , dx/dt , y , dy/dt , etc., by the inverse transformation of eq. (14):

$$(17)\quad q_n = \frac{1}{N_n}\int_0^{\ell}\Big\{\mu\Big[X_n^*(\frac{dx}{dt} + s_n x) + Y_n^*(\frac{dy}{dt} + s_n y)\Big]$$

$$+ j_T\Big(\Theta_n^*(\frac{d\theta}{dt} + s_n\theta) + \Phi_n^*(\frac{d\phi}{dt} + s_n\phi)\Big)$$

$$+ \Omega j_P(\Theta_n^*\phi - \Phi_n^*\theta)\Big\}dz + \sum_k \{X_n^*, Y_n^*, \Theta_n^*, \Phi_n^*\}_k\left(\frac{d}{ds}\,\underline{\underline{Z}}\right)_{nk}\begin{Bmatrix} x \\ y \\ \theta \\ \phi \end{Bmatrix}_k$$

Hence eq. (16) may be solved for any arbitrary external excitation after which the actual rotor amplitudes can be calculated from eq. (14). The general question of the convergence of the summation is discussed in ref. 18, but in practise it is usually found that a satisfactory answer only requires a limited number of modes depending, of course, on the form of the external excitation.

RESPONSE TO FOUNDATION EXCITATION

If the rotor foundation receives a vertically upward acceleration g_x in the negative x-direction it is equivalent to an imposed acceleration loading downwards in the positive x-direction when it is understood that the rotor amplitudes are measured relative to the moving foundation. With the rotor initially at rest and $F_x = \mu g_x$ $(F_y = F_\theta = F_\phi = 0)$, the normal coordinates are found from eq. (16):

$$q_n = \frac{1}{N_n} \int_0^t e^{s_n(t-\tau)} \left(\int_0^\ell \mu X_n g_x(\tau) dz \right) d\tau \tag{18}$$

For the simple case of a shock pulse, g_x may be expressed as:

$$g_x = v_0 \delta(t)$$

where v_0 is the step-jump in foundation velocity at $t = 0$ and $\delta(t)$ is the unit delta function. v_0 may be a function of z but, assuming it to be uniform along the rotor, eq. (18) is readily solved and substituted into eq. (14) to yield the final amplitude:

$$x = v_0 \sum_n e^{\lambda_n t} (\cos(\omega_n t) + i\sin(\omega_n t)) \frac{X_n}{N_n} \int_0^\ell \mu X_n^* dz \tag{19}$$

The imaginary parts cancel by pairing complex conjugate terms in the summation. The convergence is fast except near $t = 0$.

If, instead, g_x is harmonic with frequency ω , the acceleration loading, $F_x = \mu g_x$, may be represented as a complex quantity with $e^{j\omega t}$ as implied multiplier ($j = \sqrt{-1}$) and the convention that only the real part applies. Solving eq. (16) for the forced response and substituting into eq. (14) the resulting amplitude becomes:

$$x = \mathcal{R}e_j\left\{e^{j\omega t} \sum_n \frac{1}{j\omega - s_n} \frac{X_n}{N_n} \int_0^\ell \mu X_n^* g_x dz\right\} \tag{20}$$

where $\mathcal{R}e_j\{\}$ indicates the real part with respect to j . However, when s_n is a complex eigenvalue, then the complex conjugate $\bar{s}_n$ is also an eigenvalue whose associated characteristic functions are the complex conjugate of their counterparts for $s = s_n$. Pairing complex conjugate terms in i in the summation, the i-imaginary parts cancel. Hence, j may be replaced by i in eq. (20) and, dropping the real part-operator and the exponential multiplier, x becomes a complex quantity, similarly defined as previously g_x .

In the simple case where g_x is uniform along the rotor, eq. (20) yields:

$$x = H_g(\omega) g_x \tag{21}$$

where $H_g(\omega)$ is the complex transfer function:

$$H_g(\omega) = -i \sum_n \frac{1}{\omega - \omega_n + i\lambda_n} \frac{X_n}{N_n} \int_0^\ell \mu X_n^* dz + \frac{1}{\omega + \omega_n + i\lambda_n} \frac{\bar{X}_n}{\bar{N}_n} \int_0^\ell \mu \bar{X}_n^* dz \tag{22}$$

(any real eigenvalues where ω_n is zero and X_n , X_n^* and N_n are real, only contribute one term and not two as shown in the equation).

If g_x is a random excitation with spectral density

S_g , the resulting spectral density for the amplitude, S_x , is given by:

$$S_x = |H_g(\omega)|^2 S_g \tag{23}$$

The mean-square value of the amplitude, $\langle x^2 \rangle$, is computed as:

$$\langle x^2 \rangle = \frac{1}{2\pi}\int_0^\infty S_x d\omega = \frac{1}{2\pi}\int_0^\infty |H_g(\omega)|^2 S_g d\omega \tag{24}$$

For white noise excitation S_g is independent of ω such that:

$$\langle x^2 \rangle = S_g \frac{1}{4\pi}\int_{-\infty}^\infty H_g(\omega)\overline{H_g(\omega)}d\omega \tag{25}$$

The integral may be evaluated from the residue theorem. Assuming a stable system where all λ_n-values are negative, only the roots from $H_g(\omega)$ lie in the upper half of the ω-plane and contribute to the residues. Hence:

$$\langle x^2 \rangle = - S_g \sum_n Im\left\{ \left(\frac{X_n}{N_n}\int_0^\ell \mu X_n^* dz\right) \sum_m \left[\frac{1}{\omega_n+\omega_m-i(\lambda_n+\lambda_m)} \frac{X_m}{N_m}\int_0^\ell \mu X_m^* dz + \frac{1}{\omega_n-\omega_m-i(\lambda_n+\lambda_m)} \frac{\bar{X}_m}{\bar{N}_m}\int_0^\ell \mu \bar{X}_m^* dz \right]\right\} \tag{26}$$

In this form it is implied, as also in eq. (22), that all complex eigenvalues are only counted as single roots. For any real eigenvalue there is only one term in the m-summation while the term in the n-summation should be multiplied by ½ .

$\langle x^2 \rangle$ is the variance of x . The square root of the variance, the root-mean-square value, defines the standard deviation:

$$\mathrm{rms}(x) = \sqrt{\langle x^2 \rangle} \tag{27}$$

If the probability distribution for g_x is Gaussian, this will also be true for the distribution of x which, then, is completely defined (assuming zero mean).

RESPONSE TO RANDOM MASS UNBALANCE

Probably the most common source of rotor excitation is mass unbalance where the local mass center is offset by an eccentricity, $\varepsilon = \varepsilon(z)$, from the axis of rotation. If ε has the components ε_c and ε_s along orthogonal directions, fixed in the rotating shaft, the unbalance forces per unit length become:

(28)
$$F_x = \Omega^2\mu(\varepsilon_c\cos(\Omega t) - \varepsilon_s\sin(\Omega t))$$
$$F_y = \Omega^2\mu(\varepsilon_c\sin(\Omega t) + \varepsilon_s\cos(\Omega t))$$

where Ω is the angular speed of rotation. Adopting complex notation as previously discussed, these forces may also be written as:

(29)
$$F_x = \Omega^2\mu\varepsilon \qquad F_y = -\,i\Omega^2\mu\varepsilon = -\,iF_x$$

where:

(30)
$$\varepsilon = \varepsilon_c + i\varepsilon_s$$

In many applications the main contribution to the mass unbalance comes from heavy components such as wheels, impellers, couplings, etc.. Summing over these components with mass m_k and eccentricity ε_k , the rotor amplitude is computed from:

(31)
$$x = \sum_k H_{xk}(\Omega)\Omega^2 m_k \varepsilon_k$$

where x and $H_{xk}(\Omega)$ are complex like ε. The transfer functions $H_{xk}(\Omega)$ may be computed from eq. (22) with $\mu = \delta(z-z_k)m_k$ or directly from a forced response calculation as, for example, given in refs. 8, 9 and 12. Considering the ε_k's to be statistically determined with common variance, $<|\varepsilon|^2>$, the corresponding variance of the amplitude becomes:

$$<|x|^2> = <|\varepsilon|^2> \sum_k \Omega^4 m_k^2 |H_{xk}(\Omega)|^2 \tag{32}$$

or, in terms of root-mean-square values:

$$\frac{\mathrm{rms}(|x|)}{\mathrm{rms}(|\varepsilon|)} = \frac{\sqrt{<|x|^2>}}{\sqrt{<|\varepsilon|^2>}} = \Omega^2 \sqrt{\sum_k m_k^2 |H_{xk}(\Omega)|^2} \tag{33}$$

If the probability distribution for ε_k can be considered Gaussian, $\mathrm{rms}(|x|)$ gives the standard deviation for x.

When the total rotor mass is M, the eccentricity of its mass center, ε_{CG}, may be expressed by:

$$\frac{\mathrm{rms}(|\varepsilon_{CG}|)}{\mathrm{rms}(|\varepsilon|)} = \frac{1}{M}\sqrt{\sum_k m_k^2} \tag{34}$$

By combining eqs. (33) and (34) the probable amplitude response can be expressed directly in terms of the probable rotor mass eccentricity.

RESPONSE TO FIXED UNBALANCE

When the unbalance distribution is known, the amplitude response is computed more efficiently and more accurately by forced response methods (for example ref. 8) than by means of the characteristic functions and normal coordinates. By identifying the constituents of the response, however, the latter method may offer some advantages, for example in balancing studies, which will be explored in the following.

The unbalance forces caused by the mass eccentricity ε have been derived in eqs. (28) and (29). In addition, there may occur unbalance moments caused by an angular eccentricity, ψ, between the principal mass axis and the axis of rotation. When ε is continuous, ψ equals $d\varepsilon/dz$. Adopting the complex notation of eq. (29), the unbalance moments per unit length are:

$$(35) \qquad F_\theta = \Omega^2 (j_T - j_P)\psi \qquad\qquad F_\phi = -\, iF_\theta$$

With the known forces and moments, eq. (16) may be solved for the harmonic response after which the amplitudes are obtained from eq. (14):

$$(36) \qquad x = \sum_n \left\{ \frac{1}{i\Omega - s_n} \frac{X_n}{N_n} \int_0^\ell (X_n^* F_x + Y_n^* F_y + \Theta_n^* F_\theta + \Phi_n^* F_\phi)\, dz + \frac{1}{i\Omega - \bar{s}_n} \frac{\bar{X}_n}{\bar{N}_n} \int_0^\ell (\bar{X}_n^* F_x + \bar{Y}_n^* F_y + \bar{\Theta}_n^* F_\theta + \bar{\Phi}_n^* F_\phi)\, dz \right\}$$

The justification of using i as the imaginary unit for both response and eigenvalues is discussed in connection with eq. (20).

Substituting for F_x, F_y, F_θ and F_ϕ from eqs. (29) and (35), eq. (36) may be written as:

$$(37) \qquad x = \sum_n \left\{ \frac{2\Omega^2}{i(\Omega - \omega_n) - \lambda_n} \frac{X_n}{N_n} \int_0^\ell (\mu \xi_n^* \varepsilon + (j_T - j_P)\alpha_n^* \psi)\, dz + \frac{2\Omega^2}{i(\Omega + \omega_n) - \lambda_n} \frac{\bar{X}_n}{\bar{N}_n} \int_0^\ell (\mu \bar{\eta}_n^* \varepsilon + (j_T - j_P)\bar{\beta}_n^* \psi)\, dz \right\}$$

where the following auxiliary modal functions have been introduced:

(38)

$$\text{forward precessional modal functions: } \xi_n = \tfrac{1}{2}(X_n + iY_n)$$
$$\alpha_n = \tfrac{1}{2}(\Theta_n + i\Phi_n)$$

$$\text{backward precessional modal functions: } \eta_n = \tfrac{1}{2}(X_n - iY_n)$$
$$\beta_n = \tfrac{1}{2}(\Theta_n - i\Phi_n)$$

The corresponding adjoint functions are:

(39)

$$\xi_n^* = \tfrac{1}{2}(X_n^* - iY_n^*) \qquad \alpha_n^* = \tfrac{1}{2}(\Theta_n^* - i\Phi_n^*)$$
$$\eta_n^* = \tfrac{1}{2}(X_n^* + iY_n^*) \qquad \beta_n^* = \tfrac{1}{2}(\Theta_n^* + i\Phi_n^*)$$

Whereas the original characteristic functions, as for example X_n and Y_n, describe elliptical whirl orbits, the precessional modal functions define circular orbits. An elliptical orbit may be obtained as the sum of two circular orbits, one with forward precession and one with backward precession, as is also found from eq. (38):

(40) $$X_n = \xi_n + \eta_n \qquad Y_n = -i\xi_n + i\eta_n$$

For the ξ-orbit it is seen that its y-component is equal to $(-i)$ times its x-component, proving it to be a circular orbit with forward precession in analogy to the unbalance forces, eq. (29). By the same argument the η-orbit is found to be circular with backward precession. Because the unbalance force is a rotating force it can only do work on the forward precessional whirl component as also demonstrated by eq. (37).

If the coordinate transformation, defined by eq. (38), is introduced into the homogeneous part of eqs. (1) to (3), the equations reduce to:

$$(41)\quad \begin{Bmatrix} s_n^2\mu\xi_n \\ s_n^2\mu\eta_n \\ (s_n^2 j_T - i s_n\Omega j_P)\alpha_n \\ (s_n^2 j_T + i s_n\Omega j_P)\beta_n \end{Bmatrix} + \sum_k \delta(z-z_k)(\underline{\underline{Z}}_n^{(w)})_k \begin{Bmatrix} \xi_n \\ \eta_n \\ \alpha_n \\ \beta_n \end{Bmatrix}_k = - \begin{Bmatrix} \frac{dV}{dz}\xi n \\ \frac{dV}{dz}\eta n \\ V_{\xi n} - \frac{dM}{dz}\xi n \\ V_{\eta n} - \frac{dM}{dz}\eta n \end{Bmatrix}$$

$$(42)\quad \begin{aligned} \frac{d\alpha_n}{dz} &= \frac{1}{EI} M_{\xi n} e^{i\gamma} & \frac{d\beta_n}{dz} &= \frac{1}{EI} M_{\eta n} e^{-i\gamma} \\ \alpha_n &= \frac{d\xi_n}{dz} + \frac{1}{\kappa GA} V_{\xi n} e^{i\gamma} & \beta_n &= \frac{d\eta_n}{dz} + \frac{1}{\kappa GA} V_{\eta n} e^{-i\gamma} \end{aligned}$$

The elements of the 4 by 4 impedance matrix, $\underline{\underline{Z}}^{(w)}$, are:

$$(43)\quad \begin{aligned} (\underline{\underline{Z}}^{(w)})_{\xi\xi} &= \tfrac{1}{2}(Z_{xx}+Z_{yy}-i(Z_{xy}-Z_{yx})) \\ (\underline{\underline{Z}}^{(w)})_{\xi\eta} &= \tfrac{1}{2}(Z_{xx}-Z_{yy}+i(Z_{xy}+Z_{yx})) \\ (\underline{\underline{Z}}^{(w)})_{\eta\xi} &= \tfrac{1}{2}(Z_{xx}-Z_{yy}-i(Z_{xy}+Z_{yx})) \\ (\underline{\underline{Z}}^{(w)})_{\eta\eta} &= \tfrac{1}{2}(Z_{xx}+Z_{yy}+i(Z_{xy}-Z_{yx})) \end{aligned}$$

and analogously for the remaining elements. The impedances Z_{xx} , Z_{xy} , etc., are defined through eq. (5).

The equations for the adjoint modal functions are the same as eqs. (41) and (42) except that $(\underline{\underline{Z}}_n^{(w)})_k$ is replaced by its transposed matrix. The sign of Ω and γ is not changed.

It is seen that the only possible coupling between the forward and backward precessional modal functions is in the supports. In the special case of isotropic supports where $Z_{xx} = Z_{yy}$ and $Z_{yx} = - Z_{xy}$, there are two independent sets of eigenvalues and associated modal functions.

When the supports are not isotropic, the coupling between the modal functions may be expressed by means of coupling impedance matrices. Thus, the equations for the backward precessional modal functions (the second and the fourth equation in eq. (41)) can be solved for the η_n-β_n-values at the supports in terms of the ξ_n-α_n-values at the supports. Thereby η_n and β_n can be eliminated from the equations for the forward precessional modal functions (the first and the third equation in eq. (41)), resulting in a coupling impedance matrix $\underline{\underline{Z}}_n^{(f)}$ of dimension 2k where k is the number of supports. As the support reaction moments usually can be ignored, the dimension of the matrix is just k in most cases.

On this basis an orthogonality relationship for the forward precessional modal functions can be established in analogy to eq. (10):

$$(44)\quad \int_0^{\ell} ((s_n+s_m)(\mu\xi_m^*\xi_n+j_T\alpha_m^*\alpha_n) - i\Omega j_P\alpha_m^*\alpha_n)dz$$

$$+ \{\{\xi_m^*,\alpha_m^*\}_1,--,\{\xi_m^*,\alpha_m^*\}_k,--\}\frac{1}{s_n-s_m}(\underline{\underline{Z}}_n^{(f)} - \underline{\underline{Z}}_m^{(f)})\begin{Bmatrix}\begin{Bmatrix}\xi_n\\ \alpha_n\end{Bmatrix}_1\\ \vdots\\ \begin{Bmatrix}\xi_n\\ \alpha_n\end{Bmatrix}_k\end{Bmatrix} =$$

$$= \begin{cases} 0 & \text{if} \quad n \neq m \\ \tfrac{1}{2}N_n & \text{if} \quad n = m \end{cases}$$

where:

$$(45)\quad \left[\frac{1}{s_n-s_m}(\underline{\underline{Z}}_n^{(f)}-\underline{\underline{Z}}_m^{(f)})\right]_{n=m} = \left(\frac{d}{ds}\,\underline{\underline{Z}}^{(f)}\right)_n$$

An analogous relationship, but with a different coupling matrix, applies for the backward precessional modal functions.

Returning to eq. (37) it may be written in abbreviated form as:

$$x = \sum_n \frac{\Omega^2}{i(\Omega-\omega_n)-\lambda_n} \frac{X_n}{\sqrt{N_n}} U_n \tag{46}$$

where:

$$U_n = \frac{2}{\sqrt{N_n}} \int_0^{\ell} (\mu\xi_n^{*}\varepsilon+(j_T-j_P)\alpha_n^{*}\psi)\,dz \tag{47}$$

Although not specifically shown, eq. (46) may be considered to include the terms deriving from the conjugate roots by proper definitions of U_n, X_n and N_n from eq. (37).

Comparing eqs. (44) and (47) it is apparent that it is in general not possible to obtain distributions for ε and ψ which will cause all U_n-values, except one, to vanish. Only when there is no support coupling and no support damping, and gyroscopic moments can be ignored, can distributions be found that will excite just one mode, namely by setting ε and ψ equal to ξ_n and α_n, respectively. This is the case for the classical conservative system and forms the rationale behind the various modal balancing techniques, see for example refs. 14, 15, 16 and 17.

When the conditions for a conservative system are not satisfied, the amplitude response, even at the critical speeds, contains components from several modes (theoretically, all modes contribute). As modal separation is not possible it is necessary to balance the modal components simultaneously. To illustrate, it is assumed that the shaft amplitude is measured at a number of locations, $z = z_r$ $(r=1,2,--)$, and at several rotational speeds Ω_q $(q=1,2,--)$. From eq. (46) is found:

(48) $$x_q(z_r) = \sum_n \frac{\Omega_q^2}{i(\Omega_q-\omega_n)-\lambda_n} \frac{X_n(z_r)}{\sqrt{N_n}} U_n$$

which may be written in matrix form as:

(49) $$\underline{\underline{C}}\ \underline{U} = \underline{X}$$

where $\underline{U}$ and $\underline{X}$ are column vectors with elements U_n and $x_q(z_r)$, respectively, and $\underline{\underline{C}}$ is a matrix of dimension $(rq)\cdot n$ whose elements are the coefficients to U_n in eq. (48). If $rq = n$, the equation can be solved directly for $\underline{U}$. Otherwise, when $rq > n$, a least square minimizing procedure may be employed to yield:

(50) $$\underline{U} = (\bar{\underline{\underline{C}}}^T \underline{\underline{C}})^{-1} \bar{\underline{\underline{C}}}^T \underline{X}$$

where $\bar{\underline{\underline{C}}}^T$ is the transposed, complex conjugate matrix of $\underline{\underline{C}}$.

To balance the rotor, correction weights w_p are inserted in a number of planes, located at $z = z_p$ $(p=1,2,--)$. The weights are determined from eq. (47) as:

$$U_n + \frac{2}{\sqrt{N_n}} \int_0^{\ell} \delta(z-z_p)\xi_n^* w_p dz = 0$$

or:

(51) $$\sum_p \frac{2\xi_n^*(z_p)}{\sqrt{N_n}} w_p = - U_n$$

In matrix form this equation can be written as:

(52) $$\underline{\underline{D}}\ \underline{W} = - \underline{U}$$

where $\underline{W}$ is a column vector with elements w_p, and $\underline{\underline{D}}$ is a $n\cdot p$ matrix whose elements are the coefficients to w_p in eq. (51). The equation may either be solved directly for

$\underline{W}$ when $n = p$, or by the procedure of eq. (50) when $n > p$:

$$\underline{W} = - (\bar{\underline{\underline{D}}}^T \underline{\underline{D}})^{-1} \bar{\underline{\underline{D}}}^T \underline{U} \tag{53}$$

The method, of course, is only valid when the amplitude response can be represented sufficiently accurately by a finite number of modes. Furthermore, the eigenvalues and the associated characteristic functions are actually speed dependent which means that the U_n-values are, strictly speaking, not constant quantities as otherwise assumed. Unless this effect is small the method does not apply.

Apart from these considerations, the method is seen to be based on the calculated eigenvalues and characteristic functions which, in all probability, will deviate from those of the actual rotor system. To make the method practical, therefore, will require that means be found to carry out adjustments based on measurements.

APPLICATION OF THE ANALYSIS BY NUMERICAL CALCULATIONS

To demonstrate the application of the preceding analysis, sample calculations are performed for a compressor rotor, supported in two identical oil lubricated journal bearings. The shaft has an overall length of 2.62 m , a bearing span of 2.05 m , and a mass of 635 kg with the center of gravity almost midway between the bearings. Overhung at one end is the coupling while the other overhung end carries the thrust collar. The bearings have a diameter of 127 mm , a length of 38 mm and a clearance ratio of 0.0014 . The oil viscosity which varies with speed, is 16.1 centipoise at the supply temperature of $50^{\circ}C$.

Fig. 1 shows the transient response caused by a foundation shock pulse. In this particular case the rotor is supported in pivoted shoe bearings and operates at 5000 rpm . Based on eqs. (18) and (19), two cases are con-

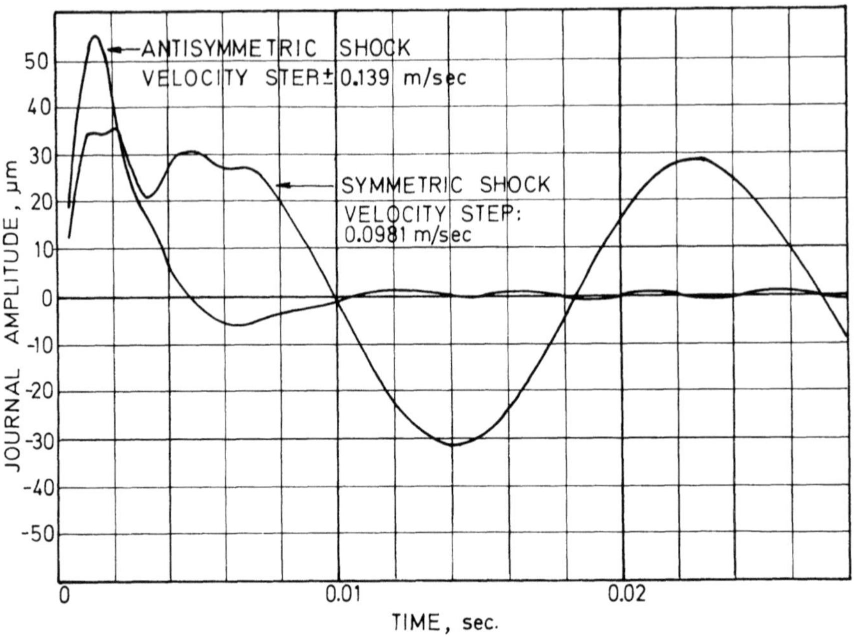

Fig. 1. Transient response of compressor rotor, operating at 5000 rpm in pivoted 5-shoe bearings. Vertical amplitude in bearing from foundation velocity step.

sidered: a symmetric shock where the two bearing pedestals receive the same upward step-jump in velocity of 0.0981 m/sec (equivalent to a 10 g shock of 1 millisecond duration), and a antisymmetric shock where the velocity step-jump is 0.139 m/sec , but in opposing directions at the pedestals. The magnitudes of the two velocity step-jumps are such that the initially imparted kinetic energy is the same in both cases. The quite different response behaviour in the two cases is explained by the close symmetry of the rotor system where the symmetric shock primarily excites the even modes while the odd modes dominate the antisymmetric shock response.

The figure shows the vertical amplitude in one of the bearings as function of time. The calculation is based on the first 14 eigenvalues (14 pairs of complex conjugate roots) but, apart from the first half cycle of the response where the influence of the higher modes is felt, it is the fundamental modes which govern the response. As the computed

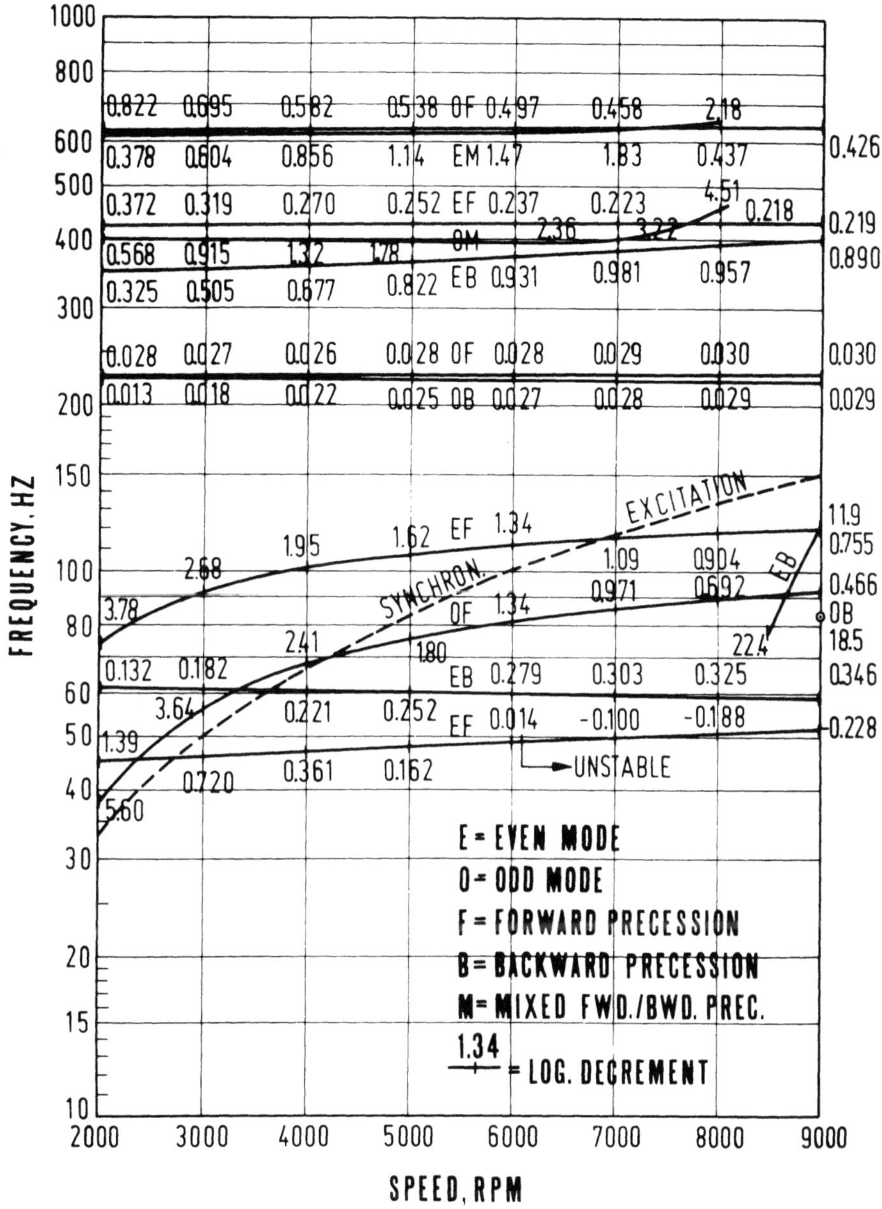

Fig. 2. Eigenvalues for compressor rotor in cylindrical journal bearing.

amplitude in general is less than 40 per cent of the 89 μm radial clearance in the bearing the assumption of linearity should be valid.

The pivoted shoe bearing is only used in this first case, and in the remaining calculations the rotor is supported in plain cylindrical journal bearings. The eigenvalues are shown in Fig. 2 where the ordinate gives the frequency in Hz (ω_n in eq. (8)) while the associated damping exponent, λ_n, is indicated through the logarithmic decrement δ :

$$\delta = -\frac{2\pi\lambda_n}{\omega_n} \tag{54}$$

Because of the speed dependent support impedances and the gyroscopic moments in the rotor, the eigenvalues become functions of the rotational speed. The intersections between the eigenvalue curves and the synchronous mass unbalance excitation, given by the dotted curve, define the damped critical speeds as:

1st. critical, forward precession:
2750 rpm (45.83Hz) , δ=0.857
1st. critical, backward precession:
3637 rpm (60.62Hz) , δ=0.208
2nd. critical, forward precession:
4180 rpm (69.61Hz) , δ=2.27
3rd. critical, forward precession:
6840 rpm (114.0Hz) , δ=1.12

As the modal whirl orbits are not circular, the designation of forward or backward precession only serves to identify the predominant component. All four critical speeds could be excited by mass unbalance but, because of the large damping exponent, the 2nd. critical has no response peak. It may be noted that the rotor becomes unstable at 6100 rpm.

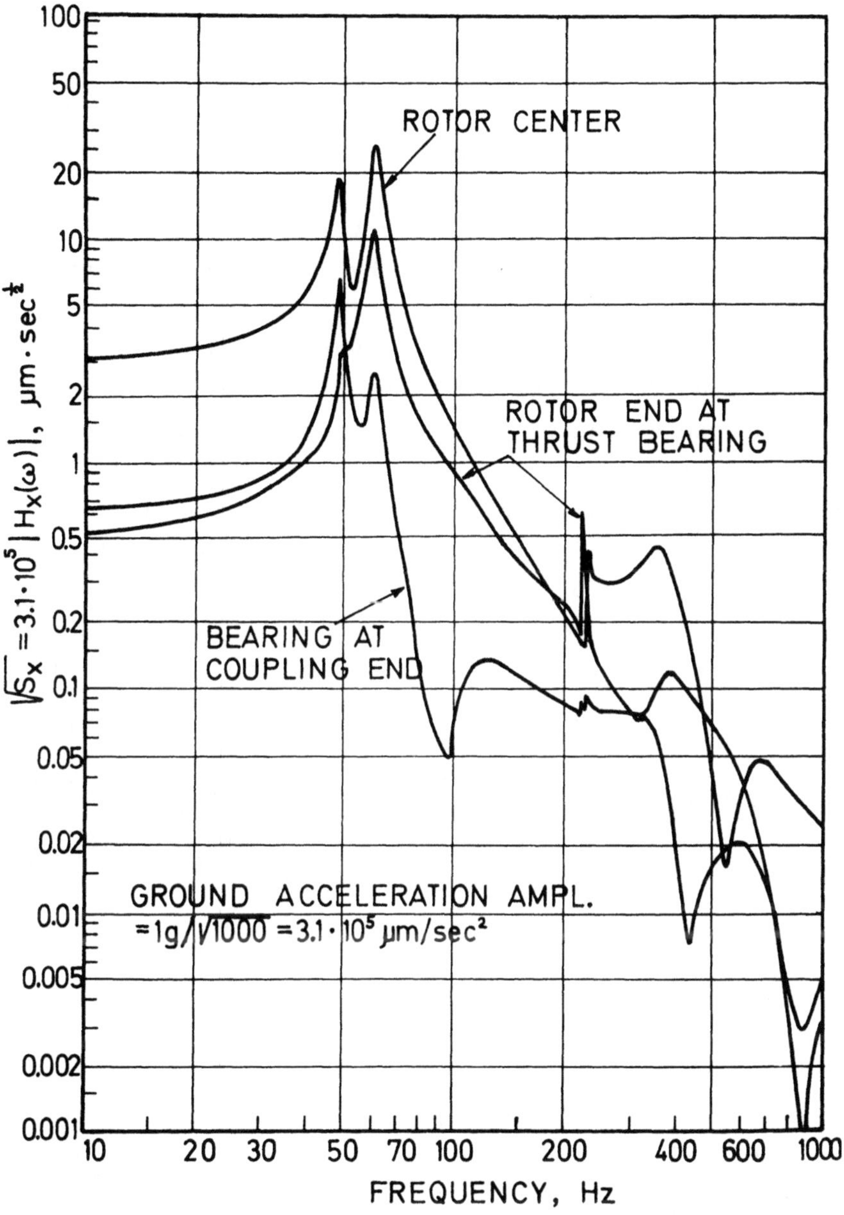

Fig. 3. Amplitude spectral density distribution for compressor rotor, operating at 5000 rpm in cylindrical journal bearings. Excitation: random foundation acceleration with uniform spectral density, 0 to 1000 Hz , and rms-value of 9.81 m/sec^2 .

Most of the eigenvalues can be identified in the response spectrum shown in Fig. 3. Running at 5000 rpm the rotor is excited by a random, vertical acceleration from the foundation with a root-mean-square value of $9.81\ m/sec^2$ (=1g) and a uniform spectral density from 0 to 1000 Hz . The resulting spectral density of the rotor amplitude, S_x , is computed from eq. (23) and plotted by its square root value for three locations as indicated in Fig. 3.

A direct integration of S_x over the specified frequency range from 0 Hz to 1000 Hz according to eq. (24) yields the variance, $\langle x^2 \rangle$, of the rotor amplitude. The square root of the variance is the root-mean-square value of the amplitude, rms(x) , and it is shown for the entire rotor in Fig. 4. It defines the standard deviation for the distribution of the amplitude and, assuming the random process to be Gaussian, the probability that the amplitude exceeds this value, is 31.7 per cent. Similarly,

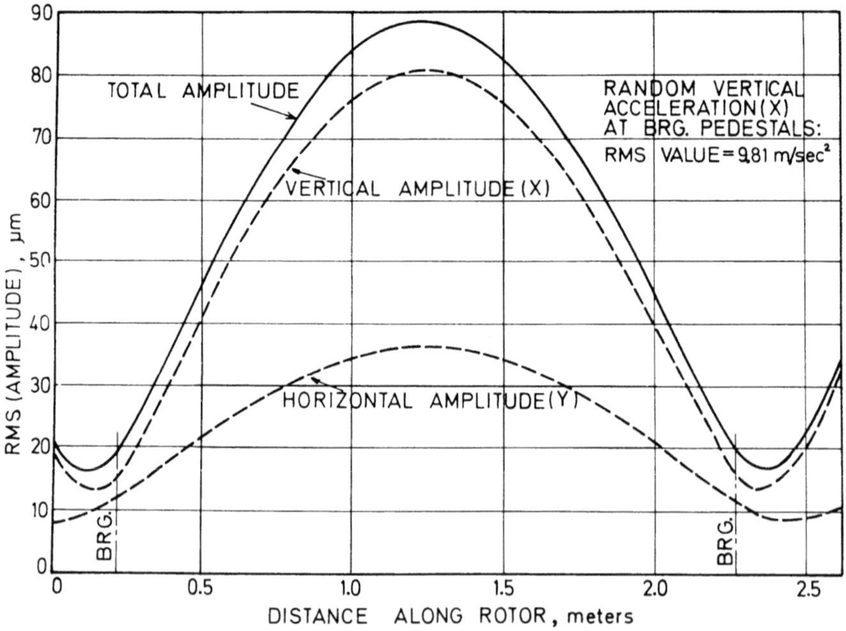

Fig. 4. Random vibration response of compressor rotor, operating at 5000 rpm in cylindrical journal bearings. Excitation: random foundation acceleration with uniform spectral density, 0 to 1000 Hz , and rms-value of $9.81\ m/sec^2$.

the probabilities for exceeding twice or three times the rms-value are 4.5 per cent and 0.26 per cent, respectively.

Rather than integrating S_x directly to obtain Fig. 4, the calculation may also be performed by means of the residue theorem as given by eq. (26). With the rotor operating at 5000 rpm, there are 16 eigenvalues below 1000 Hz, namely the 12 complex eigenvalues shown in Fig. 2 (the root close to 1000 Hz is left out), and four real roots out of which two roots are so heavily damped that they may be ignored. Summing over these eigenvalues according to eq. (26), the resulting root-mean-square values of the amplitude only differ by a few per cent from the values already obtained in Fig. 4 as would be expected by inspecting Fig. 3.

A probability interpretation also applies to Fig. 5 which shows the root-mean-square value of the rotor amplitude, caused by a random mass unbalance. The unbalance is confined to the components mounted on the shaft, namely 8 impellers located between the bearings, the coupling outboard one bearing, and the thrust collar outboard the other bearing. These ten components account for 54 per cent of the total rotor mass. Their mass eccentricities have the same probability distribution with some specified rms-value (standard deviation), and the resulting shift of the rotors center of gravity, computed from eq. (34), is found to get an rms-value equal to 0.173 times the assigned rms-value for the component mass eccentricities.

Conversely, if the rotor were to be balanced such that its residual mass eccentricity would be known to have an rms-value of, say, 2μm, then the corresponding rms-value of the component mass eccentricities would be 2/0.173 = 11.5 μm. Hence for this particular case the ordinate values in Fig. 5 should be multiplied by 11.5 to yield the actual root-mean-square values for the rotor amplitudes, thereby establishing the probability distributions.

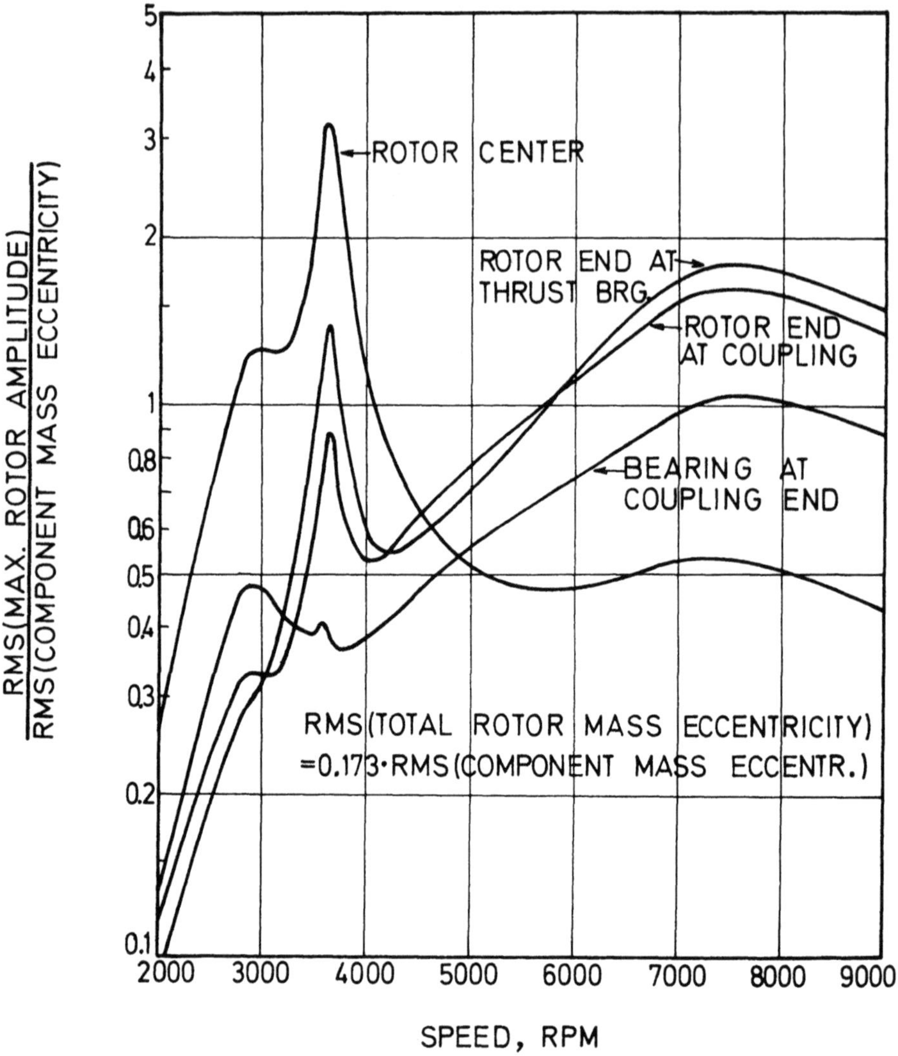

Fig. 5. Unbalance response of compressor rotor in cylindrical journal bearings, caused by random mass unbalance. Root-mean-square value of max.amplitude determines probability distribution of response.

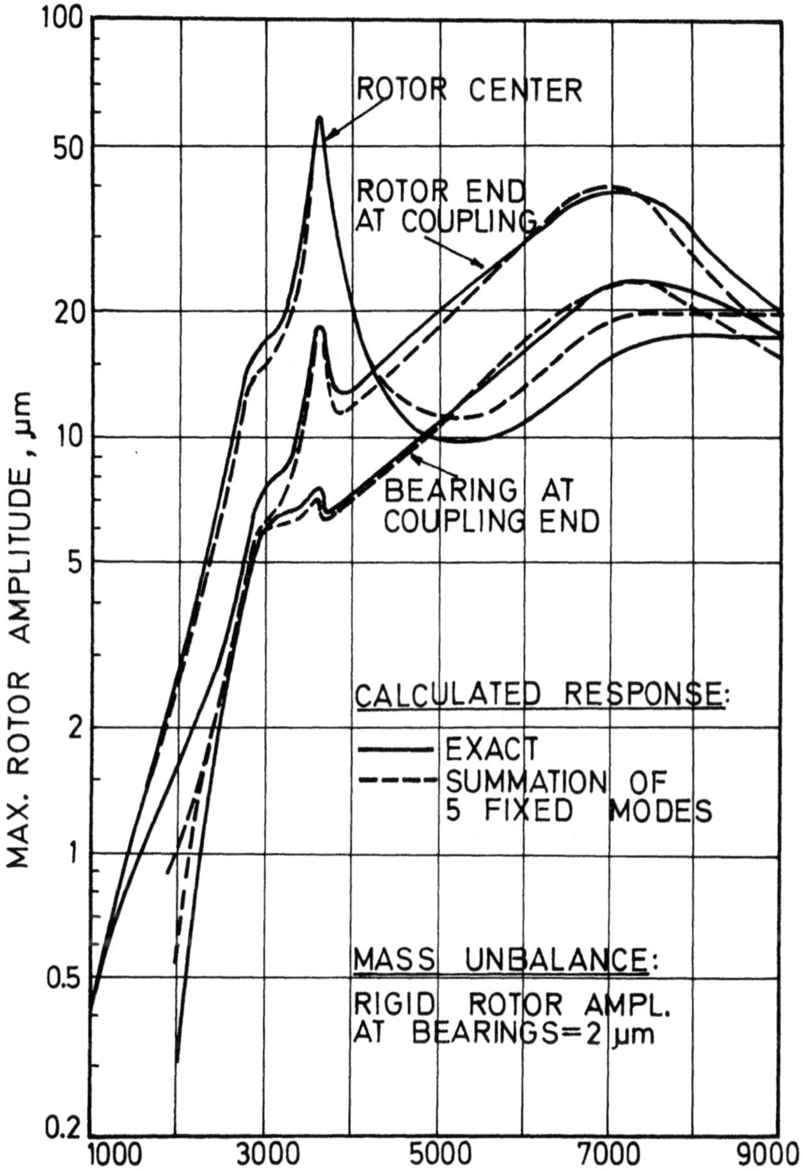

Fig. 6. Unbalance response of compressor rotor in cylindrical journal bearings, caused by specified mass unbalance. Comparison of approximate modal calculation with exact calculation.

If, instead of being random, the mass unbalance is actually known, the resulting amplitude is readily calculated (for example ref. 8). An example is given by the full line curves in Fig. 6 for the amplitude as a function of speed at three locations on the rotor. The chosen mass unbalance is such that its distribution assures even excitation of the previously determined four critical speeds in the speed range, and such that it causes a residual amplitude of 2μm at the bearing locations in a low speed, dynamic balancing machine. From this point of view it is interesting to note the similarity between this figure and Fig. 5.

Whereas the theoretically exact results are shown in full line in Fig. 6, the curves in dotted line are approximations, computed from a truncated modal summation using eq. (46). The five contributing modes are the four previously listed critical speed modes and, in addition, the forward precessional component of the fourth critical, see Fig. 2. In carrying out the calculation, however, the change in the eigenvalues and characteristic functions with speed is disregarded, and it is this approximation, rather than the inherent truncation error, which explains the deviations from the exact results.

The eigenvalues and characteristic functions on which the approximate calculations are based, are those obtained at the damped critical speeds. In this way, Fig. 6 reveals the extent to which it is possible to represent the amplitude response by such a finite set of fixed modes and, thereby, the chances for a successful application of some form of modal balancing, as for example discussed in connection with eqs. (48) to (53).

CONCLUSION

As demonstrated by the numerical examples, the presented analysis is readily applied to a variety of rotor response cases, including transient excitation, random excitation and mass unbalance response. Any practical rotor geometry may be accomodated as may also a wide range of support conditions. The computational time involved in carrying out the calculations is short, of the order of minutes on a modern digital computer.

Available experimental results in general confirm the principal validity of the basic rotor system model. There is, however, a need for further improvements in the existing data for bearing impedances, and for that purpose the present analysis may be of assistance.

REFERENCES

[1] MEIROVITCH, L.: "Analytical Methods in Vibrations", MacMillan Company, 1967.

[2] TIMOSHENKO, S.: "Vibration Problems in Engineering", 3rd. ed., D. Van Nostrand Co., Princeton, N.J., 1955.

[3] DIMENTBERG, F.M.: "Flexural Vibrations of Rotating Shafts", Butterworths, London, 1961.

[4] BISHOP, R.E.D.: "Vibration of Rotating Shafts", Journal of Mechanical Engineering Science, Vol. 1, No. 1, 1959, pp. 50-65.

[5] GLADWELL, G.M.L. and BISHOP, R.E.D.: "The Vibration of Rotating Shafts Supported in Flexible Bearings", Journal of Mechanical Engineering Science, Vol. 1, No. 3, 1959, pp. 195-206.

[6] LUND, J.W.: "Stability and Damped Critical Speeds of a Flexible Rotor in Fluid Film Bearings", Jrnl. of Engineering for Industry, Trans. ASME, Series B, Vol. 96, No. 2, May 1974, pp. 509-517.

[7] LUND, J.W.: "Modal Response of a Flexible Rotor in Fluid Film Bearings", Jrnl. of Engineering for Industry, Trans. ASME, Series B, Vol. 96, No. 2, May 1974, pp. 525-533.

[8] LUND, J.W. and ORCUTT, F.K.: "Calculations and Experiments on the Unbalance Response of a Flexible Rotor", Jrnl. of Engineering for Industry, Trans. ASME, Series B, Vol. 89, No. 4, Nov. 1967, pp. 785-796.

[9] RUHL, R.L. and BOOKER, J.F.: "A Finite Element Model for Distributed Parameter Turborotor Systems", Jrnl. of Engineering for Industry, Trans. ASME, Series B, Vol. 94, No. 1, Feb. 1972, pp. 126-132.

[10] PEDERSEN, P.T.: "On Forward and Backward Precession of Rotors", Ingenieur-Archiv 42, 1972, pp. 26-41.

[11] MORTON, P.G.: "Analysis of Rotors Supported upon many Bearings", Journal of Mechanical Engineering Science, Vol. 14, No. 1, 1972, pp. 25-33.

[12] BLACK, H.F.: "Calculation of Forced Whirling and Stability of Centrifugal Pump Rotor Systems", Journal of Engineering for Industry, Trans. ASME, Series B, Paper no. 73-DET-131.

[13] ONO, K. and TAMURA, A.: "On the Vibration of Horizontal Shaft Supported in Oil Lubricated Journal Bearings", Bulletin of JSME, Vol. 11, No. 47, 1968, pp. 813-824.

[14] FEDERN, K.: "Grundlagen einer Systematischen Schwingungsenstörung Wellenelastischer Rotoren", VDI-Berichte, Vol. 24, 1957, pp. 9-25.

[15] KELLENBERGER, W.: "Should a Flexible Rotor Be Balanced in N or (N+2) Planes", Jrnl. of Engineering for Industry, Trans. ASME, Series B, Paper No. 71-Vibr-55.

[16] PARKINSON, A.G. and BISHOP, R.E.D.: "Vibration and Balancing of Rotating Continuous Shafts", Journal of Mechanical Engineering Science, Vol. 3, No. 3, 1961, pp. 200-213.

[17] PARKINSON, A.G.: "Balancing of Flexible Shafts Rotating in Massive Flexible Bearings", Journal of Mechanical Engineering Science, Vol. 15, No. 6, 1973, pp. 430-438.

[18] LEIPHOLZ, H.: "On a Hybrid Type of Galerkin's Equation", Report No. 82, Solid Mechanics Division, University of Waterloo, Ontario, May 1971.

[19] CRANDALL, S.H. (ed.): "Random Vibration", Technology Press of The Massachusetts Institute of Technology, 1959.

[20] KIRK, R.G. and GUNTER, E.J.: "Transient Response of Rotor-Bearing Systems", Jrnl. of Engineering for Industry, Trans ASME, Series B, Vol. 96, No.2, May 1974, pp. 682-693.

[21] SHEN, F.A.: "Transient Flexible-Rotor Dynamic Analysis Part 1: Theory", Jrnl. of Engineering for Industry, Trans. ASME, Series B, Vol. 94, No. 2, May 1972, pp. 531-538.

[22] TESSARZIK, J.M., CHIANG, T. and BADGLEY, R.H.: "The Response of Rotating Machinery to External Random Vibration", Jrnl. of Engineering for Industry, Trans. ASME, Series B, Vol. 96, No. 2, May 1974, pp. 477-489.

The Derivation of Bearing Characteristics by Means of Transient Excitation Applied Directly to a Rotating Shaft

P.G. Morton

G.E.C. Power Engineering Ltd., Stafford Laboratory, UK.

ABSTRACT - The characteristics of any linear dynamical system can be deduced from the response of that system to known transient forcing applied at strategic points.

A technique has been developed in which the various characteristics of rotating systems have been so deduced. The technique, which involves applying transient forcing directly to the rotating system, is described, and both its scope and limitations are discussed.

The technique is particularly applicable to the measurement of stiffness and damping coefficients of hydrodynamic bearings under functioning conditions, and results are presented and compared with those using other techniques.

NOTATION

B	dimensional bearing damping coefficient
b	non-dimensional bearing damping coefficient
C	radial clearance of bearing
E	dimensional bearing stiffness coefficient
e	non-dimensional bearing stiffness coefficient
F_{0Y}	force vector, position 0, direction Y
$G(\omega)$	frequency response function for filter and recording equipment
$I_{AB}(\omega)$	impedance referring to bearing positions A and B (shaft only)
i	$\sqrt{-1}$
$R(\omega)$	receptance (shaft only)
ΔT	time for applied force to collapse
W	vertical steady load on bearing
$X(t)$	transient displacement function
${}_Y X_A(\omega)$	horizontal shaft displacement vector, measured at point A, due to forcing in Y direction
$x(\omega)$	horizontal oil film displacement vector

$Y(\omega)$	vertical shaft displacement vector
$y(\omega)$	vertical oil film displacement vector
τ	truncation period
ω	frequency
ω_n	lowest natural frequency of loading system
Ω	running speed
$(\cdot)$	differentiation with respect to time

1. INTRODUCTION

In any rotating system the supports exert a considerable influence, both upon the steady-state and the unstable characteristics. Even when the primary cause of instability arises from other sources - internal shaft damping, lack of rotational symmetry of the shaft, steam whirl or fluid seals, for example - the bearings can either mitigate or aggravate such phenomena. Recognition of the role played by bearings in rotor-dynamics has resulted in a great deal of research effort being devoted to the derivation of bearing data, using both theoretical and experimental methods. The simplest and most useful output of this research has been in the form of linearised "stiffness" and "damping" coefficients, of which 32 are relevant to lateral shaft vibration problems. For many practical cases the 8 translational coefficients suffice to characterise the hydrodynamic bearing contribution, although in very flexible shafts embodied in systems with low damping, the pitching and yawing "self-damping" terms have some effect in limiting the response to unbalance at the lower critical speeds.

Although theoretically-derived bearing properties have been useful in the qualitative analysis of linear and non-linear shaft whirling behaviour (Refs. [1] and [2], for example), the quantitative analysis of large hydrodynamic bearings of non-circular profile has proved to be very unreliable. Difficulties associated with mathematical modelling include: Description of the flow at oil inlets and at other discontinuities in the profile, establishment of the correct viscosity profile having regard to oil recirculation, and precise definition of the cavitation boundaries. Results obtained from small-scale models (Ref. [3]), although an improvement on mathematical modelling, are inadequate, particularly in terms of oil temperature (and hence viscosity) profile. Even in full-scale rig testing (Ref. [4]) the correct environmental conditions relating to heat transfer, distortion and bearing alignment cannot always be reproduced.

In view of the foregoing, there is a strong case for extracting bearing data directly from rotating machines under operating conditions. In the case of a static structure with reasonable access, resonance testing would fulfil this purpose. It is possible, of course, to force the shaft of a rotating machine by shaking a bearing support, but this is ineffective in that much of the force is diverted by the support itself, and the residue is very inefficiently transferred to the shaft, except over certain very narrow frequency bands*: furthermore, it is generally difficult to obtain the static coefficients by this technique, and in the special case of a two-bearing shaft, impossible. On the other hand, applying a steady vibration by external means directly to the 80 ton shaft of a large generator rotating at 3,000 r.p.m. is hardly a practical proposition. Another alternative - that of measuring the response of the shaft to unbalance - yields only a limited amount of data.

A transient technique for extracting both the rotating system characteristics and the 8 translational bearing coefficients is described in the present paper. This technique has been used by the author and his colleagues for several years on machines with rotors varying in weight between a

---

* The maximum force transmission through the oil film at any bearing r occurs at frequencies corresponding to resonances of the shaft with the bearing support at r only, infinitely stiff, but the oil film functioning normally.

few hundred pounds and 100 tons supported on bearings ranging from 3" to 21" diameter.

2. PRACTICAL CONSIDERATIONS

The technique involves the application to a rotating shaft of a transient force vector whose co-ordinates are stationary in space. The basic requirement is that this transient shall contain significant harmonic components within a prescribed frequency range. Otherwise there are, in principle, no restrictions on the "shape" of the transient. Where the structure to be excited is stationary, the production of a force transient is relatively straightforward. DOWNHAM and WOODS (Ref. [5]), for instance, describe the use of a range of devices, including a captive bolt gun (humane killer) and an instrument pendulum to excite the pedestals of turbine compressor sets for process plant. Such methods are not practicable where a moving target such as a rotating shaft is involved. Ideally, the requirements are that the transient forcing should be repeatable, should not require

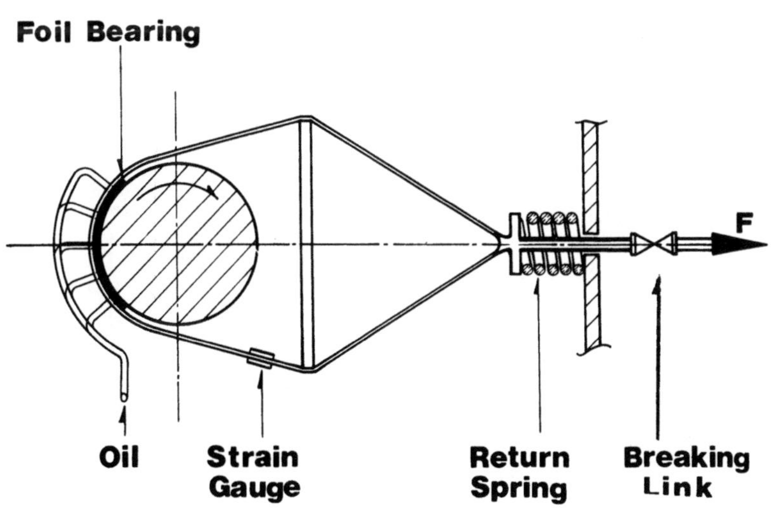

Fig. 1. Loading Mechanism.

measurement over its time span, and, in addition to containing components of a suitably high frequency, should enable the static properties of the system to be easily extracted.

The method is shown in Fig. 1, where a force is slowly applied in the required direction by means of an externally pressurised narrow bearing attached to a notched link. The link breaks at a certain force level, and the collapse rate depends on the "transmissibility" of the loading system.

The following points required consideration during the development of the technique:

1. The pressurised bearing had to be narrow (usually 1"), so that the load could be applied in some of the axially restricted conditions available in practical machines. A

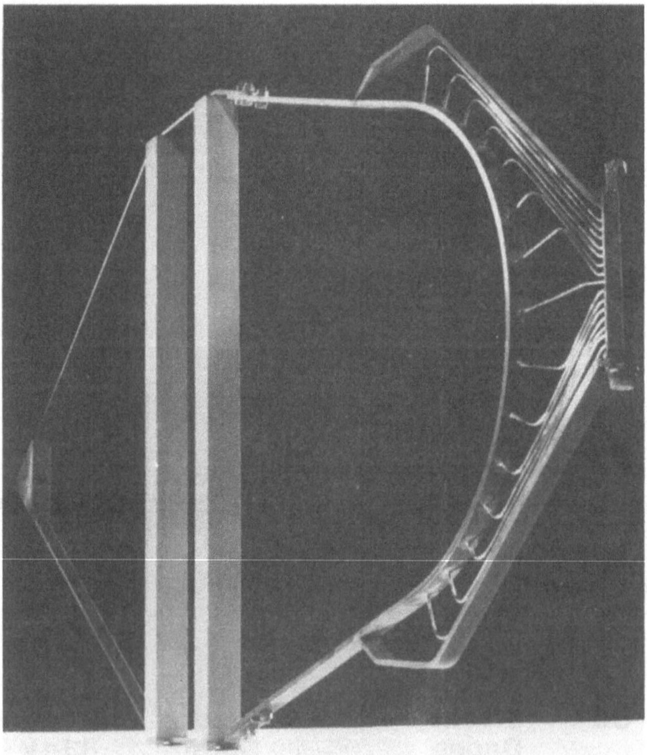

Fig. 2. Construction of Large Diameter Loading Device.

low flexural stiffness for some of the larger diameter bearings, combined with multi-inlet high pressure oil feeding a circumferential groove (the loaded arc being about 120°), ensured a reasonably uniform pressure distribution. A phosphor-bronze liner provided a safeguard against welding in the event of local oil film breakdown. Mean pressures of 400 p.s.i. were not usually exceeded, so that, in a large machine with a 20" dia. shaft, available loads of about 4 tons were normally achieved. Fig. 2 shows a typical foil bearing for use in tests on large turbo-machines.

2. The link had to be cheap and quickly replaceable, to fracture cleanly without excessive elongation and to break at a predictable load. Round mild steel bar with a circumferential 1/32" root radius, 90° V-notch reducing the diameter to ½ was found to be quite adequate, and gave a repeatability of within ±1½% .
3. The loading mechanism had to be capable of applying horizontal and vertical loads, to cause the bearing to completely clear the shaft on unloading, and to be of high flexibility in order to tolerate steady-state shaft vibrations during loading. In addition, the inertia of the system between the breaking-link and the shaft had to be low to obtain a rapid collapse of the load. The load included dead-weights and levers, hydraulic jacks and cranes, depending upon the application, using opposing springs to clear the bearing.
4. The actual time at which the load collapsed at the shaft had to be very accurately recorded. A strain gauge on the foil bearing gave a clear indication of this event. The time delay between the breaking of the notched link and complete removal of load from the shaft was invariably less than a mili-second.

A typical example of the practical application of the method is shown in Fig. 3. Here, a medium sized generator is being excited vertically, but the technique has been applied

Fig. 3. Vertical Excitation of a Medium Sized Generator.

to a wide range of machines.

The transient responses to the loads were usually measured across the oil film at the bearing centres in two directions, using eddy current displacement transducers. Ideally, measurements are required at all bearings, wherever the force is applied, if the true characteristics of any given bearing are to be deduced. In cases where the load application is close to a bearing, however, a sufficiently accurate answer may be obtained by simply measuring the displacement at that bearing.

In addition to measurements across the bearing oil film, absolute shaft motion at the bearing position has to be known. In the absence of direct measurements, bearing housing measurements were taken and added to the oil film readings.

3. ANALYSIS TECHNIQUE

The force is a mirror image of a step function (see Fig. 4), and the corresponding response at any bearing and in any direction is a displacement, starting from a steady initial value and decaying to zero in a finite time. The first requirement is to obtain cross-receptance plots appropriate to

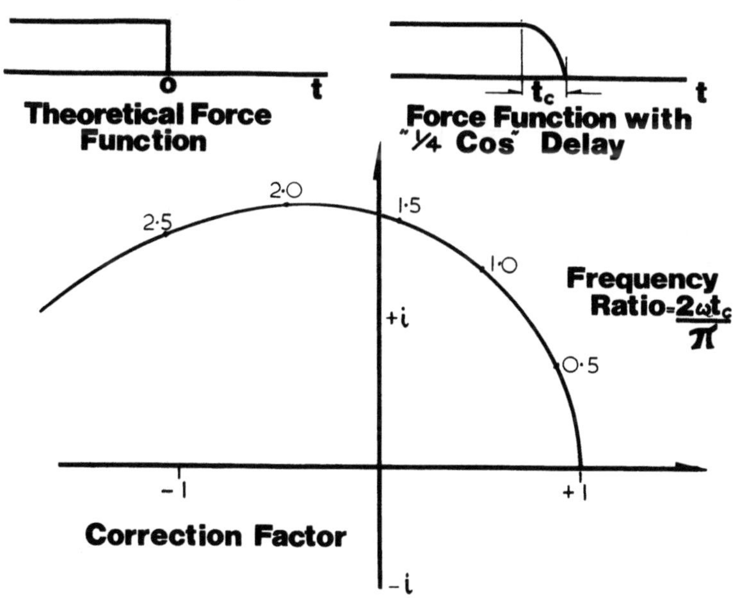

Fig. 4. Correction Factor Due to Delay in Decay of Force.

the force and displacement locations. Receptance as defined in Ref. [6] is the (generally) frequency-dependent complex number defining the quotient of a sinusoidal displacement and force.

The derivation of steady-state information from transient data is, in principle, a well-established procedure. In the present context, the choice of a step function for the force simplifies the analysis, and, in theory, all that would be required to produce a receptance plot would be to operate with the Fourier Transform as follows:

$$R(\omega) = \frac{X_0}{F_0} - \frac{i\omega}{F_0} \int_0^\infty X(t) e^{-i\omega t} \, dt \tag{1}$$

In the case of the bearing supports, where velocity or acceleration measurements have necessarily to be taken, the formula becomes:

$$R(\omega) = \frac{1}{F_0} \int_0^\infty \dot{X}(t) \, e^{-i\omega t} \, dt \quad \text{(velocity)} \tag{2}$$

or

$$R(\omega) = \frac{i}{\omega F_0} \int_0^\infty \ddot{X}(t) \, e^{-i\omega t} \, dt \quad \text{(acceleration)} \tag{3}$$

The static receptance or flexibility may be found directly from the first two formulae, appearing as

$$\frac{X_0}{F_0} \tag{4}$$

or

$$\left\{ - \int^\infty \frac{\dot{X}(t) \, dt}{F_0} \right\} \quad \text{respectively .} \tag{5}$$

In the case of the acceleration formula, however, Expression (3) becomes indeterminate at zero frequency, and static values have to be found by extrapolation.

<u>3.1 Possible Errors</u>. The practical case differs from the ideal case in the following important respects:

1. The force takes a finite time to decay. Assuming that the fracture of the breaking link is instantaneous, there is a resonant system, including the foil bearing oil film, between link and shaft. With a relatively light loading system, then the step will be modified to a $\frac{1}{4}$ cosine, no overshoot being physically possible. The modifying function in the frequency plane is shown in Fig. 4. Although the modulus of the receptance is not substantially changed, the phase error is significant, being of the order of 0.01 radian per cent of the ratio between receptance frequency and loading system frequency. The modifying factor is given by the expression:

$$(6) \qquad \frac{1}{\left[1 - \dfrac{i\omega}{\left(\frac{\pi}{2\Delta T}\right)^2 - \omega^2}\left\{i\omega + \frac{\pi}{2\Delta T}\, e^{-i\omega\Delta T}\right\}\right]}$$

Now ΔT is the period of a $\frac{1}{4}$ of a cycle of the lowest resonant frequency of the loading system. The modification, expressed in terms of the forcing frequency as a proportion of this resonant frequency, can be expressed thus:

$$(7) \qquad \frac{1}{\left[1 + \dfrac{\frac{\omega}{\omega_n}}{1 - \left(\frac{\omega}{\omega_n}\right)^2}\left\{\frac{\omega}{\omega_n} - \sin\frac{\omega\pi}{2\omega_n} - i\cos\frac{\omega\pi}{2\omega_n}\right\}\right]}$$

2. Invariably, in any sensitive measuring system, there must be "noise". Furthermore, mechanical vibrations of a random nature can be picked up by the shaft, especially when the system is running near the threshold of whirl. In general, however, most of the spurious information is of relatively

high frequency. In order to perform the integration

$$\int_0^\infty X(t)\ e^{-i\omega t}\ dt$$

the signal must be sampled. If the sampling frequency is less than twice any frequency component contained in $X(t)$, "aliasing" occurs, causing the frequency spectrum to be badly distorted (see Ref. [7] for a complete description of this well-known phenomenon). Having decided the maximum frequency at which receptances are required, an electrical filter has to be used to eliminate unwanted higher frequencies. The characteristics of most low-pass filters used, made it necessary to sample at about 10 times the cut-off frequency, cut-off being defined as the 3 dB attenuation level. It is obviously essential to allow for the filter characteristic in the analysis.

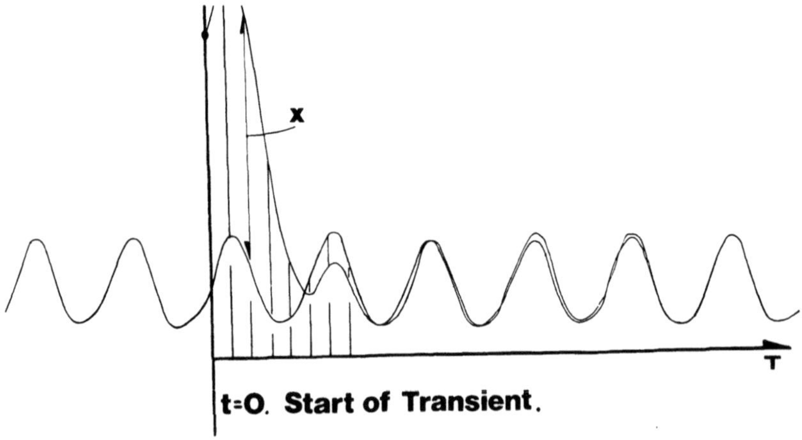

Fig. 5. Substraction or Steady State from Transient.

The frequency response function, $G(\omega)$, of the filter, together with the recording system, is derived by applying an electrical step function to the filter prior to a test. The apparent receptance plot is modified by dividing by the frequency response function of the filter and recording system in order to find the true receptance.

3. A periodic signal of synchronous fundamental frequency is superimposed on the transient, due mainly to unbalance but also to non-isotropic rotor gravity forcing. It is essential to remove this signal if Fourier transform methods are to be used. Several electronic techniques are available for substracting the pre-transient from the post-transient signals, but, with care, this can be done manually, as shown in Fig. 5.
4. Rapid decay of the transient indicates both a physically and an analytically well-behaved machine. When the decay time is long - typically near the instability threshold - truncation is a problem, since it produces spurious loops on the receptance plot. The frequency spacing of these loops is at intervals of $1/\tau$, where τ is the truncation period, so that they are recognisable if truncation is inevitable, under conditions of zero system damping, for example.

3.2. Evaluation of Bearing Characteristics. Cross-receptance plots between a forcing vector at position '0' on the shaft and oil film displacements at all bearings are obtained using the technique already described. Thus ${}_{Y}X_{A}/F_{OY}$, say, refers to the measured receptance for a forcing at '0' in direction 'Y' and a response at bearing 'A' in the direction 'X' . The problem is to eliminate the shaft mechanics from the results, and so deduce the receptances (or impedances) of the oil films alone.

Whereas bearing characteristics are not amenable to theoretical treatment, the mechanics of undamped isotropic free-free shafts can be readily modelled mathematically, and with great accuracy (Ref. [8]).

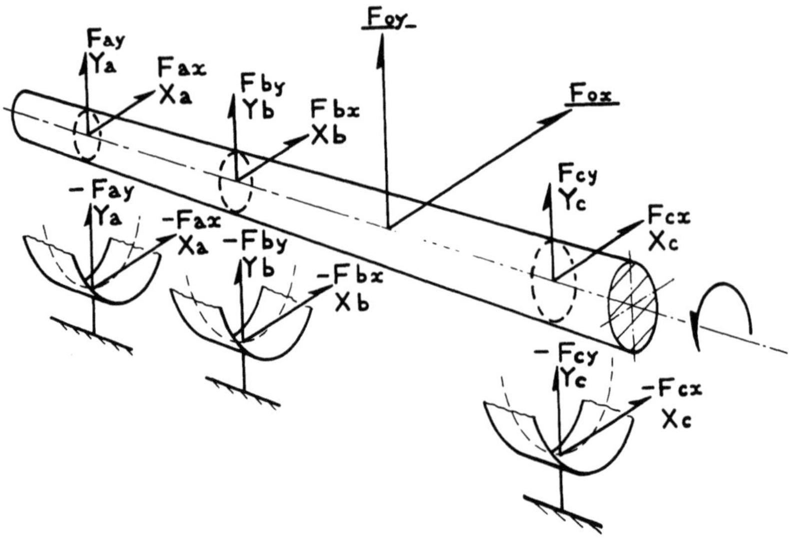

Fig. 6. Forces on Shaft and Bearings.

From Fig. 6, shaft equilibrium dictates that:

$$(8)\qquad \left[\begin{array}{c|c} R & 0 \\ \hline 0 & R \end{array}\right] \left\{\begin{array}{c} F_X \\ \hline F_Y \end{array}\right\} + \left\{\begin{array}{c} F_{OX} \vdots R_{OA} \\ F_{OX} \vdots R_{OB} \\ \hline F_{OY} \vdots R_{OA} \\ F_{OY} \vdots R_{OB} \end{array}\right\} = \left\{\begin{array}{c} X \\ \hline Y \end{array}\right\}$$

where R and X refer to bearing positions.

Whence

$$(9)\qquad \left[\begin{array}{c|c} I & 0 \\ \hline 0 & I \end{array}\right] \left[\left\{\begin{array}{c} X \\ \hline y \end{array}\right\} - \left\{\begin{array}{c} F_{OX} \vdots R_{OA} \\ F_{OX} \vdots R_{OB} \\ \hline F_{OY} \vdots R_{OA} \\ F_{OY} \vdots R_{OB} \end{array}\right\} \right] = \left\{\begin{array}{c} F_X \\ \hline F_y \end{array}\right\}$$

where $I = R^{-1}$.

At position 'A' only:

$$\left[\begin{array}{c|c} (I_{AA}, I_{AB}, \ldots) & 0 \\ \hline 0 & (I_{AA}, I_{AB}, \ldots) \end{array}\right] \left\{\begin{array}{c} X_A - F_{OX} R_{OA} \\ X_B - F_{OX} R_{OB} \\ \vdots \\ \hline Y_A - F_{OY} R_{OA} \\ Y_B - F_{OY} R_{OB} \\ \vdots \end{array}\right\} = \left\{\begin{array}{c} F_{AX} \\ \hline F_{AY} \end{array}\right\} \tag{10}$$

The bearing equation is:

$$\left[E + i\,B\right] \left\{\begin{array}{c} x_A \\ y_A \end{array}\right\} = - \left\{\begin{array}{c} F_{AX} \\ F_{AY} \end{array}\right\} \tag{11}$$

where E and $i\,B$ are dimensional impedances.

Eliminate F_{AX}, F_{AY} and consider the shaft forced separately with F_{OX} (response ${}_X X$, ${}_X Y$) and with F_{OY} (response ${}_Y X$, ${}_Y Y$) .

Combining these conditions into a single matrix equation, then:

$$\left[\begin{array}{c|c} (I_{AA}, I_{AB}, \ldots) & 0 \\ \hline 0 & (I_{AA}, I_{AB}, \ldots) \end{array}\right] \left[\begin{array}{c|c} \left\{\begin{array}{c} R_{OA} F_{OX} - {}_X X_A \\ R_{OB} F_{OX} - {}_X X_B \\ \vdots \end{array}\right\} & - \left\{\begin{array}{c} {}_Y X_A \\ {}_Y X_B \\ \vdots \end{array}\right\} \\ \hline - \left\{\begin{array}{c} {}_X Y_A \\ {}_X Y_B \\ \vdots \end{array}\right\} & \left\{\begin{array}{c} R_{OA} F_{OY} - {}_Y Y_A \\ R_{OB} F_{OY} - {}_Y Y_B \\ \vdots \end{array}\right\} \end{array}\right]$$

$$= \left[E + i\,B\right] \left[\begin{array}{cc} {}_X x_A & {}_Y x_A \\ {}_X y_A & {}_Y y_A \end{array}\right] \tag{12}$$

But

$$[E + i\,B] \left\{\begin{array}{c} x \\ y \end{array}\right\} = \frac{W}{C} \left[e + i \frac{\omega}{\Omega} b\right] \left\{\begin{array}{c} x \\ y \end{array}\right\} \tag{13}$$

where e and b are non-dimensional coefficients.

Hence the real and imaginary parts of

$$\frac{C}{W}\left[\begin{array}{c|c} (I_{AA}, I_{AB}, \ldots) & 0 \\ \hline 0 & (I_{AA}, I_{AB}, \ldots) \end{array}\right] \left[\begin{array}{c|c} \begin{Bmatrix} R_{OA} - \dfrac{X^X{}_A}{F_{OX}} \\ R_{OB} - \dfrac{X^X{}_B}{F_{OX}} \\ \vdots \end{Bmatrix} & \begin{Bmatrix} \dfrac{Y^X{}_A}{F_{OY}} \\ \dfrac{Y^X{}_B}{F_{OY}} \\ \vdots \end{Bmatrix} \\ \hline \begin{Bmatrix} \dfrac{X^Y{}_A}{F_{OX}} \\ \dfrac{X^Y{}_B}{F_{OX}} \\ \vdots \end{Bmatrix} & \begin{Bmatrix} R_{OA} - \dfrac{Y^Y{}_A}{F_{OY}} \\ R_{OB} - \dfrac{Y^Y{}_B}{F_{OY}} \\ \vdots \end{Bmatrix} \end{array}\right]$$

$$\cdot \begin{bmatrix} \dfrac{X^x{}_A}{F_{OX}} & \dfrac{Y^x{}_A}{F_{OY}} \\ \dfrac{X^y{}_A}{F_{OX}} & \dfrac{Y^y{}_A}{F_{OY}} \end{bmatrix}^{-1} \qquad (14)$$

are e and ω/Ω respectively.
ω is appropriate to the receptance: Ω is shaft speed.

It will be remarked that the oil film displacement vectors are different from those defining the shaft. This is because pedestals are interposed between the oil film and earth. Since the oil film can be assumed to be essentially massless, the forces F_{AX}, F_{AY}, etc. are transmitted through it without modification. In order therefore to obtain pedestal impedances, the same formulae are used, replacing oil film displacement vectors with those measured on the pedestal.

The system eigenvalues may be abstracted from any of the receptance plots, using the techniques of KENNEDY and PANCU (Ref. [9]).

The influence of the shaft in Expression (14) is represented by:

$$\begin{bmatrix} (I_{AA}, I_{AB}, \ldots) & 0 \\ 0 & (I_{AA}, I_{AB}, \ldots) \end{bmatrix} \begin{bmatrix} \begin{Bmatrix} R_{OA} - \dfrac{X^X{}_A}{F_{OX}} \\ R_{OB} - \dfrac{X^X{}_B}{F_{OX}} \\ \vdots \end{Bmatrix} & \begin{Bmatrix} \dfrac{Y^X{}_A}{F_{OY}} \\ \dfrac{Y^X{}_B}{F_{OY}} \\ \vdots \end{Bmatrix} \\ \begin{Bmatrix} \dfrac{X^Y{}_A}{F_{OX}} \\ \dfrac{X^Y{}_B}{F_{OX}} \\ \vdots \end{Bmatrix} & \begin{Bmatrix} R_{OA} - \dfrac{Y^Y{}_A}{F_{OY}} \\ R_{OB} - \dfrac{Y^Y{}_B}{F_{OY}} \\ \vdots \end{Bmatrix} \end{bmatrix}$$

(15)

If the forcing occurs very near to a bearing, this shaft modification matrix becomes:

$$\begin{bmatrix} 1 & 0 \\ 0 & 1 \end{bmatrix} - \begin{bmatrix} (I_{AA}, I_{AB}, \ldots) & 0 \\ 0 & (I_{AA}, I_{AB}, \ldots) \end{bmatrix}$$

(16)

$$\begin{bmatrix} \dfrac{1}{F_{OX}} \begin{Bmatrix} X^X{}_A \\ X^X{}_B \\ \vdots \end{Bmatrix} & \dfrac{1}{F_{OY}} \begin{Bmatrix} Y^X{}_A \\ Y^X{}_B \\ \vdots \end{Bmatrix} \\ \dfrac{1}{F_{OX}} \begin{Bmatrix} X^Y{}_A \\ X^Y{}_B \\ \vdots \end{Bmatrix} & \dfrac{1}{F_{OY}} \begin{Bmatrix} Y^Y{}_A \\ Y^Y{}_B \\ \vdots \end{Bmatrix} \end{bmatrix}$$

If the displacements at all other bearings become insignificant, and the pedestal stiffness is high, then the shaft modification becomes:

$$(17)\qquad \begin{bmatrix} 1 & 0 \\ 0 & 1 \end{bmatrix} - \begin{bmatrix} I_{AA} & 0 \\ 0 & I_{AA} \end{bmatrix} \begin{bmatrix} \dfrac{X^{x}{}_{A}}{F_{OX}} & \dfrac{Y^{y}{}_{A}}{F_{OY}} \\[2ex] \dfrac{X^{y}{}_{A}}{F_{OX}} & \dfrac{Y^{y}{}_{A}}{F_{OY}} \end{bmatrix}$$

which means that the actual shaft modification simply involves subtracting I_{AA} from the diagonal terms of the inverse of the measured receptance matrix.

At low frequences, the impedances become terms in the static stiffness matrix, and in a 2-bearing shaft this simply means that no correction has to be applied.

<u>3.3. The Influence of Internal Shaft Damping</u>. The preceding section has assumed that the shaft is undamped. The influence of internal shaft damping is easily allowed for, provided this damping can be expressed in terms of mofifications to the shaft receptances.

The receptances of a stationary free-free damped shaft are generally all complex, say $({}_R R - i\,{}_I R)$, for all bearing positions as well as OA, OB, etc. In effect, internal damping couples vertical and horizontal shaft displacements. It can easily be shown that, for a damped isotropic shaft, Eqn. (8) becomes:

$$\left[\begin{array}{c|c} {}_R R & {}_I R \\ \hline -{}_I R & {}_R R \end{array}\right] \left\{\begin{array}{c} F_X \\ \hline F_Y \end{array}\right\} + \left[\begin{array}{ccc|ccc} {}_R R_{OA} & & 0 & {}_I R_{OA} & & 0 \\ & {}_R R_{OB} & & & {}_I R_{OB} & \\ 0 & & \ddots & & & \ddots \\ \hline -{}_I R_{OA} & & 0 & {}_R R_{OA} & & 0 \\ & -{}_I R_{OB} & & & {}_R R_{OB} & \\ 0 & & \ddots & 0 & & \ddots \end{array}\right] \left\{\begin{array}{c} F_{OX} \\ F_{OX} \\ \hline F_{OY} \\ F_{OY} \end{array}\right\} = \left\{\begin{array}{c} X \\ \hline Y \end{array}\right\}$$

(18)

Thereafter the processes depicted in Eqns. (9) to (12) are carried out as already described.

The calculation of the receptances of a damped beam is difficult, mainly because of problems in mathematically modelling any practical beam. If the shaft is available prior to assembly in the machine, the receptances can, of course, be found by standard resonance testing techniques.

3.4. Use of Derived Coefficients. Now e and b (functions of eccentricity ratio) derived from Expression (14) are the non-dimensional stiffness and damping coefficients connecting the force and displacement vectors in the following manner:

$$\begin{Bmatrix} F_x \\ F_y \end{Bmatrix} = \frac{W}{C} \left[\begin{bmatrix} e_{xx} & e_{xy} \\ e_{yx} & e_{yy} \end{bmatrix} + \frac{i\omega}{\Omega} \begin{bmatrix} b_{xx} & b_{xy} \\ b_{yx} & b_{yy} \end{bmatrix} \right] \begin{Bmatrix} x \\ y \end{Bmatrix} \qquad (19)$$

where x, y relate to an elliptical orbit at a frequency ω.

The eigenvalues of $(e + i(\omega/\Omega)\, b)$ are influenced not only by the eccentricity ratio but by the frequency parameter ω/Ω. For a limited range of journal attitudes there exist unique values of ω/Ω at which one eigenvalue possesses no imaginary part. This means that, for the appropriate orbit or eigenvector, the bearing abstracts no energy from the dynamical system of which it forms a part. Consequently, provided no other damping is coupled into the system, the threshold of oil whirl is then reached. The stabilising effect of the bearing increases with the magnitude of ω/Ω, and at any journal attitude there exists a value of this parameter, above which the bearing will always provide positive damping (see Fig. 7).

Physically real eigenvalues of $(e + i(\omega/\Omega)\, b)$ can be considered as pure isotropic spring stiffnesses suitably non-dimensionalised. A fairly descriptive term would be perhaps "the zero damping stiffness" of the bearing. For most bearings this value varies very little over the relevant range of eccentricity ratios.

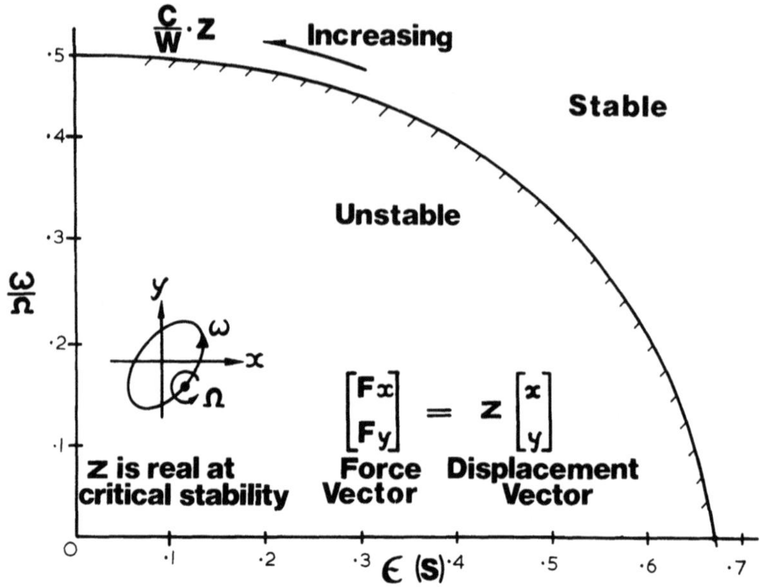

Fig. 7. Bearing Stability Chart.

4. EXPERIMENTAL WORK

4.1. Model Work. The technique was initially developed on a rotor which weighs around 1,400 lb, has a lowest critical speed at 7.8 Hz, and is supported on 4" dia. bearings. For the purpose of evaluation, circular bearings were chosen on the grounds that, in the laminar region at least, published work was available on both theoretical and experimental coefficients, and, further, that solutions of the Reynold's equation were most likely to be correct for this simple case.

In order to eliminate excessive bending in the very flexible rotor, the force resultant was arranged to occur at the centre of a bearing, straps being disposed each side of the bearing and loaded in parallel. Since the investigation was mainly conserned with linearised coefficients, small load perturbations of the order of 100 lb were applied in most of the tests. Tests were carried out, using a wide range of oils at various temperatures, the viscosities ranging from 0.4×10^{-6} to 10×10^{-6} Reyns. Journal loci and the

8 bearing coefficients were experimentally determined. The stability boundary and the appropriate "zero damping" stiffnesses were derived from both experimental and theoretical coefficients, and compared with the direct measurements of the onset of whirl.

The journal locus of Fig. 8 produced covering eccentricity ratios from 0.2 to 0.96. At the lowest eccentricity ratio the minimum film thickness was 0.00015" (radial clearance 3.75" x 10^{-3}) , so that accurate alignment was a protracted and tedious procedure. The consistency of the locus implies that here the journal attitude may be completely defined by eccentricity. The connection between eccentricity ratio and Sommerfeld Number, based on outlet oil tem-

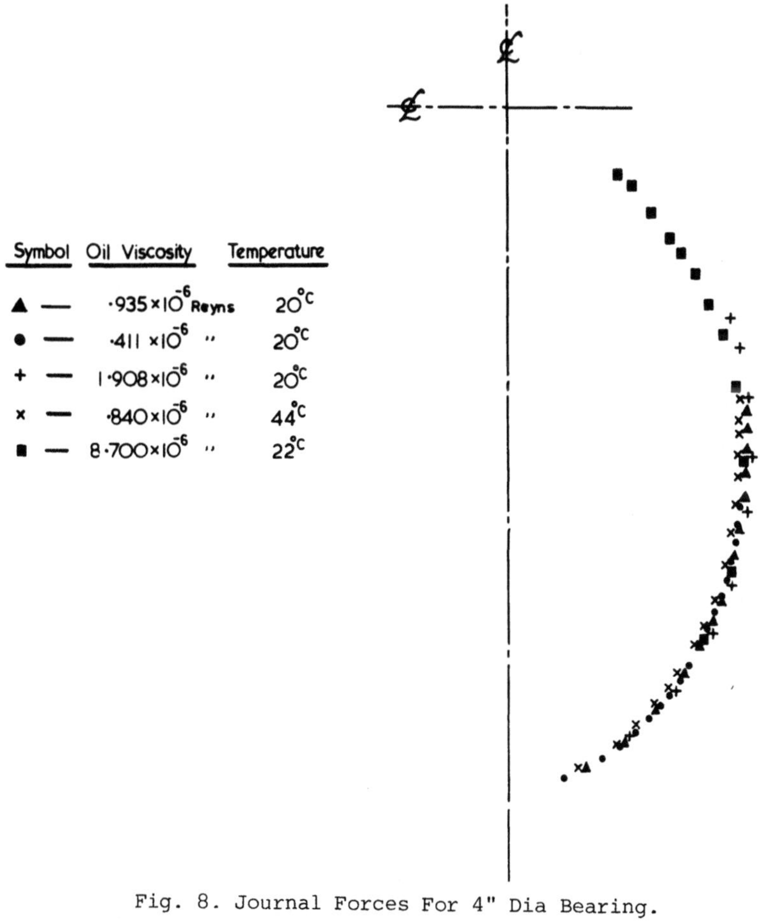

Fig. 8. Journal Forces For 4" Dia Bearing.

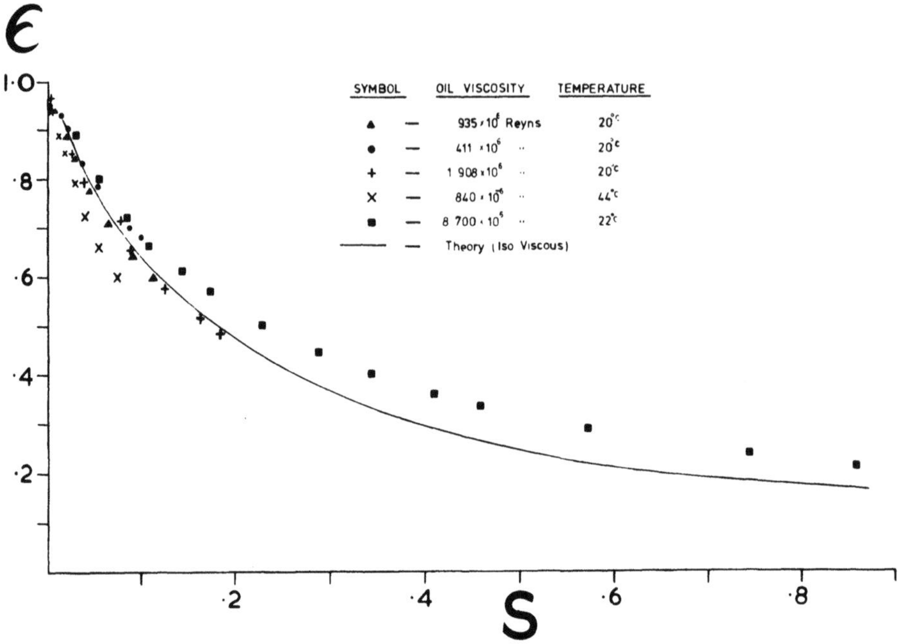

Fig. 9. Relationship between Sommerfeld Number and Eccentricity Ratio for 4" dis bearing.

perature (Fig. 9) is not so consistent, the divergence occurring mainly in the hot and higher viscosity oils. In the latter case, the greater power loss results in some divergence from the isoviscous situation, and the choice of outlet oil viscosity is probably a wrong one.

Figs. 10 and 11 plot the measured stiffness and damping coefficients against eccentricity ratio, and compare them with theoretical coefficients reported by SMITH (Ref. [10]). These results are somewhat scattered, since the receptance matrix is ill-conditioned. Further, coefficients for eccentricity ratios of over 0.85 are suspect, in view of the susceptibility of the minimum film thickness to misalignment. A point of special interest is the cross-damping coefficients which are, in theory, reciprocal. Few researchers have ever managed to conclusively establish reciprocity experimentally, and clearly the author is no exception.

The results have been processed (without any smoothing)

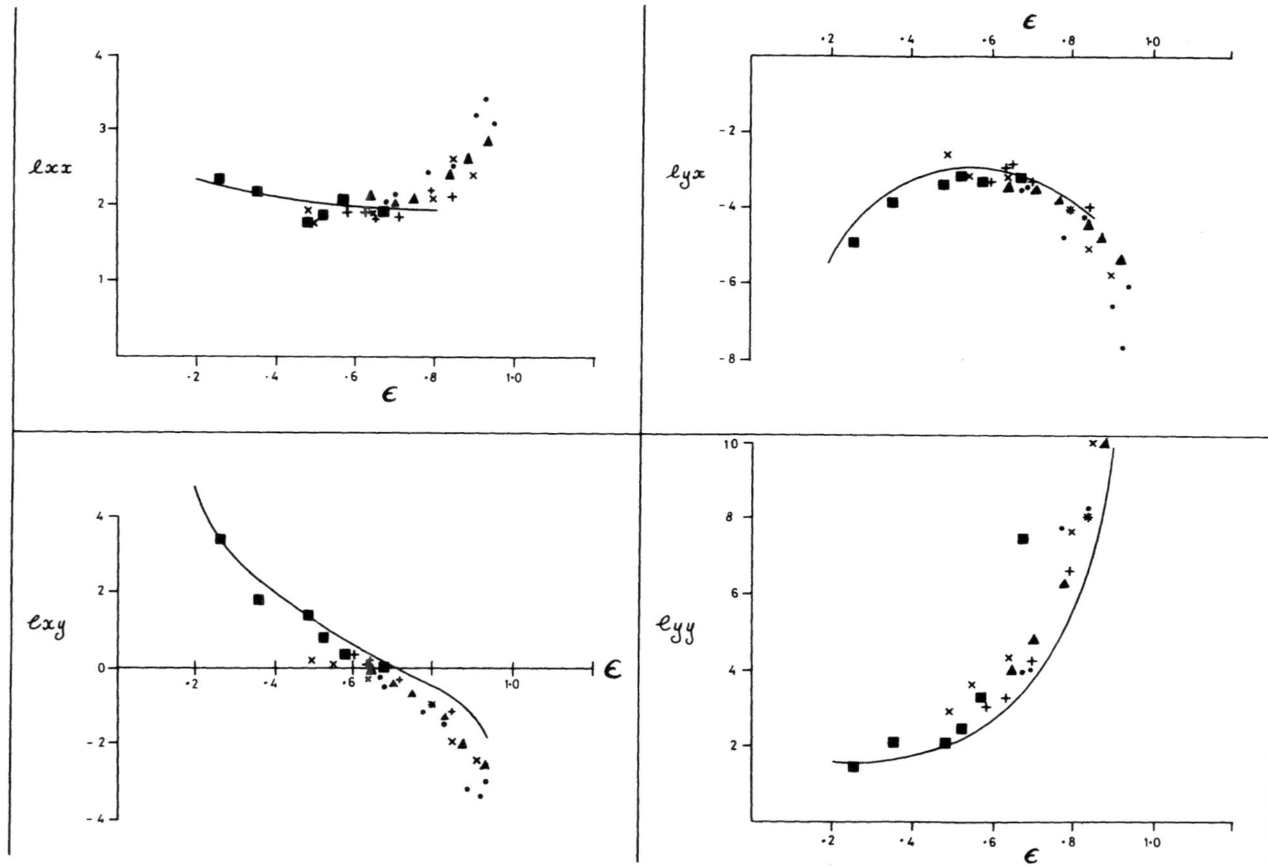

Fig. 10. Non Dimensional "Stiffness" Coefficients.

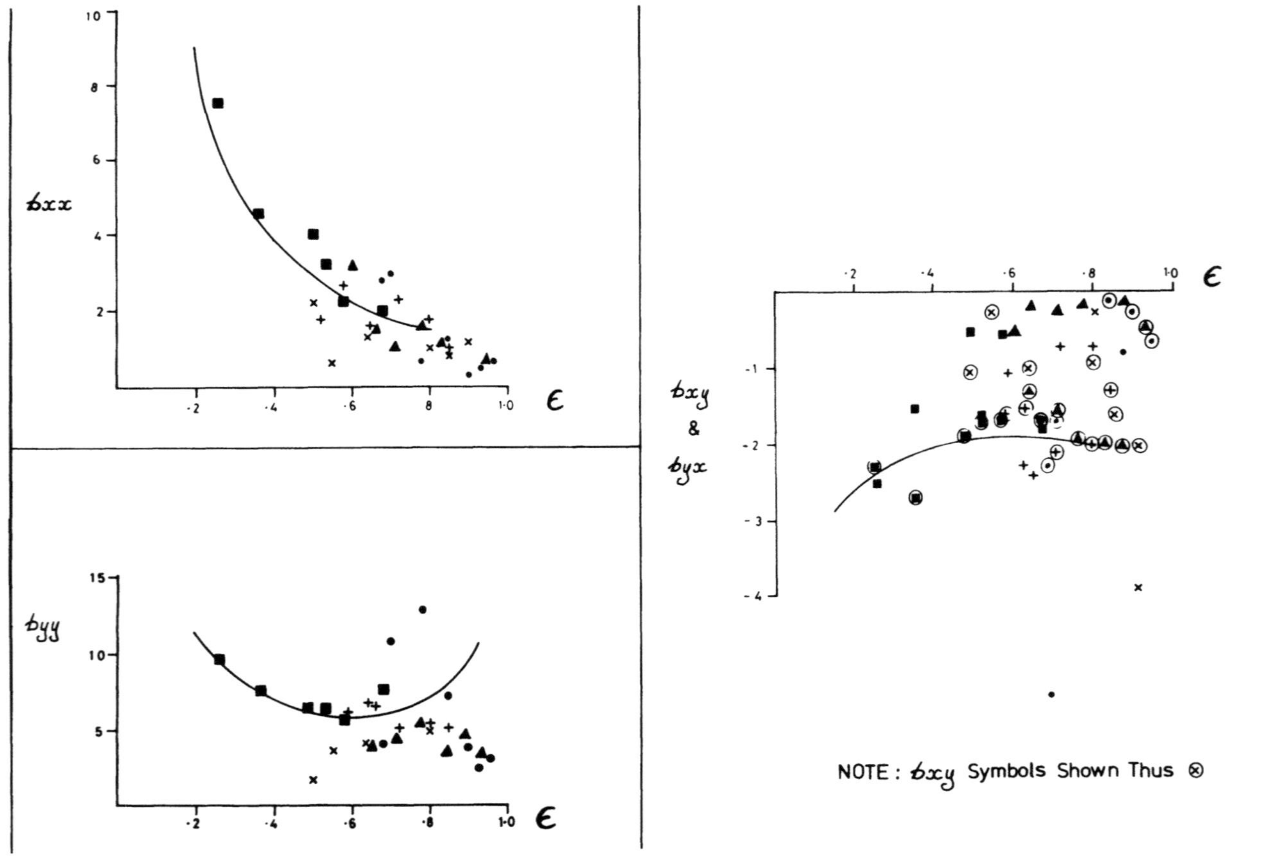

Fig. 11. Non-dimensional "Damping" Coefficients.

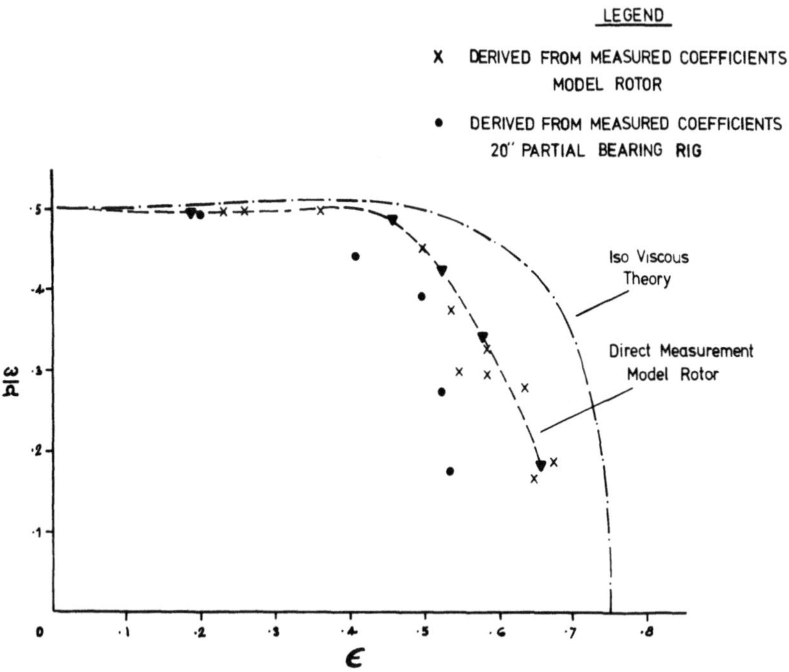

Fig. 12. Stability Chart. Comparison of Theoretical and Experimental Results.

to produce the stability chart of Fig. 12. Points derived from the coefficients are confirmed very closely by a curve established directly by test under identical conditions. The fact that this curve does not agree with the theoretical stability boundary does not invalidate the argument that the coefficients must necessarily be accurate, since they predict observed behaviour. It is of general interest that theory seems to under-estimate stability.

4.2. Tests on Large Machines. Large machines have proved to be hardly more difficult to test using the transient technique than the model. A large selection of machines has been investigated, with bearings ranging from 6" dia. to 20" dia., in both shop test berths and at site.

In a preliminary series of tests, the transient technique was used on a large test bearing in the test of Fig. 13. Although this was a case in which the non-rotating bush was forced against a rotating shaft, and is therefore not a

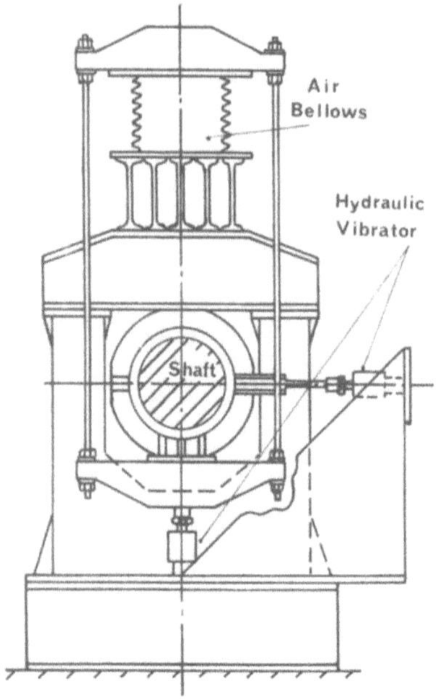

Fig. 13. Rig for the Dynamic testing of Large Bearings.

complete example of the technique under discussion, the results were of value, in that a direct comparison between steady-state forcing and transient methods was possible. A typical receptance curve for a 20" dia. partial bearing, uncorrected for structural elements, is compared with shake test results in Fig. 14. The resonances shown are those of the test bush and shaft, both weighing several tons, "bouncing" on their oil films and structure.

An example of traces from a typical test on a medium-sized generator in the overspeed pit prior to balancing is shown in Fig. 15, together with a derived receptance plot and the frequency and Q factors for the system at the

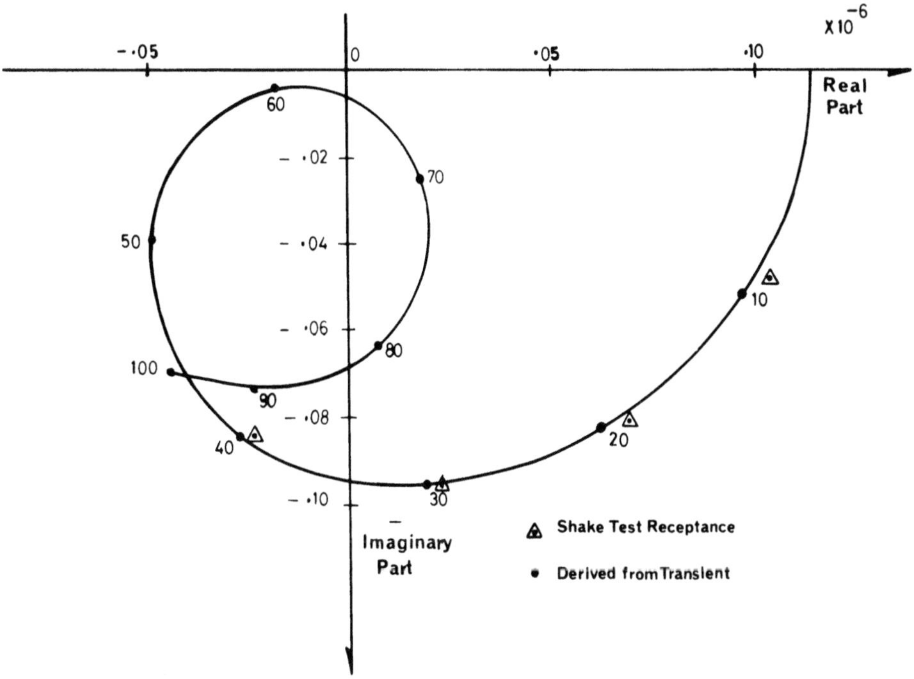

Fig. 14. Comparison of Bearing Receptance derived from transient and sinusoidal forcing.

particular running speed.

Apart from their use in deriving the bearing coefficients, such receptance curves contain information on the natural frequencies and damping of the system at a given running speed. The proximity to an unstable whirl threshold may be deduced from the level of damping at the lowest natural frequency, and the critical speeds are found by interpolation between the results at several running speeds.

5. COMMENTS AND CONCLUSIONS

One outcome of the tests, both on the model and on machines such as large induction motors, with bearings of up to 6" dia. and fairly small clearances, was to demonstrate the very powerful influence of even the smallest degree of misalignment, particularly in bearings with large L/D ratios.

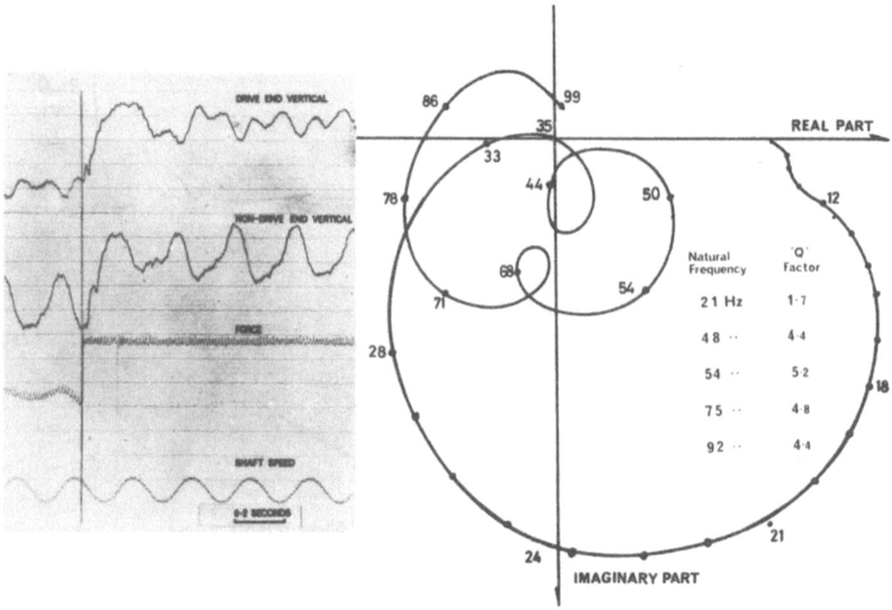

Fig. 15. Typical Test Results from a Medium Sized Generator.

The model results of Fig. 10 show that, at ecentricity ratios above 0.8, the direct stiffness coefficients climb very rapidly. This is undoubtedly solely due to contact between the shaft and the bearing. Experiments in which larger misaligments were deliberately introduced resulted in stiffness over a large proportion of the running range being significantly increased, without any apparent ill effects in the running behaviour. Having regard to distortions, settlement, etc., likely to occur under normal running conditions, predictions of the bearing characteristics at high eccentricity are not likely to be of much practical use.

On the other hand, tests on some large bearings upwards of 16" dia. have shown that at very high eccentricity ratios the stiffness coefficients are considerably lower than those obtained theoretically or by testing small bearings. It is possible that air entrainment, resulting in significant oil compressibility, could have a large effect on thin load-bearing oil films.

Testing covered bearings with Reynolds Numbers ranging from a maximum number of 265 in the case of the model tests described in the present paper, to 3,600 in the case of a large turbo-type bearing. In the former case, the critical Reynolds Number was around 1,300 , and for the large bearing this value lay in the region of 1,100 , so that oil flow in the model bearing was laminar, whereas the large bearing ran well into the turbulent regime. The main influence of turbulence was to increase the load-carrying capacity of the oil film, thus reducing the eccentricity ratio. For comparable eccentricity ratios, the influence of turbulence on stability did not seem significant. This means that, for comparable Sommerfeld Numbers, turbulent bearings were less stable than those running in the laminar regime.

Observations of the system eigenvalues, and calculations based on bearing coefficients derived from transient testing suggest that predictions of the onset of resonant whirl using theoretical coefficients are generally pessimistic. It is clear that much more detailed investigation into the flow patterns at inlets and at discontinuities in the oil film, into the effect of air entrainments, of turbulence, and of bearing distortion under operating conditions, and into cavitation conditions, is required before theoretical coefficients can be used with confidence.

Regarding the technique itself, this is still being refined, mainly in connection with the processing of the measurements. Practical questions concerning the separation of periodic and transient signals are being investigated.

The main concern, however, is the full utilisation of the recorded transient to obtain the system characteristics. The present paper is occupied primarily with the extraction of steady-state bearing characteristics, so that emphasis has been placed on deducing the steady-state system response. The extraction of system eigenvalues from response data is a somewhat approximate procedure, and it would be preferable to arrive at these eigenvalues by a more direct method. Such a method has just been developed, and will be the subject of a further paper.

ACKNOWLEDGEMENTS - The author gratefully acknowledges the help of his colleagues in the Mechanical Division of the Stafford Laboratory of G.E.C. Power Engineering Limited in preparing this paper. Thanks are due to the General Electric Company for permission to publish this paper.

REFERENCES

[1] LUND, J.W., and STERNLICHT, B.: Rotor bearing dynamics with emphasis on attenuation. Amer. Soc. Mech. Engrs., Paper 61-WA-68, 1961.

[2] HOLMES, R.: Oil whirl characteristics of a rigid rotor in 360° journal bearings. Proc. Inst. Mech. Engrs., 177, No. 11, 291, 1963.

[3] HAGG, A.C. and SANKEY, G.O.: Elastic and damping properties of oil film journal bearings for application to unbalance vibration calculations. J. Appl. Mech., Trans. Amer. Soc. Mech. Engrs., 196, 78, 302.

[4] MORTON, P.G.: Measurement of the dynamic characteristics of a large sleeve bearing. Jnl. of Lubrication Technology, Trans. Amer. Soc. Mech. Engrs., Series F, Vol. 93, 1, 143-150, 1971.

[5] DOWNHAM, E. and WOODS, R.: The rationale of monitoring vibration on rotating machinery on continuously operating process plant. Amer. Soc. Mech. Engrs., Paper No. 71-Vibr-96, 1971.

[6] BISHOP, R.E.D. and JOHNSON, D.C.: The Mechanics of Vibration - 1960. Cambridge University Press.

[7] MURPHY, G.J.: Control Engineering. Van Nostrand, 1959, pp. 230-232.

[8] MORTON, P.G.: On the dynamics of large turbo-generator rotors. Proc. Inst. Mech. Engrs., 1965-66, 180 (Pt. 1), 295.

[9] KENNEDY, C.C. and PANCU, C.D.P.: Use of vectors in Vibration measurement and analysis. Jnl. Aeronaut. Sci., 1947, 14, (11), 603.

[10] SMITH, D.M.: Dynamic characteristics of turbine journal bearings. Proc. Lubrication and Wear Convention, Bournemouth, 1963, Paper 8, 72. (Institution of Mechanical Engineers, London).

Nonlinear Excited and Self-Excited Precessional Vibrations of Symmetrical Rotors

Agnieszka Muszyńska

Institute of Fundamental Technological Research of the Polish Academy of Sciences, Warsaw, Poland

ABSTRACT - In the paper the symmetrical rotor systems having nonlinear properties are considered. Various origins of the nonlinearities are discussed (internal and structural damping, nonlinear elastic and damping characteristics of elastic supports and dampers, nonlinear aero and hydrodynamical forces in plain bearings or in seals of hydraulic machines, electrodynamical forces, geometrical nonlinearities of systems). On the grounds of a general symmetrical nonlinear mathematical model it is shown the occurrence of excited, self-excited and combined vibrations. It is found also that the equilibrium position (the center of vibrations) is displaced.

Some numerical examples demonstrate shapes of the amplitude-frequency-phase-rotational speed curves.

1. INTRODUCTION

Every mathematical model describing motion of rotors presents an abstract approach to reality. Due to the complexity of phenomena occuring during motion of a real system, the last one is usually simplified by neglecting a number of less important factors. Such an idealization is necessary not only for actual computation to be feasible, but also for the obtainment of simple and convenient mathematical relations ensuring the required accuracy of solution, which is easy for physical discussion.

From the mathematical point of view, the most often adapted mathematical models of rotors are the linear ones (linear differential equations). In spite of their numerous advantages, the linear models are not able to explain many phenomena and we face of necessity of taking into account the nonlinear properties of the systems.

On the other hand, from the geometrical point of view,

the most often applied models of rotors are the symmetrical ones, i.e. the models, in which isotropic properties in any direction perpendicular to the axis of rotation are kept. Describing only one class of possible real cases, the symmetrical models allow to explain however, many physical phenomena occuring in rotating systems and they serve as a basis for further generalizations.

In the present paper the symmetrical rotor systems having nonlinear properties are considered. (We shall say "rotor", "rotor system" meaning by it an adequate model of the rotor). Various origins of the nonlinearities are discussed. On the grounds of a general symmetrical nonlinear mathematical model it is shown the occurence of excited, self-excited and combined vibrations. It is found also that the equilibrium position of the rotors is displaced.

Some numerical examples demonstrate shapes of the amplitude-frequency and phase-frequency curves for particular cases of nonlinear functions.

2. MATHEMATICAL MODELS OF SYMMETRICAL ROTORS

2.1. General Assumptions. Several models of symmetrical rotors will be considered. We shall examine lumped models with two degrees of freedom and continuous ones. For lumped models the discussion will be concerned with two cases: small flexural deformations of the mass-less shafts with one disc of mass m, eccentrically mounted in the middle of the length ℓ of the shafts, with disregarding gyroscopic effects, and the second: small deviations of rigid rotors mounted in flexible bearings.

Let us introduce few general assumptions: 1° - The shafts have constant circular cross-section of a radius $r(S \equiv \pi r^2)$; 2° - Bearings and supports are isotropic in any two directions perpendicular to the shaft axis; 3° - The relation between the torque imparting motion to the shaft and the moment of resistance is such as to give the rotor a constant angular velocity ω , transient processes are not considered; 4° - The torsional and longitudinal rigidities of the shafts are perfect; 5° - The external damping is linear (b, $\tilde{b}$ - coefficients of the external damping); 6° - The gravitational forces are taken into account (g - gravitational acceleration).

2.2. The Rotating Shaft With Nonlinear Elastic and Internal Damping Properties. The Lumped Case. Let us consider the motion of a rotor (Fig. 1), taking into account the nonlinear rheological model of the shaft material [1 - 11]:

$$\sigma = E\varepsilon + \tilde{\mu}\dot{\varepsilon} + f(\varepsilon,\dot{\varepsilon}) \; , \; E = \text{const} \; , \; \tilde{\mu} = \text{const} \; , \; f \in C^o \tag{2.1}$$

The function $f(\varepsilon,\dot{\varepsilon})$ does not contain any linear term.

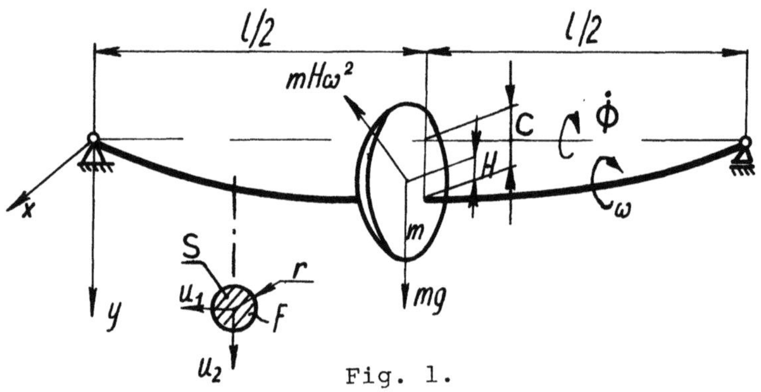

Fig. 1.

The equation of motion of the shaft is as follows:

$$\begin{aligned} m\,\ddot{z} + b\,\dot{z} + \mu(\dot{z}-i\omega z) + k\,z + z\,\Big[F_1(|z|, \tfrac{d|z|}{dt}, \omega - \dot{\phi}) - \\ - i\,F_2(|z|, \tfrac{d|z|}{dt}, \omega - \dot{\phi})\Big] = m\,H\,\omega^2 e^{i\omega t} + i\,m\,g \;, \end{aligned} \tag{2.2}$$

where

$$z = x + i\,y \; , \; i = \sqrt{-1} \; , \; |z| = \sqrt{x^2+y^2} \; , \; \phi = \arctan\frac{y}{x} \; , \tag{2.3}$$

$$\mu = \frac{48\tilde{\mu}I}{\ell^3} \; , \; k = \frac{48EI}{\ell^3} \; , \; I = \frac{\pi r^4}{4} \; , \tag{2.4}$$

$$F_w(|z|, \frac{d|z|}{dt}, \omega - \dot{\phi}) =$$

(2.5)

$$= \frac{\ell^5}{432|z|^4} \int_{-h}^{h} d z_1 \int_{-\sqrt{h^2-z_1^2}}^{\sqrt{h^2-z_1^2}} z_w \, f[z_1, (\frac{z_1}{|z|} \frac{d|z|}{dt} + (\omega-\dot{\phi}) z_2)] dz_2 ,$$

$$\lim_{|z|\to 0} F_w (|z|, \frac{d|z|}{dt}, \omega-\dot{\phi}) = \text{const}, \quad w = 1, 2, \quad h = \frac{12r|z|}{\ell^2}$$

The functions F_1 , F_2 have been obtained from the following expression of the force of elasticity:

$$\frac{4}{\ell} \int_S u_w \, \sigma \, d S , \quad w = 1,2,$$

with taking into account the internal non-elastic resistance according to the rheological model (2.1). These functions are nonlinear and they couple the equations (2.2) one to another (for scalar real variables x, y). For the assumed constant angular speed of the shaft ω the functions describing nonlinear visco-elastic properties of the shaft material have the special arguments as combinations of the variables x and y . (For the rest of notations - see Fig. 1).

2.3. The Rotating Shaft With Nonlinear Elastic and Internal Damping Properties. The Continuous Case. Let us consider the shaft of the length ℓ with continuously distributed mass (Fig.2). We shall take into account the rotary inertia, transverse shear and the gyroscopic effect [12],[13]. We shall assume that the rheological model of the shaft material is described by the relation (2.1). The equation of motion of the shaft is [14]:

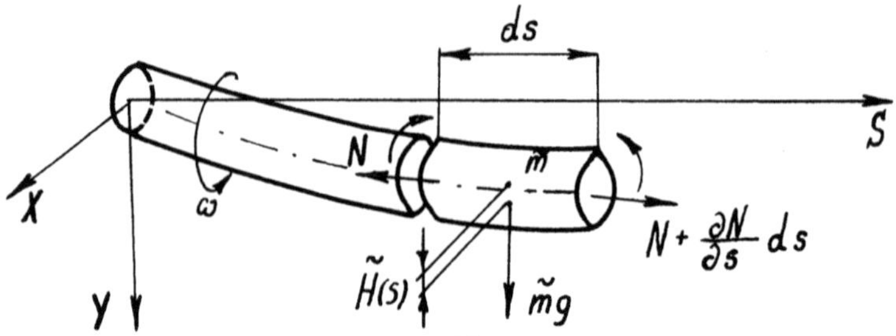

Fig. 2.

$$\tilde{m}\,\frac{\partial^2 Z}{\partial t^2} + EI\,\frac{\partial^4 Z}{\partial s^4} + \tilde{\mu} I\left(\frac{\partial^5 Z}{\partial s^4 \partial t} - i\omega\,\frac{\partial^4 Z}{\partial s^4}\right) + \frac{I\tilde{m}}{\tilde{k}GS}\left(\frac{\tilde{m}}{S}\,\frac{\partial^4 Z}{\partial t^4} -\right.$$

$$\left. - E\,\frac{\partial^4 Z}{\partial s^2 \partial t^2}\right) + i\,\frac{2\tilde{m}I\omega}{S}\left(\frac{\partial^3 Z}{\partial s^2 \partial t} - \frac{\tilde{m}}{\tilde{k}GS}\,\frac{\partial^3 Z}{\partial t^3}\right) -$$

(2.6)
$$- \frac{\tilde{m}I}{S}\,\frac{\partial^4 Z}{\partial s^2 \partial t^2} + \frac{\partial}{\partial s}\left[N(s)\,\frac{\partial Z}{\partial s}\right] + \tilde{b}\,\frac{\partial Z}{\partial t} + \frac{\partial^2}{\partial s^2}\left\{\frac{\partial 2_Z}{\partial s^2}\right.$$

$$\left.\left[\tilde{F}_1\left(\rho,\,\frac{\partial \rho}{\partial t}\,,\,\omega - \frac{\partial \tilde{\phi}}{\partial t}\right) - i\,\tilde{F}_2\left(\rho,\,\frac{\partial \rho}{\partial t}\,,\,\omega - \frac{\partial \tilde{\phi}}{\partial t}\right)\right]\right\} =$$

$$= \tilde{m}\,\tilde{H}(s)\,\omega^2\,e^{i\omega t} + i\,\tilde{m}\,g\,,$$

where

$$Z = X(s,t) + i\,Y(s,t)\,, \quad \rho = \left|\frac{\partial^2 Z}{\partial s^2}\right|\,,$$

(2.7)
$$\tilde{\phi} = \arctan\left(\frac{\partial^2 Y}{\partial s^2} \Big/ \frac{\partial^2 X}{\partial s^2}\right)\,, \quad \tilde{k} = \text{const}\,, \quad G = \text{const}\,,$$

$$\tilde{F}_w\left(\rho,\,\frac{\partial \rho}{\partial t}\,,\quad \omega - \frac{\partial \tilde{\phi}}{\partial t}\right) =$$

$$= \frac{r^4}{n^4}\int_{-n}^{n} d\,z_1 \int_{-\sqrt{n^2 - z_1^2}}^{\sqrt{n^2 - z_1^2}} z_w\,f\left[z_1,\,\left(\frac{z_1 r}{n}\,\frac{\partial \rho}{\partial t} + \left(\omega - \frac{\partial \tilde{\phi}}{\partial t}\right) z_2\right)\right] dx_2\,,$$

$$\lim_{u \to 0} \tilde{F}_w\left(u,\dot{u},\omega - \frac{\partial \tilde{\phi}}{\partial t}\right) = \text{const},\ w = 1,2,\ n = r\rho\,.$$

The function $\tilde{H}(s)$ represents the eccentricity of the centre of gravity of the shaft element at the distance s ;

$\tilde{N}(s)$ is an axial force (Fig. 2). The nonlinear functions $\tilde{F}_1, \tilde{F}_2$ are of the same character as F_1, F_2 in (2.2).

We shall seek for solutions of the Eq. (2.6) applying the Galerkin's method. Let us consider such a shape of the deflected shaft and the boundary conditions that for the first mode we have

$$(2.9)\quad Z(s,t) = z(t)\sin\frac{\pi s}{\ell}\,,\quad z(t) = x(t) + i\,y(t)\,.$$

Substituting (2.9) into (2.6), then multiplying all terms by $\sin\frac{\pi s}{\ell}$ and integrating them within limits $[0,\ell]$ we have:

$$(m+m_1)\ddot{z} + b\,\dot{z} + \mu(\dot{z}-i\omega z) + k\,z + z\left[F_1\left(|z|, \frac{d|z|}{dt}, \omega-\dot{\phi}\right) - i\,F_2\left(|z|, \frac{d|z|}{dt}, \omega-\dot{\phi}\right)\right] + \chi(\overset{....}{z} - 2\,i\omega\,\overset{...}{z}) - 2\,i\omega\delta\dot{z} =$$

$$(2.10)\qquad = m\,H\,\omega^2 e^{i\omega t} + i\,m\,g\,,$$

where

$$(2.11)\quad \begin{aligned} m &= \frac{\tilde{m}\ell}{2}\,,\quad m_1 = \delta\left(1+\frac{E}{\tilde{k}G}\right),\quad \delta = \frac{\pi^2 mI}{2\ell S}\,,\\ b &= \frac{\tilde{b}\ell}{2}\,,\quad \mu = \frac{\pi^4\tilde{\mu}I}{2\ell^3}\,,\quad \chi = \frac{\tilde{m}^2 I\ell}{2\tilde{k}GS^2}\,,\\ k &= \frac{\pi^4 EI}{2\ell^3} - \frac{\pi^2}{\ell^2}\int_0^\ell N(s)\cos^2\frac{\pi s}{\ell}\,ds\,,\quad I = \frac{\pi r^4}{4} \end{aligned}$$

$$F_w\left(|z|, \frac{d|z|}{dt}, \omega-\dot{\phi}\right) =$$

$$= \frac{\pi^4}{\ell^4}\int_0^\ell \tilde{F}_w\left(\frac{\pi^2}{\ell^2}|z|\sin\frac{\pi s}{\ell}\,,\ \frac{\pi^2}{\ell^2}\frac{d|z|}{dt}\sin\frac{\pi s}{\ell}\,,\ \omega-\dot{\phi}\right)\sin^2\frac{\pi s}{\ell}\,ds\,,$$

$$(2.12)\qquad w = 1,2\,,\quad \phi = \arctan\frac{y(t)}{x(t)}$$

The coefficients m_1 is composed of two terms. The first one is related to the rotary inertia, the second one - to the transverse shear deformation of the shaft. The coefficient χ is connected with the transverse shear $(\chi \overset{\cdots\cdot}{z})$ and gyroscopic effects $(2\chi i\omega \dddot{z})$, the coefficient δ with the gyroscopic effect.

2.4. The Shafts With Elastic Supports. Let us consider a shaft mounted in symmetrical elastic supports (Fig.3). Let us

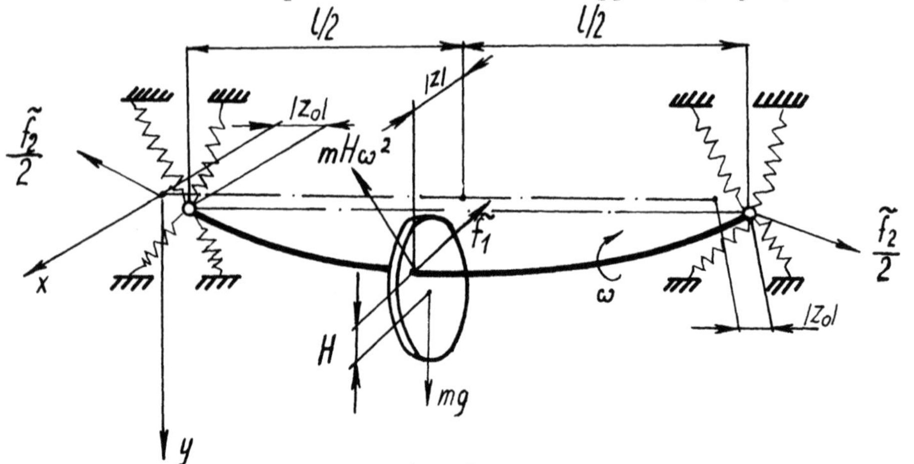

Fig. 3.

assume that the elastic forces of the shaft $(\tilde{f}_1 \equiv (|z|-|z_o|)\cdot f_1(|z|-|z_o|),\ \tilde{f}_1 \in C^1)$ and the supports $(\tilde{f}_2 \equiv |z_o|\cdot f_2(|z_o|),\ \tilde{f}_2 \in C^1)$ are nonlinear $(|z| = \sqrt{x^2+y^2}$ - displacement of the disc, $|z_o| = \sqrt{x_o^2 + y_o^2}$ - deflection of the supports).

The equation of motion of the rotor takes the form:

$$m\ddot{z} + b\dot{z} + \mu(\dot{z} - i\omega z) + kz + zF_3(|z|) = mH\omega^2 e^{i\omega t} + img , \tag{2.13}$$

where

$$z = x + iy ,$$

$$F_3(|z|) + k \equiv \frac{(|z|-|z_o|)\cdot f_1[|z|-|z_o|(|z|)]}{|z|} \tag{2.14}$$

(F_3 does not contain any constant term) and the function $|z_o|(|z|)$ can be obtained from the following relation [15]:

$$(|z| - |z_o|)\cdot f_1(|z| - |z_o|) = |z_o|\cdot f_2(|z_o|) . \tag{2.15}$$

The appearance of the nonlinear function f_2 can be caused either by nonlinear restoring forces of flexible supports [9], [10], [15]-[29] or elastic restoring force due to the deformation of ball bearing races [30]-[38].

<u>2.5. The Rigid Rotor With Liquid Tank in Susceptible Supports</u>. Let us investigate a symmetrical rigid shaft of mass M , the axial moment of inertia I_z of which is much smaller than the equatorial moment of inertia I (Fig. 4).

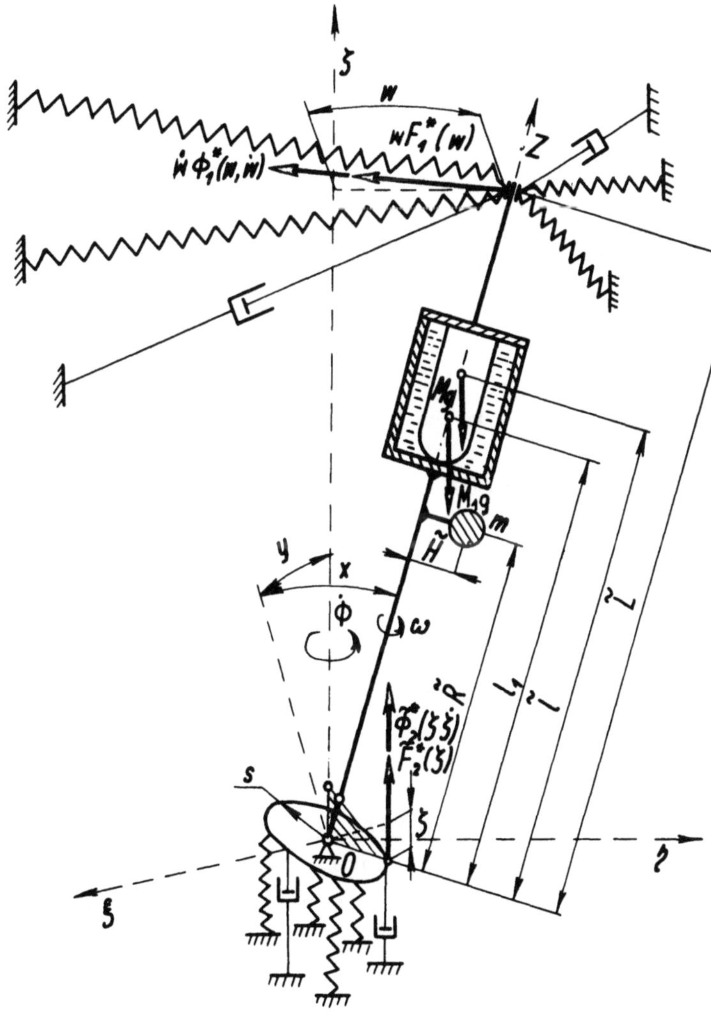

Fig. 4.

On the shaft a tank partially filled with liquid of mass M_1 is mounted. A mass m located out of the rotational axis represents the eccentricity of the mass distribution ($m \ll M + M_1$). The shaft rotates at the constant angular velocity ω and precesses in symmetrical visco-elastic supports. The elastic and damping forces of the supports $wF_1^*(w)$, $\zeta F_2^*(\zeta)$, $\dot{w}\phi_1^*(w,\dot{w})$, $\dot{\zeta}\phi_2^*(\zeta,\dot{\zeta})$ are nonlinear functions of the class C^1 (w - displacement of the upper end of the rotor, ζ - displacement of the lower one;
$\lim_{u\to 0} \frac{\dot{u}\phi_j^*(u,\dot{u})}{u} = \text{const}, \quad j = 1,2$).

Under these conditions of deflection of the shaft, the liquid produces a resistance proportional to the relative angular velocity $\omega - \dot{\phi}$ of the shaft. The equation of motion of the rotor can be written as follows [39]:

$$(m+m_1)\ddot{z} + b\,\dot{z} + \mu(\dot{z}-i\omega z) + k\,z + z\left[F_4(|z|) + F_5\left(|z|, \frac{d|z|}{dt}\right)\right] - 2\,i\omega\delta\,\dot{z} = m\,H\omega^2\,e^{i\omega t}, \tag{2.16}$$

where

$$m_1 = I - m, \quad \delta = \frac{I_z}{2I}, \quad H = \tilde{H}\,\tilde{R},$$

$$F_4(|z|) + k \equiv \tilde{L}^2 F_1^*(w) + s^2 F_2^*(\zeta) - (M\,l + M_1\,\ell_1)g,$$

$$F_5\left(|z|, \frac{d|z|}{dt}\right) \equiv \frac{1}{|z|}\left[\tilde{L}\phi_1^*(w,\dot{w}) + s\,\phi_2^*(\zeta,\dot{\zeta})\right],$$

$$w = \tilde{L}\,|z|, \quad \zeta = s\,|z|.$$

In the case under consideration, the coefficient μ represents the influence of the moving liquid in the tank partially filled [10], [40].

<u>2.6. The elastic Shafts Rotating Within a Radially Flexible Annulus Having a Radial Clearance</u>. The simple model of the symmetrical shaft whirling within a radially flexible annulus having a radial clearance is shown in Fig. 5. The annulus may represent a bearing with a radial clearance [41] - [43], a seal of a hydraulic machine [44]-[48], a dry friction damper of the lateral vibration of the shaft [49]-[52]

or at last, a rotor-stator radial rub in rotating machinery [18], [53], [54] .

Let us assume that the internal damping force of the shaft is linear (with the coefficient μ) and the elastic force of the shaft is a sum of linear and nonlinear terms $|z|[k+F_6(|z|)](F_6 \in C^1)$. The elastic characteristics of the supports of two rings are $\phi_w^* \equiv u_w \phi_w(u_w)$, $\phi_w \in C^1$, $w=1,2$ (u_1, u_2 are displacements of the centres of rings). η is the coefficient of dry friction.

The equation of motion of the shaft is as follows [49]:

$$m \ddot{z} + b \dot{z} + \mu(\dot{z}-i\omega z) + k z + z [F_6(|z|) + F_7(|z|) - i F_8(|z|)] = m H\omega^2 e^{i\omega t} + i m g \quad , \tag{2.18}$$

where

$$z = x + i y \quad ,$$

$$F_7(|z|) = \sum_{w=1}^{2} \frac{u_w\phi_w(u_w)}{|z|^2} \sqrt{|z|^2 - \frac{\eta^2(r_w-r)^2}{a_w^2 (1+\eta^2)}} \quad ,$$

$$F_8(|z|) = \pm(1+\frac{1}{\eta^2})^{-\frac{1}{2}} \sum_{w=1}^{2} \frac{u_w\phi_w(u_w)}{a_w|z|^2} (r_w - r) \quad , \tag{2.19}$$

$$u_w(|z|) = \sqrt{a_w^2|z|^2 - \frac{\eta^2(r_w-r)^2}{1+\eta^2}} \mp \frac{r_w-r}{\sqrt{1+\eta^2}} \quad , \quad w = 1,2 \; ;$$

a_1, a_2 are constant coefficients depending on ℓ_1, ℓ_2 (Fig.5) and on the shape of the deflected line of the shaft.

<u>2.7. The Shafts With Geometrical Nonlinearity and One Nonlinear Axial Bearing</u>. Let us examine a rotating shaft presented in Fig. 6. We shall take into consideration the linear rheological model of the shaft material (Eq. (2.1) with $f(\varepsilon,\dot{\varepsilon}) = 0$) and the geometrical nonlinearity of the deflected line of the shaft [10]. We assume that the axial force N in the right-hand bearing is

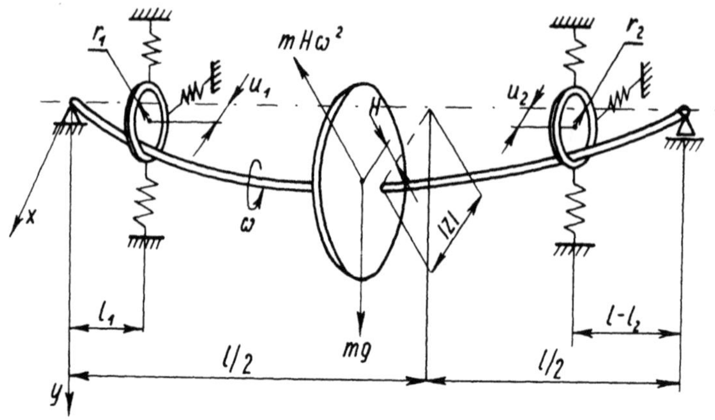

Fig. 5.

$$N(s,t) = -\psi_1[w(l,t)] - \psi_2[w(l,t)\ ,\ \dot{w}(l,t)]\ ,$$

(2.20)

$$0 \le s \le \ell\ ,\quad \psi_1 \in C^1\ ,\quad \psi_2 \in C^1\ ,$$

(2.21) $$w(s,t) = s - \int_0^s \sqrt{1-\left(\frac{\partial(|Z|)}{\partial s}\right)^2}\, d s \simeq \frac{1}{2}\int_0^s \left(\frac{\partial(|Z|)}{\partial s}\right)^2 d s\ .$$

Let us accept that the mass of the rotor is concentrated at the middle of the shaft:

(2.22) $$\tilde{m}(s) = \begin{cases} 0 & o \le s < \frac{\ell}{2}\ , \\ m^* & \frac{\ell}{2} \le s \le \frac{\ell}{2} + \delta^* \qquad \lim\limits_{\delta^* \to 0} m^*\delta^* = m \\ 0 & \frac{\ell}{2} + \delta^* < s \le \ell\ , \end{cases}$$

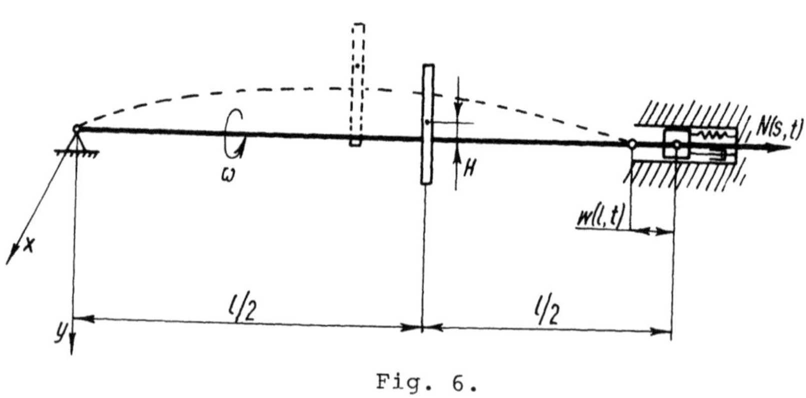

Fig. 6.

(δ^* sufficiantly small quantity).

The equation of motion of the rotor is [10], [55]:

$$\tilde{m}\frac{\partial^2 Z}{\partial t^2} + \frac{\partial^2}{\partial s^2}\left\{E\,I\,\frac{\partial^2 Z}{\partial s^2}\left[1-\left(\frac{\partial(|Z|)}{\partial s}\right)^2\right]^{-\frac{1}{2}}\right\} + \mu I\frac{\partial^5 Z}{\partial s^4 \partial t} -$$

(2.23)

$$- i\tilde{\mu}\omega I\frac{\partial^4 Z}{\partial s^4} + \frac{\partial}{\partial s}\left(N\frac{\partial Z}{\partial s}\right) + \tilde{b}\frac{\partial Z}{\partial t} = \tilde{m}\,H\omega^2 e^{i\omega t} + i\,m\,g \quad ,$$

where

$$Z = X + i\,Y \quad , \quad |Z| = \sqrt{X^2 + Y^2} \; .$$

We shall apply the Galerkin's method. Let us take the solutions in the form (2.9). From the Eq. (2.23), with taking into account (2.9), (2.20), (2.22) we obtain:

$$m\,\ddot{z} + b\,\dot{z} + \mu(\dot{z}-i\omega z) + k\,z + z\left[F_9(|z|) + F_{10}\left(|z|, \frac{d|z|}{dt}\right)\right]$$

(2.24)

$$= m\,H\omega^2 e^{i\omega t} + i\,m\,g \quad ,$$

where

$$b = \frac{\tilde{b}\ell}{2} \quad , \quad \mu = \frac{\pi^4\tilde{\mu}I}{2\ell^3} \quad , \quad k = \frac{\pi^4 EI}{2\ell^3} \quad , \quad I = \frac{\pi r^2}{4}$$

(2.25)

$$F_9(|z|) = \frac{\pi^4 EI}{\ell^3}\sum_{n=1,3,..}\frac{1}{n+3}\left(\frac{1\cdot3\cdot5\cdots n}{1\cdot2\cdot3\cdots\frac{n+1}{2}}\right)^2\left(\frac{\pi|z|}{2\ell}\right)^{n+1} ,$$

$$F_{10}\left(|z|, \frac{d|z|}{dt}\right) = \frac{\pi^2}{2\ell}\left[\psi_1\left(\frac{\pi^2|z|^2}{4\ell}\right) + \right.$$

$$\left. + \psi_2\left(\frac{\pi^2|z|^2}{4\ell} \quad , \quad \frac{\pi^2|z|}{2\ell}\frac{d|z|}{dt}\right)\right] .$$

2.8. The Shafts Rotating in Plain Bearings. Let us consider the symmetrical one disc rotor performing rotary motion in two plain bearings (Fig. 1). The equation of motion of the system can be presented as follows:

$$m \ddot{z} + b \dot{z} + k z + z \left[F_1\left(|z|, \frac{d|z|}{dt}, \omega - \dot{\phi}\right) - \right.$$

$$(2.26) \qquad \left. - i F_2 \left(|z|, \frac{d|z|}{dt}, \omega - \dot{\phi}\right)\right] = m H\omega^2 e^{i\omega t} + i m g ,$$

$$z = x + i y ,$$

where F_1, F_2 are nonlinear functions describing the hydrodynamical or aerodynamical forces in the bearings. With certain approximation we can assume that the forces due to oil pressure are symmetrical with respect to the fixed axes x, y and they depend on the displacement of journal centers and its radial and angular velocities [8], [9], [56-64] . The coefficient k represents the linear flexibility of the shaft and the oil film in the bearings ($F_1 \in C^1$, $F_2 \in C^1$) .

The similar mathematical model can be considered for rotors supported in gas bearings [65], [66] .

2.9. Other Symmetrical Nonlinear Forces in Rotor - Bearing Systems. Structural damping is the phenomenon of energy dissipation accompanying motion in the case of flexible and rigid couplings (e.g. the effect of rotary friction inherent in shrink fits). The action of the structural damping forces in rotating shafts constructed of elements resembles closely the action of the internal damping forces. Very often it is even stronger. During lateral vibration of the shaft the parts mounted on the shaft produce friction forces due to microslips on contacting surfaces. These forces can change the sign according to the direction of deformation and they depend on relative deformation and velocity of the relative deformation of the contacting parts. Mathematically they can be expressed as nonlinear functions similar as for the internal damping forces [10], [67], [68].

Forces of the similar character arise also when shafts of electrical machines rotate in electromagnetic field. Apart from the torque, a core is subject to action of forces caused by its motion relative to the field. These forces are generally connected with losses for magneto-elastic hysteresis and eddy-current effects. A general mathematical description of them can be accepted in the similar form as in the case of the internal damping forces [10], [68], [70].

3. PRECESSIONAL VIBRATIONS OF THE SYMMETRICAL ROTORS

3.1. General Remarks. In the previous chapter several models of symmetrical rotor systems have been considered. The equations of motion (the mathematical models) have been obtained (Eqs. (2.2), (2.10), (2.13), (2.16), (2.18), (2.24), (2.26)). The most general mathematical model of them, which covers all other considered ones is presented by Eq. (2.10). The Eq. (2.10) of the complex variable z (or two coupled equations of two real variables x, y) is nonlinear. The nonlinear functions F_1, F_2 have arguments as combinations of variables, describing the symmetrical displacement, the linear velocity of the displacement and the relative angular velocity (sometimes only one or two arguments). It has been assumed solely that the functions F_1, F_2 and their first derivative are continous in all their arguments. The functions F_1, F_2 of various special forms may be connected with the nonlinear elastic and internal damping characteristics of the shaft, with nonlinear characteristics of the elastic supports and dampers, with the nonlinear aerodynamical or hydrodynamical forces in plain bearings or in seals of hydraulic machines, with structural damping, with electromagnetic forces and/or some geometrical nonlinearities. Independently of their origin and their physical nature and - what follows - forms of these nonlinear functions, they have one important common feature: the arguments of the functions are a special combination of variables. This feature results from the symmetry of the systems.

We shall analyse some particular steady motions of the shafts. The solution $z = 0$ (if it exists) represents the pure rotating motion of rotors without any lateral vibration.

The periodic solution with frequency equal to the angular velocity of rotation ω describes flexural excited vibrations of the shafts - the synchronous precession of the shafts due to the inertia forces of unbalanced masses. We shall discuss this case, assuming that the gravitational forces are negligible.

We shall analyse too, the nonsynchronous precession of the shaft - the periodic self-excited lateral vibrations with frequencies different from ω. The periodic self-excited vibrations arise due to nonlinear nonconservative forces. We shall consider this case assuming that the gravitational forces are negligible and that the balance of the rotor mass is perfect.

Then we shall examine the influence of the gravitational forces on changes of the state of the pure rotating motion. In this case we shall also assume that the rotating masses are perfectly balanced.

Obviously these additional assumptions change forms of motions of the rotors under consideration. At the end we shall mention about the solving of the problem in which the common action of all factors are taken into account.

As an illustration of the theoretical considerations numerical examples will be presented. We shall take the nonlinear functions (Eq. (2.10)) in the following form*:

$$F_1\left(|z|, \frac{d|z|}{dt}, \omega-\dot{\phi}\right) = c_1|z|^2 + c_2\left[3\left(\frac{d|z|}{dt}\right)^2 + |z|^2(\omega-\dot{\phi})^2\right] \tag{3.1a}$$

$$+ 3\,c_3|z|\frac{d|z|}{dt} + c_4\left[\frac{1}{|z|}\left(\frac{d|z|}{dt}\right)^3 + |z|\frac{d|z|}{dt}(\omega-\dot{\phi})^2\right] ,$$

---

*The functions F_1, F_2 were obtained from the relations (2.8), (2.12), where the function $f(\varepsilon,\dot{\varepsilon})$ (Eq. (2.1)) had been given in the form

$f(\varepsilon,\dot{\varepsilon}) = \sum\limits_{u=0}^{4}\sum\limits_{w=0}^{4} a_{uw}\varepsilon^u\dot{\varepsilon}^w$ $(u + w \leq 4)$; a_{uw} = const. This function represents the first terms of the power series developing of any nonlinear C^1- class function. After integration (expressions (2.5) or (2.8)) the terms with even powers get the zero value.

$$F_2(|z|, \frac{d|z|}{dt}, \omega-\dot{\phi}) = (\omega-\dot{\phi}) \left\{ 2\, c_2 |z| \frac{d|z|}{dt} + c_3\, |z|^2 + \right.$$

$$\left. + c_4 \left[|z|^2 (\omega-\dot{\phi})^2 + \left(\frac{d|z|}{dt}\right)^2 \right] \right\}^{*} . \tag{3.1b}$$

In table 1 there are given data for numerical examples. The units for the coefficients are expressed in any system LMT (L - length, M - mass, T - time). There are considered soft and hard characteristics of elasticity ($c_1 = -100$, $c_1 = 50$). The coefficients k, m, H, b are fixed for all cases.

The numerical calculations have been made on the digital computer ODRA 1204 (Algol language).

Case	k $[\frac{M}{T^2}]$	c_1 $[\frac{M}{L^2T^2}]$	χ $[MT^2]$	m $[M]$	2δ $[M]$	b $[\frac{M}{T}]$	c_2 $[\frac{M}{L^2}]$	c_3 $[\frac{M}{L^2T}]$	μ $[\frac{M}{T}]$	c_4 $[\frac{MT}{L^2}]$	m_1 $[M]$	mH $[LM]$	mg $[\frac{LM}{T^2}]$
(1)	1600	-100	0.005	1	0	1	-	-	-	-	0	1	-
(2)	"	50	0	"	"	"	-	-	-	-	"	"	-
(3)	"	-100	"	"	"	"	-100	-1	250	-50	"	-	-
(4)	"	50	0.1	"	"	"	70	"	"	"	0.1	-	-
(5)	"	-100	-	-	-	-	-100	-1	"	-50	-	-	9.81

Table 1.

---

* $c_1 = 3\, h\, a_{30}$, $c_2 = h\, a_{12}$, $c_3 = h\, a_{21}$, $c_4 = 3\, h\, a_{03}$, $h = \frac{\pi^9 r^6}{64 \ell^7}$;

c_1 is related with nonlinear cubic elasticity, c_4 with nonlinear (cubic) internal damping, c_2, c_3 stand at mixed cubic terms [14], [71].

The expression ($a_{03}\, \dot{\varepsilon}^2 + a_{12}\, \varepsilon\dot{\varepsilon} + a_{21}\, \varepsilon^2$ can be interpreted as a "nonlinear coefficient" of the internal damping.

3.2. Synchronous Precession of the Rotor - the Excited Shaft Lateral Vibrations. In the case of neglecting the last term in the Eq. (2.10), it has an exact particular solution

$$z(t) = A\, e^{i(\omega t + \gamma)} \tag{3.2}$$

describing the steady synchronous precession of the shaft - the excited lateral vibrations due to the unbalanced state of the rotating masses. Constant amplitude of precession A may be obtained from the algebraic equation $(A > 0)$:

$$\left[k + F_1(A, 0, 0) - \chi\omega^4 + (2\delta - m - m_1)\omega^2\right]^2 + \\ + [F_2(A, 0, 0) - b\omega]^2 = \left(\frac{mH\omega^2}{A}\right)^2 . \tag{3.3}$$

The constant phase γ results from the relation:

$$\gamma = \arctan \frac{F_2(A,0,0) - b\omega}{k + F_1(A,0,0) - \chi\omega^4 + (2\delta - m - m_1)\omega^2} . \tag{3.4}$$

Eq. (3.3) describes the relation $A = A(\omega)$ - the main resonance amplitude - frequency relation. For any given values of parameters there can exist several solutions A of the Eq. (3.3) [1], [4], [39], [77]. This means that several particular solutions (3.2) with different amplitudes and phases appear for a given frequency ω. From the arguments of the functions F_1, F_2 only the first remains. It describes the amplitude of the shaft deflection. The relations (3.3), (3.4) do not include the coefficient μ, which describes the linear internal damping properties of the shaft. During the synchronous precession the longitudinal fibres of the shaft do not move with respect to each other, the deflected line of the shaft remains unchanged and only the elastic properties exert the influence on the amplitude of the lateral vibrations.

As an example the nonlinear functions F_1, F_2 have been choosen in the form (3.1). For the coefficients drawn up in Table 1 the relations $A = A(\omega)$ and $\gamma = \gamma(\omega)$ have been plotted (Figs. 7 and 8).

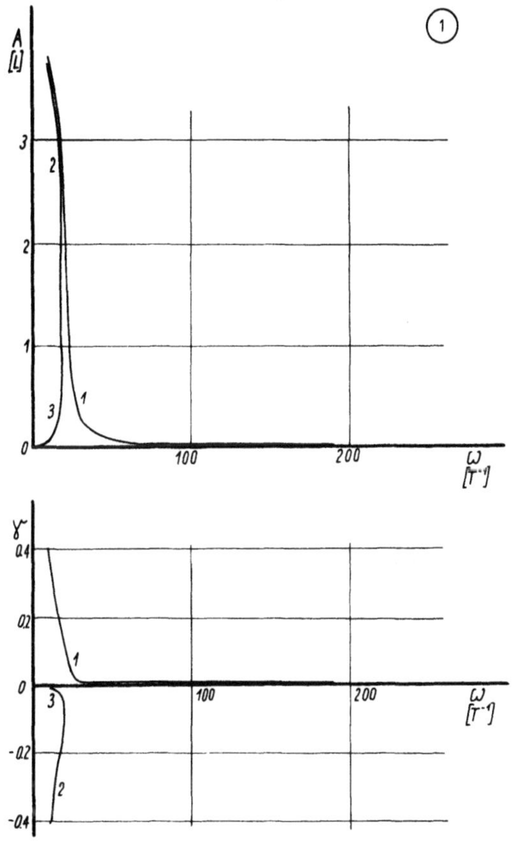

Fig. 7.

3.3. Nonsynchronous Precession of the Rotor - the Self-excited Shaft Lateral Vibrations. In the case of neglecting last two terms in Eq. (2.10) it has the exact particular solution

$$z(t) = B\, e^{j\alpha t} \qquad (3.5)$$

describing self-excited vibrations - stady nonsynchronous precession of the shaft. The constant amplitude B and the constant frequency α of precession may be found from the set of two algebraic equations:

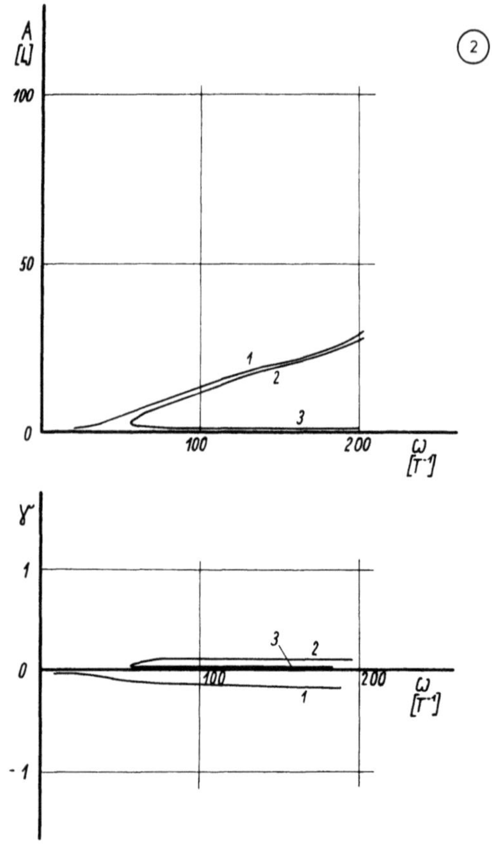

Fig. 8.

$$B[k + F_1(B,0,\ \omega-\alpha) + \chi\alpha^3(\alpha - 2\omega) + 2\delta\alpha\omega -$$

$$(3.6)\qquad - (m + m_1)\alpha^2] = 0\ ,$$

$$B[b\alpha - F_2(B,0,\omega-\alpha) + \mu(\alpha-\omega)] = 0\ .$$

For a fixed set of values of parameters the Eqs. (3.6) can have several solutions (B,α) [2,4,14,71] .

As an example let us take again the nonlinear functions F_1, F_2 in the form (3.1) and for some values of parameters (Table 1) let us calculate the amplitudes B and frequencies α from the Eq. (3.6), treating the rotational speed of the rotor ω as a parameter.

In Figs. 9, 10 the curves $B(\omega)$, $\alpha(\omega)$ have been plotted. The lack of place does not allow to show more examples.

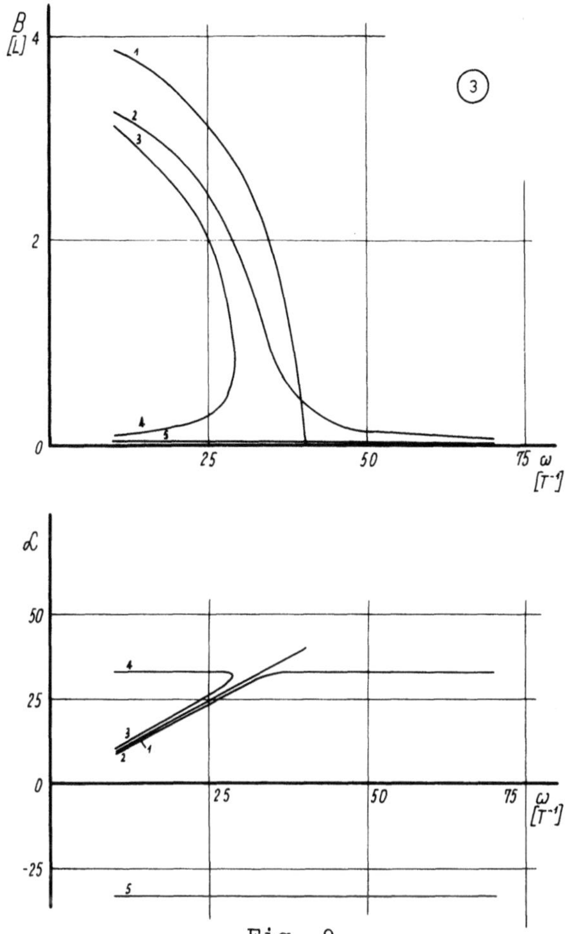

Fig. 9.

Changes of values of the parameters bring a great variety of shapes of figures. It means that the self-excited regimes can occur in different regions of values of ω .

3.4. Position of equilibrium: Influence of the Gravitational force. For horizontal heavy rotors the presence of remarkble force of gravity causes a deflection of shafts. During rotation the deflected line of the shaft remains unchanged but the longitudinal shaft fibres experience periodical stresses. Let us conventionally call the deflected line of the shaft at rest - the position of equilibrium. The shaft bends in the vertical plane. During rotation in the case of pure elasticity (without internal damping) the deflected line

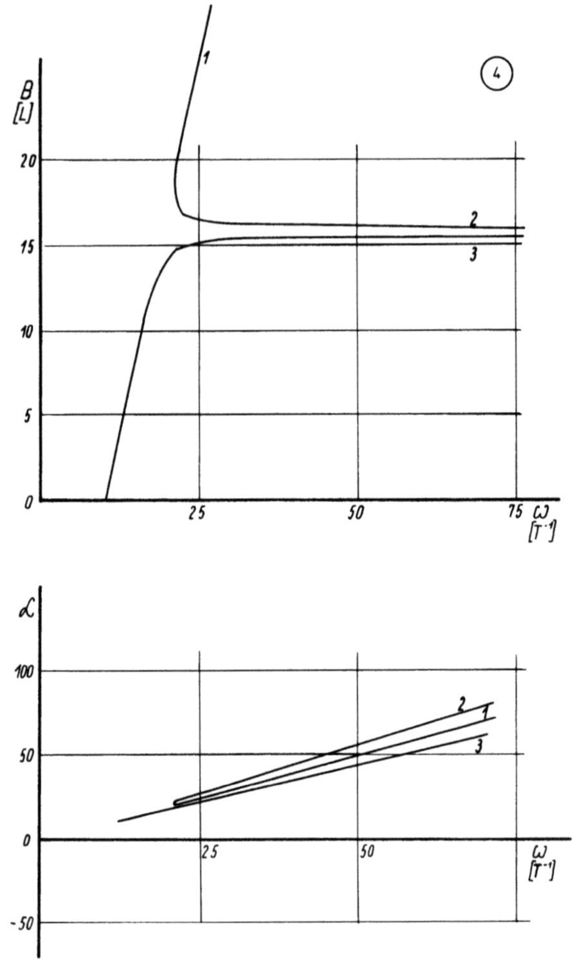

Fig. 10.

remains in the vertical plane. But if there exist some non-conservative forces, the position of it changes [9], [72].

Let us assume that the rotor is perfectly balanced ($H \equiv 0$). Then Eq. (2.10) has an axact particular solution

$$z(t) = C\,e^{i\theta} \quad , \tag{3.7}$$

where C is a constant amplitude of deflection of the center of the shaft line (Fig. 1), θ is a constant angle between the horizontal plane and the plane of the deflected shaft line. The amplitude C and the angle θ may be calculated from the following relations [72]:

$$(3.8) \qquad [k + F_1(C,0,\omega)]^2 + [\mu\omega + F_2(C,0,\omega)]^2 = \left(\frac{mg}{C}\right)^2$$

$$(3.9) \qquad \theta = \frac{k+F_1(C,0,\omega)}{-\mu\omega-F_2(C,0,\omega)}$$

An example of relations $C(\omega)$, $\theta(\omega)$ for the parameters given in the Table 1 has been plotted in the Fig. 11.

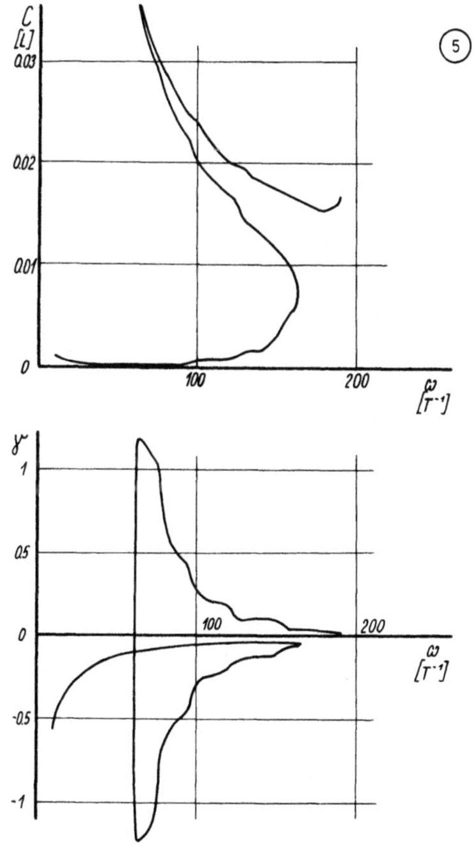

Fig. 11.

3.5. Some Remarks About Stability. The stability of solutions (3.2), (3.5), (3.10) can be analysed following the method used in [2,14,39,71,73-76]. The results of the analysis say in general that the stable solutions alternate with the unstable ones. It can be well seen in Figs. 7, 8 - the middle amplitude gives an unstable solution, the extreme ones describe the stable solutions. In the case of action of the gravitational force the extreme equilibrium positions are stable, the middle ones are unstable (Fig. 11).

It is well known that when the speed of rotation exceeds the first critical value, the pure rotating motion may turn out to be unstable. The simple linear theory explains this phenomenon. For the nonlinear cases considered here, we have found too that the zero solution $z = 0$ (the pure rotating motion of undeflected shafts in the case of the perfect balance and neglecting action of the gravitational forces) is stable up to the value of the first critical speed. When the rotational speed ω exceeds that value, the pure rotating motion becomes unstable, the periodical self-excited vibrations arise [71],[77]. The backward self-excited procession is generally unstable (Fig. 9 - the regime 5). Also in the case of the hard characteristic of elasticity the regime with the greatest amplitude $(\alpha \simeq \omega)$ is unstable (Fig. 10 - the regime 1).

The external damping always exerts a stabilizing effect. The amplitudes of vibration decrease and some modes disappear [71].

In practical cases of rotating shafts, even some stable solutions might not be attained when the stable and unstable regimes have very close amplitudes or due to the influence of factors, which have not been taken into account in the mathematical models.

3.6. Combined Vibrations of Rotors. We have analysed some particular solutions of rotors in the cases of some simplifications of the considered mathematical models. The question arises what will occur when all factors considered above separately are present together. We have found that in certain regions of values of parameters the systems show tendency to self-excited vibrations.

This has been proved with the assumption that the rotor masses are perfectly balanced. In practice some unbalance always exists and the forces of inertia exert influence on motion. The excited vibrations should interfere in the self-excited ones. If also the gravitational force is present, the resulting motion will be still more complex. Let us call this motion - the combined vibrations. We shall seek for them in the following form [78]:

$$z(t) = B\, e^{i\alpha t} + u(t)\, e^{i\alpha t} \quad , \tag{3.10}$$

where B, α are the amplitude and frequency, respectively found from the Eqs. (3.6); $u(t) = u_1(t) + i\, u_2(t)$ represent some disturbing complex function. We substitute (3.10) into the Eq. (2.10) and we take (3.6) into account. Restricting ourselves to the linear terms only, we get equation for u :

$$\begin{aligned} &(m + m_1)(\ddot{u} + 2\, i\alpha\dot{u}) + \dot{u}(b + \mu - 2\, i\, \delta\omega) + \\ &+ \left[B \frac{\partial\Phi}{\partial|z|} u_1 + B \frac{\partial\Phi}{\partial\left(\frac{d|z|}{dt}\right)} \dot{u}_1 - \frac{\partial\Phi}{\partial(\omega-\dot{\phi})} \dot{u}_2 \right]_{z=B,\ \dot{\phi}=\alpha} + \\ &+ \chi \left[\ddddot{u} - 2\, i\, \dddot{u}(\omega-2\alpha) + 6\alpha\ddot{u}(\alpha-\omega) + 2\alpha^2\, i\, \dot{u}(3\omega-2\alpha) \right] = \\ &= m\, H\, \omega^2\, e^{i(\omega-\alpha)t} + i\, m\, g\, e^{-i\alpha t} \quad , \end{aligned} \tag{3.11}$$

where

$$\Phi = F_1\left(|z|, \frac{d|z|}{dt}, \omega-\dot{\phi}\right) - i\, F_2\left(|z|, \frac{d|z|}{dt}, \omega-\dot{\phi}\right) . \tag{3.12}$$

The Eq. (3.11) is linear with constant coefficients, then solving does not present any problem. The combined vibrations (3.10) at the first approximation will be composed of constant terms, of terms containing trigonometrical functions: $\sin\alpha t$, $\cos\alpha t$, $\sin\omega t$, $\cos\omega t$ (with the frequency of the self-excited and excited vibrations) and of terms containing products of trigonometrical and exponential functions:

$$\begin{matrix}\sin\\ \cos\end{matrix}\ \alpha t \exp(k_s t) , \tag{3.13}$$

where k_s, s = 1,2,..., 7 are characteristic roots of the homogenous equation (3.11) (k_s can be complex; one of the roots is equal to zero [78]). Dependently on signs of the real parts of the roots k_s, the terms (3.13) play a significant or nonsignificant role in the expression of combined vibrations. If even one of the roots has a positive real part, then the combined vibrations are unstable. If all real parts of the roots are non-positive, then the dominant part of the combined vibrations consists of a constant term and trigonometrical terms with frequencies α and ω. The last ones may form a quasiperiodic function.

Rotor vibrations similar to those described above were observed in practice [8], [79], [80] and were simulated on analog computers as well [8], [61], [81], [82].

The next approximation for the combined vibrations should be made with taking the nonlinear terms into account. Solving of the problem can be made real with the help of approximative methods.

4. CONCLUSIONS

The ideal behaviour of any rotor system is to perform the pure rotational motion, around a proper axis, without any displacements or lateral (torsional and/or longitudinal) vibrations. There exist, however, numerous reasons disturbing this pure rotational motion.

In order to ensure reliable operation, modern design

of machines should include accurate prediction of all factors, which might disturbe the rotational motion. Having known the origins of the disturbing factors and the regions of their greatest influence, their action can be often significantly limited by applying some simple measures.

In the present paper the steady excited and self-excited vibrations of the nonlinear symmetrical rotor systems have been considered. The excited vibrations appearing in the form of the synchronous precession are caused by unbalances of the rotating masses. The amplitudes of that precession depend on the nonlinear elastic properties of the systems (for linear systems it refers to the first critical speed). A decrease of the amplitudes at operating speeds can be attained by change of values of the restoring and damping forces (e.g. by supplying the rotor system with susceptible supports [9], [15-22].

The self-excited vibrations may be caused by numerous reasons. They may arise due to the internal and structural damping forces, hydro- or aerodynamic forces or electrodynamic forces. Under conditions of high rotational speed and light loading the rotor system can be capable of precessing in its bearings at a rate different from the rotational speed. This nonsynchronous precessional motion causes alternating stresses in the shafts and can lead to destruction of all rotor system if the amplitudes exceed an admitted threshold. If the rotor damping and balancing requirements are met, it is possible only to safely pass through the rotor critical speed, whereas the occurence of the nonsynchronous precession will limit the operating speed of the rotor. The problem of decrease of amplitudes of the precession of this kind is closely associated with reduction of forces causing the self-excited vibrations (e.g. plain bearings of special shapes). This problem, however, is far from being completely solved.

The common action of the gravitational and nonconservative forces causes the displacement of the plane of the deflected shaft line. In the case of nonlinear nonconservative forces there can exist several positions of this plane. It can be dangerous in the case of slow change of the rotational speed, because it can lead to a jump phenomenon, which will significantly disturb the motion.

In the case in which all the factors considered above appear, the occurence of combined vibrations - composed of the displaced excited, self-excited and transient vibrations has been shown.

In all the cases under consideration the exact mathematical relations were obtained. The numerical examples exhibit the possible shapes of the amplitude-frequency-phase-rotational speed curves. The presented method may be used to identify the nonlinear conservative and nonconservative forces in the rotor systems.

REFERENCES

[1] MUSZYŃSKA, A.: O tlumieniu wewnetrznym w walach wirujacych, Zagadnienia Drgań Nieliniowych - Nonlinear Vibration Problems, 11, 1970.

[2] MUSZYŃSKA, A.: An Analysis of the Motion of a Shaft With Non-Linear Characteristic of Elasticity and Damping, Biul. PAN, ser. Sci. Techn., Vol. 16, 7, 1968.

[3] MUSZYŃSKA, A.: O precesji wirnika przy uwzglednieniu nieliniowych wlasnosci reologicznych materialu walu, Proc. of the V-th Conf. of Dynamic of Maschines, Liblice, Czechoslovakia 1968.

[4] MUSZYŃSKA, A.: Vibrations nonlinéaires des arbres tournants, Conférences, Acad. Pol. des Sci., Contre Sci, á Paris, Fasc. 101, 1973.

[5] GENIN, J.: Effect of Nonlinear Material Damping on Whirling Shafts, Appl. Scientific Research, A, v.15, 1, 1965.

[6] GENIN, J. & MAYBEE, J.S.: External and Material Damped Three Dimensional Rotor System, Int. Journ. of Non-Linear Mechanics, v. 5, 2, June 1970.

[7] BRONIAREK, C. & OSIŃSKI, Z.: Pewien przypadek drgań gietnych walu wirujacego przy nieliniowej charakterystyce materialu, Arch. Bud. Maszyn, t. 10, 4, 1963.

[8] GUNTER, E.J.: Dynamic Stability of Rotor - Bearing Systems, NASA SP-113, U.S. Government Printing Office, Washington 1966.

[9] TONDL, A.: Some Problems of Rotor Dynamics, Prague 1965.

[10] BOLOTIN, V.V.: Non-conservative Problems of the Theory of Elastic Stability, Moscow 1961. (English Translation, Pergamon Press, 1963).

[11] GOROBOV, V.A.: Unsteady vibrations of turbine shafts during transition through critical zones and the stability of their motion at super-critical speeds, Dynamika Strojov, 2, Bratislava 1966. (In Russian).

[12] DIMENTBERG, F.M.: Flexural vibrations of rotating shafts, Moscow 1959. (In Russian).

[13] TIMOSHENKO, S.: Vibration Problems in Engineering, D. van Nostrand Comp., Inc., Toronto, New York, London, 1955.

[14] MUSZYŃSKA, A.: Ruch walu o nieliniowych wlasnosciach sprezystych i tarcia wewnetrznego, IFTR Reports, 26, 1973.

[15] MUSZYŃSKA, A.: O ruchu sprezystego walu wirujacego w podatnych lozyskach, Rozpr. Inz., v. 17, 3, 1969.

[16] KEL'ZON, A.S. & PRYADILOV, V.I.: Removal of dangerous vibrations in high-speed vertical shafts, Izvestiya AN USSR, Mech. 6, 1965. (In Russian).

[17] ISAEV, R.I.: A study of the critical speed of a rotor under variable support stiffness, Strength and Dynamics of Aircraft Engines, 1, Mashinostroyene, 1964. (In Russian).

[18] SERGEEV, S.I.: Damping of mechanical vibrations, Moscow 1959. (In Russian).

[19] GROBOV, V.A.: Asymptotic methods of investigating flexural vibrations of turbine shafts, Moscow 1961. (In Russian).

[20] ZACRZHEVSKII, M.V.: Forced Vibrations of an unbalanced rotor on non-linearly elastic supports, Problems of Dynamics and Strength 14, Zinatne, Riga, 1967. (In Russian).

[21] GRIGOR'EV, N.V.: Dynamic damping of critical states of rotating turbine shafts, Vibrations in turbomachines Izvestiya AN USSR, 1956. (In Russian).

[22] ZAITMAN, M.F.: On the use of elastic supports in flexible rotors, Mashinovedenie, 1, 1967. (In Russian).

[23] OVCHAROVA, D.K.: Nonlinear vibrations of a flexible rotor on spring-mass supports, Dynamics and Strength of Machines, Collection of Interdepartimental Scientific and Technical reports (RMNT), 16, 1972. (In Russian).

[24] BORISOV, D.S. & DIMENTBERG, F.M.: On limiting the deflection of a non-linearly elastic rotating shaft in the post-critical range, Mashinovedenie, 2, 1972. (In Russian).

[25] DUBKO, N.A. & ZACRZHEVSKII, M.V.: Investigation of forced vibrations of a rigid unbalanced rotor on non-linearly elastic symmetric suspension, Problems of Dynamics and Strength, 26, Zinatne, Riga 1973. (In Russian).

[26] LACTOVETSKII, L.E.: Flexural vibrations of a rotor on non-linearly elastic supports subjected to two harmonic forces. Investigation on Vibrations, Strength and Constructional details of Aircraft Engines, Trans. Moscow Aviation Inst. (MAI), 180, 1968. (In Russian).

[27] BILLET, R.A.: Experimental Studies of the Effects of a Nonlinear Bearing Stiffness of a Shaft Whirl, Instn. Mech. Engrs. Appl. Mech. Convention, paper 29, 1966.

[28] BILLET, R.A.: Effects of Symmetrical Nonlinear Bearing Flexibility of Shaft Whirl, Journ. Mech. Engng. Sci., V. 8, No. 2, 1966.

[29] LEVY, S.: Critical Speed Damper, Trans. ASME, E, V.30, 3, 1963.

[30] STANKIEWICZ, E.: Wplyw nieliniowej podatnosci jednej z podpor na obroty krytyczne i postać ugiecia wirujacego walu, Arch. Bud. Maszyn V. 11, 3, 1964.

[31] TAMURA, A.: On the Vibrations Caused by Ball Diameter Differences in a Ball Bearing, Bull. JSME, V. 11, 44, 1968.

[32] GRIGOR'IV, N.V.: Nonlinear vibrations of machine and building elements, Moscow 1961. (In Russian).

[33] KEL'ZON, A.S. & YANVAREV, N.V.: Influence of the axial tension of a ball bearing on the transverse vibrations of a rigid shaft, Strojnicky Casopis, V. 22, 2, 1971. (In Russian).

[34] KEL'ZON A.S. & TSIMANSKII Yu.P.: On the dynamics of a rigid shaft on non-linearly elastic trunnions, Mashinovedenie, 4, 1972. (In Russian).

[35] POZNYAK, E.L.: Nonlinear vibrations of unbalanced vertical rotors on trunnions, Mashinovedenie, 1, 1971. (In Russian).

[36] GOLOSKOKOV, E.G. & OVCHAROVA, D.K.: Flexural vibrations of a horizontal rotor in the post-critical zone, Dynamics and Strength of Machines, Collection of RMNT, 15, 1972. (In Russian).

[37] BERGER, E.G. & KEL'ZON, A.S.: Free and forced vibrations of a simply supported rigid rotor with a non-linear restoring force, Proc. 5th Int. Conf. Non-linear Vib., 1969, V. 3, Kiev, 1972. (In Russian).

[38] ZHURAVLEV, V.F.: Dynamics of a rotor on non-ideal ball bearing, Izvestia AN USSR, Mech. Tver. Tela (Solid Mechanics), 5, 1971. (In Russian).

[39] MUSZYŃSKA, A.: On the precession of a high-speed partially filled centrifuge on non-linear bearings, Zagadnienia Drgań Nieliniowych, Non-linear Vibration Problems, 9, 1968. (In Russian).

[40] EPISHEV, L.V.: On the dynamic instability of a rotor partially filled with liquid, Scientific Communications of HighSchools Mashinostroenie and Pribovoatroenie, 2, 1959. (In Russian).

[41] MORRIS, J.: The Impact of Bearing Clearances on Shaft Stability, a New Approach to shaft Balancing, Aircr. Engng., V. 29, Dec. 1957.

[42] BILLET, R.A.: Shaft Whirl Induced by Dry Friction, Engineer, 220, No. 5727, 1965.

[43] EHRICH, F.F. & OCONNOR, J.J.: Stater Whirl with Rotors in Bearing Clearance, ASME WA/MD-8, Paper 66, 1966.

[44] ALFORD, J.S.: Protecting Turbomachinery From Self-Excited Rotor Whirl, Trans. ASME, SFA J. Engng. Paper, 4, Oct. 1965.

[45] BLACK, H.F.: Synchronous Whirling of a Shaft Within a Radially Flexible Annulus Having Small Radial Clearance, Symp. Vibr. in Hydr. Pumps and Turbines, Proc. Instn. Mech. Engrs, 181, 65, 1966-67.

[46] BLACK, H.F.: Effects of Hydraulic Forced in Annular Pressure Seals on the Vibrations of Centrifugal Pump Rotors, Journ. Mech. Engng. SCI., V. 11, 2, 1969.

[47] MARTSINKOVSKII, V.A. & RUDIS, M.A.: On the dynamics of hydromachine rotors, Theory of Machines and Mechanisms 98-99, Nauka, 1964. (In Russian).

[48] MARTSINKOVSKII, V.A.: Hydrodynamics and Strength of Centrifugal pumps, Mashinostronie, Moscow 1970. (In Russian).

[49] MUSZYŃSKA, A.: O tlumieniu drgań gietnych walow wirujacych za pomoca sprezysto-ciernych tlumikow, Archiwum Budowy Maszyn, V. 20, 1, 1973.

[50] PICHUGIN, D.F.: Analysis of the work of a dry friction damper used to reduce the vibrations of a shaft during transition through critical speeds, Izvestuja VUZOV, Aviation Technology, 1, 1958. (In Russian).

[51] EPISHER, L.V.: Investigation of the vibrations of a separator as a gyroscope taking into account dry friction in the elastic support. Chem. Mashinostroennie, 6, 1960. (In Russian).

[52] POZNYAK, E.L.: Damping of forced flexural vibrations of flexible rotors by dry frictional forces, Dynamics of Machines, Mashinostroenie, 1969. (In Russian).

[53] BLACK, H.F.: Interaction of a Whirling Rotor With a Vibrating Stator Across a Clearance Annulus, Journ. Mech. Ingng. Sci., V. 10, No. 1, 1968.

[54] EHRICH F.F.: The Dynamic Stability of Rotor-Stator-Radial Rubs in Rotating Machinery, Paper ASME, Vibr. 56, 6, 1969.

[55] MUSZYŃSKA, A.: On certain dynamic models of a rotor, Zagadnienia Drgań Nieliniowych. Nonlinear Vibration Problems, 10, 1969.

[56] RENTZEPIS G.M. & STERNLICH, B.: On the Stability of Rotors in Cylindrical Journal Bearings, Trans. ASME, D, V. 84, 4, 1962.

[57] JENNINGS, U.D. & OCVIRK, F.W.: The Simulation of Bearing Whirl on an Electronic Analog Computer, Trans. ASME, D, V. 84, Dec. 1962.

[58] OLIMPEV, V.I.: Free and forced vibrations of rotors in journal bearings, Kotlotrubostroenie (Boiler Construction), Proc. TsKTI (Central Boiler Research Inst.), 44, 1964. (In Russian).

[59] PORITSKY, H.: Rotor Stability, 5th US Congr. of Appl. Mech., Minneapolis, 1966.

[60] EHRICH, F.F.: The Influence of Trapped Fluids on High Speed Rotor Vibration, Trans. ASME, B, V. 89, 4, 1967.

[61] KIKUCHI, K.: Analysis of Unbalance Vibration of Rotating Shaft System With Many Bearings and Disks, Bull. ISME, V. 13, 61, 1970.

[62] KEL'ZON, A.S. & YAKOVLEV, V.I.: Transition through self-excited vibration zone of a rigid rotor in journal bearings, Doklady AN USSR, Vol. 193, A. 1970. (In Russian).

[63] KEL'ZON A.S. & YAKOVLEV, V.I.: Transition of a vertical rotor in journal bearings through self-excited vibration zone, Izvestiya VUZOV (Higher Tech. Inst.), Radio Physics, 14, 7, 1971. (In Russian).

[64] KEL'ZON, A.S. & YAKOVLEV, V.I.: Influence of the local component of lubricant inertia forces on the transition of a vertical rotor in journal bearings through self-excited vibration zone, Strojnicki Casopis, Vol. 23, 2, 1972. (In Russian).

[65] LARSON, H.H. & RICHARDSON, H.H.: A Preliminary Study of Whirl Instability for Prassurized Gas Bearings, Trans. ASME, D, V. 84, 4, 1962.

[66] TONDL, A.: On Self Excited Vibrations of Rotors Superted in Gas Bearings, Zagadnienia Drgań Nieliniowych-Nonlinear Vibration Problems, 9, 1968.

[67] PANOVKO, Ya.G.: Internal friction during vibration of elastic systems, Moscow 1960. (In Russian).

[68] KHRONIN, D.V.: Theory and calculation of vibrations in engines of space ships, Moscow 1970. (In Russian).

[69] POZNYAK, E.L.: Stability of a rotating iron core in a magnetic field, Izvesterja AN USSR, OTN Energetics and Automatic, 3, 1959. (In Russian).

[70] KELLENBERGER, W.: Der magnetische Zug in Turbogenerator-Rotoren als Ursache einer Unstabilität des mechanischen Laufes, Arch. Elektrotechn., V. 50, 4, 1966.

[71] MUSZYŃSKA, A.: Analiza drgań samowzbudnych wirnika o dwóch stopniach swobody, Rozprawy Inzynierskie, V. 21, 4, 1973.

[72] MUSZYŃSKA, A.: On motion of the Horizontal Shaft with Nonlinear Elastic and Damping Properties, Zagadnienia Drgań Nieliniowych, Nonlinear Vibration Problems, 12, 1971.

[73] GENIN, J. & MAYBEE, J.S.: Stability in the Three Dimensional Whirling Problem, Int. Journ. Nonlinear Mechanics, V. 4, 1969.

[74] GENIN, J. & MAYBEE, J.S.: Further Remarks on the Theory of Three Dimensional Whirling, Ins. Journ. of Non-Linear Mech., V. 6, 5, Oct. 1971.

[75] WEHRLI, C.: Stabilitätsbetrachtungen im Ljapunowschen Sinn bei rotierenden, einfach besetzen Wellen unter periodischer Belastung, Z. angew. Math. Phys., V. 18, 2, 1967.

[76] BROMMUNDT, E.: Stability Investigation by an Energe Method for a System With Gyroscopic Terms, Proc. VII Conf. on Machine Dynam., Gliewice, Poland, 1971.

[77] MUSZYŃSKA, A.: On rotor Dynamics (Survey), Zagadnienia Drgań Nieliniowych, Nonlinear Vibration Problems, 13, 1972.

[78] MUSZYŃSKA, A. Drgania kombinowane symetrycznego wirnika, IFTR Reports, 47, 1973.

[79] BROZGUL' L.I.: Investigation of the dynamics of cantilevered turbine rotors, Vestnik Mashinostroenie, 3, 1959. (In Russian).

[80] KUSHUL' M.Ya.: Self-excited vibrations of rotors, Moscow 1963. (In Russian).

[81] ANIKEEV, G.I.: Unsteady processes in the self-excited vibrations of rotors. Inzh. Zhurnal Mekh. Tverg. Tela, 1, 1968. (English Translation Mechanics of Solids).

[82] ANIKEEV, G.I.: Simulation of almost periodic vibrations of a rotor, Mashinovedenie, 2, 1970. (In Russian).

The Balancing of Flexible Rotors

A.G. Parkinson

Department of Mechanical Engineering
University College London, Torrington Place, London, UK

INTRODUCTION

During the last fifteen years or so much interest has been directed at the vibration and whirling of flexible rotors. Such systems, which include, for example, steam turbine-generator sets for electrical power generation and aircraft jet engines, give rise to many and varied problems of practical importance in a wide range of industries. Many of the practical problems are a consequence of the development of machines with rotors which are much more flexible than earlier designs. For example, the alternator rotor in a modern turbo-generator set is likely to have a service speed in excess of its third or fourth, flexural, critical speed, whereas at one time only the first critical speed would be below the operating speed.

A common source of rotor whirling is unbalance and it is a matter which usually demands practical attention. Safe and satisfactory operation of rotating shafts can generally be ensured by suitable balancing procedures. The present "state of the art" of balancing is reviewed and several of the recently developed balancing processes are compared.

EQUATION OF MOTION

Figure 1 illustrates a thin slice of a horizontal shaft which is transverse to the undeflected horizontal axis of the bearings, denoted by OX . Let y and z represent the deflections of the mid point E of the slice parallel

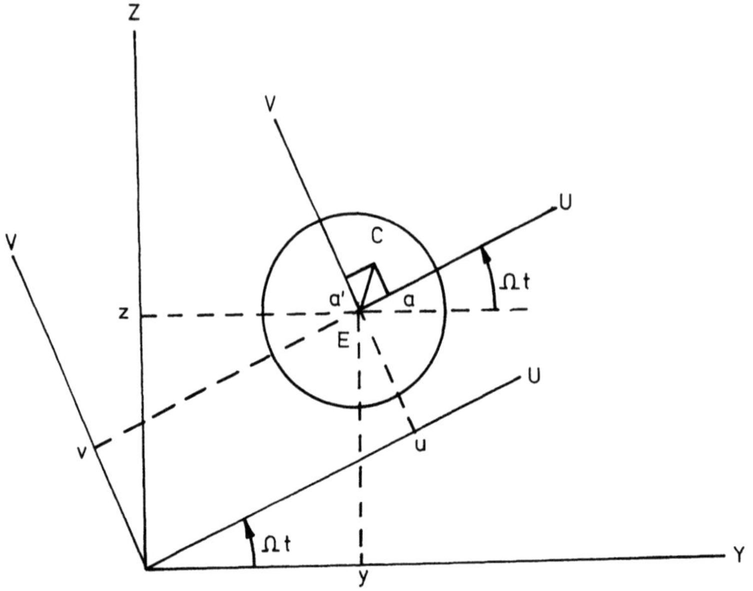

Fig. 1

to the axes OY and OZ respectively, which are orthogonal to OX . For simplicity OY is selected to be horizontal and OZ is directed vertically upwards. The shaft is assumed to have a circular cross section (that is it has axial symmetry), but it may be non-uniform along its length, so that its mass per unit length

$$\rho A = \rho A(x)$$

The equation of motion of the shaft can be expressed in terms of the complex coordinate

(1) $$\xi = y + iz$$

in the form*

---

* The effects of gravity have been omitted from this linear equation, as they are not relevant to the subject matter of this paper.

(2) $$\rho A\frac{\partial^2\xi}{\partial t^2} + b\frac{\partial\xi}{\partial t} + \frac{\partial^2}{\partial x^2}\left[EI\ \frac{\partial^2\xi}{\partial x^2}\right] = \Omega^2\rho A\bar{a}(x)e^{i\Omega t}$$

The term $\bar{a}(x)$ represents the unbalance of the shaft and is determined by the eccentricities, $a(x)$ and $a'(x)$, of the mass centre C of the thin slice resolved along the orthogonal directions EU and EV which rotate with the shaft at the steady speed Ω. Thus

(3) $$\bar{a}(x) = a(x) + ia'(x)$$

Finally the term $b\frac{\partial\xi}{\partial t}$ represents in an approximate manner the damping of the rotor.

Equation (2) is used in a variety of ways. Thus the response of the shaft to a particular unbalance can be calculated by specifying the value of $\bar{a}(x)$ and using digital or analogue computation. Such calculations, to be successful, however, demand adequate data for the bearing characteristics, which may be available from experiment or occasionally from theory. In addition the equation permits a primarily qualitative investigation in terms of a modal analysis into the nature of unbalance response. In this way it is possible to devise techniques whereby unbalance components can be deduced from observations of rotor response and corrected. The theory can also be employed in the calculation of natural frequencies (critical speeds) and principal modes which form the basis of many balancing procedures, provided once more the bearing characteristics are known. This paper is concerned with the use of these equations for balancing purposes and adopts a modal treatment which is most readily achieved by using the classical vibration approach of principal modes and principal coordinates.

To simplify the mathematics for the discussion of balancing we will introduce a rotating complex coordinate vector η such that

$$\eta = u + iv = \xi e^{-i\Omega t} \tag{4}$$

where u and v represent the deflections of the shaft parallel respectively to the rotating axes OU and OV (see figure 1).

The response to unbalance can then be expressed (see ref. [1]) as the particular integral of equation (2) in the form,

$$\eta = \sum_{r=1}^{\infty} N_r(\Omega)\Omega^2 \bar{a}_r e^{-i\zeta_r} \phi_r(x) \tag{5}$$

where

$$N_r(\Omega) = [(\omega_r^2 - \Omega^2)^2 + 4\mu_r^2\omega_r^2\Omega^2]^{-\frac{1}{2}} \tag{6}$$

$$\zeta_r = \tan^{-1}\left[\frac{2\mu_r\omega_r\Omega}{\omega_r^2 - \Omega^2}\right] \tag{7}$$

and

$$\bar{a}_r = \frac{1}{Z}\int_0^{\ell} \rho A\bar{a}(x)\phi_r(x)dx \tag{8}$$

In equation (5) $\phi_r(x)$ represents the rth principal mode of the shaft, when it is vibrating as an undamped beam without rotation $(\Omega = 0)$ and ω_r is the corresponding natural frequency, such that

$$\int_0^{\ell} \rho A\phi_r(x)\phi_s(x)dx = \begin{cases} 0 & (r \neq s) \\ Z & (r = s) \end{cases} \tag{9}$$

The relatively simple solution (5) can be deduced, provided that the damping in the rotating system is such that the response in the principal modes is uncoupled.

What are the principal characteristics of the forced vibration due to unbalance? Essentially the rotor response consists of a combination of components in the individual principal modes. The component in the rth mode is caused by the component of unbalance, $\bar{a}_r\phi_r(x)$, in the rth mode which is defined in equation (8). When the rotor speed Ω is close to one of the critical speeds (or natural frequencies) $\omega_1, \omega_2, \ldots$ the rotor undergoes a resonance. Thus when $\Omega \sim \omega_r$ the amplitude of whirling is large and most of the response is in the rth mode, that is it is associated with the rth term in equation (5). Moreover, even though the total unbalance eccentricity $\bar{a}(x)$ may vary in radial direction around the rotor along its length, the modal component $\bar{a}_r\phi_r(x)$ lies in one diametral plane through the shaft axis EX and thus rotates about this axis as a plane curve as the shaft rotates. The plane of $\bar{a}_r\phi_r(x)$ may, however, differ from mode to mode. Similarly the response at all points along the shaft in the rth mode lies in one diametral plane through the bearing axis which rotates about OX with the speed Ω .

Balancing is a process whereby small masses are attached to the shaft to modify the initial unbalance distribution $\bar{a}(x)$.* Suppose that a small mass m_j is attached to the shaft at $x = c_j$ at a radius r_j , so that in the transverse plane at $x = c_j$ the radius through the point of attachment makes an angle γ_j with the direction EU (see figure 2). The unbalance distribution is modified by the attachment of this mass and becomes $\bar{\alpha}(x) = \alpha(x) + i\alpha'(x)$, where

* This is the process which we are going to discuss, although sometimes material is removed from the shaft to modify the mass distribution.

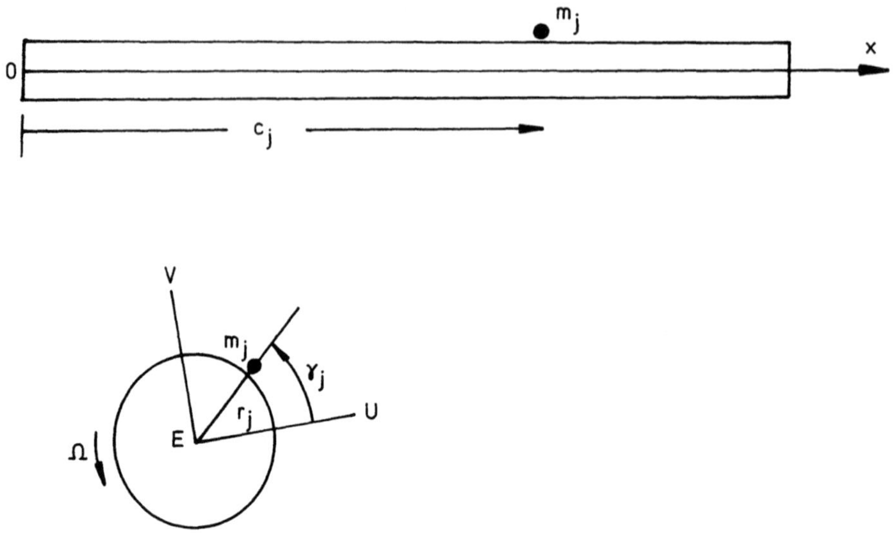

Fig. 2

$$(10)\quad \rho A\bar{\alpha}(x) = \rho A\bar{a}(x) + m_j r_j e^{i\gamma_j}\delta(x-c_j) = \rho A\bar{a}(x) + \bar{U}_j\delta(x-c_j)$$

where

$$(11)\qquad \bar{U}_j = m_j r_j e^{i\gamma_j}$$

and $\delta(x-c_j)$ is a delta function defined as,

$$(12)\qquad \delta(x-c_j) = 0 \quad , \quad (x \neq c_j) \quad ; \quad \int_0^{\ell} \delta(x-c_j)dx = 1$$

Consequently the modal components of unbalance are also modified and, according to equation (8), become

$$(13)\qquad \bar{\alpha}_r = \frac{1}{Z}\int_0^{\ell} \rho A\bar{\alpha}(x)\phi_r(x)dx$$

$$= \bar{a}_r + \frac{1}{Z}\int_0^{\ell} \bar{U}_j\delta(x-c_j)\phi_r(x)dx$$

$$= \bar{a}_r + \frac{\bar{U}_j\phi_r(c_j)}{Z}$$

This result is fundamental to the development of balancing procedures and also to a discussion of the relative merits of the various techniques.

If the only unbalance on the shaft is the mass m_j at $x = c_j$ which we have just discussed, then the whirling of the shaft is given by

(14) $$\eta^* = \frac{\bar{U}_j}{Z} \sum_{r=1}^{\infty} N_r(\Omega)\Omega^2 \phi_r(x)\phi_r(c_j)e^{-i\zeta_r}$$

This expression can be further simplified by introducing the polar receptance α_{xh} which has been devised by GLADWELL and BISHOP [2]. The receptance is a response function which relates the displacement of the shaft at x to a rotating force at $x = h$. If the force is rotating with the shaft at the speed Ω (as would be so for a force due to the discrete balance), then the receptance can be written in the notation of the present paper as

(15) $$\alpha_{xh} = \frac{1}{Z} \sum_{r=1}^{\infty} N_r(\Omega)\phi_r(x)\phi_r(h)e^{-i\zeta_r}$$

Consequently the displacement of the shaft due to the single, unbalance mass m_j can be expressed as

(16) $$\eta^* = \bar{U}_j \Omega^2 \alpha_{xc_j}(\Omega)$$

where the dependence of the receptance on Ω is indicated explicitly.

At least three methods for balancing flexible rotors have been developed during the last twenty years. Two methods depend directly on the modal analysis outlined in the present section and upon the predictions of equation (5). The third method does not depend explicitly on modal analysis, although, as we shall see, this process may be expedited by a modal interpretation of the rotor vibration. The principles of the three methods are outlined below for systems with axial symmetry. They also form the background of a proposed International Standard which is at present under development [3].

MODAL BALANCING

In this method the modal components of unbalance $\bar{a}_1, \bar{a}_2, \bar{a}_3, \ldots$ are detected and corrected in turn, starting with the first, by the attachment of suitable correction masses. In many instances the unbalance component in the rth mode, for example, can be detected directly from vibration readings at a single speed Ω near ω_r, if the magnification factor $N_r(\Omega)$ is sufficiently high at this speed for the contribution of other modes to the vibration to be negligible. Thus

$$(17) \qquad \eta \doteq \eta_r = N_r(\Omega)\Omega^2 \bar{a}_r e^{-i\zeta_r} \phi_r(x)$$

for $\Omega \doteq \omega_r$. The simple use of a trial mass will then give enough information to determine the magnitude of $\bar{a}_r$ and its angular location around the rotor relative to the axes EUV. If other modes are significant at speeds near ω_r, then the vibration component, η_r, in the rth mode must be isolated from the other modes by more refined techniques (see ref. [4]).

The use of a trial mass in this way is simply illustrated. Let the vibration of the shaft in the rth mode at some point $x = x_i$ and at some speed $\Omega = \Omega_r \doteq \omega_r$ be $\eta_r(x_i, \Omega_r)$. Let the corresponding modal component be $\eta_{rj}(x_i, \Omega_r)$ when a trial correction $\bar{U}_j^*$ is attached to the rotor at $x = c_j$. Then the effect of the correction U_j^* alone may be found by the simple vector subtraction+

$$(18) \qquad \eta_{rj}^*(x_i, \Omega_r) = \eta_{rj}(x_i, \Omega_r) - \eta_r(x_i, \Omega_r)$$

+ This may easily be performed by a geometric construction.

The magnitude and angular location of $\bar{a}_r$ may then be determined by comparing the initial modal component of vibration, namely

$$\eta_r(x_i\,,\Omega_r) = N_r(\Omega_r)\Omega_r^2 e^{-i\zeta_r}\phi_r(x_i)\bar{a}_r \tag{19}$$

with that due to the trial correction alone, that is

$$\eta_{rj}^*(x_i\,,\Omega_r) = N_r(\Omega_r)\Omega_r^2 e^{-i\zeta_r}\phi_r(x_i)\,\frac{\phi_r(c_j)\bar{U}_j^*}{Z} \tag{20}$$

Thus in terms of complex vibration vectors

$$\bar{a}_r = \frac{\eta_r(x_i\,,\Omega_r)}{\eta_{rj}^*(x_i\,,\Omega_r)}\quad\frac{\phi_r(c_j)\bar{U}_j^*}{Z} \tag{21}$$

It should be noted that in using this expression to determine a value of $\bar{U}_j$ sufficient to annul $\bar{a}_r$, it is not necessary to know either $\phi_r(c_j)$ or Z . It is only required to adjust $\bar{U}_j$ to ensure that the resultant modal vibration $\eta_{rj}(x\,,\Omega_r)$ is zero for all points x along the shaft. It should also be observed that in calculating $\eta_{rj}^*(x_i\,,\Omega_r)$, we are really determining

$$\bar{U}_j^* \; X\left[\text{rth. term in series for } \alpha_{x_i c_j}(\Omega_r)\right]$$

One other important feature of modal balancing must be mentioned. At any stage in the balancing procedure one must ensure that any corrections used to balance a particular mode do not upset the balance already achieved in the lower modes. This requirement can be readily met because of the orthogonality, or strictly linear independence, (see ref. [1]), of the mode shapes or characteristic functions. For example, assume that the first $(s-1)$ modes have been balanced. That is

$$\bar{\alpha}_1 = 0 = \bar{\alpha}_2 = \bar{\alpha}_3 = \ldots\ldots\ldots\ldots = \bar{\alpha}_{s-1} \tag{22}$$

but $\bar{\alpha}_s \neq 0$, where $\bar{\alpha}_r$ represents the resultant unbalance in the rth mode at this stage in the balancing process. Moreover $\bar{\alpha}_s$ has been found to have a magnitude $|\bar{\alpha}_s|$ and to lie in the diametral plane through EX at an angle θ_s with the plane EXU . Then at least s corrections, say $\bar{U}_1, \bar{U}_2, \bar{U}_3, \ldots, \bar{U}_s$ must be used to correct $\bar{\alpha}_s$. They must all be attached to the rotor in the plane at an angle θ_s with EXU and so chosen that

$$(23)\qquad \begin{bmatrix} \phi_1(c_1) & \phi_1(c_2) & \cdots & \phi_1(c_s) \\ \phi_2(c_1) & \phi_2(c_2) & \cdots & \phi_2(c_s) \\ \cdot & \cdot & \cdots & \cdot \\ \phi_{s-1}(c_1) & \phi_{s-1}(c_2) & \cdots & \phi_{s-1}(c_s) \\ \phi_s(c_1) & \phi_s(c_2) & \cdots & \phi_s(c_s) \end{bmatrix} \begin{bmatrix} \bar{U}_1 \\ \bar{U}_2 \\ \cdot \\ \bar{U}_{s-1} \\ \bar{U}_s \end{bmatrix} = -Z \begin{bmatrix} 0 \\ 0 \\ \cdot \\ 0 \\ \bar{\alpha}_s \end{bmatrix}$$

It should be noted that as the corrections and the unbalance component are all confined to one plane through EX , the above equation could be rewritten in terms of real quantities, not complex ones, by replacing $\bar{U}_j$ by $m_j r_j$ $(j = 1,2,\ldots\ldots,s)$ and α_s by $|\bar{\alpha}_s|$.

Alternatively, although this is not a technique which would usually be adopted in practice, but it does afford an easier comparison with method 3 below, the first s modes of the shaft could be balanced simultaneously, by choosing a different set of corrections $\bar{U}_1, \bar{U}_2, \ldots\ldots, \bar{U}_s$ such that

$$(24)\qquad \begin{bmatrix} \phi_1(c_1) & \phi_1(c_2) & \cdots & \phi_1(c_s) \\ \phi_2(c_1) & \phi_2(c_2) & \cdots & \phi_2(c_s) \\ \cdot & \cdot & \cdots & \cdot \\ \phi_{s-1}(c_1) & \phi_{s-1}(c_2) & \cdots & \phi_{s-1}(c_s) \\ \phi_s(c_1) & \phi_s(c_2) & \cdots & \phi_s(c_s) \end{bmatrix} \begin{bmatrix} \bar{U}_1 \\ \bar{U}_2 \\ \cdot \\ \bar{U}_{s-1} \\ \bar{U}_s \end{bmatrix} = -Z \begin{bmatrix} \bar{a}_1 \\ \bar{a}_2 \\ \cdot \\ \bar{a}_{s-1} \\ \bar{a}_s \end{bmatrix}$$

In general, of course, the components $\bar{a}_1, \bar{a}_2, \ldots\ldots\ldots, \bar{a}_s$ are not confined to one diametral plane in the rotor, so that the orientations $\gamma_1, \gamma_2, \ldots\ldots, \gamma_s$ may differ along the rotor.

It must be realised that modal balancing is not a technique for correcting all the components of unbalance, $\bar{a}_r$ $(r=1,2,3,\ldots\ldots)$. The step by step process is continued only for enough modes to ensure that the vibration, or possibly bearing forces (see ref. [5]), comply with appropriate tolerances *within the operational speed range of the rotor*. Due to the resonant nature of the response in the modes, unbalance components in modes with critical speeds beyond the maximum operating speed of the rotor are not likely to contribute greatly to the vibration of the rotor within its operational speed range. Generally, if the maximum speed of the shaft is in the vicinity of the nth critical speed, then only the first n modes of unbalance require correction. Sometimes, however, if this approach is adopted, residual vibration at the maximum speed may be too large, in which case a final "average" correction for all the higher modes $\bar{a}_{n+1}, \bar{a}_{n+2}, \ldots\ldots$ at this speed must also be made using at least n+2 balance masses to avoid upsetting the balance in the first n modes (see ref. [6]).

MODAL BALANCING PRECEDED BY RIGID BODY BALANCING

Some balancing engineers advocate balancing the rotor as a rigid body, that is correcting the "static" and "dynamic" unbalance, before attempting modal balancing. In the present notation the static unbalance can be expressed as

$$\Omega^2 \int_0^{\ell} \rho A \bar{a}(x)\,dx = Z\bar{a}_S^{*}\Omega^2 \qquad (25)$$

and the dynamic unbalance is

$$\Omega^2 \int_0^{\ell} \rho A x \bar{a}(x)\,dx = Z\bar{a}_D^{*}\Omega^2 \tag{26}$$

They represent the resultant force and couple on the rotor, if it were to rotate without experiencing any bending deflection.

The rigid body balancing is conducted in a conventional manner although not usually in a special low speed facility. Then the shaft is balanced modally by the technique outlined for method 1, with the proviso that at each step in the process, the selection of corrections for a particular mode must not only ensure that the previously balanced lower modes are not upset, but also that the rigid body balance is not disturbed. Although the method is a step by step process, it can be illustrated in principle, by showing the requirements for balancing up to the sth mode simultaneously. For this purpose at least $s+2$ corrections, $\bar{U}_1, \bar{U}_2, \ldots\ldots, \bar{U}_{s+2}$, are needed which are related to the initial unbalance components as follows,

$$\begin{bmatrix} 1 & 1 & \cdots & 1 \\ x & x & \cdots & x \\ \phi_1(c_1) & \phi_1(c_2) & \cdots & \phi_1(c_{s+2}) \\ \phi_2(c_1) & \phi_2(c_2) & \cdots & \phi_2(c_{s+2}) \\ \cdot & \cdot & \cdots & \cdot \\ \phi_{s-1}(c_1) & \phi_{s-1}(c_2) & \cdots & \phi_{s-1}(c_{s+2}) \\ \phi_s(c_1) & \phi_s(c_2) & \cdots & \phi_s(c_{s+2}) \end{bmatrix} \begin{bmatrix} \bar{U}_1 \\ \bar{U}_2 \\ \bar{U}_3 \\ \bar{U}_4 \\ \cdot \\ \bar{U}_{s+1} \\ \bar{U}_{s+2} \end{bmatrix} = -Z \begin{bmatrix} \bar{a}_S^{*} \\ \bar{a}_D^{*} \\ \bar{a}_1 \\ \bar{a}_2 \\ \cdot \\ \bar{a}_{s-1} \\ \bar{a}_s \end{bmatrix} \tag{27}$$

The general observations, including the results of balancing and the number of modes which require attention, are the same as those outlined for modal balancing.

INFLUENCE COEFFICIENT METHOD

A third method for balancing has been developed primarily in the United States which does not depend, at least in principle, on the insight provided by the modal analysis (see, for example, RIEGER [7]). The method embodies the use of "influence coefficients" and is outlined below.

Assume that the vibration of the rotor can be measured at n points, $x_1, x_2, \ldots\ldots, x_n$ along its length and that the vibration is observed at m suitable shaft speeds $\Omega_1, \Omega_2, \ldots\ldots, \Omega_m$. In addition suppose that balance corrections can be made at q points $x = b_1, b_2, \ldots\ldots, b_q$ along the rotor. The first steps in the procedure are to evaluate the influence coefficients. Let the vibration at $x = x_i$ for the speed Ω_k of the rotor in its initial unbalanced state be $\eta\ (x_i, \Omega_k)$. If a trial correction $\bar{U}_j^*$ is attached to the shaft at $x = b_j$, then the vibration at $x = x_i$ for $\Omega = \Omega_k$ is modified to $\eta_j(x_i, \Omega_k)$. If these two vibrations are measured, the associated influence coefficient $a_{ij}^{(k)}$ can be calculated from

$$(28) \qquad a_{ij}^{(k)} = \frac{\eta_j(x_i, \Omega_k) - \eta(x_i, \Omega_k)}{\bar{U}_j^*}$$

For balancing purposes these coefficients are determined for $i = 1,2,\ldots,n$, $j = 1,2,\ldots\ldots, q$ and $k = 1,2,\ldots\ldots,m$.

The influence coefficients which have been computed in this way can be used to formulate an influence coefficient matrix $\underset{\sim}{A}$ of the form

$$\underset{\sim}{A} = \begin{bmatrix} a_{11}^{(1)} & a_{12}^{(1)} & \cdots & a_{1q}^{(1)} \\ a_{21}^{(1)} & a_{22}^{(1)} & \cdots & a_{2q}^{(1)} \\ \cdot & \cdot & \cdots & \cdot \\ a_{n1}^{(1)} & a_{n2}^{(1)} & \cdots & a_{nq}^{(1)} \\ a_{11}^{(2)} & a_{12}^{(2)} & \cdots & a_{1q}^{(2)} \\ \cdot & \cdot & \cdots & \cdot \\ a_{n1}^{(2)} & a_{n2}^{(2)} & \cdots & a_{nq}^{(2)} \\ \cdot & \cdot & \cdots & \cdot \\ a_{n1}^{(m)} & a_{n2}^{(m)} & \cdots & a_{nq}^{(m)} \end{bmatrix} \tag{29}$$

The object of this method is to ensure that the vibration of the shaft at the points x_i $(i = 1,2,\ldots\ldots,n)$ for the speeds Ω_k $(k = 1,2,\ldots\ldots,m)$ is zero. As the initial vibration of the shaft at all these points is known, the necessary corrections $\bar{U}_1 , \bar{U}_2 , \ldots\ldots , \bar{U}_q$ can be calculated using the influence coefficient matrix, thus

$$\begin{bmatrix} \bar{U}_1 \\ \bar{U}_2 \\ \cdot \\ \cdot \\ \cdot \\ \bar{U}_q \end{bmatrix} = - [\underset{\sim}{A}]^{-1} \begin{bmatrix} \eta(x_1 , \Omega_1) \\ \eta(x_2 , \Omega_1) \\ \cdot \\ \cdot \\ \eta(x_n , \Omega_1) \\ \eta(x_1 , \Omega_2) \\ \cdot \\ \cdot \\ \eta(x_n , \Omega_2) \\ \cdot \\ \cdot \\ \eta(x_n , \Omega_m) \end{bmatrix} \tag{30}$$

The elements of $\underset{\sim}{A}$ and the vibration vectors are complex numbers, so that the corrections can be calculated as complex numbers which, for $\bar{U}_j$, for example, represent m_j , r_j , and the necessary angular location γ_j of the correction at $x = x_j$.

This procedure, in which $\tilde{A}^{-1}$ is calculated, is, of course, only valid in the relatively simple manner outlined in equation (30), if the matrix $\tilde{A}$ is square and non-singular. Thus the number of corrections q must be such that

$$q = m \times n \tag{31}$$

For example, if only one balancing speed is selected $(m = 1)$, then the number of corrections must equal the number of vibration measuring locations $(q = n)$. Conversely, if the vibration is only measured at one point $(n = 1)$, then the number of balancing speeds must equal the number of correction planes $(q = m)$.

If more than the minimum necessary data is acquired $(m \times n > q)$ then the data may be modified. Two alternative approaches are usually suggested. Either sufficient rows of $\tilde{A}$ are eliminated to produce a square matrix or the extra data is retained and a least squares technique is used to achieve an optimum balance. Of course care must be exercised in selecting the appropriate correction planes, measuring planes and balancing speeds to ensure that the influence coefficient matrix is not singular.

A word of caution should be entered at this stage. In the form in which the influence coefficient technique is usually advocated, all the corrections are calculated simultaneously on the basis of equation (30). The step by step approach of the other two methods is not usually adopted. In these circumstances, if any of the balancing speeds exceeds the first critical speed of the rotor, care must be taken, when running the rotor through a critical speed, that the vibration does not exceed permissible levels for the safe running and control of the rotor. This problem may require the use of preliminary corrections to permit the operation of the shaft through any difficult critical speeds, or it may result in the method being modified to a step by step approach.

SOME COMMENTS ON THE THREE METHODS

The first two methods described above are closely related. Their relative merits have been the subject of much discussion (and argument!). Recent reviews of this debate may be found in references [5] and [8] and it is not proposed to repeat this discussion here. It is more profitable to compare the first two approaches, which are both essentially modal techniques, with the influence coefficient method.

A comparison of the methods is assisted, if the elements used in the explanation of the third method are rewritten in terms of the notation used in the modal analysis. The influence coefficient, which is defined in equation (28), represents the vibration due to a discrete unit correction alone, so that, from equations (14) and (16), we can write

$$a_{ij}^{(k)} = \frac{1}{Z} \sum_{r=1}^{\infty} N_r(\Omega_k)\Omega_k^2 \phi_r(b_j)\phi_r(x_i)e^{-i\zeta_r} \tag{32}$$

or

$$a_{ij}^{(k)} = \Omega_k^2 \alpha_{x_i b_j}(\Omega_k) \tag{33}$$

These results can also be derived directly from the definition (28) by substituting the appropriate response expressions for $\eta_j(x_i, \Omega_k)$ and $\eta(x_i, \Omega_k)$, as we are using a linear theory. Thus the influence coefficients can be expressed in terms of the shaft receptances. Moreover, if we simplify the notation by writing

$$\alpha_{x_i b_j}(\Omega_k) = \alpha_{ij}(\Omega_k) \tag{34}$$

the influence coefficient matrix can be rewritten in terms

of block matrices as:

$$\underset{\sim}{A} = \begin{bmatrix} \underset{\sim}{A}_1 \\ \underset{\sim}{A}_2 \\ \cdot \\ \cdot \\ \cdot \\ \underset{\sim}{A}_m \end{bmatrix} \tag{35}$$

where

$$\underset{\sim}{A}_k = \Omega_k^2 \begin{bmatrix} \alpha_{11}(\Omega_k) & \alpha_{12}(\Omega_k) & \cdots & \alpha_{1q}(\Omega_k) \\ \alpha_{21}(\Omega_k) & \alpha_{22}(\Omega_k) & \cdots & \alpha_{2q}(\Omega_k) \\ \cdot & \cdot & \cdots & \cdot \\ \alpha_{n1}(\Omega_k) & \alpha_{n2}(\Omega_k) & \cdots & \alpha_{nq}(\Omega_k) \end{bmatrix} \tag{36}$$

$$(k = 1,2,\ldots,m)$$

This is a form of receptance matrix of order $n \times q$.

The first observation which should be made in any comparison of the methods is that, although all three methods are intended to produce appropriately low levels of vibration in a rotor, they set different detailed objectives. Consider a shaft for which the maximum operating speed is Ω_o , such that

$$\omega_s < \Omega_o < \omega_{s+1} \tag{37}$$

The principal aim of the first two methods is to ensure that the net unbalance $\bar{\alpha}(x)$ after balancing satisfies the requirements* that

---

* As has been remarked earlier, attempts may also be made to balance the $s+1$ and $s+2$ modes.

(38) $$\bar{\alpha}_1 = 0 = \bar{\alpha}_2 = \ldots = \bar{\alpha}_s$$

In addition the second method requires that

(39) $$\begin{cases} \int_0^{\ell} \rho A \bar{\alpha}(x)\,dx = 0 \\ \int_0^{\ell} \rho A x \bar{\alpha}(x)\,dx = 0 \end{cases}$$

Thus the residual vibration after balancing has the form

(40) $$\eta = \sum_{r=s+1}^{\infty} N_r(\Omega)\Omega^2 \bar{\alpha}_r e^{-i\zeta_r} \phi_r(x)$$

where $\bar{\alpha}_r$ is given by

(41) $$\bar{\alpha}_r = \bar{a}_r + \frac{1}{Z} \sum_{j=1}^{N} \bar{U}_j \phi_r(c_j)$$

and $N(\geq s)$ † is the number of corrections used. There is no guarantee that the residual vibration at any point along the rotor vanishes for $\Omega < \Omega_o$, but it is likely to be very small and within any relevant tolerances.

Conversely the influence coefficient method attempts to ensure that the vibration is zero* at a number of points $x = x_1, x_2, \ldots, x_n$ along the rotor at some speeds $\Omega = \Omega_1, \Omega_2, \ldots, \Omega_m$, where $\Omega_m \leq \Omega_o$. If corrections are chosen in accordance with equation (30) (assuming that $q = m \times n$), then the residual vibration can be written as

---

† For method one the minimum value of N is likely to be between s and $s+2$, whereas for method two this value is probably between $s+2$ and $s+4$ (see ref. [5]).

* Strictly this process is referred to by its advocates as "the exact point-speed flexible rotor balancing method" (see ref. [9]).

$$\eta = \sum_{r=1}^{\infty} N_r(\Omega)\Omega^2 \left\{ \bar{a}_r + \frac{1}{Z} \sum_{j=1}^{q} \bar{U}_j \phi_r(b_j) \right\} e^{-i\zeta_r} \phi_r(x) \tag{42}$$

such that

$$\eta(x_i , \Omega_k) = 0 \qquad (i = 1,2,\ldots,n \ , \ k = 1,2,\ldots,m) \tag{43}$$

One cannot, therefore, assume automatically that the vibration is low at all speeds $\Omega \leq \Omega_o$. Indeed the possibility of resonance at critical speeds is very likely unless the balancing speeds are carefully (or fortunately) chosen. Just as with the other methods, the influence coefficient method makes no claim to balance the shaft completely, that is $\bar{\alpha}(x) \neq 0$. Indeed in this case one cannot even be sure that conditions (38) are satisfied.

Intuitively, using the insight afforded by modal analysis, one would expect the effectiveness of the influence coefficient method to be improved for this example by selecting $m = s$ balancing speeds, such that $\Omega_k \doteq \omega_k$ $(k = 1,2,\ldots,s)$ and also by using $q = s$ corrections $\bar{U}_1 , \bar{U}_2 , \ldots , \bar{U}_s$. In these circumstances the vibration need only be measured at one point x_1 $(n = 1)$, although the modal interpretation would be aided if more information were available concerning the general nature of the vibration at the balancing speeds. In addition suppose that the response at the speed $\Omega_k \doteq \omega_k$ is predominantly in the kth mode, as is very common in systems with light damping. The relevant influence coefficients for $i = 1$ can then be approximated by writing

$$a_{1j}^{(k)} = \Omega_k^2 \alpha_{1j}(\Omega_k) \doteq \frac{1}{Z} N_k(\Omega_k)\Omega_k^2 \phi_k(b_j)\phi_k(x_1) e^{-i\zeta_k} \tag{44}$$

Consequently the matrix $\underset{\sim}{A}_k$ in equation (36) becomes

$$
\begin{aligned}
\underset{\sim}{A}_k &= \Omega_k^2 \left[\alpha_{11}(\Omega_k) \;\; \alpha_{12}(\Omega_k) \ldots\ldots \alpha_{1s}(\Omega_k) \right] \\
&\doteq \frac{1}{Z} N_k(\Omega_k) \Omega_k^2 \phi_k(x_1) e^{-i\zeta_k} \left[\phi_k(b_1) \;\; \phi_k(b_2) \ldots\ldots\ldots\ldots \phi_k(b_s) \right]
\end{aligned} \tag{45}
$$

In addition, if the vibration of the rotor at $\Omega_k \doteq \omega_k$ is largely in the kth mode, the initial response vector at $x = x_1$ can be approximated by

$$
\eta(x_1, \Omega_k) \doteq \eta_k(x_1, \Omega_k) = N_k(\Omega_k) \Omega_k^2 \bar{a}_k \phi_k(x_1) e^{-i\zeta_k} \tag{46}
$$

The corrections $\bar{U}_1, \bar{U}_2, \ldots, \bar{U}_s$ can be calculated from equation (30) for $q = s = m$, $n = 1$, so that

$$
\begin{bmatrix} \bar{U}_1 \\ \bar{U}_2 \\ \cdot \\ \cdot \\ \cdot \\ \bar{U}_s \end{bmatrix} = - [\underset{\sim}{A}]^{-1} \begin{bmatrix} \eta(x_1, \Omega_1) \\ \eta(x_1, \Omega_2) \\ \cdot \\ \cdot \\ \cdot \\ \eta(x_1, \Omega_s) \end{bmatrix} \tag{47}
$$

Therefore, if the matrix $\underset{\sim}{A}$ is developed through equation (35) and the approximation (45) and the approximations (46) are also employed, the necessary corrections can be determined approximately in the form

$$
\begin{bmatrix} \bar{U}_1 \\ \bar{U}_2 \\ \cdot \\ \cdot \\ \cdot \\ \bar{U}_s \end{bmatrix} = - Z \begin{bmatrix} \phi_1(b_1) & \phi_1(b_2) & \ldots & \phi_1(b_s) \\ \phi_2(b_1) & \phi_2(b_2) & \ldots & \phi_2(b_s) \\ \cdot & \cdot & \ldots & \cdot \\ \cdot & \cdot & \ldots & \cdot \\ \cdot & \cdot & \ldots & \cdot \\ \phi_s(b_1) & \phi_s(b_2) & \ldots & \phi_s(b_s) \end{bmatrix}^{-1} \begin{bmatrix} \bar{a}_1 \\ \bar{a}_2 \\ \cdot \\ \cdot \\ \cdot \\ \bar{a}_s \end{bmatrix} \tag{48}
$$

Thus the corrections predicted approximately by the influence coefficient technique in equation (48) in these circumstances are identical to the exact values suggested by modal balancing in equation (24).

If the influence coefficient technique is employed in this way, that is by selecting balancing speeds at shaft speeds for which the vibration is large, it is effectively applied at the critical speeds in the operating speed range of the rotor (see ref. [9] and [10]). It then closely resembles modal balancing, except that in principle it is not a step by step process, but a computerised, simultaneous method. Indeed, if there is interference from other modes at the critical speeds, then the influence coefficient program could be applied to appropriately isolated modal components by using a technique like the one outlined in ref. [4]. In assessing the advantage of a one stage computer solution to a balancing problem, however, the reader should remember that high initial vibration levels at some critical speeds may not permit the safe acquisition of all the necessary data at all the balancing speeds without the need for preliminary balancing. For this reason the influence coefficient technique may also have to be performed in a step by step manner.

If the balancing speeds are chosen to be fairly remote from the critical speeds, the close relationship between the influence coefficient method and modal balancing is not tenable. In these circumstances, however, it seems doubtful whether a rotor, which initially has a general, unknown unbalance and is corrected by the third method, will have suitable low vibration levels at any critical speeds within its speed range. The vibration levels are those predicted by equation (42) which indicates the possibility of resonance at the critical speeds.

The influence coefficient method will not meet with this difficulty, of course, if the rotor is known to only have discrete unbalances in say the n planes $x = x_1, x_2, \ldots, x_n$ and if it is also corrected in these planes. Then, provided that the balancing speeds and the measuring planes are well chosen, so that the influence coefficient matrix $\underset{\sim}{A}$ is non-singular and can be inverted, equation (47) will predict the corrections which will remove the unbalance entirely. The rotor will subsequently experience no vibration over its entire speed range. Of course, modal balancing is also likely to result in "perfect" balance for a simple situation like this.

If the influence coefficient technique can be used in the simple way in which it has been outlined and all the balancing corrections calculated in one computation using simply acquired vibration data, then undoubtedly it permits the use of fairly low trained personnel. In the mass production of rotors this is a valuable asset, but many flexible rotors are not mass produced. For these rotors the quest for a method which can be used with little thought by low grade staff seems to the author to be misguided. There is much to be gained by employing a suitable qualified engineer, who can take advantage of some understanding and physical insight into the modal character of shaft vibration due to unbalance. Certainly an understanding of the modal interpretation is likely to increase the success of the influence coefficient method. In this way the balancing process is also likely to be efficient, although it represents a different form of efficiency.

The balancing of flexible rotors is a subject of practical importance. The three methods which have been discussed have some characteristics in common and, if used with care and preferably some insight, will balance flexible rotors. It is hoped that some of the mathematical relationships which have been presented here will permit a further assessment of the relative merits of the three techniques. Ideally such assessments require practical tests with production rotors, although some progress is possible by means of numerical examples. Too much reliance should not, however, be placed on the latter, as their predictions are very sensitive to the assumed unbalance distributions.

REFERENCES

[1] PARKINSON, A.G.: "Balancing of flexible shafts rotating in massive flexible bearings", J. Mech. Engng. Sci., Vol. 15, 430, 1973.

[2] GLADWELL, G.M.L. and BISHOP, R.E.D.: "The receptances of uniform and non-uniform rotating shafts", J. Mech. Engng. Sci., Vol. 1, 78, 1959.

[3] Revised Draft International Standard (or Technical Report) for the mechanical balancing of flexible rotors. ISO/TC108/SC1/WG2 (Sec.-1) 4, 1974.

[4] BISHOP, R.E.D. and PARKINSON, A.G.: "On the isolation of modes in the balancing of flexible shafts", Proc. Inst. Mech. Engrs., Lond., Vol. 177, 407, 1963.

[5] BISHOP, R.E.D. and PARKINSON, A.G.: "On the use of balancing machines for flexible rotors", Trans. ASME, J. Engng. Industry, Vol. 94 (B2), 561, 1972.

[6] PARKINSON, A.G. and BISHOP, R.E.D.: "Residual vibration in modal balancing", J. Mech. Engng. Sci., Vol. 7, 33, 1965.

[7] RIEGER, N.F.: "Flexible rotor-bearing system dynamics, III Unbalance response and balancing of flexible rotors in bearings", ASME, New York, USA, 1973.

[8] KELLENBERGER, W.: "Should a flexible rotor be balanced in N or (N+2) planes?", Trans. ASME, J. Engng. Industry, Vol. 94 (B2), 548, 1972.

[9] TESSARZIK, J., BADGLEY, R.H. and ANDERSON, W.J.: "Flexible rotor balancing by the exact-point speed influence coefficient method", ASME paper 71-Vibr-91, 3rd Vibration Conference Toronto, Canada, 1971.

[10] RIEGER, N.F. and BADGLEY, R.H.: "Flexible rotor balancing of a high-speed gas turbine engine", Society of Automotive Engineers, Paper 72041, 1972.

Some Recent Computer Studies on the Stability of Rotors in Fluid-Film Bearings

Neville F. Rieger
Charles B. Thomas, Jr.

Rochester Institute of Technology
Mechanical Engineering Department
Rochester, New York 14623, USA

ABSTRACT - Several computer studies of rotor system stability are presented. The rotor systems include rigid and flexible rotors, fluid-film bearings, gas seals, and structural hysteresis. Equations for the stability of a general two bearing rotor system model were obtained, and several rotor systems used by previous investigators were calculated. Close correlation between test and calculated rotor threshold speeds was observed. A comparison of digital and analog techniques for linear rotor systems with various destabilizing factors was made, using a specific rotor system test case. Close correlation between all analog and digital results was observed. A short bearing theory for non-linear forces was developed and programmed for the EAI Pacer 600 hybrid computer. Published results for rigid rotor stability tests were used for program verification. Results from all cases are given and discussed in the paper.

1. INTRODUCTION

The stability of rotor systems is influenced by the bearings, shaft seals, impeller seals, rotor stiffness and hysteresis properties, and by the foundation dynamics. Each of these effects may be described analytically, and with the digital computer it is possible to study the combined effect of such factors on the rotor system stability.

Digital computers are widely used in studies of rotor stability problems. Several commercial stability programs are presently available which represent the rotor as a mass-elastic beam of variable cross section, supported in bearings having linear dynamic properties. Recently, programs with non-linear bearing formulations have become available, with provision for non-linear synchronous and non-synchronous forces applied to the rotor.

Good correlation has been reported between machine threshold speeds and the linear bearing analysis predictions, and with various other instability conditions using the non-linear bearing approach. At present, the precise conditions under which one approach is superior to the other have not been defined. It is, however, obvious that post-threshold rotor behavior such as bounded orbits can only be predicted rationally using the non-linear bearing force representation. Computing costs are necessarily greater using a non-linear bearing program because of the point-by-point orbit calculations involved, whereas the linear programs computer threshold speeds as eigenvalues of a determinent. A comparative review of digital computer stability programs in use on the United States has been given by RIEGER [1].

Analog computer studies of rotor stability have also been made, but analog techniques have not found widespread acceptance as either a research tool or in design practice. Analog computers are less available in the sizes required for rotor analysis, but they are well-suited to the non-linear aspects of the rotor stability problem, and orbit data is obtained directly. Commercial hybrid computers are available which overcome the analog machines size limitation*. Problem setup time on a hybrid machine has also been reduced to the point where a general orbital stability program for rotor-bearing systems is possible. At the present, however, there does not appear to have been previous hybrid computer work on stability analysis of rotor systems.

This paper describes a number of recent studies of rotor system stability problems in which each of the above methods, digital, analog, and hybrid were used. These studies provide insight into the relative abilities of each technique to predict the onset of unstable rotor behaviour when exposed to destabilizing influences present in the system,

* For example, the Pacer 600 machine made by Electronics Associates Incorporated, West Long Branch, New Jersey, USA.

as mentioned above. First, digital computer predictions of whirl threshold speeds were compared with other published test results to verify the linear analysis. The analog results were then compared with the verified digital results for a variety of unstable operating conditions, to check the accuracy and functioning of the analog analysis procedure.

Finally, effectiveness of hybrid computer methods for studying the stability characteristics of rotor systems was studied for several rotor systems, and compared to results achieved with linearized studies of such systems.

2. PREVIOUS WORK

Comprehensive surveys of rotor system stability have been made by TONDL [2], RIEGER [3], STERNLICHT and RIEGER [4], and GUNTER [5,6]. Analytical contributions have been made by LUND [7,8], by GUNTER [9], and by GIBERSON [10]. Experimental investigations of rotor system stability have been made by NEWKIRK and TAYLOR [11], NEWKIRK and LEWIS [12], PINKUS [13], MORRISON [14], TONDL [15], and RAO [16].

Both linear and non-linear analyses of rotor stability are currently being made. In both approaches the rotor is considered to be massive and linearly elastic. Linear bearing analysis has been successful in predicting whirl instability threshold speeds for small journal motions about the static equilibrium position; see LUND [8], PAN and STERNLICHT [17], and DUCKWORTH and RIEGER [18]. A comprehensive study of this procedure has recently been made by LUND [8].

The linear method predicts a threshold speed beyond which the rotor will become unstable and then whirl with increasing amplitude in time, until otherwise restrained. In practice, this does not always occur: the rotor may form a stable bounded orbit at speeds beyond the threshold speed, if such a speed exists. Such so-called bounded whirl motions can be predicted using a non-linear representation of the bearing forces. Any rotor motions which arise may be predicted depending on the completeness of the non-linear bearing force representation. This type of analysis is usually more time consuming as the equations of motion must be integrated in time around the whirl path.

Analog studies of rotor stability have been made by JENNINGS and OCVIRK [19] who studied rigid rotor motions in nonlinear bearings. No sharply defined instability threshold speed was found, but the orbit size grew from small to large (bounded) over certain speed ranges. GUNTER [5] described flexible rotor motions in linear, four-coefficient bearings by both analog and digital methods, and gave parametric curves for the stability threshold conditions. Holmes [20] and MICHELL et al [21] studied a rigid rotor in short and long linear and nonlinear bearings. Analog orbits were compared with test rotor orbits. Good correlation of results was obtained in a highly unstable situation (i.e., 360 degree film).

3. LINEAR BEARING SYSTEM STUDIES

The rotor system model in Figure 1 was devised to study:

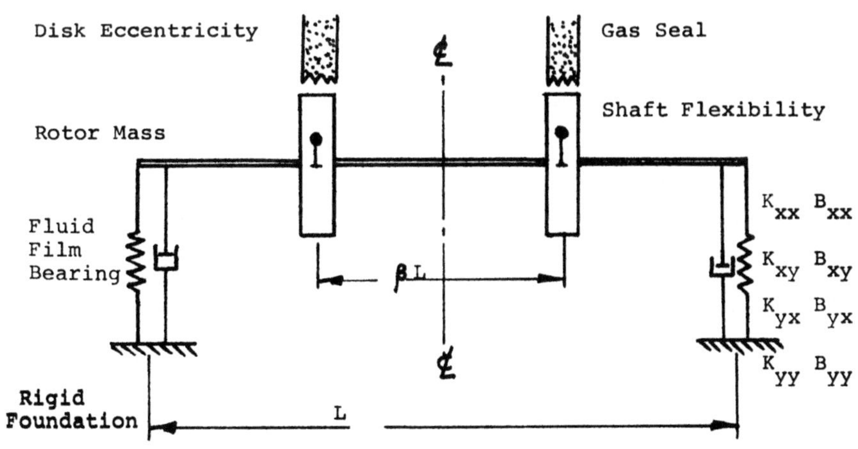

Fig. 1. Rotor System.

(a) effectiveness of the linear bearing representation for predicting instability threshold speeds which had been reported by other experimental investigators, references [2], [11], through [13], and [16].

(b) ability of the analog computer to predict the same threshold speeds obtained from digital computer analysis of an identical rotor system.

This rotor system has sufficient generality to undertake studies of instability arising from bearing, seal, and structural hysterisis effects on the rotor. Most experimental rotors described in the literature may be simulated with this model. Details of the analytical rotor system are as follows:

Rotor: This consists of a flexible shaft of specified transverse stiffness carrying two massive rigid disks a certain distance apart. Disk eccentricity (small) is included only in the analog equations for convenience of initiating the analog response, and was omitted from the digital study. No disk translatory or gyroscopic effects are included. The rotor is axially symmetric and is also symmetric about its mid-plane. When β is non-zero the system has two whirl modes, and one mode when β is zero. The shaft has structural hysteresis damping effects proportional to the disk radial velocity in rotating coordinates.

Bearings: The eight coefficient representation of bearing dynamic properties shown in Figure 2 was used for this part of the study i.e., linear bearings, equations (A1.2). Both bearings are identical and are located at the ends of the shaft. Suitable coefficient values were chosen to match the bearings used in the experimental investigations, using the data given by LUND in reference [22].

Gas Seals: Forces proportional to disk dynamic radial displacement were used, equation (A1.3). In the absence of established coefficient values for such seals, a parametric study of the effect of this coefficient on stability was made.

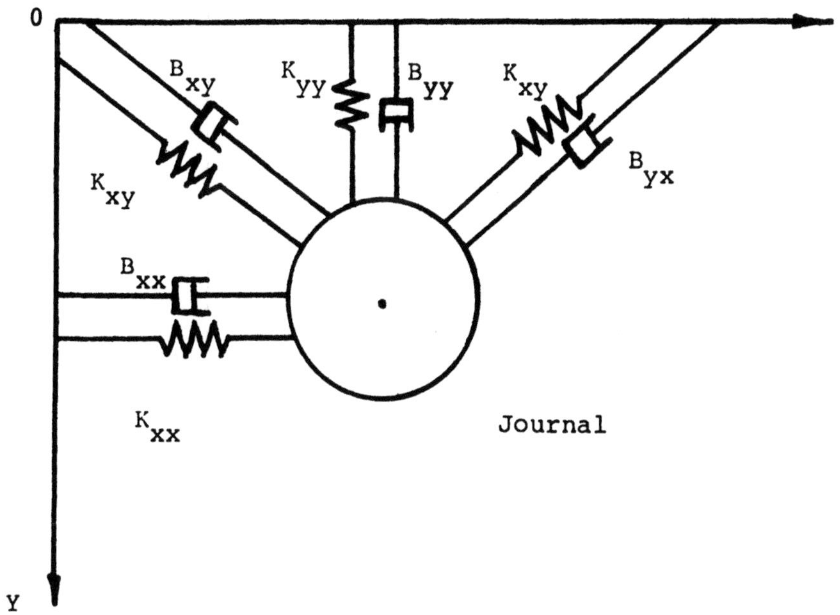

Fig. 2. Bearing Linear Dynamic Force Representation.

Pedestals and Foundations: A rigid level support system beneath the bearings was used.

The theory for linear motions of such a system is given in Appendix 1 of this paper.

(a) Comparison of theoretical and experimental values for threshold speeds. Bearing-induced rotor instability may be conveniently studied by omitting the gas seal term and the structural hysteresis term from the stability determinant, equation (A1.9). The analysis then gives the following expressions, LUND [23]:

Threshold Parameter:

$$\text{(A1.10)} \quad \kappa = \frac{(K_{xx}\omega B_{yy}+K_{yy}\omega B_{xx}) - (K_{xy}\omega B_{yx}+K_{yx}\omega B_{xy})}{K_{xx}\omega B_{yy} + K_{yy}\omega B_{xx}}$$

Frequency ratio:

$$\text{(A1.11)} \qquad \gamma^2 = \frac{(K_{xx}-\kappa)(K_{yy}-\kappa) - K_{xy}K_{yx}}{\omega B_{xx}\omega B_{yy} + \omega B_{xy}\omega B_{yx}} = (\frac{\gamma}{\omega_T})^2$$

Rotor flexibility parameter:

(A1.12) $$\kappa = \frac{1}{2} K \frac{(s\gamma)^2}{1-(s\gamma)^2}$$

Threshold speed ratio:

(A1.13) $$s = \frac{\omega_T}{\omega_n}$$

The $K_{xx}...B_{xx}...$ are the bearing dynamic coefficients, ω is the rotational speed rad/sec, ω_n is the rigid bearing critical speed rad/sec, and ν is the system unstable whirl frequency rad/sec. The above expressions contain ν and ω_{th} as unknowns.

A computer program was written to give N_{th} in terms of the rotor parameters, K, M with $\beta = 0$ (lowest instability mode). Variation of the eight bearing coefficients was conveniently incorporated into the program by curve-fitting the bearing coefficient charts. A complete description of this work is given in reference [18].

Table 1 gives details of the experimental rotor-bearing systems used by previous investigators. The threshold speed was calculated for each of these systems using the above expressions. Where details were insufficient, e.g., descriptions of the bearing surface, oil supply pressure, outlet temperature, etc. reasonable assumed values were taken.

Correlation of predicted threshold speeds with the experimentally observed values was quite close, and within the usual range of practical threshold speed variation. It can be concluded that where valid data on bearing geometry and operating conditions are available, the threshold speed of smooth-running simple rotors in fluid film bearings can be predicted to design accuracy using linear bearing theory. Common sources of prediction errors are: viscosity variations, bearing clearance, misalignment, and inlet pressure effects on the fluid film.

	ROTOR SYSTEM PARAMETERS									THRESHOLD SPEED	
	D	L	C	R	W	μ	L1	F1	D1	Experiment	Cal.
SOURCE	IN.	IN.	IN.	IN.	LBS	$X10^{-6}$	IN.	RPM	IN.	RPM	RPM
Pinkus-1956	2	1	.0025	1	θ4	9.425	32	2500	2	5000-7500	6200
Pinkus-1956	2	2	.0025	1	β2	9.425	46	6100	2	12,000	13,200
Newkirk-1956	2	1	.002	1	96	3.45	36	1600	2	3375	3600
Newkirk-1956	2	1	.004	1	96	5.24	36	1580	2	4200	4100
Newkirk-1956	2	2	.004	1	48	2.57	42	1580	2	3275	2950
Tondl-1964	1.55	1.55	.008	.75	62.7	3.25	40	1000	1.38	3800	3850
Tondl-1964	1.55	1.55	.008	.75	30.3	3.25	40	2500	1.38	4900	4900
Rao-1970	1	1	.002	.5	4.87	7.7	20	1500	.5	3000-4000	4000

Table 1. Details of Experimental Rotor Systems.
Comparison of Threshold Speeds Observed and Calculated.

(b) Comparison of digital and analog predictions of instability threshold speeds. It is shown in Appendix 1 that the system stability determinant including bearing, seal and hysteresis terms may be written as:

$$(A1.8) \qquad A_1S^6 + A_2S^5 + A_3S^4 + A_4S^3 + A_5S^2 + A_6S + A_7 = 0$$

The A_i coefficients of the polynomial are complex algebraic expressions. Full details are given in reference [27].

A computer program was written to solve the stability polynomial for the lowest complex eigenvalue. Values of M, K, G, and H were held constant for given cases. The program will increment on speed, and corresponding values of the speed-dependent bearing coefficients are selected from an input table. Instability occurs when the real part of the eigenvalue changes sign from negative to positive.

The compressor shown in Figure 3 was used as a sample

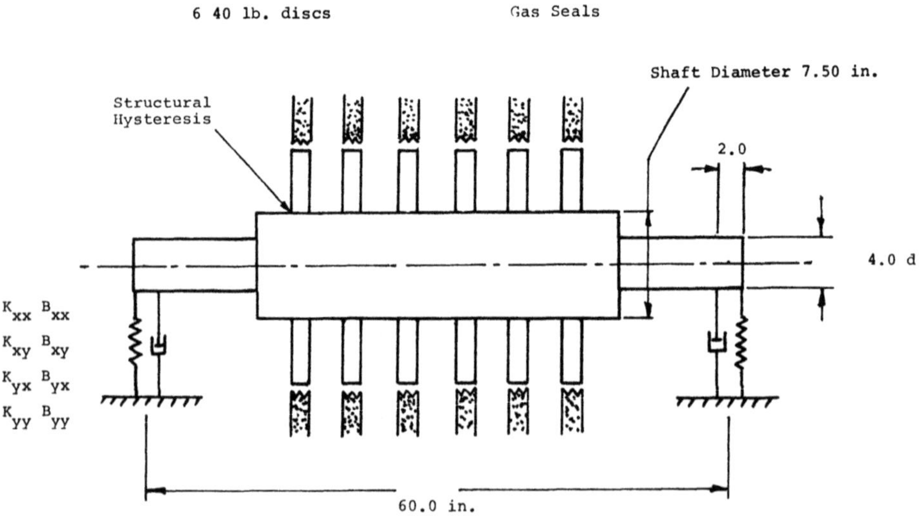

Fig. 3. Compressor Rotor in Bearings with Seal and Structural Hysteresis Effects.

calculation. The rotor weighs 1000 lb ($M = 1.39$ lb sec^2/in per bearing) and operates in two 4.0 diameter by 2.0 in. long plain cylindrical bearings of diametral clearance 0.004 in. Oil of viscosity 22 cP at 110 deg. F. and 4.84 cP at 200 degree F is supplied to each bearing at a rate of 1.25 gal/min at operating speed under 20 lb/in^2 gage pressure at 120 degree F . This rotor system has been discussed in detail in reference [24].

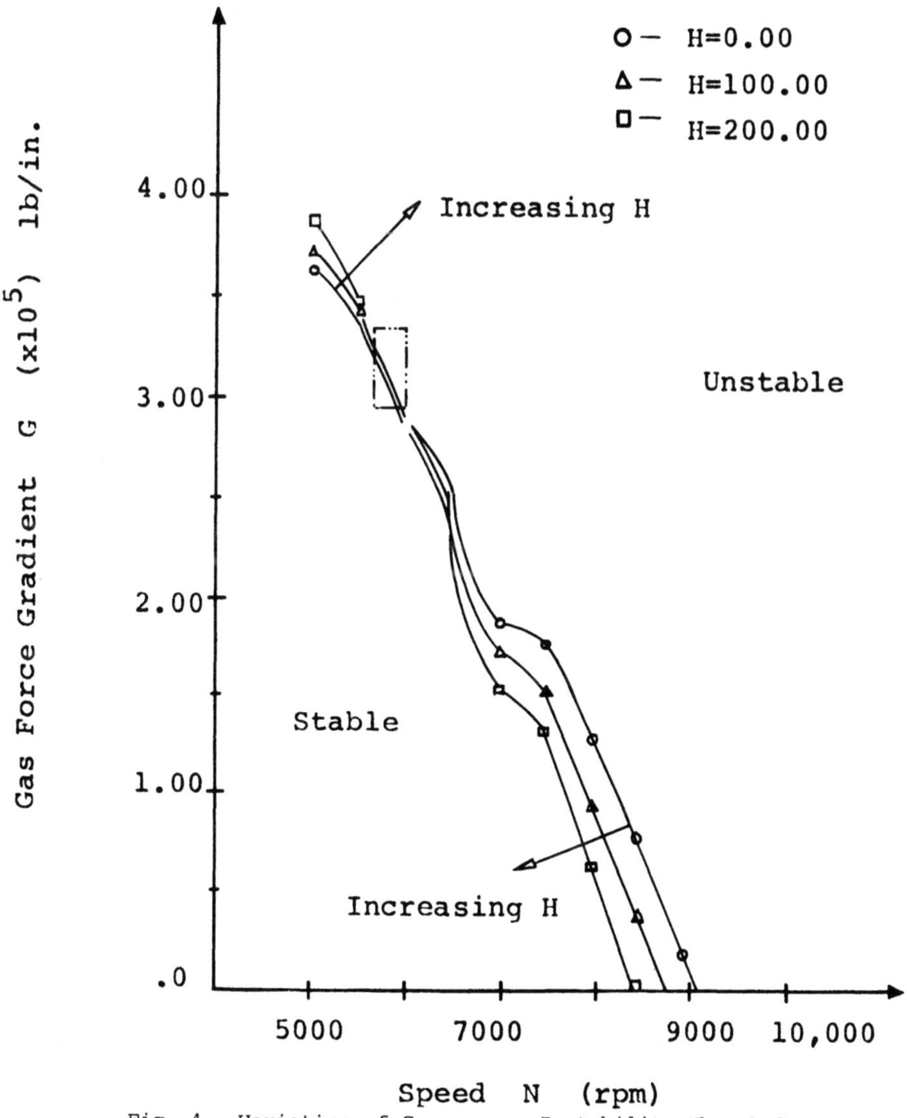

Fig. 4. Variation of Compressor Instability Threshold Speed with Gas Seal Force and Structural Hysteresis.

Parametric studies of the relative significance of bearing, seal, and structural hysteresis effects on the compressor rotor system stability were made. The system threshold speed without seals or hysteresis (i.e., bearing instability) was first determined to be 9200 rpm with the rotor whirling in its translatory mode. The gas seal coefficient was then introduced and increased in value. A decrease in threshold speed was observed. Lastly, the effect of increasing the structural hysteresis was studied, with gas seal forces operating. For known threshold values, the value of H was added and new threshold speeds were found. The results of these studies are shown in Figure 4 in which the system threshold speed N_T is plotted against gas seal coefficient G, with structural hysteresis coefficient H as parameter. It is evident that the threshold of instability is depressed with increase of the gas seal coefficient. The effect of hysteresis is mixed: below the system fundamental critical speed (5850 rpm) it raises the threshold speed, and above 5850 rpm it depresses the threshold speed, as anticipated from previous experience. Structural hysteresis has a minor effect on system instability threshold compared with gas seal effects, where the rotor system incorporates fluid-film bearings.

An analog stability study of the compressor rotor system was also undertaken. The analog circuit is shown in Figure 5. Bearing coefficients corresponding to each speed were introduced by hand adjustment of circuit resistors. The gas seal coefficient G and the hysteresis coefficient H were held constant for each condition and the corresponding threshold speed was found by observing orbit growth or decay at selected speeds.

Sample whirl orbits are shown in Figures 6 and 7. Figures 6a and 6b show the journal orbits at 9100 rpm and at 9200 rpm , below and above the instability threshold speed (9200 rpm) for the case $G = 0$ and $H = 0$, i.e., no gas seal or shaft hysteresis effects. The influence of gas seals

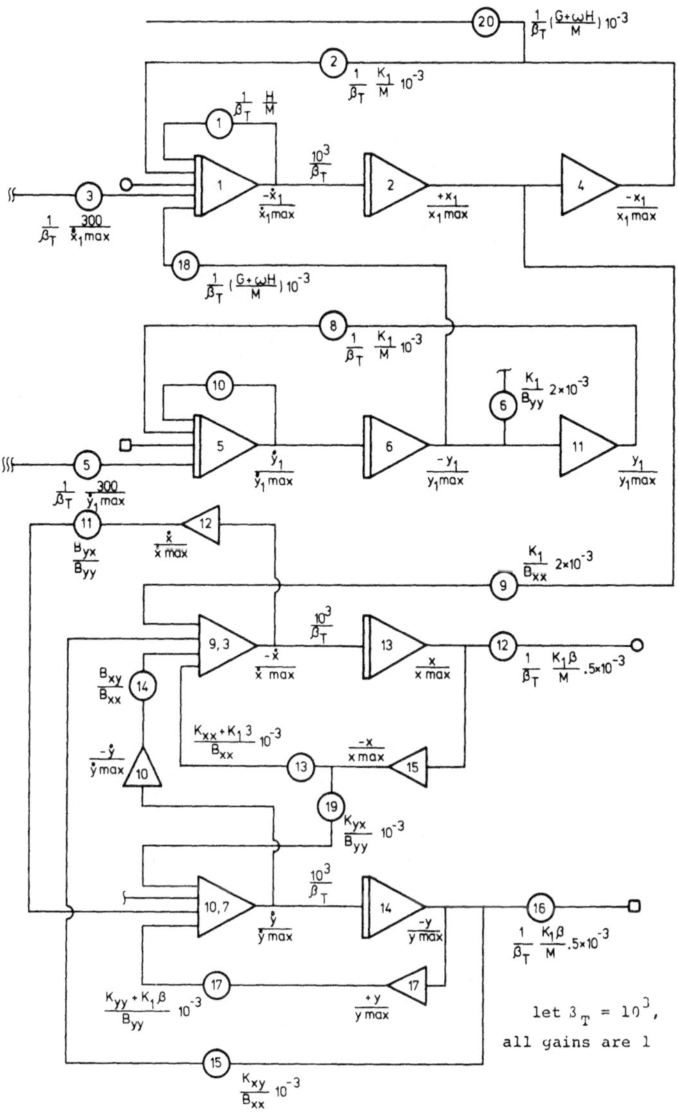

Fig. 5. Linear Analog Circuit.

$G = 1.50 \times 10^5$ lb.in and hysteresis $H = 200$ lb.in are shown in Figures 7a and 7b below and above the threshold speed.

The analog orbits show that the tendency towards instability is a steadily developing trend. Below the threshold speed the rotor takes increasingly longer periods

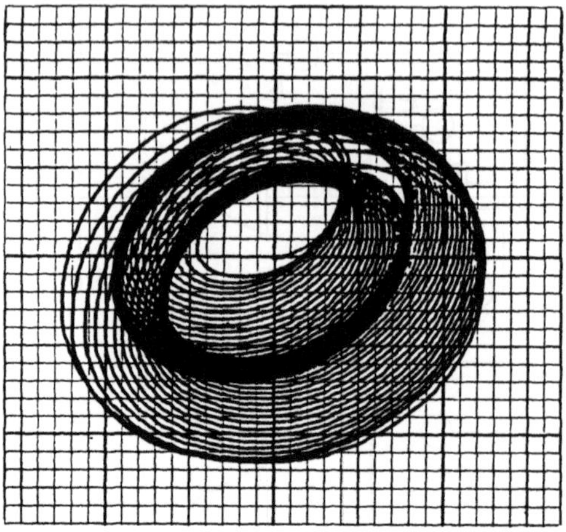

(a) N = 9100 rpm. Below Threshold Speed.

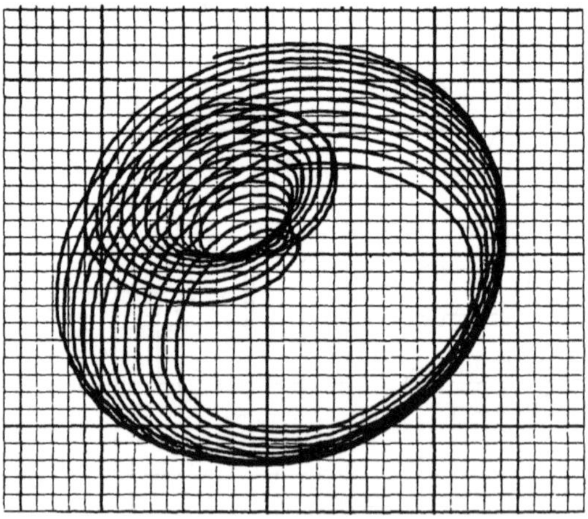

(b) N = 9200 rpm. Slightly Above Threshold Speed.

Fig. 6. Journal Whirl Orbits for G = 0 and H = 0 .
Linear Short Bearing Analysis.

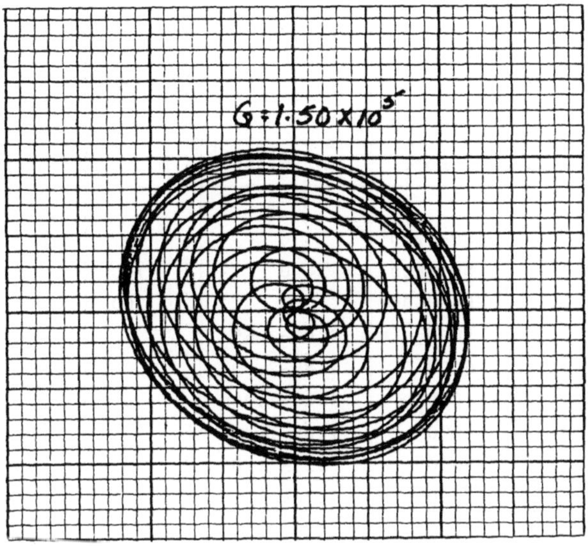

(a) G = 1.50 10^5 lb/in. Below Threshold Speed.

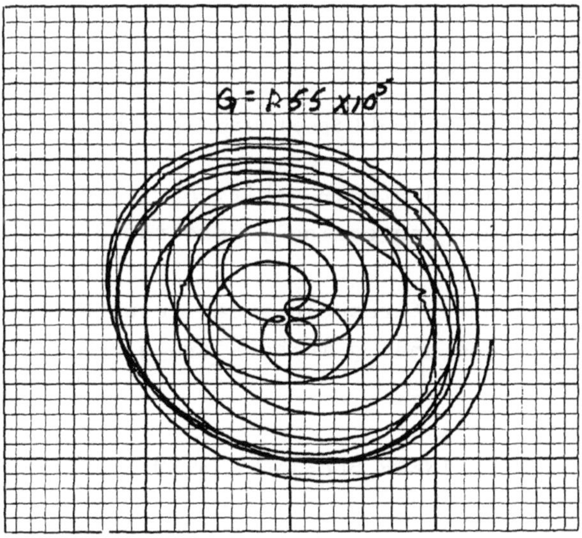

(b) G = 1.55 10^5 lb/in. Above Threshold Speed.

Fig. 7. Journal Whirl Orbits Where H = 200 and N = 7000 rpm.

Comparison of Analog and Digital Threshold Speed Values					
Analog			Digital		
G	H	Threshold Speed	G	H	Threshold Speed
lb/in x10^5	lb sec/in	rpm.	lb/in. x10^5	lb sec/in	rpm.
0	0	9100-9200	0	0	9100-9200
0	100	8700-8800	0	100	8800-8900
0	200	8500-8600	0	200	8500-8600
3.60-3.65	0	5000	3.60-3.65	0	5000
2.85-2.90	0	6000	2.85-2.90	0	6000
1.85-1.90	0	7000	1.85-1.90	0	7000
1.20-1.25	0	8000	1.25-1.30	0	8000
.15- .20	0	9000	.15- .20	0	9000
3.70-3.75	100	5000	3.70-3.75	100	5000
2.80-2.85	100	6000	2.85-2.90	100	6000
1.75-1.80	100	7000	1.70-1.75	100	7000
.90- .95	100	8000	.90- .95	100	8000
3.85-3.90	200	5000	3.85-3.90	200	5000
2.75-2.80	200	6000	2.80-2.85	200	6000
1.50-1.55	200	7000	1.50-1.55	200	7000
.60- .65	200	8000	.60- .65	200	8000

Table 2. Variation of Threshold Speed with Compressor System Effects.

of time to establish a stable orbit. Above the threshold, orbit growth becomes increasingly more rapid as speed is further increased.

A comparison of threshold speed results obtained by both the digital and analog methods are given in Table 2.

4. NON-LINEAR BEARING SYSTEM STUDIES

Linear bearing is unable to determine whether a rotor will whirl in a bounded orbit beyond the instability threshold speed. To study such motions, the non-linear short-bearing theory described in Appendix 2 was developed. Equations (A2.10) and (A2.11) for the radial and tangential fluid-film force components include the effects of the rotor whirl motions about the static equilibrium position. These equations were coupled with the rotor system equations (A2.13) as shown. The complete system equations include the effects of gas seals, rotor hysteresis and rotor unbalance, as described for the linear bearing study. A hybrid computer program was developed to obtain orbital solutions for the equations. Two test cases as follows were used to study the functioning of this program.

(a) Rigid rotor system, Figure 8. Based on the test set-up used by MITCHELL [21], this case was examined first in an attempt to verify program functioning.

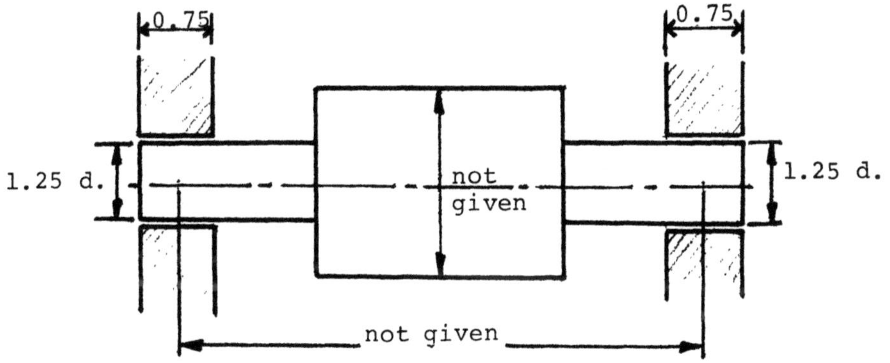

Rotor weight: W = 8.094 ln.
Bearing clearance: C = 0.0039 in., radial
Oil inlet pressure: 0.5 in.Hg. to 6.0 in.Hg.
Bending critical: 70,000 rpm

Fig. 8. Rigid Rotor used for Hybrid Calculation. Original from Mitchell et.al. [25].

(b) Compressor rotor system described in the linear bearing study of this paper, but with non-linear bearing forces.

The non-linear force expressions (A2.10, A2.11) apply between the limits θ_1 and θ_2 in the circumferential direction, as shown in Figure 9. For programming simplicity, a 180 degree fluid-film extending from the minimum thickness to the maximum thickness locations was considered. An EAI Pacer 600 hybrid computer was used. Part of the hybrid analog circuit for the general rotor system with non-linear bearing forces is shown in Fig. 10. Full circuit details and description of the software development has been given by OKUNSEINDE and MASLO [25]. Man-machine interface is achieved with a Hybran Operations Interpreter (HOI) linked to a Fortran IV function generator and display subroutines. An initial parameter listing is displayed on the CRT, and specific values may be introduced or changed in response to the display questions. The HOI then sets the analog coefficients

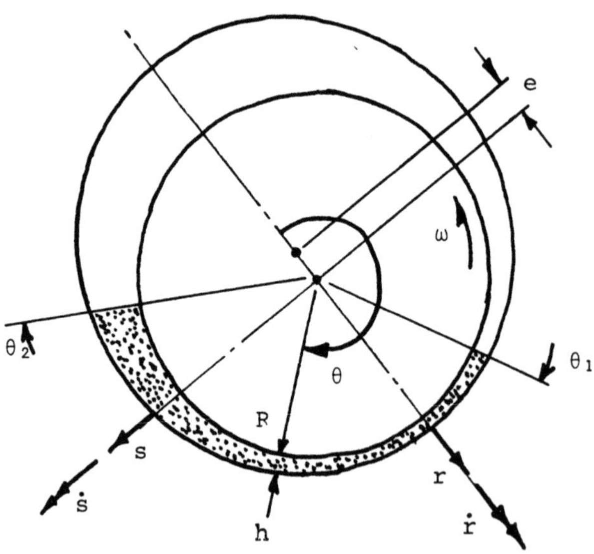

Fig. 9. Bearing Fluid Film Geometry.

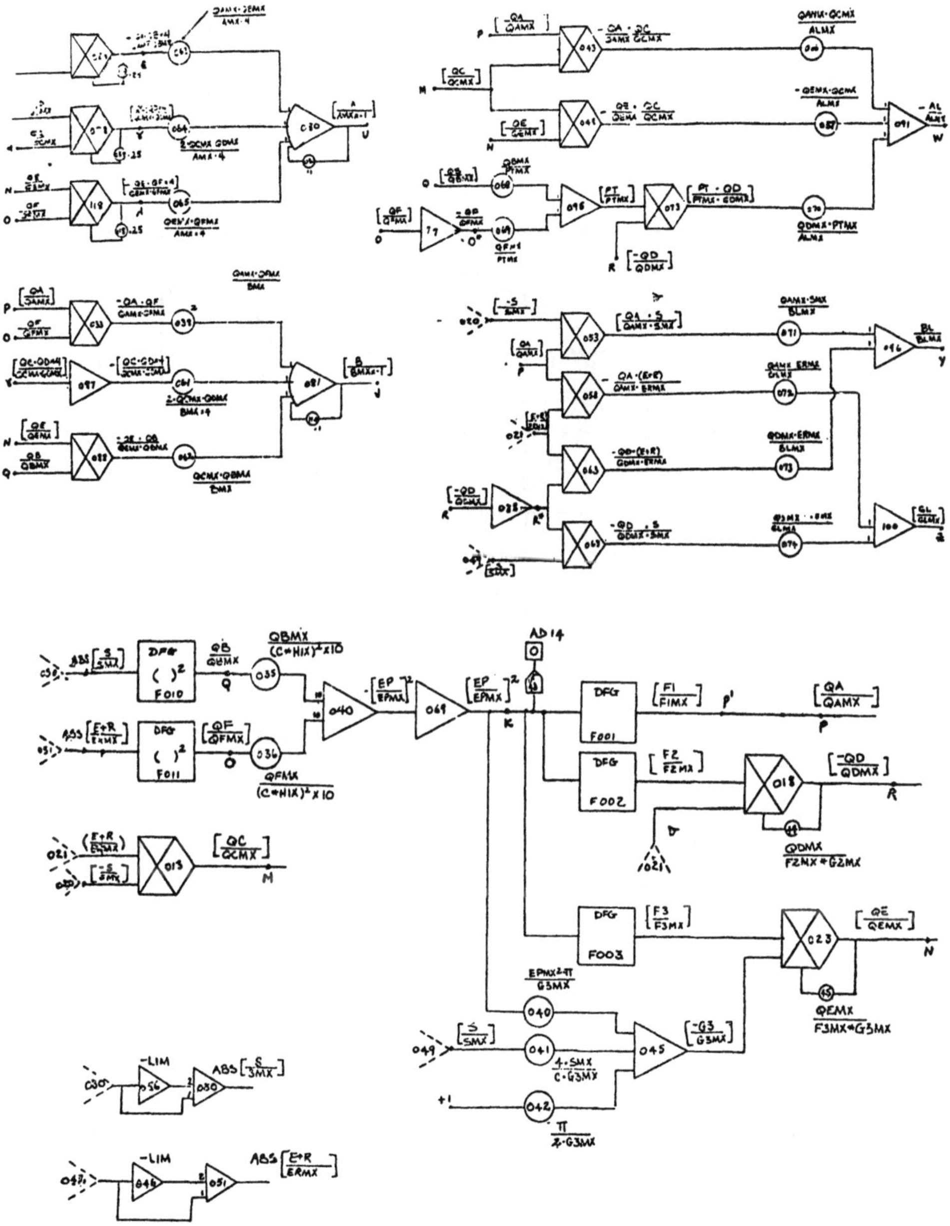

Fig. 10. Hybrid Computer Circuits for Rotor System with Non-linear Bearings.

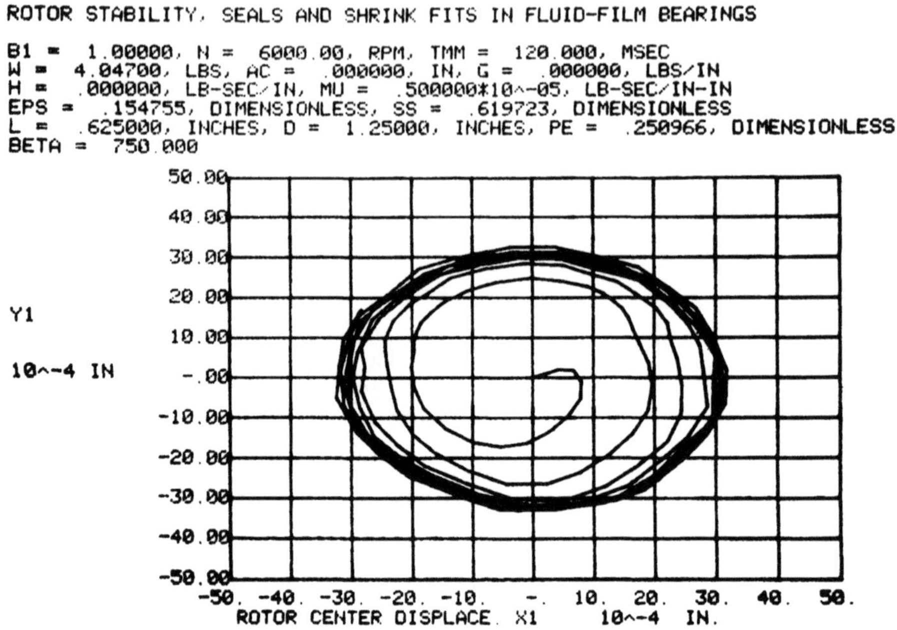

Fig. 11. Rigid Rotor Orbits. Hybrid Computer.
N = 6000 rpm, G = 0 , H = 0 . Stable.

for the desired run and puts the hybrid computer into 'operate' mode. Figures 11 through 17 show typical input data and orbits for the two test cases. Output may be whirl orbits at the journal or at the rotor mass, or plots of dynamic eccentricity vs time. Two plots may be displayed simultaneously on the CRT , as a console dialog option.

(a) Rigid rotor in non-linear bearings. Figure 11 shows the journal orbit at 6000 rpm with oil viscosity $\mu = 5.10^{-6}$ $lb.sec/in^2$. Orbits taken at 1910, 3000, and 6000 rpm were stable in each instance, but more time was required to form a stable orbit as speed increased, indicating a trend towards instability. The linear threshold speed for this case as obtained from curves by LUND [26] is $N_T = 7700$ rpm, for $L/D = 0.5$. Orbits at 6000, and 10,000 rpm are shown in Figures 12 and 13 for the same rotor with $\mu = 3.10^{-6}$ $lb.sec/in^2$. These orbits are also stable and bounded. With-

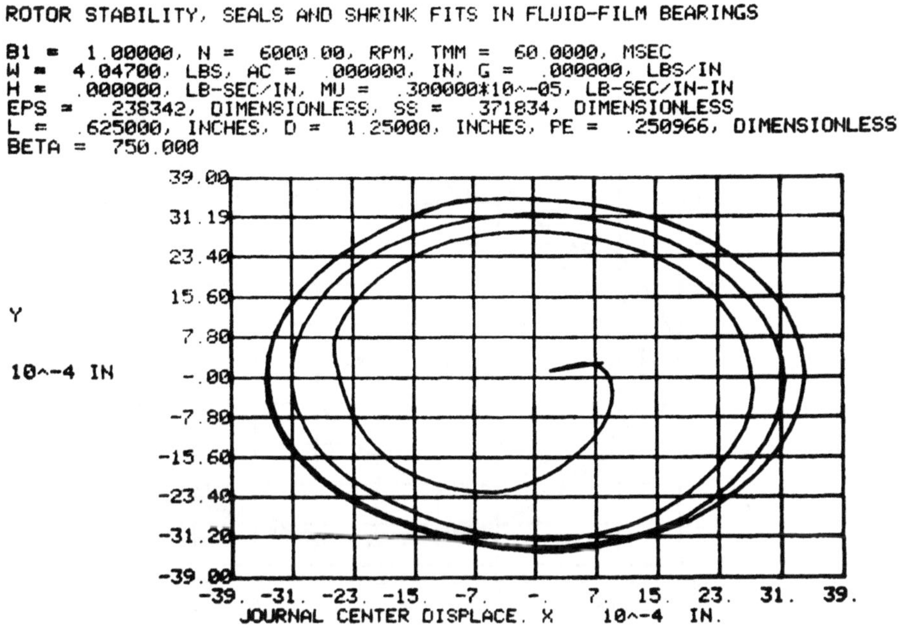

Fig. 12. Rigid Rotor Orbits. Hybrid Computer.
N = 6000 rpm, G = 0 , H = 0 . Stable.

in this range the orbits tended to grow with speed until the journal touches the bearing wall at about 10,000 rpm using the hybrid program.

There is similarity between these results and those obtained by Mitchell. It should be borne in mind that the bearing used by Mitchell was a flooded 360 degree bearing which is inherently unstable. Also, the viscosity values stated above are somewhat lower than Mitchell's which by inference were around 10.10^{-6} lb.sec/in^2.

There is similarity between these results and those obtained by Mitchell. It should be borne in mind that the bearing used by Mitchell was a flooded 360 degree bearing which is inherently unstable. Also, the viscosity values stated above are somewhat lower than Mitchell's which by inference were around 10.10^{-6}lb.sec/in^2.

Table 3 summarized Mitchell's operating conditions. A comparison with calculated linear threshold speeds applicable

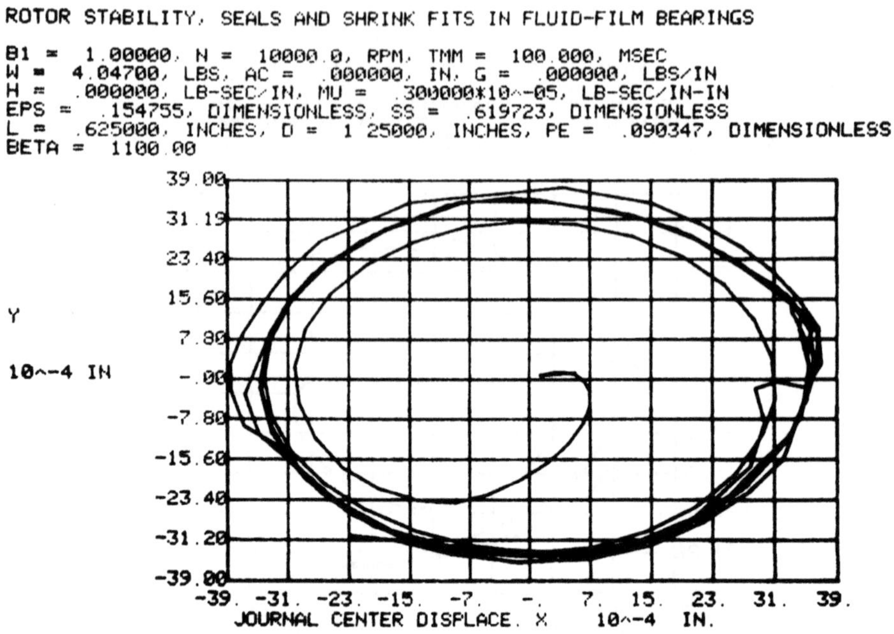

Fig. 13. Rigid Rotor Orbits. Hybrid Computer
N = 10,000 rpm, G = 0 , H = 0 . Unstable.

SELECTED CONDITIONS, MITCHELL EXPERIMENT [25]			Comparable Plain Cylindrical Bearing Operating Conditions	
Load Parameter $P = c\omega^2/g$	Operating Eccentricity, ε	Operating Speed N	Threshold Speed N_T*	Operating Condition
2.5	0.6	1910	8550	Stable
0.86	0.4	3260	7780	Stable
0.86	0.6	3260	8850	Stable
0.56	0.4	4040	7780	Stable
0.32	0.5	5350	8165	Stable
0.32	0.4	5350	8550	Stable
0.20	0.4	6740	7780	Stable
0.20	0.6	6740	8550	Stable

*Based on curves by LUND [31] Fig. 6

Table 3. Summary of Operating Conditions in Mitchell Experiment and Comparable Plain Cylindrical Bearing Conditions.

to Mitchell's rotor is also given, had plain cylindrical bearings been used in Mitchell's test. Such conditions are reasonably close to those computed in present study.

The hybrid results confirm that whirl orbits are influenced in size and form by oil viscosity (i.e., damping). This feature is not apparent in the results obtained by linear threshold theory. The hybrid results show that a lightly loaded rotor may experience non-linear whirling in a bounded orbit which will grow until the bearing is contacted. An increase in oil viscosity may help to stabilize the orbit to a smaller size by increasing the bearing damping under certain circumstances.

(b) Compressor rotor. Figures 14 through 17 indicate the hybrid results for the compressor rotor in non-linear bearings. Bearing size is the same as for the linear study. Oil

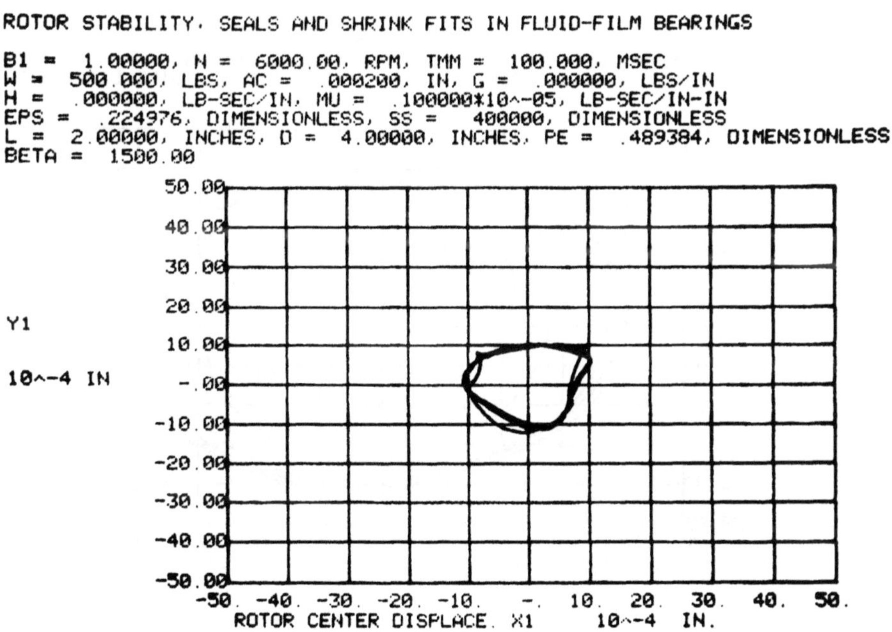

Fig. 14. Compressor Rotor Orbits. Hybrid Computer. N = 6000 rpm, G = 0 , H = 0 . Stable.

viscosity is held constant at $\mu = 1.10^{-6}$ lb.sec/in^2 in each calculation. The compressor has a mass eccentricity of 0.0002 in. in each disk (total, 3.2 ox.in). The orbit at 6000 rpm shown in Figure 14 indicates stable unbalance whirling. Orbits at 8000, 8500 (not shown) and 9000 rpm (Figure 15) are also stable. However, a slightly longer run at 9000 rpm, Figure 16, indicates a final rapid growth of the whirl orbit until the rotor hits the bearing wall. This correlates with the linear threshold speed under similar operating conditions (9200 rpm) , and the instability is clearly defined. The orbit grows steadily as the threshold speed is approached as was evident in the linear study. Figure 17 indicates the effect of a seal coefficient

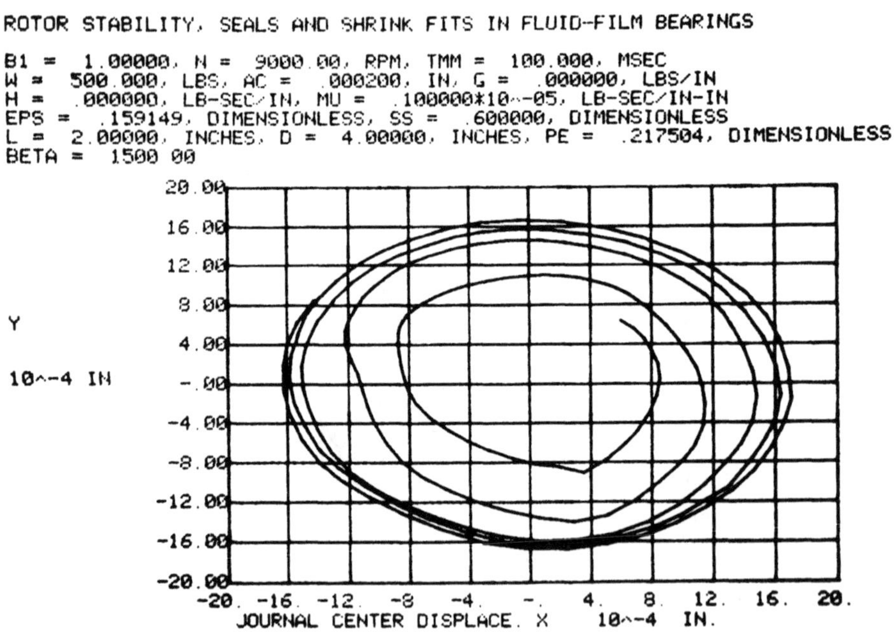

Fig. 15. Compressor Rotor Orbits. Hybrid Computer.
N = 9000 rpm. G = 0 , H = 0 . Stable.

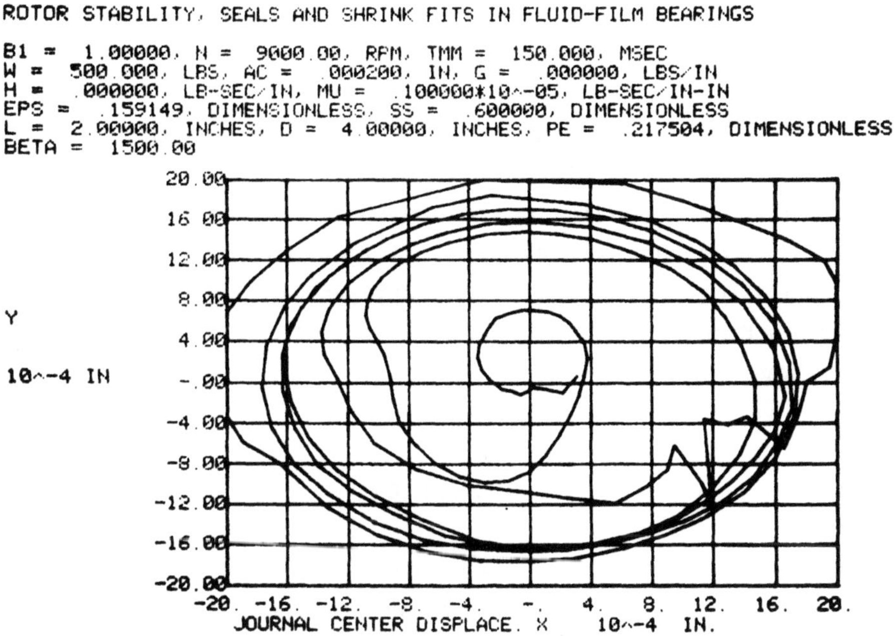

Fig. 16. Compressor Rotor Orbits. Hybrid Computer. N = 9000 rpm. G = 0 , H = 0 . Unstable.

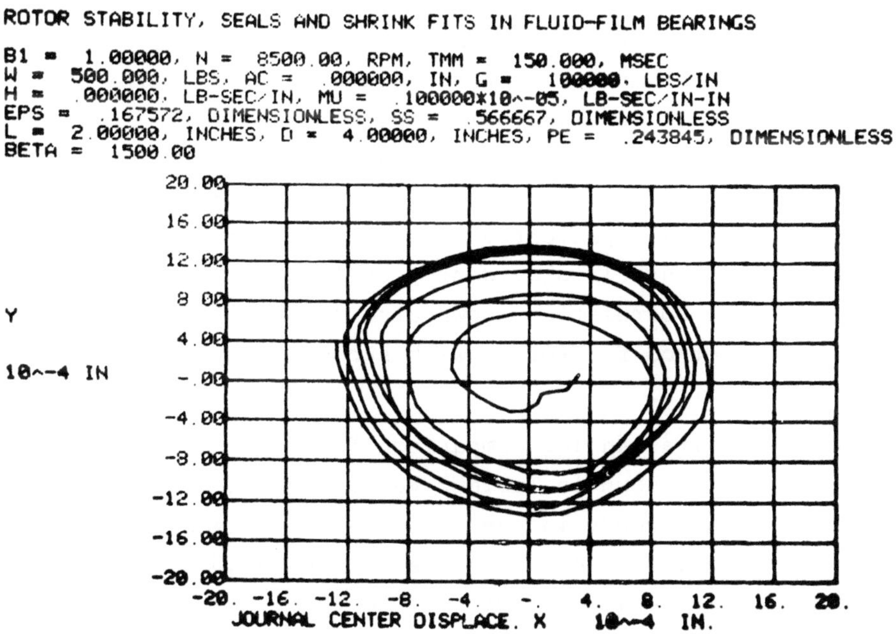

Fig. 17. Compressor Rotor Orbits. Hybrid Computer. N = 8500 rpm. $G = 1.10^5$, H = 0. Unstable.

$G = 1.10^5$ lb.in. on the system. The orbit appears to be unstable after 150 milliseconds of computation. This also agrees with the linear calculation, Figure 4, which predicts a threshold decrease to around 8500 rpm with a seal force coefficient of 1.10^5 lb/in.

The hybrid results correlate with the analog and digital results for the same systems. It appears that the 180 degree bearing represents a good approximation to the plain cylindrical bearing for the case $L/D = 0.5$, as Holmes and others have suggested.

5. DISCUSSION

(a) Comparison threshold speed results with experimental data. In all instances there is correlation between the experimental threshold speeds and the predicted threshold speeds to within +10% to -6% , and in several instances to within 2% . As actual threshold values are not usually repeatable to closer than ±5% due to the causes mentioned, the procedure demonstrated is considered to be effective. It was observed that most test rotors consist of a single central massive disk on a uniform shaft between two identical bearings. Test data indicates that such rotors will experience a pronounced threshold speed, i.e., strong oil-whip will occur. The analysis presented is valid for such rotors, and the model in Figure 1 is evidently well suited to such calculations. The correlation shown also endorses the eight-coefficient method for rotors with substantial shaft flexibility.

(b) Linear digital-analog comparison for compressor rotor. The correlation shown in Table 2 verifies both these techniques, and the trends in the results are in keeping with practical experience, e.g., decrease in threshold speed with increase in gas seal coefficient. The theory is therefore considered valid and could be applied in practice to rotor systems of the type shown in Figure 1, provided that suitable values for coefficients G and H were available. To extend this theory to rotors which have no reasonable mid-plane symmetry requires a more complex analysis, and a polynomial solution is no longer suitable.

The digital analysis required the same amount of program development as the analog analysis, although the analog analysis took a longer period of time (ratio 4:3). The digital analysis is convenient and clean-cut. To do the same work with the analog with significant parameter changes would usually require some re-scaling of the system parameters. Input effort for speed-dependent items such as the bearing coefficients is also considerable, although in return the orbit results are more information than the threshold speed alone. The parameters are readily changed once the analog machine is set up. In general, the analog computer is a more versatile and is well adapted to parametric studies, whereas the digital method is more efficient for threshold speed calculations.

(c) Hybrid computer studoes of rotors in non-linear bearings. The hybrid computer circuit is shown in Figure 10. Hybrid results are shown in Figures 11 and 17. The general correlation between these results and those of other authors confirms the functioning of the hybrid circuit. Two problems were encountered in obtaining the results. The scaling of the circuit to suit the wide range of parameters required some modifications between cases. Initially this caused overloading problems with the amplifiers. For example, rotor eight was varied from 4.08 lb/brg to 20,700 lb/brg for certain studies and it was necessary to modify the circuit to deal with these values. The small numbers for viscosity (around 1.10^{-6}) also caused problems, as the system parameters were input directly into the computer via a program keyboard. Whirl orbits occasionally contained confused or jagged sections toward the end of a run. It was not immediately clear whether the rotor itself had become unstable or whether the problems were caused by circuit overloading. Most of these difficulties were resolved by rescaling but where this was not successful the run duration was extended from 100 millisec to around 300 millisec to resolve this question. As the number of bits of information per run was usually restricted to 200, this detracted from the smoothness of long-run orbits. In general, run durations of 150 millisec was adequate to determine whether the rotor was stable or not.

The influence of initial conditions on instability appeared to be minor where the system was well below threshold speed, but some effect was noted close to the threshold speed. For example, the initial displacements r, s could be specified as program input. Large r, s values tended to make the rotor unstable. This is consistent with practical experience, as many authors have noted that instability may be initiated by a blow applied to the rotor which corresponds to large initial values.

6. CONCLUSIONS

1. Instability threshold speeds of flexible rotors in fluid-film bearings may be predicted with good accuracy using linear eight coefficient bearing theory.
2. Digital computer and analog computer methods are equally reliable and accurate in predicting threshold speeds based on linear bearing theory.
3. Analog computer development time was somewhat greater than the digital computer approach, but analog orbital results provide more information on the rotor motion than do threshold speed values alone.
4. Analog studies are convenient for studying parametric effects on system performance but parametric iterations made by hand can be time consuming, especially where substantial re-setting is involved, e.g., speed-dependent bearing coefficients.

5. Parametric changes and bearing coefficient iterations may be conveniently accomplished on a hybrid computer, which is programmed to input such changes directly. Complicated non-linear effects may be incorporated in a hybrid computer program where a large machine is available.

6. The non-linear short bearing theory described herein was found to give results which correlate well with other published results.

ACKNOWLEDGEMENTS - Grateful acknowledgement is made to the following for their contributions: R.M. Maslo and B. Okunseinde of E.A.I., West Long Branch, N.J. for the hybrid computer work; to J. Browns and A. Arslancan, R.I.T. Research Associates, for the non-linear bearing studies, and the linear system program; and to Dean Richard A. Kenyon for the use of the College of Engineering facilities at R.I.T.

8. REFERENCES

[1] RIEGER, N.F.: Rotor-Bearing Systems, Structural Mechanics Computer Programs - Surveys, Assessments and Availability, W. Pilkey, K. Salzalski, M. Schaeffer, University Press of Virginia, Charlottesville, 1974.

[2] TONDL, A.: Some Problems of Rotor Dynamics, Publishing House Czechoslovakian Academy of Sciences, Prague, 1965.

[3] RIEGER, N.F.: Survey of Previous Work, TR-45-65. Wright-Patterson AFB Aero Propulsion Lab Report, Vol. 1. Rotor Bearing Design Technology Series.

[4] STERNLICHT, B. and RIEGER, N.F.: Rotor Stability, Proc. Inst. Mech. Engrs., London Vol. 182 PT 3A. P 82. 1968.

[5] GUNTER, E.J.: Dynamic Stability of Rotor-Bearing Systems, N.A.S.A. Report, SP-113, 1966.

[6] GUNTER, E.J.: Survey of the Literature on Rotor-Bearing Stability, A.S.M.E. Flexible Rotor-Bearing Dynamics Sub-Committee Report, Machine Design Division. 1974 (in press).

[7] LUND, J.W.: Computer Program Manual for Rotor Response and Stability, AFAPL-TR-65-45 Part V Rotor Bearing Dynamics Design Technology, Air Force Systems Command, Wright-Patterson Air Force Base, May, 1965.

[8] LUND, J.W.: Stability and Damped Critical Speeds of a Rotor in Fluid-Film Bearings, A.S.M.E. Paper 73 DET 103 Design Engineering Technical Conference, Cincinatti, Ohio, September 1973.

[9] GUNTER, E.J. and DeCHOUDHURY, P.: Dynamic Stability of Flexible Rotor-Bearing Systems, Report No. ME-4040-104-700 Research Laboratory for the Engineering Sciences, University of Virginia, Charlottesville, December, 1970.

[10] GIBERSON, M.F.: Unpublished Analysis and Computer Program. Turbo Research Incorporated, West Chester, Pa. 1973.

[11] NEWKIRK, B.L. and TAYLOR, H.D.: Shaft Whirling Due to Oil Action in Journal Bearings, General Electric Review, August, 1925.

[12] NEWKIRK & LEWIS: Oil Film Whirl, An Investigation of Disturbances Due to Oil Films in Journal Bearings, Trans. A.S.M.E., January 1956, Vol. 65, 78.

[13] PINKUS, I.J.: Experimental Investigations of Resonant Whip. Trans. A.S.M.E., Vol. 87. July, 1956.

[14] MORRISON, D.: The Influence of Plain Journal Bearing on the Whirling Action of an Elastic Rotor, Proceedings of the Institution of Mechanical Engineers, Vol. 176, 1962.

[15] TONDL, A.: Notes on the Problem of Self Excited Vibrations and Non-Linear Resonances of Rotors Supported in Several Journal Bearings, Wear, Vol. 8, 1965.

[16] RAO, J.S.: Experimental Investigation on Oil Whip of Flexible Rotors, Tribology, May 1970.

[17] PAN, C.H.T. and STERNLICHT, G.: Comparison Between Theories and Experiments for the Threshold of Instability of Rigid Rotor in Self-Acting, Plain-Cylindrical Journal Bearings, Trans. A.S.M.E., Jnl. of Basic Engineering, Paper No. 63-LUBS-3, 1964.

[18] DUCKWORTH, S.J. and RIEGER, N.F.: Computerization of Threshold Speed Analysis in Flexible Rotor-Bearing Systems, Union College Independent Study Report, October, 1970.

[19] JENNINGS, U.D. and OCVIRK, F.W.: The Simulation of Bearing Whirl on an Electronic-Analog Computer, Trans. A.S.M.E. Jnl. of Engineering for Industry, Series D. pp. 503-510, 1962.

[20] HOLMES, R.: Oil Whirl Characteristics of a Rigid Rotor in Journal Bearings, Proceedings of the Institution of Mechanical Engineers, 1963 (No. 11), 291.

[21] MITCHELL, J.R., HOLMES, R., and BYRNE, J.: Oil Whirl of a Rigid Rotor in 360 degree Journal Bearings: Further Characteristics, Proceedings of the Institution of Mechanical Engineers, Vol. 180, Part 1, No. 25, 1965-66.

[22] LUND, J.W.: Design Handbook for Fluid-Film Type Bearings, AFAPL-TR-65-45, Part 3, Rotor Bearing Dynamic Design Technology, Whright-Patterson Air Force Base, Dayton, Ohio, 1965.

[23] LUND, J.W.: Discussion of paper by P.C. Warner and R.J. Thoman, Effect of 150-degree Partial Bearing on Rotor-Unbalance Vibration, A.S.M.E. Paper 63-LUB-36, (1963).

[24] RIEGER, N.F. and CUNDIFF, R.: Discussion of paper by P.G. Morton, Influence of Coupled Asymmetric Bearings on the Motion of a Massive Flexible Rotor. Proceedings I. Mechanical Engineers, Vol. p82, Part 1, No. 13, p. 255, (1967-68).

[25] OKUNSEINDE, B. and MASLO, R.M.: Hybrid Simulation of Rotor Stability with Seals and Shrink Fits in Non-Linear Fluid-Film Bearings, Summer Computer Simulation Conference, Houston, Texas, July, 1974.

[26] LUND, J.W.: Self-Excited Stationary Whirl Orbits of a Journal in a Sleeve Bearing. Ph.D. Thesis Rensselaer Polytechnic Institute, Troy, New York 1966.

[27] RIEGER, N.F., THOMAS, C.B. and ARSLANCAN, A.: Rotor Stability with Seals and Shrink Fits, in Fluid-Film Bearings, Rochester Institute of Technology, Mechanical Engineering Department, Technical Report 72 WRL R3, August 1972.

APPENDIX 1.

Stability Analysis of a Flexible Rotor System

A two mass flexible rotor in fluid-film bearings is shown in Figure 1. The disks are rigid and have no rotary or polar inertia. The shaft is massless and may be of any shape provided the transverse stiffness, K, is known or may be found from the rigid-bearing critical speed ω_n, i.e.,

$$K = M\omega_n^2 \tag{A1.1}$$

Both bearing forces are identical and linear with rotor displacement and velocity from the equilibrium position of the journal in each bearing. The eight-coefficient bearing force representation includes both direct and cross coupled effects;

(A1,2)
$$F_x = K_{xx}X + K_{xy}Y + B_{xx}\dot{X} + B_{xy}\dot{Y}$$
$$F_y = K_{yx}X + K_{yy}Y + B_{yx}\dot{X} + B_{yy}\dot{Y}$$

Gas seals are located at both disks. Forces arising from disk eccentricity with the stator clearance are given by:

$$F_S = iGr \qquad i = \sqrt{-1} \tag{A1.3}$$

where G is the gas seal coefficient and r is the whirl radius. Structural hysteresis effects which occur due to the shrink fit of the disk or the shaft and are given by:

$$F_H = H\dot{\zeta} \tag{A1.4}$$

where H is the hysteresis coefficient and $\dot{\zeta}$ is the whirl velocity in rotating coordinates. The rotor is axisymmetric and system effects are symmetric about mid-span. The equations of motion for either disk including the shaft, seal, and hysteresis effects are:

(A1.5)
$$M\ddot{X}_1 + H\dot{X}_1 + K(X_1-\beta X) + (G+\omega H)Y_1 = Ma\omega^2 \text{ Cos } \omega t$$
$$M\ddot{Y}_1 + H\dot{Y}_1 + K(Y_1-\beta Y) - (G+\omega H)X_1 = Ma\omega^2 \text{ Sin } \omega t + Mg$$

A force balance at either bearing gives:

$$K(X_1-\beta X) = K_{xx}X + K_{xy}Y + B_{xx}\dot{X} + B_{xy}\dot{Y}$$

$$K(Y_1-\beta Y) = K_{yx}X + K_{yy}Y + B_{yx}\dot{X} + B_{yy}\dot{Y}$$

To obtain a solution assume that:

(A1.6)
$$X = x_o \, e^{st}$$
$$Y = y_o \, e^{st}$$

where $s = \alpha + i\nu$

Substituting in the above expressions and simplifying gives:

$$\begin{vmatrix} \{(Ms^2+Hs+K)\Lambda^*_{xx}+(Gs+H\omega)\Lambda_{yx}-K^2\beta\} & \{(Ms^2+H\omega+K)\Lambda_{xy}+(Gs+H\omega)\Lambda^*_{yy}\} \\ \{(Ms+Hs+K)\Lambda_{yx}-(Gs+H\omega)\Lambda^*_{xx}\} & \{Ms^2+Hs+K)\Lambda^*_{yy}-(Gs+H\omega)\Lambda_{xy}-K^2\beta\} \end{vmatrix} = 0$$

(A1.7)

where

$$\Lambda^*_{xx} = K_{xx} + \beta K + sB_{xx}$$

$$\Lambda_{xy} = K_{xy} + sB_{xy}$$

$$\Lambda_{yx} = K_{yx} + sB_{yx}$$

$$\Lambda^*_{yy} = K_{yy} + \beta K + sB_{yy}$$

To obtain the roots of this determinant a convenient procedure is to expand and form the corresponding polynomial in S, viz:

$$A_1S^6 + A_2S^5 + A_3S^4 + A_4S^3 + A_5S^2 + A_6S + A_7 = 0 \quad \text{(A1.8)}$$

The coefficients A_i of this polynomial are complex algebraic expressions.

The special case where $G = 0$ and $H = 0$ (flexible rotor in eight-coefficient bearing) was originally presented by LUND [23]. The stability determinant becomes:

$$\text{(A1.9)} \quad \begin{vmatrix} \{(M_s{}^2+K)(K_{xx}+K+sB_{xx})-K^2\} & \{(Ms^2+K)(K_{xy}+sB_{xy})\} \\ \{(Ms^2+K)(K_{yx}+sB_{yx})\} & \{(Ms^2+K)(K_{yy}+K+sB_{yy})-K^2\} \end{vmatrix} = 0$$

At the stability threshold the real part of the eigenvalue vanishes $(\alpha = 0)$. Substituting $s = i\nu$ for this case and simplifying gives on separating real and imaginary parts:

$$\text{(A1.10)} \quad \kappa = \frac{(K_{xx}\omega B_{yy}+K_{yy}\omega B_{xx}) - (K_{xy}\omega B_{yx}+K_{yx}\omega B_{xy})}{K_{xx}\omega B_{yy} + K_{yy}\omega B_{xx}}$$

$$\text{(A1.11)} \quad \gamma^2 = \frac{(K_{xx}-\kappa)(K_{yy}-\kappa) - K_{xy}K_{yx}}{\omega B_{xx}\,\omega B_{yy} + \omega B_{xy}\omega B_{yx}} = \left(\frac{\nu}{\omega_T}\right)^2$$

where

$$\text{(A1.12)} \quad \kappa = \frac{1}{2} K \frac{(S\gamma)^2}{1-(S\gamma)^2}$$

and

$$\text{(A1.13)} \quad S = \frac{\omega_T}{\omega_n}$$

These expressions are cited in the main text of this paper.

APPENDIX 2.

Short Bearing Approximation for Non-Linear Forces.

Consider the plain cylindrical bearing shown in Figure 9. The fluid-film extends from θ_1 to θ_2. Displacements r and s are directed along the line of bearing centers, and at right angles to it as shown. Assuming an adequate oil supply to support full film lubrication and a length-to-diameter ratio small enough for the short bearing approximation to be valid we have from Reynolds equation:

(A2.01) $$\frac{\partial}{\partial Z}\{h^3 \frac{\partial P}{\partial Z}\} = - 6\ \mu\omega \frac{dh}{d\theta} - 12\ \mu V$$

(A2.02) $$h = c + e \cos\theta + r \cos\theta + s \sin\theta$$

(A2.03) $$V = -\dot{r} \cos\theta - \dot{s} \sin\theta$$

(A2.04) $$\frac{dh}{d\theta} = -(e+r) \sin\theta + s \cos\theta$$

Integrating 3 times with respect to Z and using the usual boundary conditions $P = 0$ at $Z = 0$, and $Z = L$ obtain:

(A2.05) $$\begin{vmatrix} W_r \\ W_s \end{vmatrix} = \frac{1}{2}\ \mu R L^3 \omega \int_{\theta_1}^{\theta_2} - \frac{\{\frac{dh}{d\theta} + \frac{2V}{\omega}\}}{h^3} \begin{vmatrix} \cos\theta \\ \sin\theta \end{vmatrix} d\theta$$

Defining

(A2.06) $$\begin{vmatrix} W_r{}^* \\ W_s{}^* \end{vmatrix} \triangleq \begin{vmatrix} W_r \\ W_s \end{vmatrix} \cdot \frac{c^2}{\frac{1}{2}\ \mu R L^3 \omega}$$

and substituting for $\frac{dh}{d\theta}$ and $\frac{2V}{\omega}$ and h

$$\text{(A2.07)}\qquad \begin{vmatrix} W_r^* \\ W_s^* \end{vmatrix} = c^2 \int_{\theta_1}^{\theta_2} \frac{(e+r+\frac{2\dot{s}}{\omega})\sin\theta+(\frac{2\dot{r}}{\omega}-s)\cos\theta}{(c+(r+e)\cos\theta+s\,\sin\theta)^3} \begin{vmatrix} \cos\theta \\ \sin\theta \end{vmatrix} d\theta$$

If the denominator is put in the form $(1 + \varepsilon\cos\beta)^3$

where

$$\text{(A2.08)}\qquad \varepsilon = \frac{\sqrt{s^2+(r+e)^2}}{c}$$

$$\text{(A2.09)}\qquad \beta = \theta + \arctan(-\frac{s}{e+r})$$

then upon changing the variable of integration to β and using Sommerfeld's substitution which may be formulated as

$$\int \frac{\sin^n\theta\cos^m\theta}{(1+\varepsilon\cos\theta)^p}\, d\theta =$$

$$(1-\varepsilon^2)^{\frac{1+n-2p}{2}} \int \sin^n\phi(\cos\phi-\varepsilon)^m(1-\varepsilon\cos\phi)^{p-m-n-1}\, d\phi$$

We obtain after considerable algebra:

$$\text{(A2.10)},\ \text{(A2.11)}\qquad \begin{vmatrix} W_r^* \\ W_s^* \end{vmatrix} = \varepsilon^{-2}(1-\varepsilon^2)^{-3/2} \left|\frac{\theta}{2} - \frac{\sin 2\theta}{4}\right|_{\theta_1}^{\theta_2} \frac{1}{c^3}\ \cdot$$

$$\begin{vmatrix} -s\{(e+r)(e+r+\frac{2\dot{s}}{\omega})-s(\frac{2\dot{r}}{\omega}-s)\} \\ (e+r)\{(e+r)(e+r+\frac{2\dot{s}}{\omega})-s(\frac{2\dot{r}}{\omega}-s)\} \end{vmatrix} + \varepsilon^{-2}(1-\varepsilon^2)^{-2}\ \cdot$$

$$\left|\varepsilon\cos\theta - \frac{\cos 2\theta}{4}\right|_{\theta_1}^{\theta_2} \frac{1}{c^3} \begin{vmatrix} ((e+r)^2-s^2)(e+r+\frac{2\dot{s}}{\omega})-2s(e+r)(\frac{2\dot{r}}{\omega}-s) \\ 2s(e+r)(e+r+\frac{2\dot{s}}{\omega})+((e+r)^2-s^2)(\frac{2\dot{r}}{\omega}-s) \end{vmatrix}.$$

$$+ \varepsilon^{-2}(1-\varepsilon^2)^{-5/2} \left| \frac{\sin 2\theta}{4} - 2\varepsilon \sin\theta + \varepsilon^2 + \frac{\theta}{2} \right|_{\theta_1}^{\theta_2} \cdot$$

$$\frac{1}{C^3} \begin{vmatrix} (e+r)\{s(e+r + \frac{2\dot{s}}{\omega}) + (e+r)(\frac{2\dot{r}}{\omega} - s)\} \\ s\{s(e+r + \frac{2\dot{s}}{\omega}) + (e+r)(\frac{2\dot{r}}{\omega} - s)\} \end{vmatrix}$$

or

$$\begin{vmatrix} W_r^* \\ W_s^* \end{vmatrix} = f_1(\varepsilon)g_1(\theta) \begin{vmatrix} J_{r1} \\ J_{s1} \end{vmatrix} + f_2(\varepsilon)g_2(\theta) \begin{vmatrix} J_{r2} \\ J_{s2} \end{vmatrix} +$$

$$f_3(\varepsilon)g_3(\theta) \begin{vmatrix} J_{r3} \\ J_{s3} \end{vmatrix} = \sum_{i=1}^{3} f_j(\varepsilon)g_i(\theta) \begin{vmatrix} J_{ri} \\ J_{si} \end{vmatrix}$$

Stabilization of Self-Excited Rotor Vibrations by an Active Damper

G. Schweitzer

Institut B für Mechanik
Technische Universität München, BRD

1. INTRODUCTION

Lateral vibrations of a rotor can become unstable due to forces normal to the deflection of the rotor. Physically these nonconservative forces are caused by internal damping or friction, by bearing characteristics or by unsymmetrical steam flow in a steam turbine [1]. Self-excited vibrations induced by such mechanisms may severely restrict the operational range of a rotor. Therefore means are looked for in order to avoid or to damp self-excited vibrations.

One method is to determine the still available constructional parameters of the rotor-bearing-system in such a way as to maximize some measure of stability of the rotor motion [2]. In this case the bearing-forces are a "passive" reaction on the motion of the rotor. The efficiency of this method depends on the physical and constructional boundaries of the range within which the system parameters can still be varied.

Another method allows a much further reaching influence on the motion of the rotor. Here an "active" bearing exerts forces, which can be specified in such a way that the motion of the rotor remains stable and the vibrations are damped. Further, the active damper reduces resonance amplitudes when the rotor runs through critical speeds. In terms of control theory the rotor is the plant and the active damper is the controller. It is the evaluation of an active damper that will be investigated here more thoroughly.

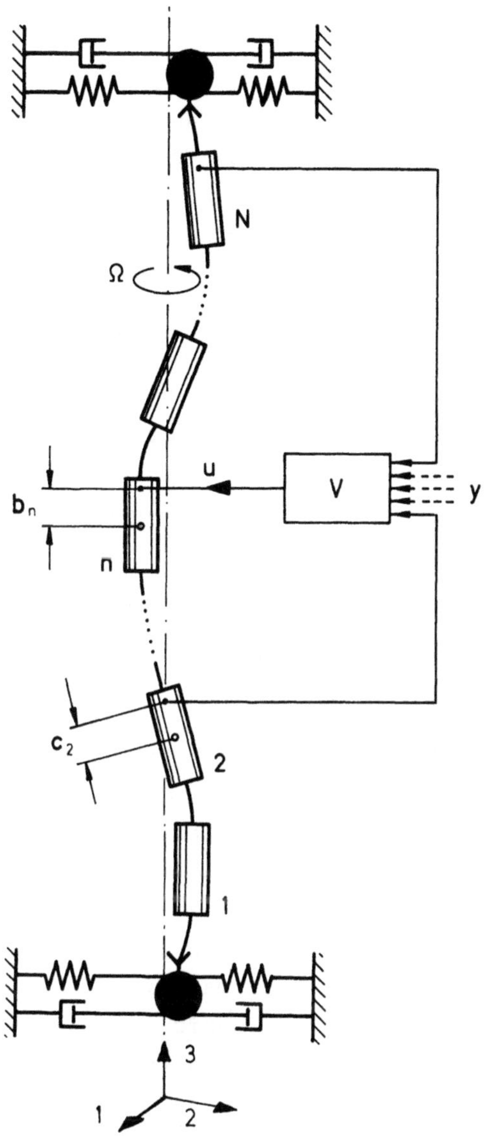

Fig. 1. Model of elastically suspended multibody rotor with an active damper.

At first the equations of motion are stated for a symmetric rotor model (Fig. 1). It consists of a string of viscoelastically interconnected rigid bodies that are elastically supported by bearings at the end of the rotor. This discrete rotor model is chosen because it makes the equations analytically tractable, because it easily incorporates the action of the discrete forces of the active damper, and because the investigation was stimulated by work done for a multistage rotor corresponding actually to Fig. 1. Though, the model still is a workable approximation of a continuous elastic rotor. The internal damping as a source of self-excitation is chosen because it is well amenable to theoretical and experimental analysis.

In order to assess the influence of internal and external damping and of the rotor velocity on the natural rotor motion a stability analysis is performed making use of matrix theorems for mechanical systems [3] and of generalized Rayleigh-quotients.

Then the active damper (Fig. 1) is introduced. It exerts forces u on the rotor that depend linearly by a gain matrix V on some selected states y of the rotor motion. In evaluating the location and the structure of the damper three questions will have to be answered. Where is the optimal point of attack for the damper forces u in order to stabilize the rotor motion most effectively? Where are the optimal positions for observing the rotor motion in order to get a maximum of information about it from the measuring signals y ? And what does the optimal control gain matrix V look like that ensures stability of the motion? The theoretical results are applied to a three stage rotor, experimental results with an electromagnetic active damper are showing its performance.

2. EQUATIONS OF MOTION

It is assumed that small deviations of the rotor from the axis of permanent rotation can be described by linearized equations of motion. Each of the equal rigid rotor elements (Fig. 1) has two rotational and two translational degrees of freedom. There are no vertical or torsional displacements, the angular velocity Ω is constant. The shaft elements are of viscoelastic material, their mass is neglected. For each rigid rotor element the basic laws of mechanics are formulated. The forces acting on a rotor element are elastic forces of the rotor shaft, and eventually there are damper forces. The elastic forces depend on the deformation of the shaft element and its velocity relative to the adjoining rotor element, and they can be eliminated by expressing them in terms of displacement variables. Similarly the equations for the linearly damped, translational motion of the elastically suspended masses of the bearings are derived [10].

The displacement variables are arranged as coordinates

of the vector

$$(2.1) \qquad \tilde{z} = \left[\ldots {}^n s_1 \ldots \vdots \ldots {}^n w_1 \ldots \Vert \ldots {}^n s_2 \ldots \vdots \ldots {}^n w_2 \ldots \right]^T$$

where ${}^n s_i$ and ${}^n w_i$ denote the displacements of the n-th rigid body in the direction of the i-axis $(i = 1,2)$ due to translation and rotation, respectively. Then the equation of motion takes the form

$$(2.2) \qquad \begin{bmatrix} I & 0 \\ 0 & I \end{bmatrix} \ddot{z} + \begin{bmatrix} \mathcal{D} & 0 \\ 0 & \mathcal{D} \end{bmatrix} \dot{z} + \begin{bmatrix} 0 & \Omega G \\ -\Omega G & 0 \end{bmatrix} \dot{z} + \begin{bmatrix} K & 0 \\ 0 & K \end{bmatrix} z + \begin{bmatrix} 0 & N \\ -N & 0 \end{bmatrix} z = \nu \; .$$

The displacement vector z follows from $\tilde{z} = Tz$. The transformation T does not change the physical character of $\tilde{z}$ but is only introduced to transform the mass matrix into the unit matrix. The submatrix $\mathcal{D}$ is termed damping matrix. It is composed of the matrices of external and internal damping

$$(2.3) \qquad \mathcal{D} = \mathcal{D}_A + \mathcal{D}_I \quad ,$$

both matrices being symmetric and positiv semidefinite. The external damping is caused by the damping of the bearing, the internal damping by the deformation of the rotating shaft elements. The gyroscopic matrix G is a diagonal matrix, partly occupied by terms C/A, where C and A are the moments of inertia of a rigid rotor element about the longitudinal and an equatorial axis, respectively. The stiffness matrix is symmetric and positiv definite. The matrix N of the nonconservative forces is due to internal damping and it holds

$$(2.4) \qquad N = \Omega \mathcal{D}_I \quad .$$

The vector ν contains the control forces of the active

damper.

The typical arrangement of the sub-matrices in (2.2) is a consequence of the rotational symmetry of the rotor-bearing-system. It can be further simplified by introducing complex variables. The displacements in the direction of the 1-axis and 2-axis are considered as the real part and the imaginary part of a vector, respectively, so that

$$(2.5)\quad \zeta = \zeta_{RE} + i\zeta_{IM}\ ,\quad z = \left[\zeta_{RE}^T\ ,\ \zeta_{IM}^T\right]^T\ ,\quad \upsilon = \left[\upsilon_{RE}^T\ ,\ \upsilon_{IM}^T\right]^T .$$

Then eq. (2.2), substituting (2.3), (2.4), (2.5) becomes complex and reads

$$(2.6)\quad \ddot{\zeta} + (\mathcal{D}_A + \mathcal{D}_I - i\Omega G)\dot{\zeta} + (K - i\Omega N)\zeta = \upsilon_{RE} + i\upsilon_{IM} .$$

It will be used subsequently for stability analysis. Another form of writing (2.2) is the state equation

$$(2.7)\qquad \dot{x} = Ax + Bu\ ,$$

$$x = \begin{bmatrix} \zeta_{RE} \\ \dot{\zeta}_{RE} \\ \hdashline \zeta_{IM} \\ \dot{\zeta}_{IM} \end{bmatrix}\ ,\quad A = \left[\begin{array}{cc:cc} 0 & I & 0 & 0 \\ -K & -\mathcal{D} & -N & -\Omega G \\ \hdashline 0 & 0 & 0 & I \\ N & \Omega G & -K & -\mathcal{D} \end{array}\right]\ ,\quad Bu = \begin{bmatrix} 0 \\ \upsilon_{RE} \\ \hdashline 0 \\ \upsilon_{IM} \end{bmatrix} .$$

This representation is familiar to control theory and will be applied to evaluate the active damper.

3. STABILITY

A homogeneous solution of (2.6) is $\zeta = q_j \exp\lambda_j t$, where q_j is an eigenvector and $\lambda_j = \delta_j + i\omega_j$ is an eigenvalue of the eigenvalue equation

$$(3.1)\qquad \left[I(\delta_j+i\omega_j)^2 + (D_A+D_I-i\Omega G)(\delta_j+i\omega_j) + (K-i\Omega D_I)\right]q_j = 0 \ .$$

By rearranging the imaginary terms eq. (3.1) reads

$$(3.2)\qquad \left[I(\delta_j+i\omega_j)^2 + \bar{D}(\delta_j+i\omega_j) + \bar{K}\right]q_j = 0 \ ,$$

where

$$\bar{D} = D_A + D_I - \frac{\Omega}{\omega_j}D_I - \delta_j\frac{\Omega}{\omega_j}G \ , \quad \bar{K} = K + \frac{\Omega}{\omega_j}(\omega_j^2+\delta_j^2)G + \delta_j\frac{\Omega}{\omega_j}D_I$$

are real symmetric matrices, and (3.2) can be understood as the eigenvalue equation of some mechanical system

$$(3.3)\qquad \ddot{\gamma} + \bar{D}\dot{\gamma} + \bar{K}\gamma = 0$$

with the real variable γ . The gyroscopic forces and the nonconservative forces of the original system (2.6) have thus been broken up into terms that can be regarded as additional damping and elastic forces. The original solution $q_j \exp\lambda_j t$ will satisfy (2.6) as well as (3.3) and if all solutions of (3.3) are stable, then the solution of (2.6) will be stable, too. According to general theorems for mechanical systems [3], connected with the names of THOMSON, TAIT and CHETAEV, the solutions of (3.3) are stable if

$$(3.4)\qquad \bar{K} > 0 \ , \quad \bar{D} \geq 0 \ ,$$

i.e., if the stiffness matrix $\bar{K}$ is positive definite and the damping matrix $\bar{D}$ is positive definite or semidefinite. Of course these matrices still contain the natural damping δ_j and the natural frequency ω_j . The natural frequency can assume positive and negative values, thus characterising - due to the physical interpretation of the complex variables - forward and backward whirl [4]. This means that a natural rotor vibration (of a conservative system) takes place in a plane that whirls with the natural frequency around the vertical axis in the same or in the opposite sense as the rotor velocity Ω . At the limit of stability for the eigenvalue λ_j the natural damping is $\delta_j = 0$ and the natural frequency is $\omega_j \neq 0$. The stability condition $\overline{K} > 0$ can not be violated, as it would require $\overline{K} = 0$ at the stability limit and hence $\omega_j = 0$, which is a contradiction as for $\omega_j = 0$ always $\overline{K} > 0$ is guaranteed. Therefore the stability condition (3.4) is fulfilled if

$$\overline{D} = D_A + D_I - D_I(\Omega/\omega_j) \geq 0 \quad . \tag{3.5}$$

This simple condition implies some meaningful physical interpretations. Internal damping D_I can only destabilize if the rotor is spinning $(\Omega \neq 0)$. The external damping D_A has to be suitably structured in order to stabilize effectively. Backward whirls $(\omega_j < 0)$ can not become unstable. Forward whirls $(\omega_j > 0)$ can only become unstable if $\Omega > \omega_j$, even if there is no external damping.

In order to show the influence of damping and spin velocity on the stability more explicity the eigenvalues of (3.1) are estimated by generalized Rayleigh-quotients. Premultiplying (3.1) by $q_j^*/(q_j^* q_j)$, i.e. the conjugate transposed and normalized eigenvector, reduces the system (3.1) of linear equations to a scalar equation

$$\lambda_j^2 + (d_{Aj} + d_{Ij} - i\Omega g_j)\lambda_j + (\nu_j^2 - i\Omega d_{Ij}) = 0 \quad . \tag{3.6}$$

The coefficient matrices are replaced by scalar quantities,

$$\frac{q_j^* D_A q_j}{q_j^* q_j} = d_{Aj} \quad , \quad \frac{q_j^* G q_j}{q_j^* q_j} = g_j \quad , \quad \frac{q_j^* K q_j}{q_j^* q_j} = \nu_j^2 \quad . \tag{3.7}$$

For any choice of q_j these Rayleigh-quotients (3.7) cover the range of value of the corresponding matrices in the complex plane. As all coefficient matrices are real, symmetric and p.d. or p.s.d. their range of value covers a piece of the positive real axis and therefore the following relations hold

$$(3.8)\qquad \begin{array}{ll} 0 \le d_{Aj} \le d_{Amax} & 0 \le d_{Ij} \le d_{Imax} \\ 0 \le g_j \le g_{max} & 0 < \nu_{min}^2 \le \nu_j^2 \le \nu_{max}^2 \ . \end{array}$$

As a consequence of the structure of the gyroscopic matrix G

$$(3.9)\qquad g_{max} = C/A \ .$$

The Rayleigh-quotient ν_j^2 of the stiffness matrix lies between the square of the lowest and the highest value of the undamped natural frequency of the elastic rotor system. For high spin velocity Ω and low damping the solution of (3.6) is approximated, retaining only terms of the highest order of Ω . Four groups of eigenvalues are obtained

$$(3.10)\qquad \lambda_q = -(d_{Aq} + d_{Iq}) + d_{Iq}/g_q + i\Omega g_q \ ,$$

$$(3.11)\qquad \lambda_p = -d_{Ip}/g_p - i\nu_p^2/g_p\Omega \ ,$$

$$(3.12)\qquad \lambda_f = -\frac{1}{2}(d_{Af} + d_{If}) + d_{If}\frac{\Omega}{\nu_f} + i\nu_f\left[1 + \frac{1}{8}\left(d_{If}\frac{\Omega}{\nu_f^2}\right)^2\right]$$

$$(3.13)\qquad \lambda_b = -\frac{1}{2}(d_{Ab} + d_{Ib}) - d_{Ib}\frac{\Omega}{\nu_b} - i\nu_b\left[1 + \frac{1}{8}\left(d_{Ib}\frac{\Omega}{\nu_b^2}\right)^2\right] \ .$$

In the first two groups the imaginary parts or the natural frequencies depend on the spin velocity Ω . For the nutation (3.10) the natural frequency tends asymptotically to $(C/A)\Omega$, for the precession (3.11) it tends to zero with increasing Ω . For the socalled pendulous vibration (3.12), (3.13) the frequency dependance on Ω is less distinct and only relevant if there is internal damping. The eigenvalues λ_p and λ_b belong to natural modes with backward whirl. The nutation and one group of the pendulous vibrations (3.12) are whirling forward.

The real parts of the eigenvalues are indicating stability properties. As already shown the backward whirling natural modes can not become unstable, only the forward whirling ones. The nutation can only become unstable if $g_q < 1$, i.e., if the pervasiveness of the gyroscopic forces is not sufficiently large, or according to (3.9) if the rotor is elongated. In any case the gyroscopic forces tend to stabilize the motion. Any forward pendulous whirl (3.12), however, will become unstable if the spin velocity Ω exceeds some limit. In this case the destabilizing effect of internal damping will even increase with higher spin velocity Ω .

4. STABILIZABILITY

The rotor motion is stabilizable if its unstable natural modes are controllable. The motion is completely controllable if there exist control forces that let any displacement of the rotor go to zero in finite time. The physically quite clearly formulated concept of controllability has been introduced into control theory by KALMAN [5]. One of the various methods for investigating controllability will be used here.

Starting with eq. (2.7) the influence of control forces on the natural modes is studied. At first (2.7) is transformed by using modal coordinates r with

$$x = \psi r \tag{4.1}$$

into a system of decoupled equations

$$\dot{r} = Jr + \psi^{-1}Bu \quad (4.2)$$

The transformation matrix ψ is the nonsingular matrix of the arranged eigenvectors of the system matrix A, and $J = \psi^{-1}A\psi$ is the Jordan form of A. If A has distinct eigenvalues λ_j then J is the diagonal matrix

$$J = \mathrm{diag}[\lambda_1, \ldots, \lambda_j, \ldots, \lambda_m] \quad (4.3)$$

Each of the modal coordinates of the vector r corresponds to a natural mode of the system.

If there exists only a single control force then the control vector u is a scalar. The vector B indicates how the action of the control force is distributed on the different variables. For a control force acting in the 1-direction on the n-th rotor element the vector Bu of (2.7) assumes the form

$$Bu = [\ldots 0 \ldots \vdots 0 \ldots 1 \ldots 0, 0 \ldots b_n k \ldots 0 \vdots \ldots 0 \ldots \vdots \ldots 0 \ldots]^T u \quad (4.4)$$

Here b_n characterizes the point of attack of the control force (Fig. 1). It is zero if the control force acts on the center of mass of the rotor element. The factor k is a constant.

Obviously the system (4.2) is controllable only if each natural mode can be influenced by the control. That means that the vector $\psi^{-1}Bu$ must not have any zero coordinates and J must not have multiple eigenvalues. Otherwise a scalar control u cannot influence all modes simultaneously. For stabilizing the rotor motion, however, it is already sufficient to control the unstable modes only. They are the more controllable the larger the amount of the control force is acting on each unstable mode. The coordinates of the control vector $\psi^{-1}Bu$ are $\theta_j^T Bu$, where θ_j^T is the j-th row vector of the matrix ψ^{-1}. Adapting the ideas of [6], the following measure of stabilizability of an unstable mode is defined

$$ST_k = \frac{\theta_k^* BB^T \theta_k}{\theta_k^* \theta_k} \quad . \tag{4.5}$$

In order to assess the simultaneous stabilizability of several unstable modes a suitable overall measure is formed by averaging the weighted individual measures

$$ST = \left[\sum_k (ST_k \cdot \omega_k \cdot \exp\delta_k)^{-1} \right]^{-1} \quad . \tag{4.6}$$

The averaging over all unstable modes weights the damping δ_k and the frequency ω_k of the individual unstable mode. In order to achieve optimal stabilizability the control (4.4) can be structured now in such a way that the measure (4.6) reaches a maximum.

For practical purposes it is suggested to exert a control force not only in the 1-direction but also an independent one in the 2-direction. The motions in the 1,3- and in the 2,3-plane are coupled by gyroscopic and non-conservative forces, but the coupling is weak. A control force in the 1-direction only (4.4) would control all natural modes, but the measures for the quality of their controllability would be very low for some modes. Therefore the active damper is represented by a control vector u with two coordinates.

The measures (4.5), (4.6) are applied to determine the optimal location b_n of an active damper for a three stage rotor, that is used for experiments, too. At the desired operational speed two unstable pendulous vibration modes exist that have to be stabilized by an active damper. The measures for stabilizability are shown in Fig. 2 as functions of the location b_n of the damper, together with the respective geometric modal shapes. The unstable fundamental mode $(\lambda_1 = 0{,}18 \pm 19{,}3i)$ is most effectively stabilized by an active bearing at the position ①, for stabilizing the other unstable mode the position ② is to

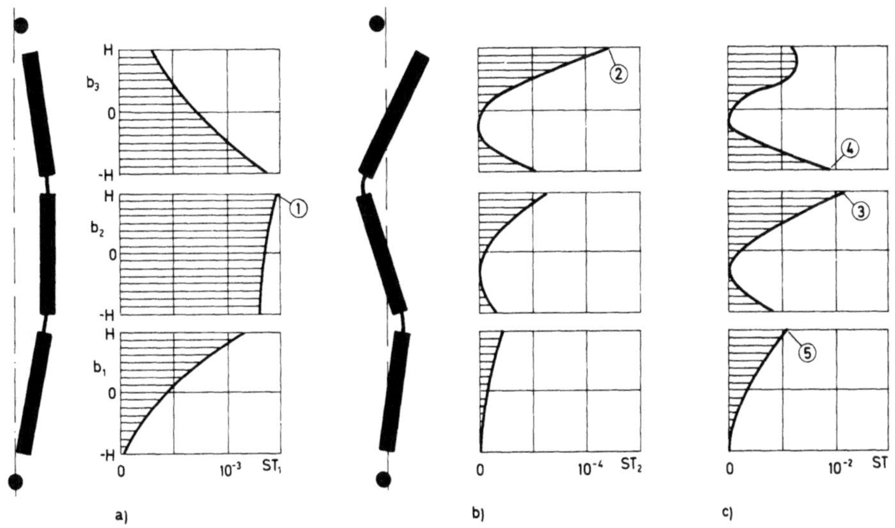

Fig. 2. Measures of stabilizability for the unstable modes of a three-stage rotor plotted as functions of the damper position. a) unstable fundamental mode ($\omega_1 = 19{,}3s^{-1}$) with measure ST_1 ; b) unstable high frequency mode ($\omega_2 = 125s^{-1}$) with measure ST_2 ; c) overall measure ST .

be preferred, and for stabilizing both modes simultaneously position ③ is offering a good compromise. If the active bearing cannot be mounted to this position due to constructional or other reasons, Fig. 2 at least points out suboptimal solutions. For the experimental verification position ④ was chosen, because the optimal position ③ was occupied by measuring devices.

5. OBSERVABILITY

It is assumed that the measured signals, represented by the output vector y , are linearly connected by the measuring matrix C to the states x of the rotor motion

$$y = Cx \quad (5.1)$$

The first principal question concerning the number of outputs that are necessary for the construction of a stabilizing controller for a multivariable system cannot yet be answered generally. A sufficient condition [7] is as follows: The number of stabilizable eigenvalues of a real vibration system is at least equal to the number of outputs, if the system is completely controllable and observable and if the control is not limited. For the stabilization of two unstable natural vibrations with two pairs of conjugate eigenvalues four outputs will be sufficient. It appears to be reasonable to choose as outputs the displacement and the velocity of the rotor measured in the 1- and 2-direction in a single plane or in several ones (Fig. 1). The optimal location for these measuring planes for getting a maximum of information about the state of motion will be determined in a similar way as the location of the damper.

The total motion consists of the superposition of all natural modes, and the output y, too, is composed of the natural modes as can be seen by substituting (4.1) into (5.1)

$$y = C\psi r = \sum_j C\Psi_j r_j \quad , \tag{5.2}$$

where Ψ_j is the j-th column vector of ψ and r_j is the j-th modal coordinate. The contribution of the k-th unstable mode to the output y is the larger the higher the amount of $\| C\Psi_k \|$ becomes. According to [6] the measure for the observability of the k-th unstable mode is defined as

$$OB_k = \frac{\Psi_k^* C^T C \Psi_k}{\Psi_k^* \Psi_k} \quad . \tag{5.3}$$

For measuring several unstable modes simultaneously the overall measure

$$OB = \left[\sum_k (OB_k \cdot \omega_k \cdot \exp\delta_k)^{-1} \right]^{-1} \tag{5.4}$$

sums up the weighted individual measures. By maximizing (5.3) or (5.4) the optimal value of c_n (Fig. 1) indicating the position of the measuring plane can be determined.

For the same three stage rotor as in chapter 4 the measures for the observability of the two unstable modes are calculated as functions of the position parameter c_n . The curves obtained are quite similar to the curves of Fig. 2. The vibrations of the experimental model are measured at the positions ③ and ⑤ of Fig. 2. A simple physical explanation of the measures for stabilizability and observability is that they are proportional to the square of the respective modal deflections. The conclusion is nearly trivial, that a natural vibration can be controlled and measured optimally at its antinode. This direct geometric interpretation is of course a consequence of the chosen correlation between state, control and output. If the control and the output are composed of physically different components then Fig. 2 may loose its clearness.

6. CONTROL GAINS

It is reasonable to let the control u linearly depend on the output y

$$u = Vy \,, \tag{6.1}$$

where V is the gain matrix. The advantage of a linear control law is that it usually works satisfactorily and that there are quite a number of methods for analysis and synthesis available. Nonlinear control laws require more specialized investigations.

Combining (2.7), (5.1), (6.1) the rotor motion is described by

$$\dot{x} = (A + BVC)x \,. \tag{6.2}$$

The structure of the matrices B and C has already been

laid down. Now the gain matrix V has to be determined so that the rotor is optimally stable at a certain operational speed. A technically useful measure for the stability reserve is the smallest relative damping

$$\text{RD} = \min_j [-\delta_j/|\lambda_j|] \quad . \tag{6.3}$$

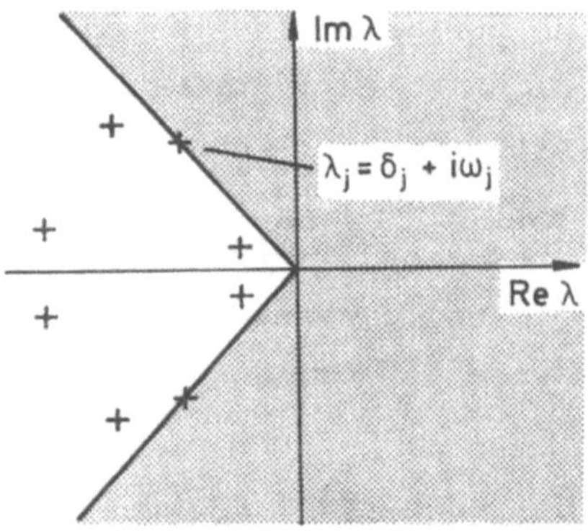

Fig. 3. Definition of relative damping.

According to Fig. 3 RD is positiv if all eigenvalues $\lambda_j = \delta_j \pm i\omega_j$ lie in the left half of the root locus plane. It is a problem of parameter optimization to determine the elements of the gain matrix V so that RD becomes a maximum; i.e. that the shaded area in Fig. 3 grows as large as possible. To find these optimal gains the method of direct search [8] was used. The number of gain elements to be determined is reduced by the fact that due to the rotational symmetry of the rotor and the damper the matrix V has the form

$$V = \begin{bmatrix} V_1 & -V_2 \\ V_2 & V_1 \end{bmatrix} \quad . \tag{6.4}$$

For the three stage rotor, mentioned before, the optimal gain was calculated. From technical reasons the control gains were limited to low values, and therefore it was not possible to stabilize the two unstable natural modes by feeding back only the four signals of a single measuring plane. By measuring the rotor vibration in two planes (Fig. 4), thus "doubling" the information about the state of the

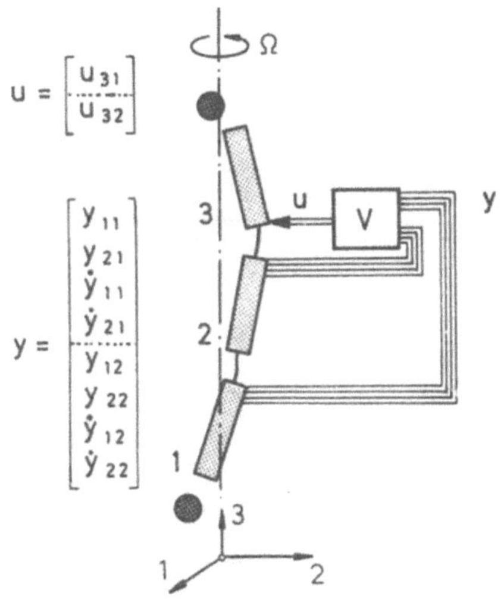

Fig. 4. Active damping of a three-stage rotor using signals from two measuring planes.

rotor, stability was achieved. Only six measuring signals, however, were really needed for the feedback, as the optimal gain matrix V contained several zero elements. The rotor motion, previously being unstable at a rotor velocity $\Omega = 190\,s^{-1}$, could thus be stabilized theoretically up to a velocity of $\Omega = 320\,s^{-1}$.

7. EXPERIMENTS

The vertically suspended rotor is composed of three tubes that are connected by elastic PVC-shafts. The overall length of the rotor is about 0.5 m, the mass of one rigid element 0.3 kg and the ratio of the inertias C/A is about 1:14. The ratio of the damping modulus E^* to Young's modulus E characterizing the viscoelastic shaft material lies between $0{,}8 \cdot 10^{-2}\,s < E^*/E < 2 \cdot 10^{-2}\,s$, and it seems to depend on the rotor velocity or the vibration frequency. This high internal damping causes the rotor to reach the stability limit already at a low spin velocity Ω. For safety reasons it is highly desirable to keep the velocity Ω of the rotor and its kinetic energy at low level, because the rotor often crashes during experiments on the verge of stability. Nevertheless, in a certain sense the rotor can be termed as a high-speed rotor as it runs through five critical speeds, though, before reaching the stability limit.

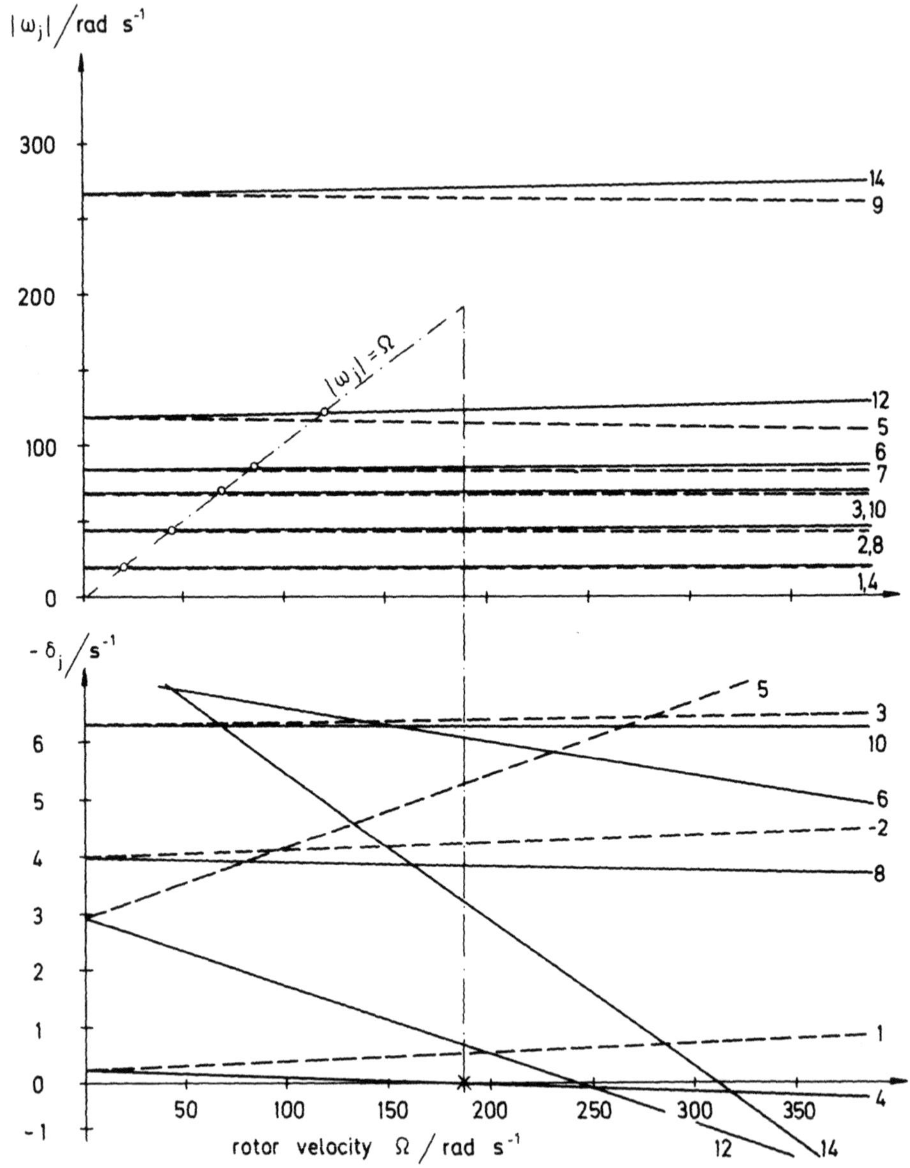

Fig. 5. Detail plot of the eigenvalues $\lambda_j = \delta_j \pm i\omega_j$ as functions of the rotor velocity Ω for a three-stage rotor. ——— eigenvalues of forward whirls; – – – eigenvalues of reverse whirls; O critical speeds.

Fig. 5 shows the calculated natural frequencies and the natural damping of the eigenvalues depending on the rotor velocity Ω . The circles in the frequency plot indicate

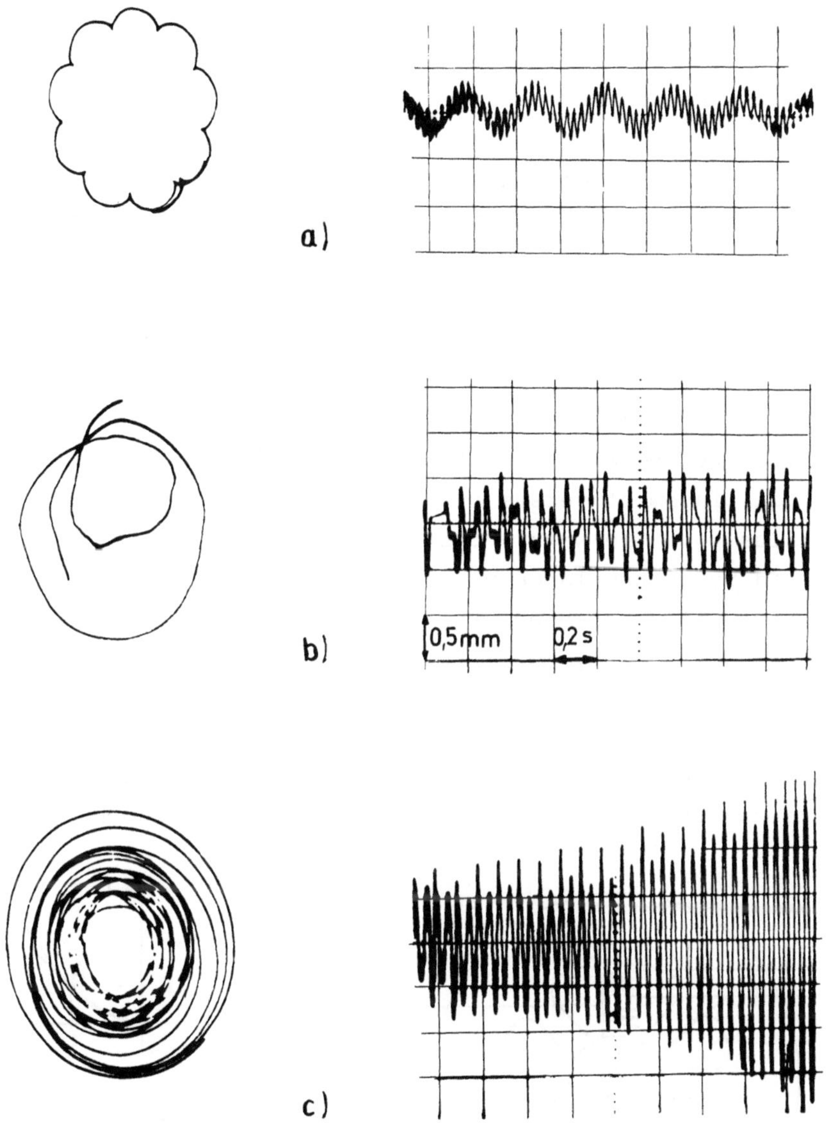

Fig. 6. Measured path of the rotor axis and time response when crossing the stability boundary. a) weakly unstable fundamental mode (ω_1 = 19,8 s^{-1} , Ω = 200 s^{-1}) ; b) onset of a higher frequency unstable mode (ω_2 = 125 s^{-1} , Ω = 220 s^{-1}) ; c) strongly unstable mode (ω_2 = 125 s^{-1} , Ω = 230 s^{-1}) .

where the rotor velocity coincides with the frequency of a forward natural whirl. Only the forward whirls are excited by the residual unbalance of the rotor and lead to resonance.

But not only these critical speeds obtainable from the frequency plot have to be taken into account. The damping of some of the natural vibrations decreases with increasing rotor velocity until the stability limit is reached at $\Omega = 190\,s^{-1}$. Here the fundamental forward whirl ($\omega_1 = 19{,}3\,s^{-1}$) becomes unstable but as the slope of its damping curve is very slight it is only a "weak" destabilization. By increasing the rotor velocity Ω further another natural vibration with a higher natural frequency ($\omega_2 = 125\,s^{-1}$) becomes "strongly" unstable. This principal behaviour at the stability limit could be verified experimentally. Fig. 6 shows measured polar plots of the rotor axis and the vibrational displacement versus time for three different rotor velocities Ω. The weakly unstable fundamental mode, modulated by an unbalance vibration with the rotor frequency Ω, is superseded step by step by a strongly unstable high-frequency mode. In order to facilitate a stable rotor motion at the desired operational speed $\Omega = 300\,s^{-1}$ an active damper was applied.

The electromagnetic damper was developed as a constructional unit, following ideas of KLIMEK [9] who had used

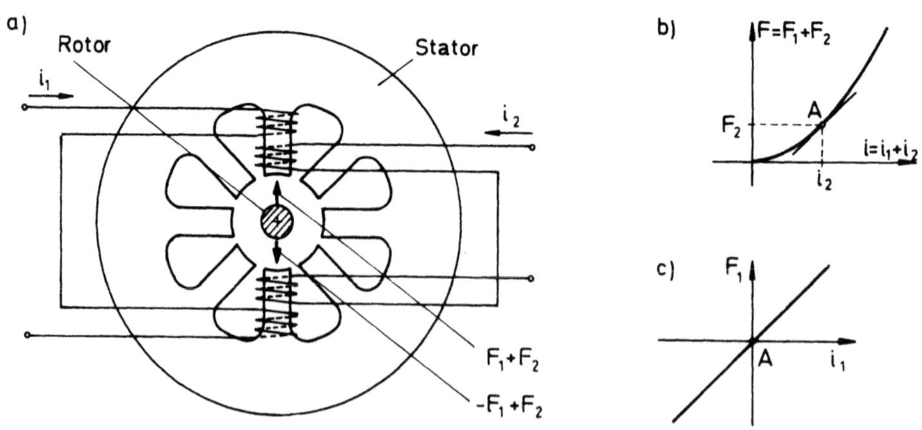

Fig. 7. Circuit diagram for electromagnetic damper. a) differential wiring; b) shifting of stationary operational point A by premagnetizing; c) control force F_1 as function of control current i_1.

the electromagnetic principle to suspend a measuring component for an inertial navigation system. Fig. 7 shows the circuit diagram for the generation of a control force F_1 in one direction. By premagnetizing with a constant current i_2, thus shifting the operational point to A, and by a differential wiring the magnetic force acting on the rotor depends linearly on the control current i_1. For a gap between the rotor and the coil of 1 mm and a constant current $i_2 = 0{,}8A$, a "stiffness" of $F_1/i_1 =$ 18 N/A and a maximal control force of 14 N is obtained. As the damper unit is able to generate torques, too, about the driving axis, it appears to be a promising future task to incorporate the functions of driving the rotor and damping its vibrations into one unit.

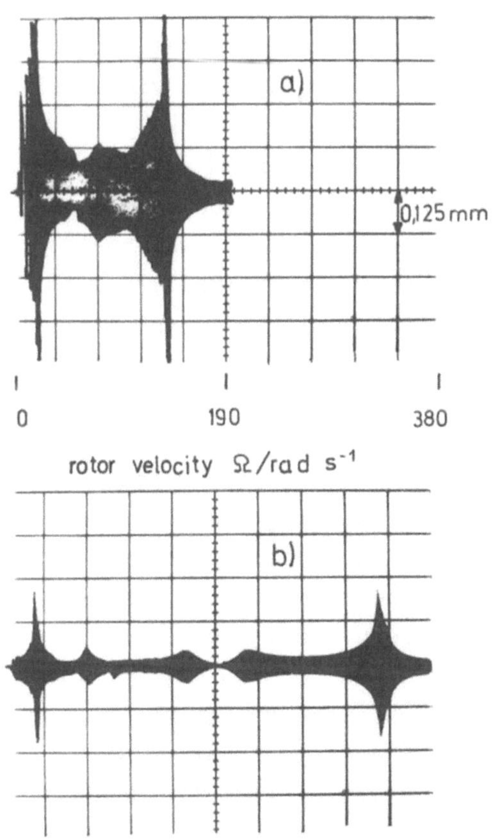

Fig. 8. Measured resonance curves of the rotor running down. a) without active damper; b) with active damper.

The active damper is mounted according to Fig. 4. With its help it is possible to pass through still another resonance at the critical speed $\Omega = 272\ s^{-1}$ (compare Fig. 5) and to shift the stability boundary to an even higher rotor velocity than the calculated one. In fact the rotation is stable within the total realizable velocity range $\Omega < 450\ s^{-1}$. An additional favourable performance of the active damper is that it reduces the resonance peaks considerably when running through critical speeds, even though the damper had not been specially designed to do that. Fig. 8 shows the running down of the uncontrolled rotor from the stability boundary ($\Omega = 190\,s^{-1}$) and the running down of the actively damped rotor from $\Omega = 380\,s^{-1}$. The damping effect is obvious.

8. CONCLUSION

It has been shown how internal damping can create self-excited vibrations of an multi-body rotor. Analytical stability conditions have been derived and the influence of the rotor velocity and the external and internal damping on the frequency and the stability of the natural vibration have been evaluated.

In order to stabilize the rotor motion an active damper was developed. Control forces are exerted on the rotor that depend on the rotor motion in such a way that the unstable, self-excited vibrations are optimally damped. Suggestions are made for the optimal position of the damper along the rotor axis as well as for the optimal choice of measuring signals that are used to create the control signals. Experiments with a three stage rotor verify the stabilizing effect of an active electromagnetic damper.

The principle of actively influencing the rotor motion is not restricted to stabilizing or damping of self-excited vibrations. It appears to be promising to develop active bearings that allow the smooth running of an unbalanced rotor. A supplementary active bearing may contribute to reducing resonance peaks when the rotor runs through critical speeds. An advantage of the active bearing with respect to the passive one is that its effectivity depends less on its mounting position. The consequences of an unfavourable arrangement of the active bearing can partially be compensated by the choice of suitable measuring signals and control gains.

REFERENCES

[1] TONDL, A.: "Some Problems of Rotor Dynamics", Chapman G. Hall, London, 1968.

[2] SCHWEITZER, G., SCHIEHLEN, W., MÜLLER, P.C., u.a.: "Kreiselverhalten eines elastisch gelagerten Rotors", Ing.-Arch., 41 (1972), S. 110-140.

[3] MAGNUS, K.: "Kreisel, Theorie und Anwendungen", Springer-Verlag, Berlin-Heidelberg-New York, 1971.

[4] PEDERSEN, P.T.: "On Forward and Backward Precession of Rotors", Ing.-Arch., 42 (1973), S. 26-41.

[5] KALMAN, R.E., HO, Y.C. and NARENDRA, K.S.: "Controllability of Linear Dynamical Systems", In Contrib. to Differential Equ., Vol. 1, Interscience, New York, 1963.

[6] MÜLLER, P.C. und LÜCKEL, J.: "Modale Masse für Steuerbarkeit, Beobachtbarkeit und Störbarkeit dynamischer Systeme", Z. Angew. Math. Mech., 54 (1974), Heft 4, T 57.

[7] DAVISON, E.J. and CHOW, S.G.: "An Algorithm for the Assignment of Closed-Loop Poles Using Output Feedback in Large Linear Multivariable Systems", IEEE Trans. on AC, Febr. 1973, pp 74-75.

[8] DRENIGK, R.F.: "Die Optimierung linearer Regelsysteme", Verlag Oldenbourg, München und Wien, 1967.

[9] KLIMEK, W.: "Ein Beitrag zur messtechnischen Anwendung der aktiven elektromagnetischen Lagerung", Deutsche Luft- und Raumfahrt. Forschungsbericht 72-30, 1972.

[10] SCHWEITZER, G.: "Stabilisierung eines schnelldrehenden Rotors". Habilitationsschrift, Techn. Univ. München, 1974.

New Field Balancing Method on Tandem Connected Multispan Flexible Rotor System

K. Shiraki and H. Kanki

Takasago Technical Institute

Mitsubishi Heavy Industries, LTD, Japan

ABSTRACT - Turbomachinery has recently been increasing in both capacity and size, especially in turbine generator systems. This results in many problems in balancing the rotors.

To solve these problems, the vibration characteristics of multi-span rotor system due to mass unbalance and initial bend according to geometrical inaccuracy of coupling have been investigated and a new field balancing procedure has been developed.

The method is based on the modal balancing principle, and uses the practical modal response circle analysis with the aid of a new vibration analyser system and a portable computer.

That is,

1. To find out the approximate axial and radial location of unbalance from the modal response circles plotted for all journal and all speed range during deceleration.
2. To choose the correction weights as effectively as possible to reduce the vibration of the unit to the excellent level for all speed range.

This method was applied for the field balancing of a 600 MW turbine generator unit and brought about excellent results on reducing the vibration of the unit and saving time.

INTRODUCTION

Turbomachinery has recently increased in capacity and size, especially in turbogenerator systems, (Fig. 1). According to this trend, not only the span of each rotor but also the number of spans increases. As a result, vibration characteristics are becoming more complex, and difficult problems arise in balancing the rotors.

To prevent multispan rotor from vibration troubles, it is most important that each single rotor be balanced accurately and independently. However, the multispan rotor has

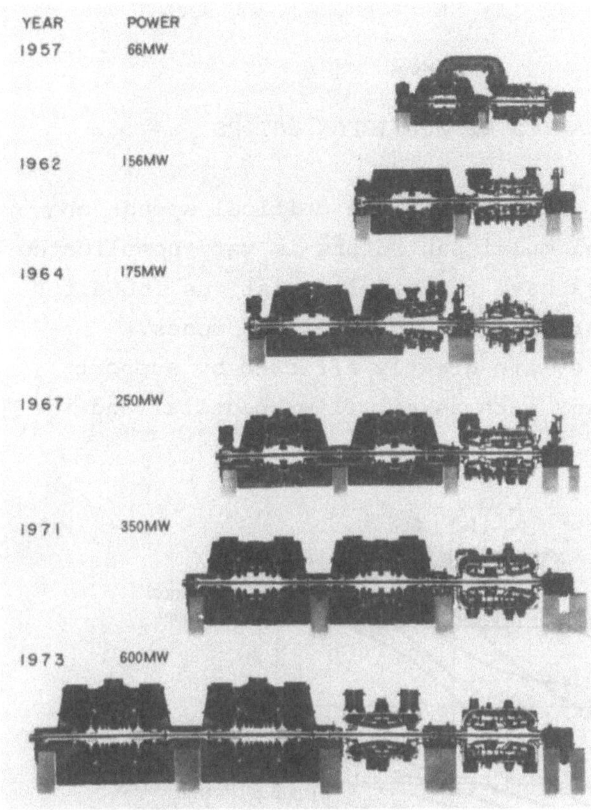

Fig. 1. Development of Steam Turbines

many causes of vibration which are not solved by the single rotor balancing.

For example, there is the effect of coupling inaccuracy and different overhang condition between single span and multispan.

To solve these problems, the vibration characteristics of multispan rotor system due to mass unbalance and coupling inaccuracy were investigated and the new field balancing method was developed.

The method is based on the modal balancing principle, and uses the practical modal response circle analysis with the aid of a new vibration analyser system and a portable computer. That is, to find out the approximate axial and radial location of unbalance from the modal circles plotted for all speeds during deceleration, and to choose the correction weights as effectively as possible to reduce the vibration of the unit to the optimum level for all speed ranges.

This method has been applied to the field balancing of 600 MW turbogenerator units and has resulted in a highly satisfactory reduction of the vibration of the unit and savings in time.

2. DYNAMICS OF MULTISPAN ROTORS

2.1. Critical Speeds. Determining the critical speeds and their modes for actual multispan rotors is very complicated and they do not always have the simple relations found for each single span rotor's critical speeds and modes.

In addition, they are greatly affected by support dynamic characteristics such as oil film, pedestal and foundations.

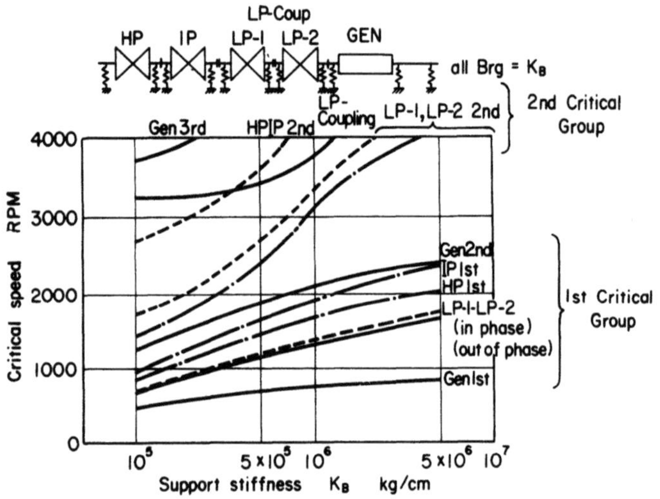

Fig. 2. Example of Calculated Critical Speeds.

Fig. 2 shows an example of calculated critical speeds versus support stiffness (for the sake of simplicity, it is assumed that the support stiffnesses are all K_B and gyroscopic moment is neglected).

The results of the analysis show the following characteristics.

1st Critical Speeds Group

- o Mode shapes are simple and relatively independent of individual span.
- o They correspond to the mode shapes of each single span rotor.
- o They are not seriously affected by support stiffness as compared to 2nd Critical Speeds.

2nd Critical Speeds Group

- o Mode shapes are very complicated.
- o They do not always correspond to the mode shapes of each single span rotor.
- o They are greatly affected by support stiffness.

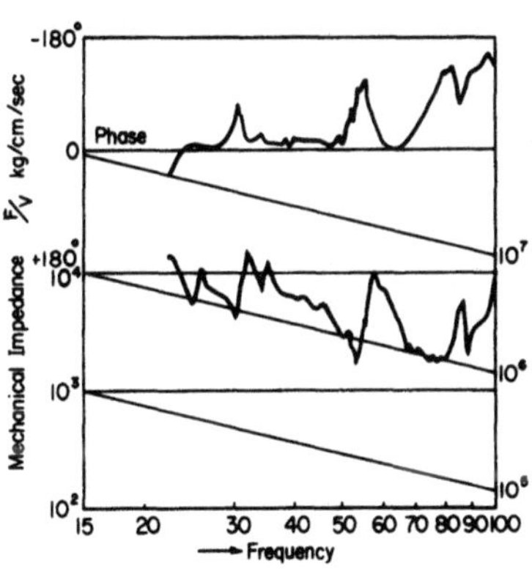

Fig. 3. Example of Measured Mechanical Impedance.

Therefore, to improve the accuracy of calculation of critical speeds, an investigation of support stiffness is essential.

The support consists of oil film, pedestal and foundation.

The oil film properties are described in the literature and many papers [1][2].

On the other hand, the dynamic properties of pedestals containing the foundation are not well known, and it is very difficult to establish the general formulation.

Therefore, the mechanical impedance of actual pedestal in the field condition was first measured and analysed.

Then the results were applied for the calculation of critical speeds and dynamic response.

The test condition is shown in Fig. 4 and the example of the measured and analysed data is shown in Fig. 3.

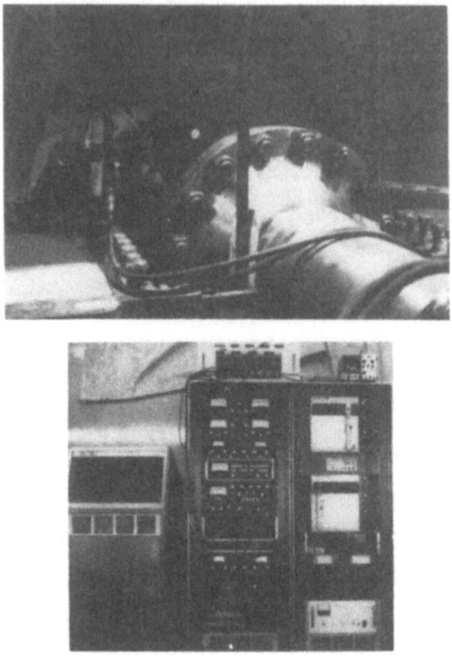

Fig. 4. Measurement of Pedestal Mechanical Impedance in the Field.

<u>2.2. Unbalance Response of Multispan Rotor.</u> Usually, large multispan rotors are coupled together with rigid couplings. The numerical calculation of unbalance response of a multispan rotor is usually carried out as described by many authors [3].

Two calculations of unbalance response are carried out.

The calculation models are shown in Fig. 5.

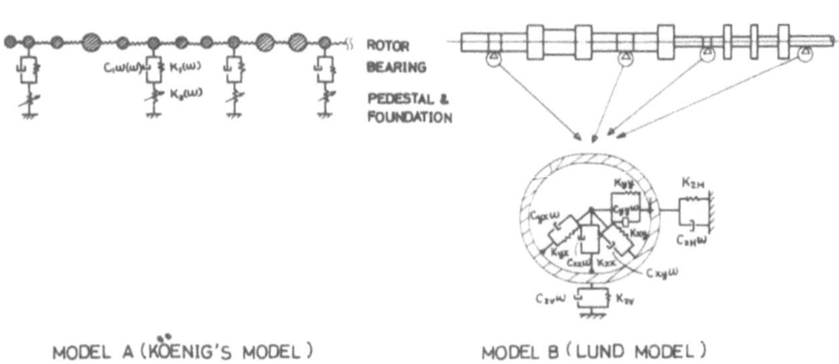

Fig. 5. Calculation Model of Multispan Rotor System.

	CASE A König Model	CASE B Lund Model
Shaft	Myklestad model (discrete model)	Prohl model (continuous model)
Oil film Data by	Symmetric non-coupled Hagg & Sankey Chart 2 coefficients (each direction)	Asymmetric Coupled Lund and Others 8 coefficients
Pedestal	Measured Data	Measured Data One degree of Freedom for each direction

Case A has the advantages of time-saving and easily comprehensible results.

Case B gives an accurate simulation of actual industrial rotor systems.

2.3. Response due to Coupling Inaccuracy. On the response and balancing of initially bent rotor, Bishop gave the modal analysis of the problem [4]. Actually, in the multispan rotor connected by rigid coupling, the problem of initial bend is very important because of unavoidable tolerance of coupling machining and shrink fit.

Thus, it is necessary to investigate the response of multispan rotor with coupling inaccuracy.

	Mass Unbalance	Inclined Coupling	Eccentric Coupling
Model	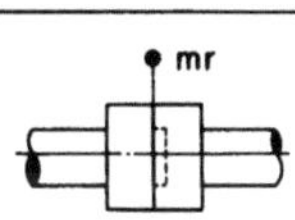	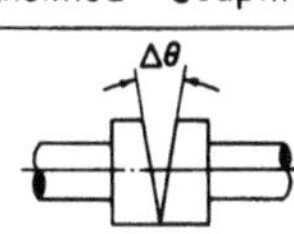	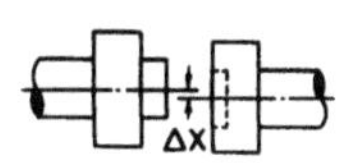
Rotor Mode at $\omega = 0$ / $\omega = \infty$	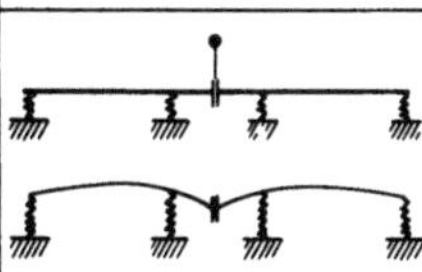	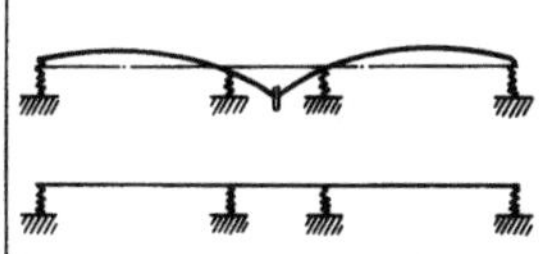	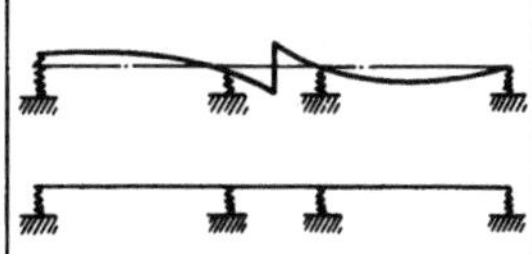
Iteration Equation by Myklestad Holtzer Method	$\theta_{n+1} = \frac{\ell^2}{2EI}\theta_n + \frac{\ell}{EI}M_n + \frac{\ell^2}{2EI}S_n$ $X_{n+1} = X_n + \ell\theta_n + \frac{\ell^2}{2EI}M_n + \frac{\ell^2}{2EI}S_n$ $M_{n+1} = M_n + \ell S_n$ $S_{n+1} = S_n + \frac{W_n}{g}X_n\omega^2 + \underset{\text{Unbalance Force}}{mr\omega^2}$	$\theta_{n+1} = \frac{\ell^2}{2EI}\theta_n + \frac{\ell}{EI}M_n + \frac{\ell^2}{2EI}S_n + \underset{\text{Coupling Inclination}}{\Delta\theta}$ $X_{n+1} = X_n + \ell\theta_n + \frac{\ell^2}{2EI}M_n + \frac{\ell^2}{2EI}S_n$ $M_{n+1} = M_n + \ell S_n$ $S_{n+1} = S_n + \frac{W_n}{g}X_n\omega^2$	$\theta_{n+1} = \frac{\ell^2}{2EI}\theta_n + \frac{\ell}{EI}M_n + \frac{\ell^2}{2EI}S_n$ $X_{n+1} = X_n + \ell\theta_n + \frac{\ell^2}{2EI}M_n + \frac{\ell^2}{2EI}S_n + \underset{\text{Coupling Eccentricity}}{\Delta X}$ $M_{n+1} = M_n + \ell S_n$ $S_{n+1} = S_n + \frac{W_n}{g}X_n\omega^2$
Characteristics	$\omega = 0$ $F = 0$ $X = 0$ $\omega = \infty$ $F = 0$ $X = 0$ Large Effect at Only Coupling Span Critical Speed	$\omega = 0$ $F \neq 0$ $X \neq 0$ $\omega = \infty$ $F \neq 0$ $X \neq 0$ Large Effect at Low Speed Critical Speed of Main Rotor	

Table 1. Unbalance Response and Response due to Coupling Inaccuracy.

The computer program is modified as shown in Table 1, to estimate the response due to any coupling inaccuracy.

Of course, this program can be used for the calculation of the response due to general initial bend. The program has been checked by means of the model test shown in Fig. 6.

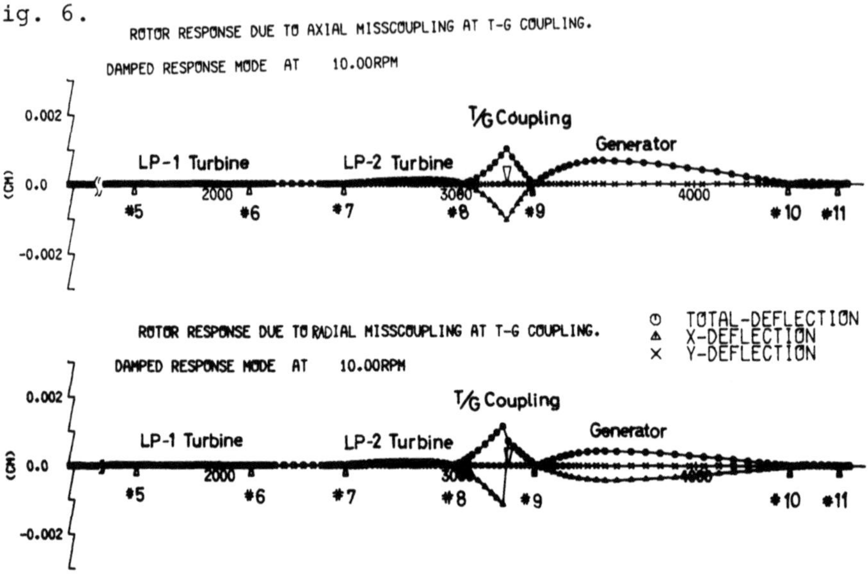

Fig. 6. Initial Bend due to T/G Coupling Inaccuracy.

Fig. 7 shows the calculated example for the large turboset with initial bend at low speed for the two cases inclination and eccentricity of coupling, respectively.

The calculated results show that the response of generator rotor at first critical speed is most marked. Therefore, the coupling of large flexible rotor such as a generator rotor must be manufactured under possible minimum tolerances.

Further, this study confirms that the vibration of a multispan rotor is greatly affected by both coupling accuracy and the quality of balancing in single span.

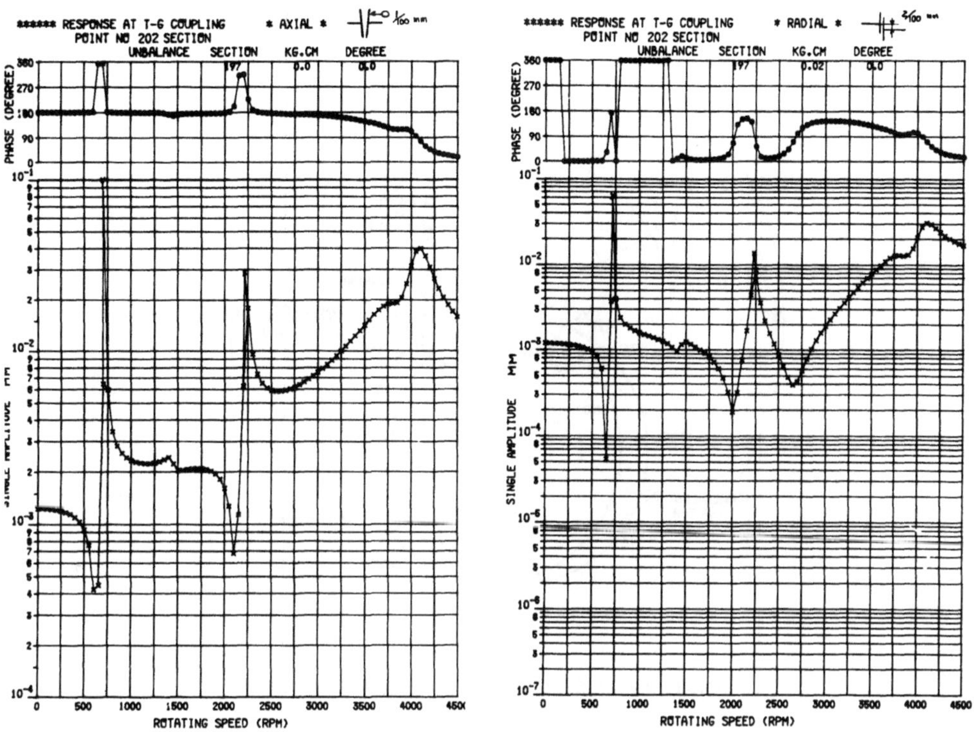

Fig. 7. The Example of Calculated Response.

3. MODAL RESPONSE CIRCLE

Modal response circle is defined as a polar plot of synchronous vibration. It is advantageous to describe the vibration near its resonance [6].

Bishop and Parkinson proposed the application of modal response circle (Kennedy & Pancu Method) to the balancing of flexible rotor about ten years ago. However, it may have been difficult to plot the modal response circle in that year.

Since then, the development of electronics has made it easy to plot modal response circle and this technique has been used in various vibration tests.

Then, the use of the modal response circle was tried

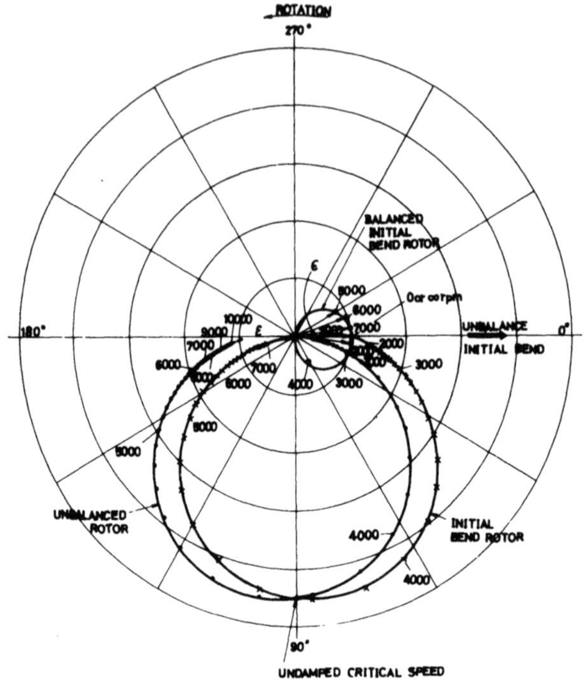

MODAL RESPONSE CIRCLE OF UNBALANCED AND INITIAL BEND ROTOR

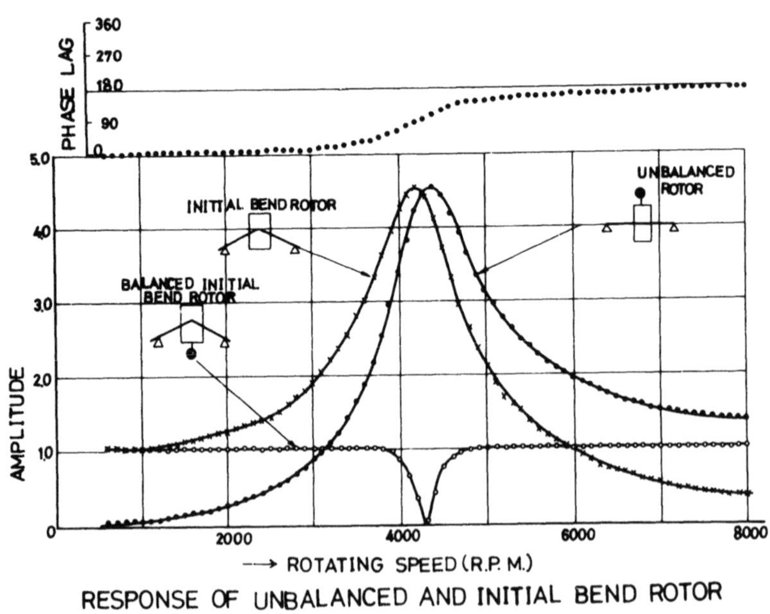

RESPONSE OF UNBALANCED AND INITIAL BEND ROTOR

Fig. 8. Response of One Degree of Freedom System.

in the analysis of the unbalance response of rotor and balancing of flexible rotor.

The important characteristics are interpreted in this section.

3.1. Modal Response Circle of One Degree of Freedom. Fig. 8(a) and 8(b) show the response of a simple rotor with one degree of freedom.

- · shows the response to unbalance,
- x shows the response to initial bend,
- o shows the residual response of initially bent rotor after modal balancing.

The figure shows the following characteristics of the response.

Case / N rotating speed	o	Critical Speed (max. response)	∞
Unbalance	o	$\varepsilon/2\zeta\sqrt{1-\zeta^2}/\phi > 90^o$	$\varepsilon/180^o$
Initial Bent	$\varepsilon/0^o$	$\varepsilon/2\zeta\sqrt{1-\zeta^2}/\phi < 90^o$	0°
Initial Bent -Unbalance	$\varepsilon/0^o$	$0/0^o$	$\varepsilon/0^o$

Table 2.

where
- ε : eccentricity
- ζ : damping ratio
- phase : phase lag

3.2. Symmetric Model. Fig. 9 shows the modal response circles of symmetric single rotor. The unbalance relates to 1/4 of span near #1 Brg. (bearing).

The figure shows that the modal response circles of 1st and 2nd critical speeds become concentric at unbalance side measuring point (exciting point response), otherwise the #2 Brg. (transfer response) shows the antiphase for lst and 2nd critical speed.

This is understood by modal analysis for two plane balance.

lst modal unbalance	↑ 0.5	↑ 0.5
second modal unbalance	↑ 0.5	↓ -0.5
the sum of modal unbalance	↑ 1.0	

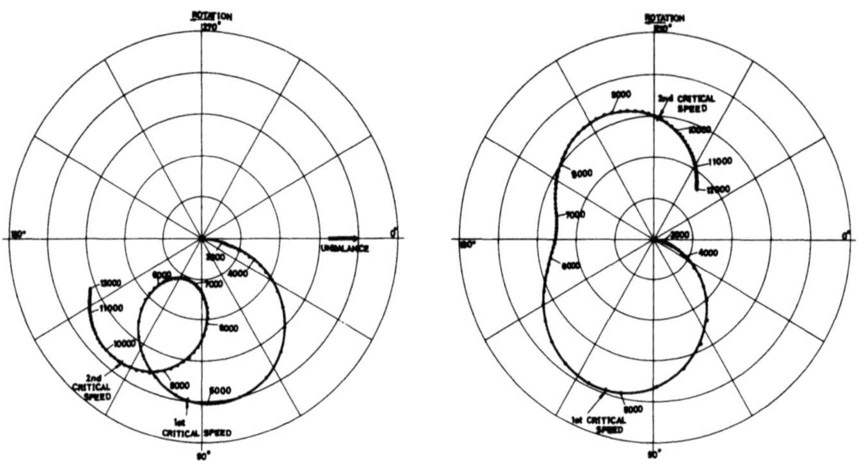

RESPONSE OF #1 JOURNAL POINT (UNBALANCE SIDE)

RESPONSE OF #2 JOURNAL POINT

Fig. 9. Response of Symmetric Single Span Rotor.

This suggests that if a proper balancing plane is selected, the modal balancing weight can usually be used for lst and 2nd critical speed (it means that the equivalent unbalance is approximately located at that point).

<u>3.3. The Response of Single Span Rotor with Overhang</u>. Fig. 10 shows the response of single span rotor with overhang for 3 kinds of unbalance location.

It is a most difficult problem to distinguish between unbalance in span and unbalance in overhang part. However, the figure shows that the unbalance located in overhang part can be easily found out by characteristics of modal response circles.

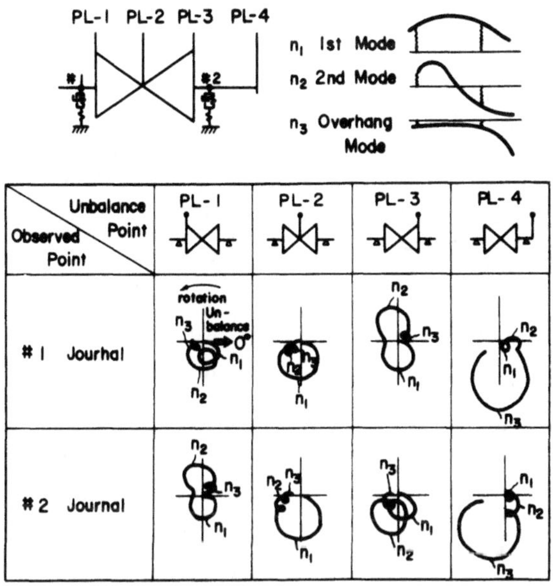

Fig. 10. Modal Response Circles of Single Rotor with Overhang.

3.4. Typical Modal Response Circles of Multispan Rotor. Fig. 11 shows the typical modal response circles of a multispan rotor consisting of turbine and generator.

The figure is composed of the above three cases of modal response circles and shows the method of finding the approximate unbalance location, or approximate common modal correction weights.

3.5. The Modal Response Circles of LP Turbine Rotors Supported by Asymmetric Brg. & Pedestal. The examples, Fig. 8-Fig. 11, show the simple result, neglecting the asymmetry and combined effect of oil film and the pedestal asymmetry.

However, a real rotor supported by the bearings having such effects, deviates from the above example. Fig. 12 shows the calculated results of modal response circle for the vertical direction, considering asymmetry and the combined effect of oil film and asymmetry of dynamic characteristics of pedestal.

Of course, they agree with the measured modal response circles of an actual turboset.

The examples show the slight difference from the simplified examples.

In practice, the effects are checked by trial weight.

The modal response circles of an asymmetric shaft are also studied, and results are described in Appendix 1.

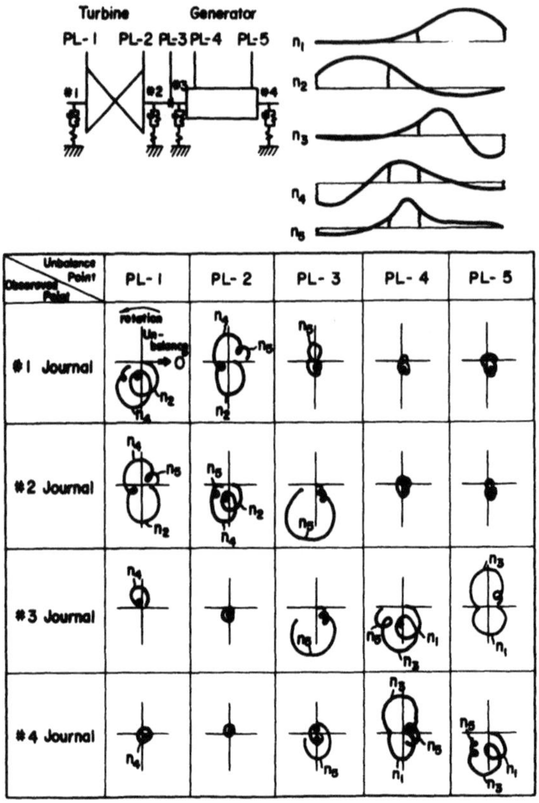

Fig. 11. Typical Modal Response Circles of Multispan Rotor.

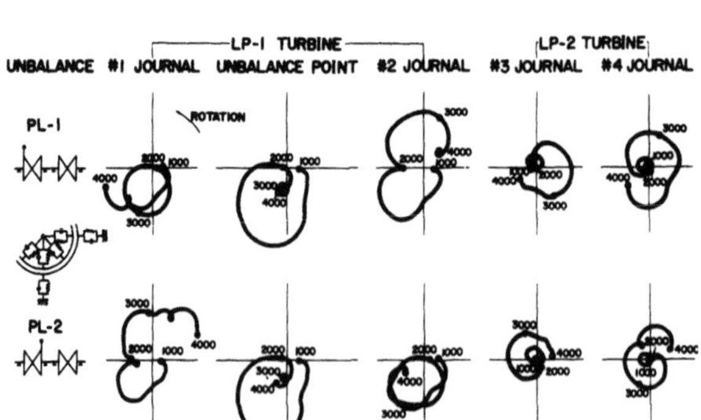

Fig. 12. Influence of Bearing Asymmetry and Couple Terms. Calculated Modal Response Circles.

4. FIELD BALANCING OF MULTISPAN ROTORS

<u>4.1. General Remarks on Field Balancing.</u> Field balancing is defined as the balancing of the rotor supported by its own bearings and casing.

The objects and necessity of field balancing are as follows.

(1) To remove the residual unbalance remaining due to unavoidable difference in vibration mode caused by different overhang conditions and support conditions.

(2) To reduce the vibration due to initial bend caused by the unavoidable coupling geometrical inaccuracy mentioned in Chap. 2.

(3) To balance the permanent thermal unbalance occuring after application of the

load, for example, relaxation of residual internal stress of rotor material and shrink fit of the coupling and the blade disc.

The field balancing of single span rotor is similar to high speed balancing and it is relatively easy.

On the other hand, the field balancing of tandem-connected multispan rotor is very difficult.

The causes of the difficulty are as follows.

(1) Complex vibration characteristics; as shown in Chap. 2, the vibration characteristics of a multispan rotor are very complex i.e. it has many critical speeds and each mode shape is very complicated.
(2) Limited time; it is desirable to save time spent on balancing in the field for reasons of economy and to reduce the risk to operate at high vibration amplitudes.
(3) Limited correction planes; sufficient numbers and locations of correction planes are not usually available in design limitation and accessibility in the field.
(4) Limited vibration measuring points; in the actual machine only shaft vibrations near its bearings are measured.

Form the above-mentioned points, it can be seen what is important in the field balancing of multispan rotors:
(1) Accurate measuring of vibration.
(2) Accurate interpretation of measured vibration.
    For example
    1. normal mass unbalance response,
    2. initial bend response,
    3. thermal unbalance response,
    4. unstable vibration.
(3) Adequate balancing procedure to reduce the vibrations to optimum level as soon as possible (usually for all speeds and for all conditions).

4.2. Problems relating to Influence Coefficient Method and Modal Balancing Method for the Field Balancing of Large Multispan Rotor.

(1) Influence coefficient method.

As described by many authors [7][8][9][10], the influence coefficient method is normally used in field balancing.

The conventional influence coefficient method is one-plane, one-speed balancing. Without a good understanding of vibration and plenty of experience this method is of limited value for the multispan rotor.

Recently, many authors, [7][9][10], reported the computerization of the generalized influence coefficient method and its efficiency.

The method gives good results for simple rotors if there are enough correction planes and a sufficient number of vibration measuring points.

However, without some special device, the method may not be adequate for the field balancing of multispan rotors such as large turboset, because of the many limitations in the field balancing as described in the above section.

The biggest limitation on this method is the accuracy of the influence coefficient. The vibration of actual rotor varies according to the operating condition. In this case, the exact influence coefficients are not usually obtained.

(2) Modal balancing.

Modal balancing can be successfully applied for many rotors.

The balancing procedure is reasonably accurate for use on the high speed balancing machine or single rotor in the field.

Modal balancing is also suitable for field balancing because of many advantages (for example, clear physical understanding of each balancing run). However, for a multispan rotor it is very difficult to apply the method precisely. The reasons are as follows.

1. The multispan rotor has many critical speeds and complex mode shapes. They cannot be exactly assumed easily.
2. So it is very difficult to find the modal correction weights to correct only the mode in question.
3. It is difficult to correct the higher mode above the rated speed.
4. It is necessary to measure the vibration at the critical speed exactly and without dangerous conditions.

4.3. Modal Response Circle Balancing Method. According to the above discussion, we tried to find a field balancing method that has the advantages of both the influence coefficient method and the modal balancing method.

The method is named the modal response circle balancing method in this paper.

The basic points of the method are as follows.

1. Modal analysis can be applied approximately in actual multispan rotor i.e. the response of any point of rotor can be expressed by the sum of each modal response.

$$y(x,\Omega) = \sum_{r=1}^{\infty} \frac{\bar{U}(x)\phi_r(x)\Omega^2 + \bar{V}(x)\phi_r(x)\omega_r^2}{(\omega_r^2 - \Omega^2 + 2i\zeta_r\omega_r\Omega)M_r} \cdot \phi_r(x)$$

where $\bar{U}(x)$: distributed mass unbalance

M_r : modal mass of r'th mode

$\bar{V}(x)$: distributed initial bent

$\phi_r(x)$: mode shape function of r'th mode

ω_r : critical speed of r'th mode

ζ_r : damping ratio of r'th mode

Ω : rotating speed

2. According to the calculated mode shapes for actual multispan rotors, the response of a multispan rotor can be approximately treated as semi-local. Semi-local means not necessarily one span [7], but some regions to suffering a relatively large effect.
3. Modal response circles are useful for an understanding of the vibration of multispan rotors and for determing the approximate unbalance distribution and common modal correction weight. (Common modal correction weight means that the weight can reduce all the mode which is concerned, and it often coincides with the approximate unbalance distribution).
4. Modal response circles are useful for finding the cause of vibration change and error of influence coefficient. Then the usefulness of influence coefficients is easily noted.
5. Reasonably accurate modal response circles are available in transient condition through critical speeds for repeatable small acceration and deceleration.
6. It is useful to search the correction weights by calculation of residual vibration using modal influence coefficient and assumed correction weight

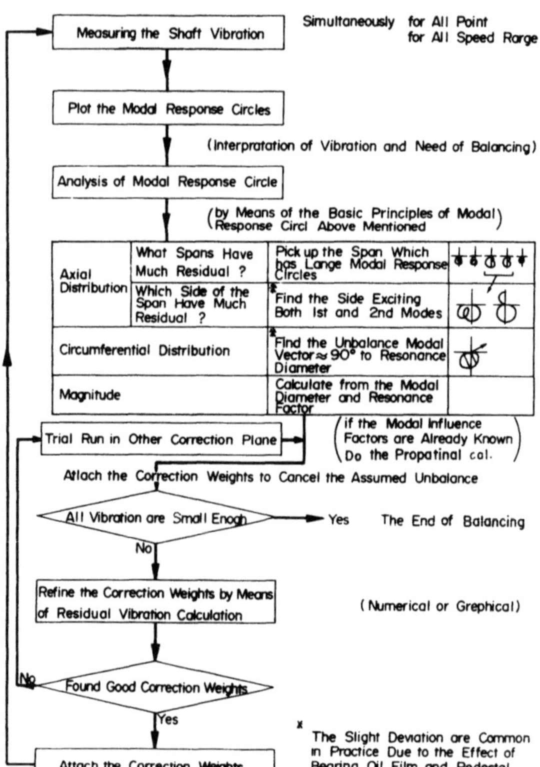

Fig. 13a. Method of modal response circle balancing.

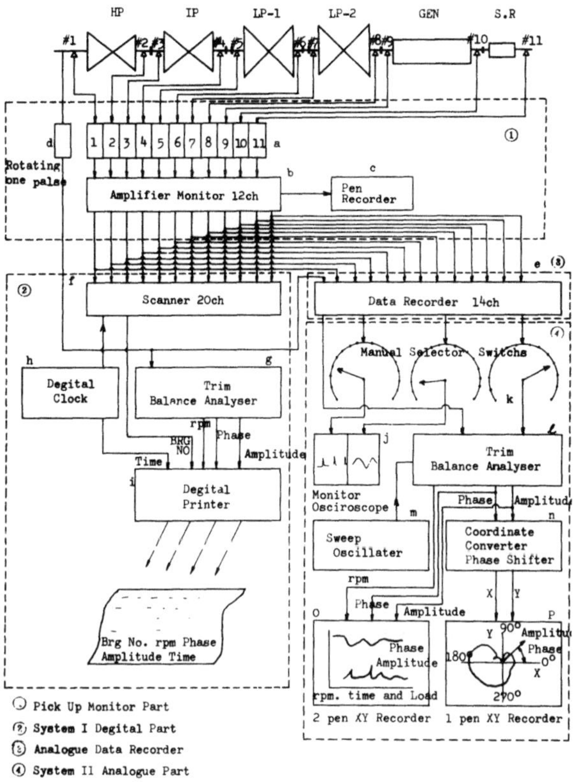

Fig. 13b. Rotor vibration analysing system.

according to modal response circle analysis. This method is not greatly affected by error of influence coefficient as to solve the simultaneous equations [8]. Of course, the calculation of residual vibration is possible semi-locally, as described in item 2.

7. It is essential to add the calculation of residual vibration at rated speed, to cancel the higher mode and to check the error of the correction weights calculation. 2, 6, 7 are explained in Appendix 2.

The practical procedure is described in Fig. 13.

The modal response circles obtained are analysed and the first trial weights are determined. The trial weights are not limited to single weight, combination of weights in suitable planes also being used (for example, static and dynamic coupled weights).

The influence coefficients are obtained from the trial run

$$[\alpha] = \frac{[A]' - [A]^{O}}{U}$$

where

$[\alpha]$: influence coefficients
$[A]^o$: vibration of 1st run
$[A]'$: vibration of trial run

and optimum correction weights in the corresponding planes are calculated by the following condition.

$$[t]_{local} = [A]^o_{local} - [\alpha]_{local}[U]_{local}$$

$[\varepsilon]_{local}$ = local residual vibration
$[A]_{local}$ = local initial vibration
$[\alpha]_{local}$ = local influence coefficients
$[U]_{local}$ = optimum weights in the planes

Determine $[U]$ under the condition

$t_{ij} < A_{ij}$ (for all elements)
$\varepsilon_{ij} < \varepsilon_a$ (as possible)

ε_a : objective residual vibration

The calculation of influence coefficients and residual vibration is easily carried out by portable computer or by graphical methods.

If the sufficient optimum weight cannot be found, it is necessary to consider the modal response circle again and repeat the trial run.

It is also possible to calculate the multiplane correction weights after obtaining the influence coefficients of several correction planes.

4.4 Vibration Measuring and Analysing System. In Fig. 14, an example of vibration measuring system for field balancing of large turboset is shown.

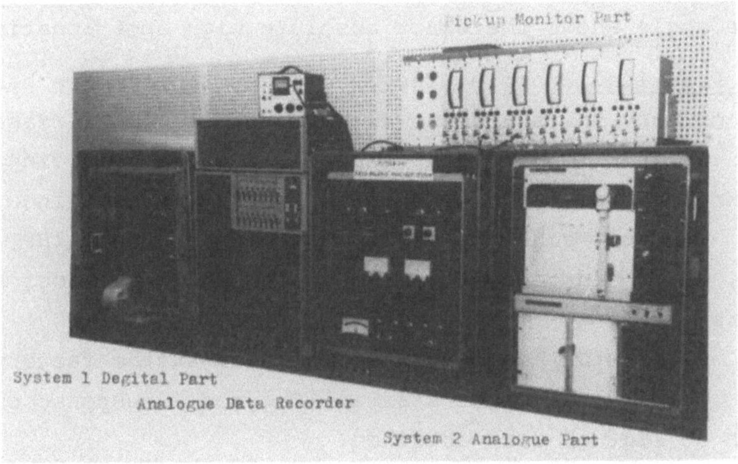

Fig. 14. Rotor Vibration Analysing System.

The functions of the system are as follows.

1. Measurement of the shaft vibration simultaneously for all measuring points and whole speed range.
2. Analysis of the phase and amplitude of the synchronous vibration and plotting the modal response circle, and monitoring and recording the amplitude and phase for all running time.
3. Analysis of the abnormal vibration when this occurs.

This system consists of the following units.

1. Vibration and phase angle reference pick-ups and monitoring unit.
2. Phase amplitude digital recording unit.
3. Analogue data recorder unit.
4. Phase amplitude continuous analysing and plotting unit.
5. Off-line calculator to calculate influence coefficient and residual vibration.

In this system, the special attention is paid to phase angle measurement to give exact phase angle for wide speed range. (For example, low natural frequency and low damping vibration pick up, and high performance tracking filter and data recorder).

The system is fully calibrated, then the measured phase angle shows the exact absolute high spot locations.

4.5. Practical Example. The method described above has been used on many units and showed good results in all cases.

In this section, the example of field balancing on 600 MW 3600 rpm tandem compound turboset is explained.

The unit consists of four turbine elements (HP, IP, LP-1, LP-2) and generator, and whole length is about 47 meters.

Each rotor had been well balanced at the factory.

When the turboset started, the modal response circles shown in Fig. 15 were obtained.

STEP 1.

Step A. Pick up the span and mode required in field balancing; The modal response circles showed the relatively large modal response for

(1) LP Turbine (near 3600 rpm)

(2) HP Turbine (1st critical speed)

(3) LP-1 Turbine (1st critical speed)

First, it was determined to balance the mode (1) to achieve the smooth operation at rated speed.

Step B. To balance the mode (1): the modal response circle was analysed and the unbalance was determined in the connecting shaft, and a trial weight was attached to the center of the connecting shaft and gave the smaller vibration at the same speed range.

Step C. To get the maximum balance effect, the residual vibration calculation using modal influence coefficients was repeated and the optimum weight in the plane was found as shown in Fig. 16; the optimum weight determined was 150 gr/0° and the results of the balance are shown with solid line in Fig. 16.

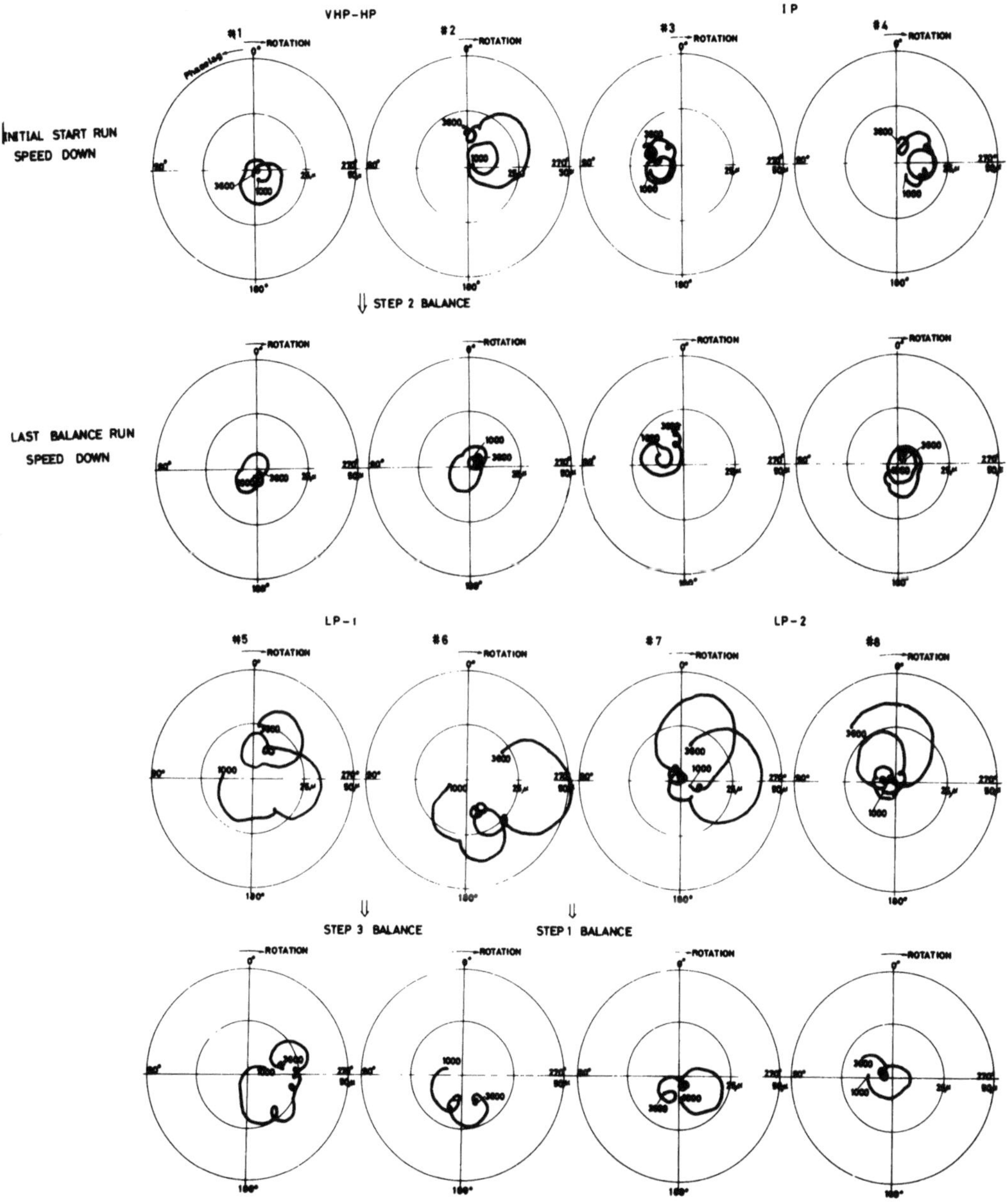

Fig. 15a. Modal Response Circles of the Turboset.

The results show that the vibrations for all speeds were reduced. The second and third steps of fine balancing were done after load condition and over-speed test.

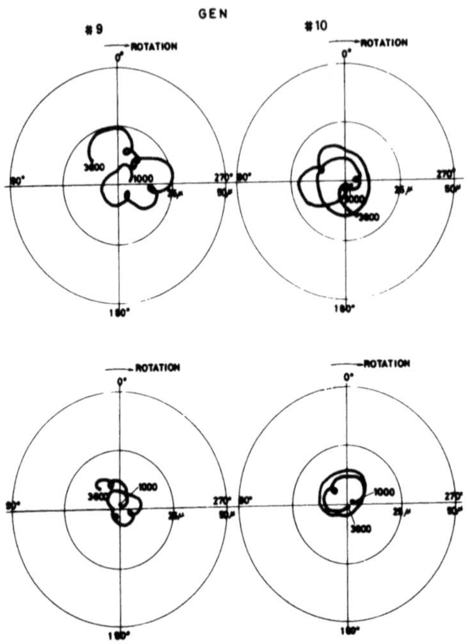

Fig. 15b. Modal Response Circles of the Turboset.

STEP 2. HP Turbine Balance.
Amplitude at the 1st critical speed of HP turbine was reduced by the modal response circle method. Balance plane was selected to center plane of HP rotor and balance weight was determined from the modal response circle. The HP turbine showed the three-dimensional mode at 1st critical speed, because of the difference of pedestal between =1 and =2.
The phase angle of unbalance was determined from both modal response circles of =1 and =2 journals, and the amount of the unbalance was determined from the data of the shop test. The determined balance weight was 208gr/310^{0}.
The effects of the balancing are shown in Fig. 17. As shown in Fig. 17, only one shot reduced the vibration to excellent level.

STEP 3. LP-1 Turbine 1st Critical Speed.
The response of LP-1 turbine on 1st critical speed was balanced by static coupled weights as shown in Fig. 18.

The modal response circle after 3 steps of balancing is shown in Fig. 15(b), and the response curves before and after balancing are shown in Fig. 19.

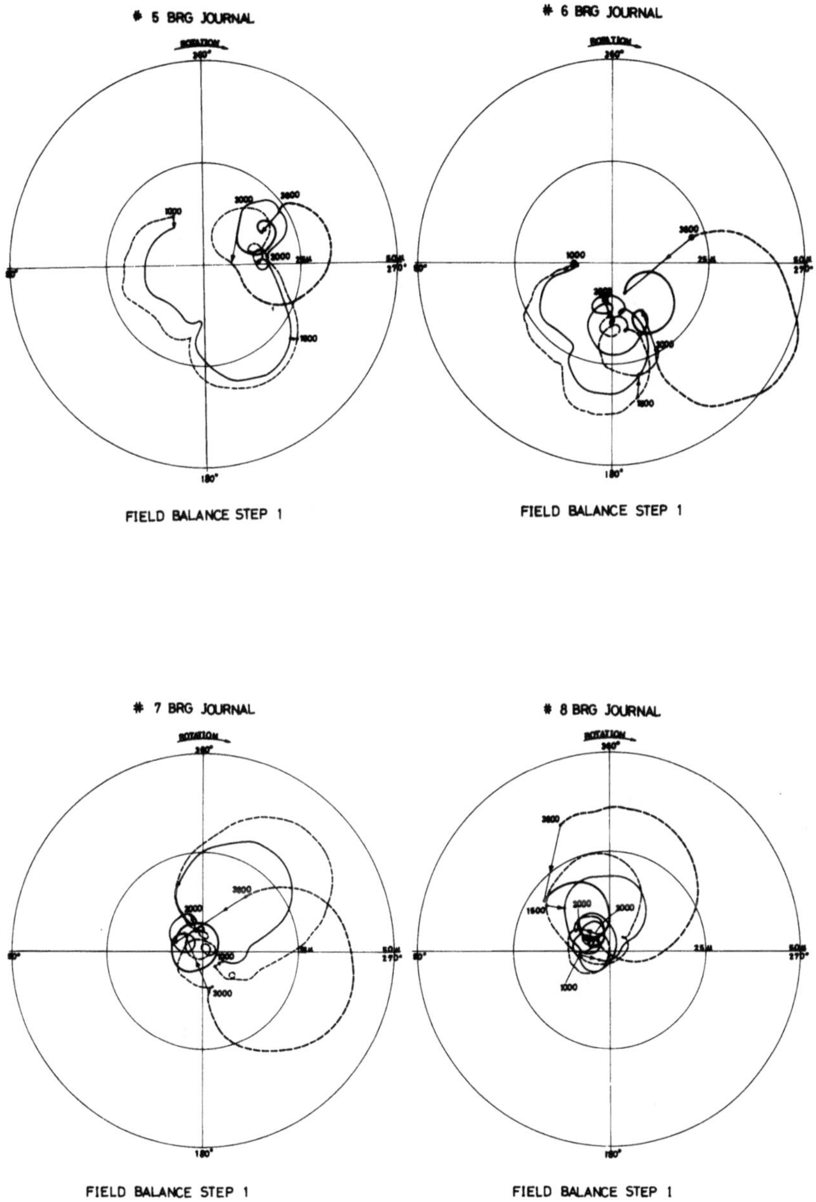

Fig. 16. Field Balancing STEP 1.

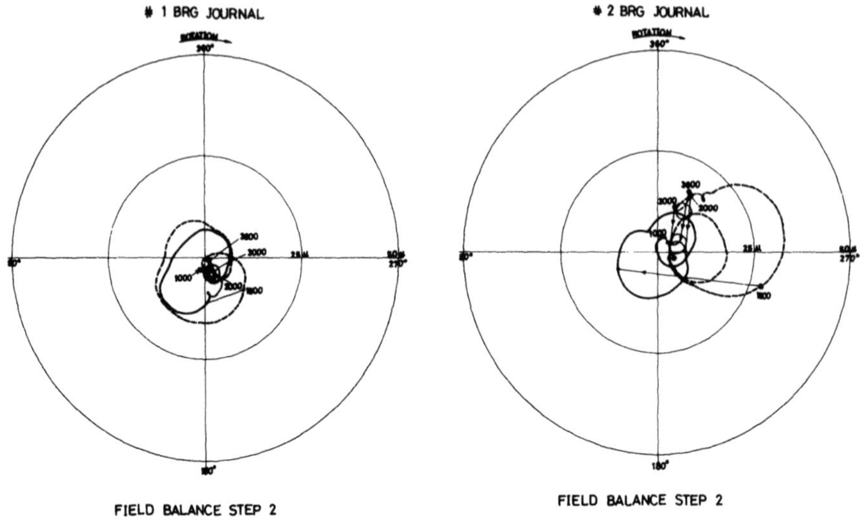

Fig. 17. Field Balancing STEP 2.

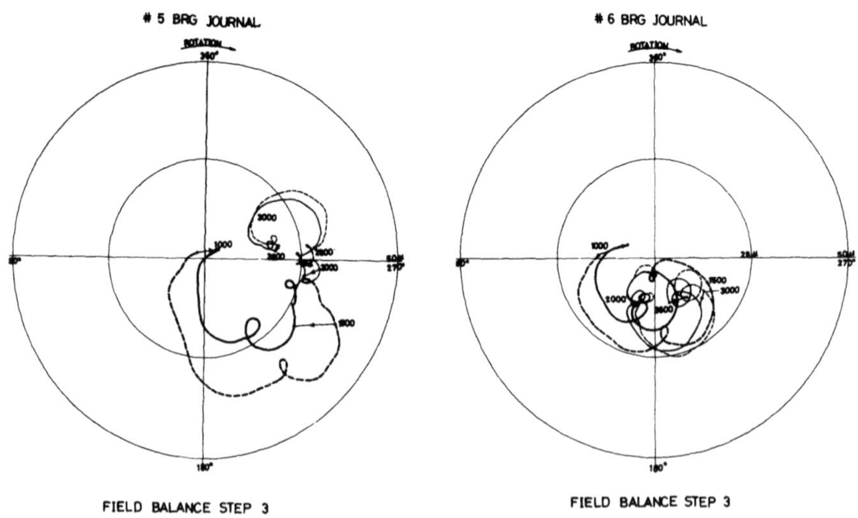

Fig. 18. Field Balancing STEP 3.

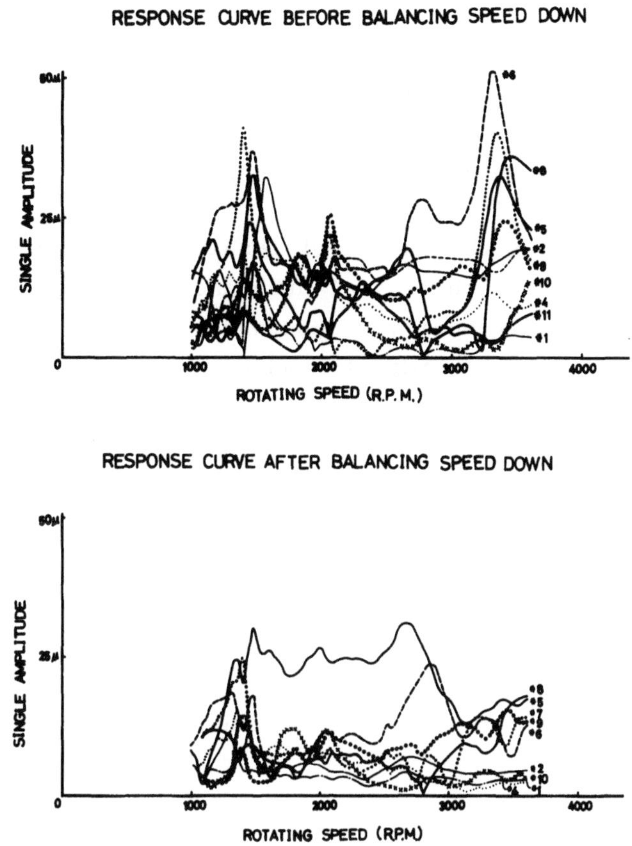

Fig. 19. Response Curve of the Turboset before and after Field Balancing.

5. CONCLUSION

In this paper, the vibration characteristics of a multispan rotor such as a large turboset and the field balancing of a multispan rotor are described.

The conclusion of this work is as follows.

1. The analysis of vibration response due to initial bend of rotor caused by coupling inaccuracy shows the importance of the tolerance of coupling machining and shrink fitting,

especially for large flexible rotors such as generator rotors.

2. The application of modal response circle to the field balancing of multispan rotor instead of conventional trial and error method makes it easy to apply modal balancing to multispan rotors.
   This method shows the excellent result as regards time-saving and controllability of unbalance vibration and gives a physically easily understandable field balance procedure. In other words, the advantage of this method lies in the fact that the trial weights of this method always act as the correction weights.

ACKNOWLEDGMENT - The authors wish to thank H. Nakane (manager of technical department of Takasago Works of their company) and Y. Shoda (submanager of manufacturing department of Takasago Works) for their kind support, encouragement and valuable advice.

They also thank the members of Takasago Technical Institute and the members of Takasago Works of their company and members of Kobe Works of Mitsubishi Electric Co., Ltd., who supported and assisted their work.

REFERENCES

[1] Hagg, A.C. and SANKEY, G.D.: "Elastic and dynamic properties of oil film journal bearings for application to unbalance vibration calculation", Trans. ASME J. App. Mech., Vol. 80, 1955.

[2] ORCUTT, F.K. and ARWAS, E.B.: "The steady-state and dynamic characteristics of full circular bearing and partial arc bearing in the laminar and turbulent flow regimes", Trans. ASME, Journal of Lubrication Technology, Vol. 89, No. 2, April 1967.

[3] LUND, J.W. and ORCUTT, F.K.: "Calculation and experiments on the unbalance response of flexible rotor", Trans. ASME, J. Engrg. Ind., 89, No. 4, Nov. 1957.

[4] BISHOP, R.E.D. and GLADWELL, G.M.L.: "The vibration and balancing of an unbalanced flexible rotor", Journal of Mechanical Engineering Science, Vol. 1, 1959.

[5] HOFFMAN, H.: "Einfluss des kupplungs verhaltens auf die laufrehe mehrfach gelagerter wellen systeme", VDI-Berichte Nv. 161, 1971.

[6] BISHOP, R.E.D. and PARKINSON, A.G.: "On the isolation of modes in the balancing of flexible shafts", Proc. Instn. Mech. Engrs., Vol. 177, No. 16, 1963.

[7] LINDSEY, J.R.: "Significant developments in method of balancing high speed rotors", ASME Paper 69-Vibr.-53.

[8] LUND, J.W. and TONNESON, J.: "Analysis and experiments of multi-plane balancing of flexible rotor", ASME Paper No. 71-Vibr.-74.

[9] GOODMAN, T.P.: "A least-squares method for computing balance corrections", ASME Paper No. 63-WA-295.

[10] TESSARZIK, J.M., BADGLEY, R.H. and ANDERSON, W.J.: "Flexible rotor balancing by the exact point-speed influence coefficient method", ASME Paper No. 71-Vibr.-91.

[11] TONDL, A.: "Some problems of rotor dynamics", Publishing House of the Czechoslovak Academy of Sciences Praugue, 1965.

[12] IWATSUBO, T., KANKI, H. and KAWAI, R.: "Vibration of asymmetric rotor through critical speed with limited power supply", Journal of Mechanical Engineering Science, Vol. 14, No. 3, 1972.

APPENDIX 1

Application of modal response circle analysis to the balancing of asymmetric shaft.

As many authors described [10], asymmetric shafts have different characteristics in their unbalance response.

One of the most important points are the difference of modal response near its critical speed due to angular location of unbalance.

The numerical example of approximate analytical solution [11] is shown in Fig. A.1.

As shown in Fig. A.1,

1. The modal response circles of stable asymmetric shaft are elliptic having their principal axis ±45 degrees directions.
2. And max. response shown in the case =-45 degrees
3. Minimum response shows in the case =+45 degrees and
4. Only above two cases the phase lag of response to unbalance show 90 degrees.

Considering the above characteristics, the following balancing procedure is very useful and gave many good results.

1. Plot the modal response circle (ellipse) in deceleration or acceleration through its critical speed.
2. Attach the balancing weight at 45° or $(45^{\circ}+180^{\circ})$ direction to coincide the principal axis of ellipse to -45° axis.
3. Attach the balancing weight at -45° or $(-45^{\circ}+180^{\circ})$ direction to eliminate the major axis of ellipse.

The example is shown in Fig. A.2.

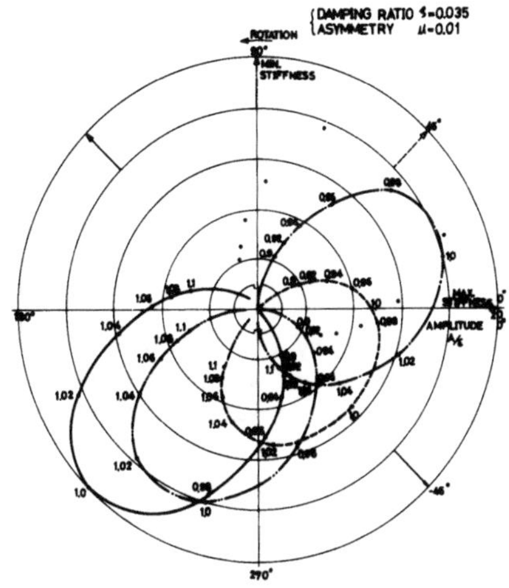

Fig. A-1. Modal Response Circle of Asymmetric Shaft.

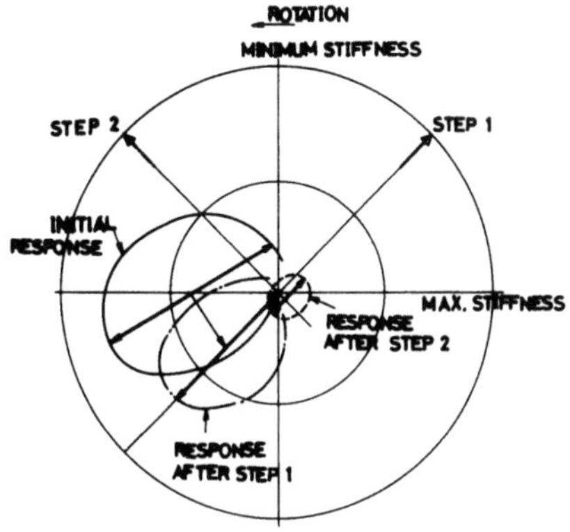

Fig. A-2. Example of Modal Balancing on Asymmetric Shaft.

Response Matrix

Speed \ Points		#1	#2	#3	#4	#5	#6	#7	#8	#9	#10	#11
Gen 1st	1	A_{11}	A_{12}	A_{13}	A_{14}	A_{15}	A_{16}	A_{17}	A_{18}	A_{19}	A_{110}	A_{111}
Lp 1st	2	A_{21}	A_{22}	A_{23}	A_{24}	A_{25}	A_{26}	A_{27}	A_{28}	A_{29}	A_{210}	A_{211}
Lp 1st	3	A_{31}	A_{32}	A_{33}	A_{34}	A_{35}	A_{36}	A_{37}	A_{38}	A_{39}	A_{310}	A_{311}
Hp 1st	4	A_{41}	A_{42}	A_{43}	A_{44}	A_{45}	A_{46}	A_{47}	A_{48}	A_{49}	A_{410}	A_{411}
Ip 1st	5	A_{51}	A_{52}	A_{53}	A_{54}	A_{55}	A_{56}	A_{57}	A_{58}	A_{59}	A_{510}	A_{511}
Gen 2nd	6	A_{61}	A_{62}	A_{63}	A_{64}	A_{65}	A_{66}	A_{67}	A_{68}	A_{69}	A_{610}	A_{611}
Lp 2nd	7	A_{71}	A_{72}	A_{73}	A_{74}	A_{75}	A_{76}	A_{77}	A_{78}	A_{79}	A_{710}	A_{711}
Lp 2nd	8	A_{81}	A_{82}	A_{83}	A_{84}	A_{85}	A_{86}	A_{87}	A_{88}	A_{89}	A_{810}	A_{811}
C . S	9	A_{91}	A_{92}	A_{93}	A_{94}	A_{95}	A_{96}	A_{97}	A_{98}	A_{99}	A_{910}	A_{911}
3600	10	A_{101}	A_{102}	A_{103}	A_{104}	A_{105}	A_{106}	A_{107}	A_{108}	A_{109}	A_{1010}	A_{1011}

The Following Local Matrix are Obtained by Neglecting the Small Component

- HP Local Matrix

	#1	#2
Hp 1st	A_{41}	A_{42}
3600	A_{111}	A_{112}

- IP Local Matrix

	#3	#4
Ip 1st	A_{53}	A_{54}
3600	A_{113}	A_{114}

- LP Local Matrix

	#5	#6	#7	#8	#9
Lp 1st	A_{25}	A_{26}	0	0_8	0
	0	0_6	A_{37}	A_{38}	0
Lp 2nd	A_{75}	A_{76}	A_{77}	A_{78}	0
	A_{85}	A_{86}	A_{87}	A_{88}	0
C . S	A_{95}	A_{96}	A_{97}	A_{98}	A_{99}
3600	A_{105}	A_{106}	A_{107}	A_{108}	A_{109}

- Gen Local Matrix

	#9	#10	#11
Gen 1st	A_{19}	A_{110}	A_{111}
Gen 2nd	A_{69}	A_{610}	A_{611}
3600	A_{109}	A_{1010}	A_{1011}

Influence Coefficient Matrix is Treated Simultaneously

Fig. A-3. The Example of Local Matrix.

The Dynamics of Synchronous Whirl in Turbine Rotors

D.M. Smith

formerly Consulting Mechanical Engineer
Turbine Generator Division
Associated Electrical Industries Limited

ABSTRACT - Synchronous whirl is a type of instability of rotors which is characterised by slow build up of transverse vibration with fundamental frequency equal to the running speed. It is not due to wander in out-balance.

The paper describes an investigation of the dynamics of occurences in a particular group of steam turbines. It was concluded that instability is due to interaction of torsional vibration of frequency 2/rev., with transverse vibration of frequency components 1 and 3/rev. Build up takes place only when torsional vibration in a suitable mode can reach high resonant magnification.

The principal mechanism of interaction in this group is unbalanced blade vibration. The torsional vibration and the blade vibration are greatly affected by nonlinear characteristics of the blade root fixing. The transverse vibration is modified by the nonlinear dynamic characteristics of the journal bearings.

The work is compared with analytical investigations by A. TONDL (1965) and by V.O. KONENKO (1969).

Reference is made to synchronous whirl in generator rotors where there must be a different mechanism of interaction.

1. SYNCHRONOUS WHIRL IN A GROUP OF STEAM TURBINES

Scope of Investigation. Synchronous whirl was experienced in this group of steam turbines during the years 1958-67. When the vibration was recognised as a type of instability in 1960, it was soon established that the source of trouble was nct out-of-balance; when speed was varied, amplitude dropped quickly above and below the sensitive speed range. The vibration was found to differ from other known types of rotor instability, which were of frequency below running speed and which grew up much more rapidly when instability occured.

Various possible causes of vibration were considered, but no satisfactory explanation was found at that time.

Troublesome cases were dealt with by empirical methods to bring sets to serviceable condition. The investigation described was made in 1969-73, on the basis of existing records and recollections, without further observations or tests.

Description of Steam Turbine Plant. The steam turbines in the group were all three-cylinder sets with HP, IP and double-flow LP turbines, and with running line solidly coupled to generator. Speed was 50 revs/sec. (3000 RPM). All the sets except two were of a common turbine frame of rating 120 MW . The sets of the common frame were not completely identical; there were variations both in blading and in mechanical build. On the earlier sets, the LP rotor was built up with separate discs shrunk on the shaft, while on the later sets, the LP rotor was a solid forging. A section of part of the running line on one of the later sets shown in Fig. 1.

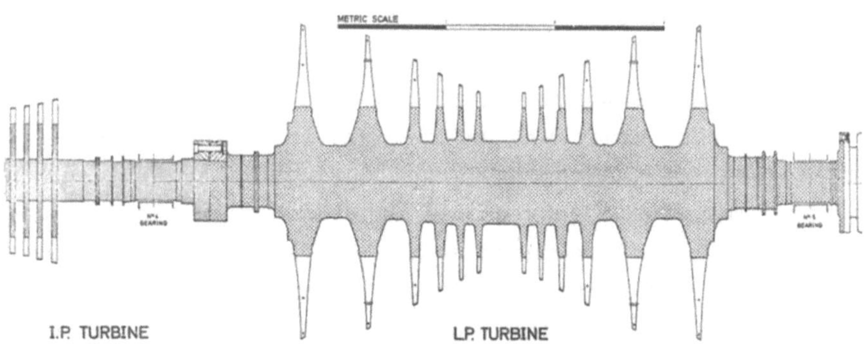

Fig. 1.

On all the sets of the common frame, there was a multi-exhaust of the Baumann type in each flow. The long blading of the last two stages was carried in riveted root fixings. On the second last stage, the two-tier blading was in packets, pinned at the bridge and laced at the outer tier. On the last stage, the blading was laced in packets, with the exception of one or two sets which had longer blading with a continuous lacing wire.

The two remaining sets - those not of the common frame - are identified as "P" and "Q" . Neither of these had Baumann exhausts. Set "P" was of much lower rating - 54 MW - but in some respects resembled the sets of the common frame. Set "Q" was of 120 MW rating, but was designed and built by another manufacturer, with every detail different from those of the remaining sets.

Experience of Vibration. Serious vibration trouble had been experienced on "P" in 1958-59 a few days after commissioning and recommissioning. The only vibration records taken were on the permanent instruments. It was assumed that the set was running too close to a critical speed; the HP rotor was machined down to reduce this critical speed. Following this change the set was serviceable, but it required careful handling, especially in starting. The vibration was not at first recognised as synchronous whirl; but symptoms of incipient synchronous whirl were found at a later date.

Synchronous whirl was experienced in 1960 on two duplicate sets of the common frame, during preliminary running for commissioning. Extensive experimental vibration observations were made on these sets. Both sets were found unserviceable in the initial build and were not synchronised. On these sets the HP rotor was machined down. This rendered the sets serviceable, but they continued to exhibit occasional symptoms of synchronous whirl.

Synchronous whirl was later observed on some sets, but not all sets, of the common frame, mostly during commissioning or shortly after commissioning. The troubles were dealt with by empirical methods, mainly bearing modifications and operating techniques. No more rotors were machined down.

The commissioning of troublesome sets was facilitated by a discovery made on "Q" , the set built by another manufacturer. This set had been in operation for a year or more without reported vibration trouble. After a priming incident (passage of wet steam through the turbine) the set developed serious synchronous whirl. One day, under pressure from customer, the erection staff tried the only change in build which they could make quickly - increasing the vertical clearance in the elliptical journal bearings. This was in the opposite direction to the change which often improves stability in cases of low-frequency whirl. Nevertheless the increase in vertical clearance improved conditions sufficiently to make "Q" serviceable. Similar results were obtained on other sets.

The occurrence of synchronous whirl on various sets was erratic, but of the same general type. Vibration grew up within a limited speed band, about 5% to 15% wide. Vibration was at first particularly marked on the LPT , and of frequency 1/rev. with harmonics, especially 3/rev. No low-frequency component below 1/rev. was present. After a short time the harmonics grew much more slowly than the synchronous component, while the mode of the synchronous component changed, becoming relatively greater on the HPT . At the higher amplitudes No. 1 bearing seemed to "lose its grip" on the journal; the journal amplitude increased with little vibration on the HP pedestal. The generator did not participate appreciably in the vibration.

When speed could be varied, the synchronous whirl dropped rapidly in amplitude outside the sensitive speed band. Amplitude generally increased very slowly within the band, maximum amplitude taking about 3 to 15 minutes to double. The vibration sometimes stabilised at a constant or slowly varying amplitude. Growth was liable to be started or restarted by a change in running conditions, particularly load or steam temperature. One set tended to fall in amplitude when brought to 90 MW from either higher or lower load.

Rebalancing of rotors, tried on several occasions, made the running conditions sometimes better and sometimes worse.

The same erratic result was obtained by dismantling and reassembling with the object of improving alignment. On nearly all sets, there was a gradual improvement in liability to vibration with passage of time.

2. THE SEARCH FOR AN EXPLANATION

Possibility of torsional-transverse interaction. When a systematic investigation of the cause of instability was initiated in 1969, particular attention was given to symptoms observed in the early stages of growth of vibration. Attention was drawn to the possibility of torsional-transverse interaction by two lines of evidence:-

(a) the presence of 3/rev. as well as 1/rev. components of transverse vibration suggested interaction with 2/rev. torsional vibration,

(b) the slow growth of transverse vibration seemed consistent with limited rate of energy exchange between torsional and transverse vibration.

Method of investigation. The basic work involved establishing and examining the equations of motion of a rotor body vibrating with small amplitude in five degrees of freedom - two transverse linear, two transverse angular and one torsional. This torsional vibration was superimposed on rotation at steady speed. The rotor body was one of a line of rotor bodies carried by a flexible mounting consisting of shaft, hydrodynamic journal bearings and bearing supports. The conditions of vibration and the properties of components of the system were assumed to fall within the limitations known to hold for a real turbine in synchronous whirl.

The transverse equations of motion, linear and angular, correspond to those of a previous investigation for rotation at constant speed (SMITH 1933) with the addition of terms arising from the torsional vibration.

The torsional-transverse and transverse-torsional excitation terms were examined for their influence on rotor motion. Investigation was made of characteristics of components which might affect vibration, including nonlinear characteristics. The results were correlated with observed phenomena in synchronous whirl. No attempt was made to solve the equations of motion of the system. The analysis is not here reproduced; a brief description of the work is given.

In the numerical assessments, the limits of amplitude in transverse vibration were based on observations of rotors in service, plus knowledge of clearances. Maximum out-of-balance was based on experience. As an example of figures used, on a representative LP turbine, with values expressed as ratios to radius of gyration, maximum amplitude of transverse vibration in service was estimated as 0.8×10^{-3}, and maximum out-of-balance eccentricity permitted as 0.2×10^{-3}. These figures give a measure of relative order of magnitude for some of the quantities involved. Throughout this paper, "order" refers to order of magnitude, not to order of frequency component.

No measurements had been made of torsional vibration, but a limit of amplitude in a known mode of vibration was provided by the observation that there had been no sign of distress due to torsional vibration in any turbine involved. In the suspected mode of torsional vibration, the maximum amplitude is unlikely to have exceed 0.ol radian and may have been far less.

In certain calculations it was found of great value to express periodic transverse displacements, forces, etc. in terms of components in forward and reverse precession. Transverse vibration can be expressed in terms of co-ordinates in Fourier series. Alternatively, each frequency component is expressible as the sum of two circular components, one of forward precession and one of reverse precession. For components of frequency n times mean speed of rotation, the component of forward precession is denoted in this paper by nF, and the component of reverse precession by nR.

Each component is characterised by specified frequency, amplitude and phase.

Publications by TONDL (1965) and by KONONENKO (1969) on torsional-transverse interaction in rotor vibration were examined. Each of these authorities has investigated the motion of rotors with out-of-balance components. Both have concluded that in certain conditions, interaction can result in high amplitude or unstable vibration within a narrow speed range.

Out-of-balance and unbalanced blade vibration. The equations of motion first examined were those for transverse and torsional motion of a rigid rotor body with fixed out-of-balance. When rotation is at constant speed, there is a direct out-of-balance excitation of frequency component 1F . When there is torsional vibration of frequency 2/rev. and of amplitude within the limitations already discussed, torsional-transverse excitation components 1R and 3F are set up. These are of smaller order than the direct out-of-balance excitation component. Torsional-transverse excitation with fixed out-of-balance was found to be incapable of explaining the vibration phenomena observed in these instances of synchronous whirl.

Examination was then made of gyroscopic excitation arising in transverse angular motion. It was found that gyroscopic terms affected the transverse angular vibration, but that torsionally-excited gyroscopic action could not explain synchronous whirl.

These conclusions did not invalidate the opinion that synchronous whirl was a consequence of torsional-transverse interaction, but led to the search for a mechanism responsive to torsional vibration which would give much larger excitation than could arise with fixed out-of-balance alone.

The only mechanism found in these turbines to give a satisfactory explanation was unbalanced blade vibration (U.B.V.). This results from irregularity of blade vibration round a blade row. It is greatest in the long blades of the last two stages of each exhaust flow of the LP rotor. The significent blade vibration frequency is 2/rev.

In turbines of the design considered, U.B.V. can account for torsional-transverse excitation of higher order than that due to fixed out-of-balance. It can also account for transverse-torsional excitation of higher order, because the torsional excitation arises from moment-of-momentum acting at the radius of blade vibration. This action permits transverse displacement of pivoting axis to be ignored in assessing torsional excitation due to U.B.V., whereas effective pivoting axis must be taken into account in assessing torsional excitation due to fixed out-of-balance.

U.B.V. due to non-uniformity of workmanship round the blade row was first examined. This appears to play a significant part in the initial stage of build up of synchronous whirl, but is inadequate to explain the substantial excitation in the later stages of whirl. U.B.V. set up through nonlinear response characteristics of blade vibration was found to give larger and more systematic excitation.

Torsional vibration of rotor. The torsional vibration of the rotor presented several problems.

It was concluded that there must be high magnification ratio of the torsional vibration, probably at least 100, and that the rotor must therefore be in resonance at frequency 2/rev. in a suitable mode of vibration with low damping. Torsional frequencies previously calculated for the rotor line included 69.8 hz for the third mode and 135.2 hz for the fourth mode, as compared with 100 hz required for 2/rev. The calculated mode of vibration of the fourth mode seemed significant, with one of the modes near the middle of the LP rotor.

The calculation of torsional frequency had been carried out assuming blading rigid on each disc. This is a reasonable approximation for the lower modes of vibration, but is grossly inaccurate for the fourth mode in which there are long heavy blades in the region of maximum torsional amplitude. It was found that torsional frequency in this mode would come down to about 100 hz if, in addition to allowing for flexibility of the blading, some allowance was made for slackening of the riveted root fastening in the last two stages. Such slackening is known to occur at speed.

This modified mode was taken as the 2/rev. torsional vibration which characterised synchronous whirl. It was assessed that there was considerable participation of the HP and IP turbine rotors, but relatively small amplitude of the generator rotor.

Damping of torsional vibration was examined. The principles of calculation used were the same as those of KER WILSON (1963). At low and moderate amplitudes, damping comes mainly from the torque-speed characteristics of steam flow in blading, plus a small contribution from the variation in bearing torque with speed. These together give in each turbine rotor a damping torque of approximately 1% of steam torque for 1% speed rise at full load condition. The resonant magnificantion ratio for the whole line in the assessed mode is about 8000 at full load and is higher at partial load.

Above a certain amplitude, however, other sources of damping arise, due to working in blade fastenings, shrink fits, and abutting joints. In the rotors here discussed, these nonlinear effects probably occur first in the riveted blade root fixings.

The problem of resonant bandwidth at high manification received much consideration. In a linear system of one degree of freedom with light damping, a magnification ratio of 100 is possible only within a total frequency range of about 1%. It was clear, however, that the high magnification, whatever it was, must be attainable over a frequency range of several percent - this was shown by the extent of the sensitive speed range in synchronous whirl, and by the observation that sets with appreciable variations in design were sensitive at the same operating speed. The solution offered, both

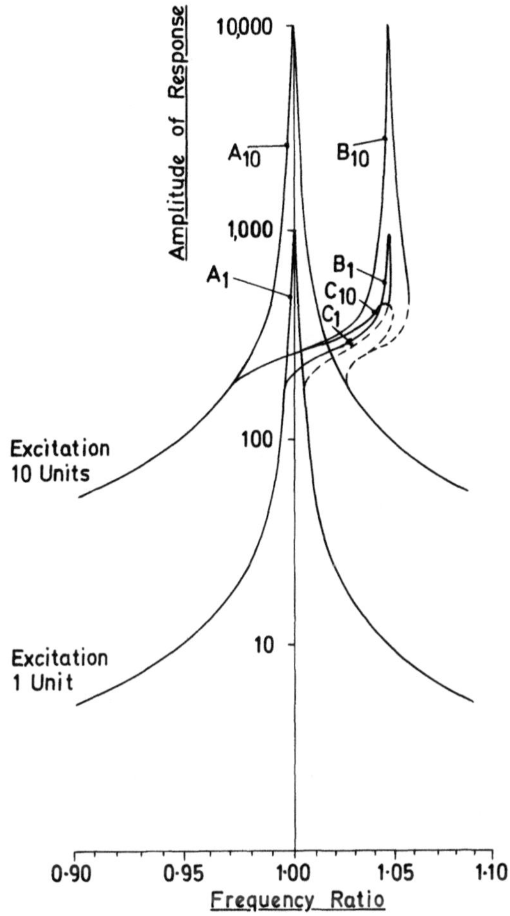

Fig. 2. Response Diagram for Linear and Nonlinear Systems.

to this problem and to the problem of "cut-off" which limits amplitude of vibration, arises from nonlinearity of the turbine blade vibration characteristics.

The nonlinearity occurs in the same blades already mentioned as having major influence on the torsional frequency - the last two stages in each flow of the LP - and is set up in the blade root fixing. There is some slackening of the root fixing due to centrifugal expansion of the disc at speed. In blade vibration, the loss of restraint is partly taken up as the amplitude of blade vibration increases, giving "hard" nonlinearity which stiffens the blade response characteristics. If the blade vibration is excited by the torsional vibration of rotor, then the combined system of rotor and blading exhibits characteristics of "hard" nonlinearity, with resonant frequency increasing as amplitude increases over a certain range.

To illustrate how nonlinearity can give rise to broadband resonance and out-off in amplitude in a system with light damping, response diagrams for simple systems of one degree of freedom are drawn in Fig. 2. The figure is plotted in logarithmic coordinates. Steady response to periodic support movement is shown for three systems, A, B and C, at two amplitudes of excitation, one ten times the other.

The three systems have the same characteristics at low amplitude of vibration but have different characteristics at high amplitude. System A is a linear system with resonant magnification ratio 1000. The response curves A_1 and A_{10} are similar; at any frequency, response is proportional to excitation.

System B exhibits hard nonlinearity in stiffness, the resonant frequency at high amplitude is 1.06 times that at low amplitude. The response curves B_1 and B_{10} are no longer similar. On each curve there is a "plateau" which, within a certain frequency range, gives response only slightly higher when excitation is multiplied by 10. The plateau is reached through a frequency "gate" of width depending on magnitude of excitation.

System C exhibits nonlinearity in both stiffness and damping. The stiffness characteristics are as for B, while damping increases nonlinearly within the same range as stiffness, but at a proportionately greater rate. The plateau and gate phenomena remain, but peak reasonance at high magnification ratio is cut off.

The response characteristics of system C represent

in some respects those which characterise the rotor torsional vibration in synchronous whirl.

The rotor vibration just discussed assumed blade vibration excited only by torsional vibration of the disc. There are other sources of blade vibration. The only other sources found significant in synchronous whirl are transverse vibration of the disc, and steam flow through the blading.

Search has also been made for sources of excitation of rotor torsional vibration. In the particular group of steam turbines here discussed, the only sources found significant are the dynamic effects from transverse interaction.

Unbalanced blade vibration due to nonlinearity. Unbalanced blade vibration (U.B.V.) is of greatest significance in the last two LP stages, in which the blades are long, heavy, lowest in fundamental blade frequency, and exposed to greater torsional amplitude of disc than in other turbine stages. The discussion concerns primarily these blades. Blade vibration is excited by rotor torsional vibration, by rotor transverse vibration, and by steam flow. The U.B.V. to be described arises from interaction between vibration excited by these sources when the blade vibration response characteristic are nonlinear.

The torsional excitation is of frequency 2/rev. This is the torsional mode excited with high magnification; there is no reason to expect any other torsional component of appreciable amplitude. This torsional excitation is of the same amplitude and phase on all blades in the same blade row.

The rotor transverse vibration, acting on the blades by tangential component of movement of the blade root, is transformed by the disc rotation. Blade excitation of frequency 2/rev. is given by the transverse vibration components 1R and 3F . The blade excitation from 1R is of the same amplitude on all blades, but of phase which travels round the blade row in the reverse direction at rate 1/rev.

3F gives 2/rev. blade excitation which is similar except that in phase it travels round the blade row in the forward direction. The rotor transverse vibration component 1F produces no excitation of blade vibration, while 3R produces blade excitation of frequency 4/rev.

The steam flow produces excitation of blade vibration and steady reaction on blade root fixing, both varying with load on turbine.

On the blades of any one blade row, the 2/rev. excitation components from torsional motion of disc and from transverse motion of disc give a resultant excitation of the various blades which is of varying amplitude and phase. This resultant is conveniently represented on a vector diagram.

Fig. 3 shows a representative vector diagram of the 2/rev. excitation on each of eight blades evenly spaced round one rotor disc. The vector OT represents the excitation component due to 2/rev. torsional vibration of the disc; it is the same for every blade. The excitation components due to the 1R and 3F transverse vibration components of the disc are compounded with OT to give the resultant excitation components Oa, Ob,...Oh for the various blades, with varying amplitude and phase as shown. These resultants include only the dynamic excitation due to disc vibration; any 2/rev. excitation from steam flow must be added.

The response of each blade to the 2/rev. excitation can likewise be shown on a vector diagram (not reproduced). When the blading is uniform and of linear vibration characteristics, the response diagram is geometrically similar to the excitation diagram. For this condition the components of rotor torsional vibration and of rotor transverse vibration, interacting through the blade vibration, are linearly independent; the transverse vibration does not influence the torsional vibration, nor does the torsional vibration influence the transverse vibration.

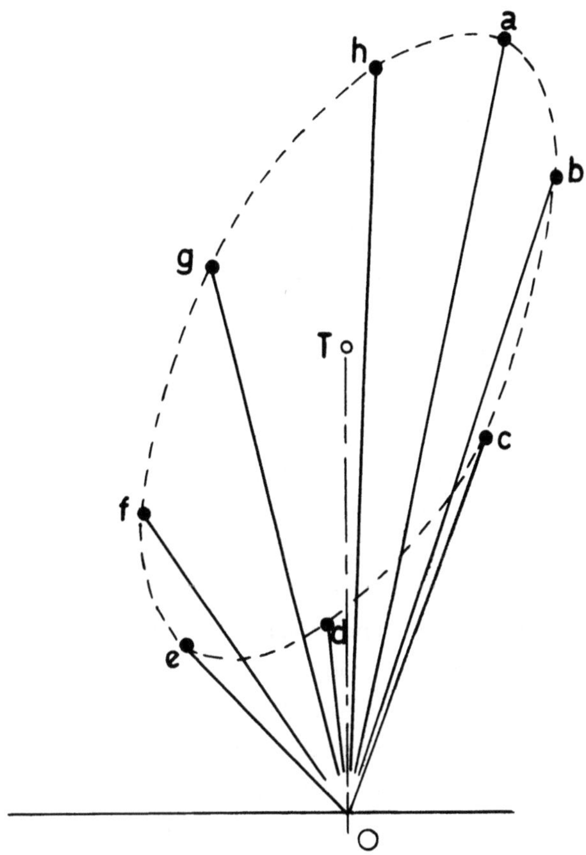

Fig. 3. Vector Representation of Blade Excitation.

When there is nonlinear blade response to the joint excitation, there is unbalanced blade vibration in which the rotor torsional vibration excites transverse vibration, and the rotor transverse vibration excites torsional vibration.

Since the nonlinear effect in blade vibration is a function of blade amplitude, the U.B.V. effect is greatest, and is nearly in phase, on neighbouring blades of the group in which blade excitation is greatest. In consequence, the blades of this group reinforce one another in excitation of rotor motion. In the same way, on adjacent discs on which rotor motion is nearly in phase, the resultant excitations reinforce one another.

This characteristic of U.B.V. set up by nonlinearity contrasts with that of U.B.V. set up by nonuniformity of workmanship of blades. The latter is irregular from one disc to another and is likely to be of random distribution round a blade row.

The U.B.V. of frequency 2/rev. is defined in rotating axes. The rotor torsional frequency which it excites is also 2/rev. but the rotor transverse vibration excited is in fixed axes and is of precessional frequencies 1R and 3F .

There is an additional consequence of nonlinear blade vibration which arises if the nonlinearity is unsymmetrical, as when slackness in blade root fixing is taken up more by movement to one side of the equilibrium position than to the other. Unsymmetrical nonlinearity causes the mean position of the blade in vibration to move away from the equilibrium position. When amplitude of the vibration varies round the blade row, this sets up temporary out-of-balance, giving 1F excitation of transverse vibration, in addition to that excited by the permanent out-of-balance of the rotor.

The excitation from steam flow, whatever its frequency, tends to increase the total amplitude of blade vibration, and influences the nonlinearity. The steady load from work done on blading does not directly excite vibration, but it affects blade root fixing conditions. It therefore affects both U.B.V. due to nonlinearity and temporary out-of-balance.

Steam temperature affects tightness of blade root fixing. Sets in synchronous whirl are often sensitive to changes both in load and in steam temperature.

The experience reported on set Q , where a turbine previously free from this trouble experienced serious synchronous whirl after a priming incident, is readily explained by the theory here put forward. The blade root fixings were loosened by thermal and mechanical shock in the priming incident. The experience on this and other sets, that in regular service the tendency to synchronous whirl diminishes with passage of time, is consistent with the known tightening of blade root fixings in steam turbines with passage of time. This tightening is due to corrosion products filling up root clearances.

Limitations of numerical calculation. The identification of those actions which play a significant part in synchronous whirl turns largely on calculation of their comparative or absolute magnitude. Reliable data is available for certain calculations, as already mentioned; but in other calculations only an upper limit is known, as for amplitude of torsional vibration; and in some cases, as with nonuniformity and nonlinearity in blade root fixings, the calculation is based wholly on guesswork.

Assumptions to facilitate approximate calculation can sometimes be made. For example, interaction in blade vibration beteen rotor torsional vibration and rotor transverse vibration is promoted when the two types vibration give comparable blade excitation. As a first approximation, the two may be assumed equal; this gives amplitude of torsional vibration when amplitude of transverse vibration is assumed. On this basis, the figure taken for maximum amplitude of transverse vibration on the reference LP rotor gives maximum amplitude of torsional vibration about 0.4×10^{-3} radians. This is 4% of the figure calculated for "safe" torsional amplitude.

If this amplitude were excited by transverse-torsional excitation from maximum fixed out-of-balance, the magnification ratio required would exceed 30×10^{3} , a figure which is clearly impossible. The alternative excitation from U.B.V. can only be very roughly guessed, but guesses which have been made suggest a torsional magnification ratio likely to be within the range 100 to 1000 . At least the lower figures within this range appear possible in practice.

These figures illistrate the extent of uncertainty in the numerical calculations.

Effect of dynamic characteristics of journal bearings. As has been mentioned, journal bearings of reduced ellipticity were found to give dimished liability to synchronous whirl. The bearings of reduced ellipticity are more nearly symmetrical in dynamic characteristics than bearings of high ellipticity, and this leads to a smaller 1R component in the rotor transverse vibration for the same out-of-balance excitation of type 1F . There is therefore less interaction.

In the turbines under consideration the running line is flexible, with three calculated critical speeds below the normal running speed. The rotor runs reasonably clear of critical speeds so long as vibration remains moderate. At the lower amplitudes all the journal bearings are running in the oscillating film regime (SMITH 1969). As journal amplitudes increase, the dynamic characteristics of the bearings become increasingly nonlinear. The bearings which are most lightly loaded (Nos. 1 and 2) enter a transition film regime, with dimished journal restraint and damping, and with increase in journal amplitude reach the rotating leading film regime. In this condition the journal orbit is guided round the bearing clearance, with low damping at the orbiting frequency.

The diminished journal restraint shows itself by fall in pedestal vibration while amplitude of journal is increasing; this is particularly noticeable on No. 1 bearing. The mode of vibration of the running line changes. The 3/rev. component reaches a limiting amplitude while the 1/rev. component keeps on increasing, particularly on the HP rotor. If the amplitude does not stabilise on the way up, the rotor reaches a critical speed condition of nonlinear resonance.

Phase relationships in torsional and transverse vibration. The torsional vibration is of frequency 2/rev. with fixed nodes in the running line, and with consequent reversal of phase across each node. The transverse precessional components which interact with this torsional vibration are 1R and 3F.

If the transverse vibrations are the same on both sides P and Q of a torsional node, then the excitations on side Q are the same as they were on side P half a revolution earlier. Hence, allowing for torsional phase reversal, both torsional excitation and transverse excitation are in opposition on sides P and Q .

If the transverse vibrations are in opposite phase on sides P and Q , then likewise both torsional excitation and transverse excitation are in opposition on sides P and Q .

If, however, the transverse vibrations on side Q are advanced in phase by one-quarter of a revolution compared with side P (which corresponds to one-half cycle of the torsional vibration), then both torsional excitation and transverse excitation are advanced in time on side Q by one-quarter of a revolution, and both excitations become additive. Similarly, the excitations are additive when the transverse vibrations on side Q are retarded by one-quarter of a revolution.

Examination of records of vibration taken on a turbine in synchronous whirl has shown that on the LP turbine, the transverse vibrations at the ends of the LP rotor were approximately in quadrature. This confirms that the synchronous whirl involved torsional-transverse interaction with torsional node in the LP rotor.

The relationship corresponds to a result obtained by TONDL (1965) for a running line of two rotor bodies, each out-of-balance. Tondl showed that maximum liability to vibration in torsional-transverse interaction was given when the out-of-balance angles were in quadrature.

3. BUILD-UP OF SYNCHRONOUS WHIRL

A summary of the explanation of build-up of synchronous whirl is given in the Abstract, and some of the actions in this build-up are discussed in 2 . In the present Section,

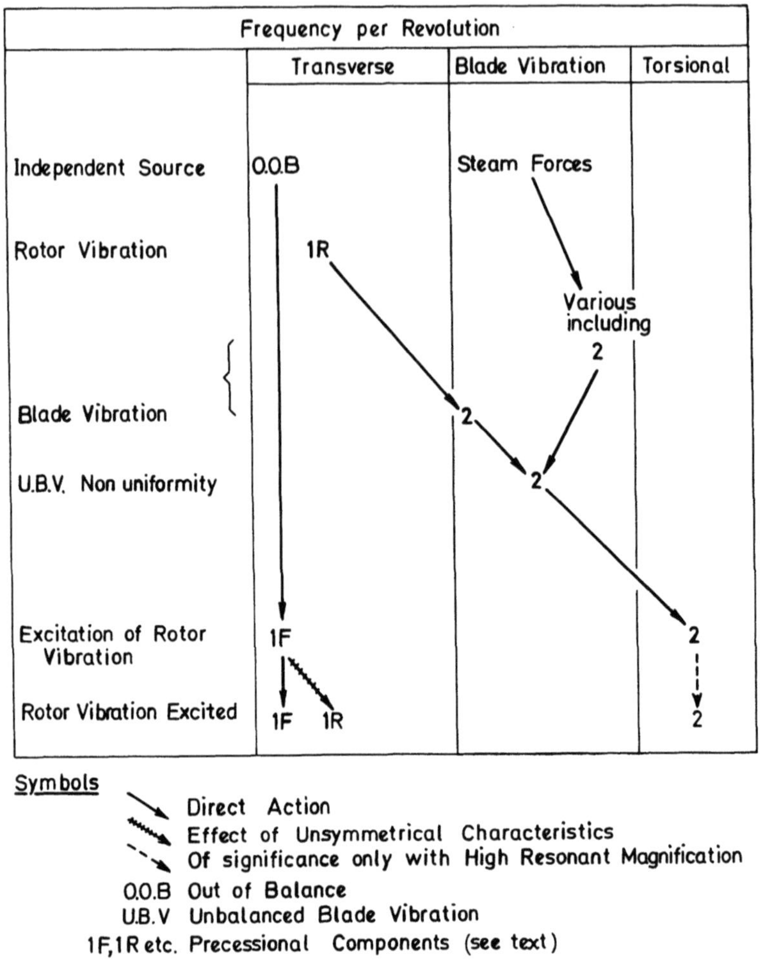

Fig. 4. Principal Actions in Initial Build up of Synchronous Whirl.

the principal actions are illustrated in two diagrams, Fig. 4 showing the actions in initial build-up, and Fig. 5 the actions in the later stages.

The sources of vibration shown in the diagrams as independent are always present present. In the early stages of self-excitation, as shown in Fig. 4, the 2/rev. unbalanced blade vibration (U.B.V.) excited by transverse vibration of rotor and unbalanced by non-uniformity of blading in a row,

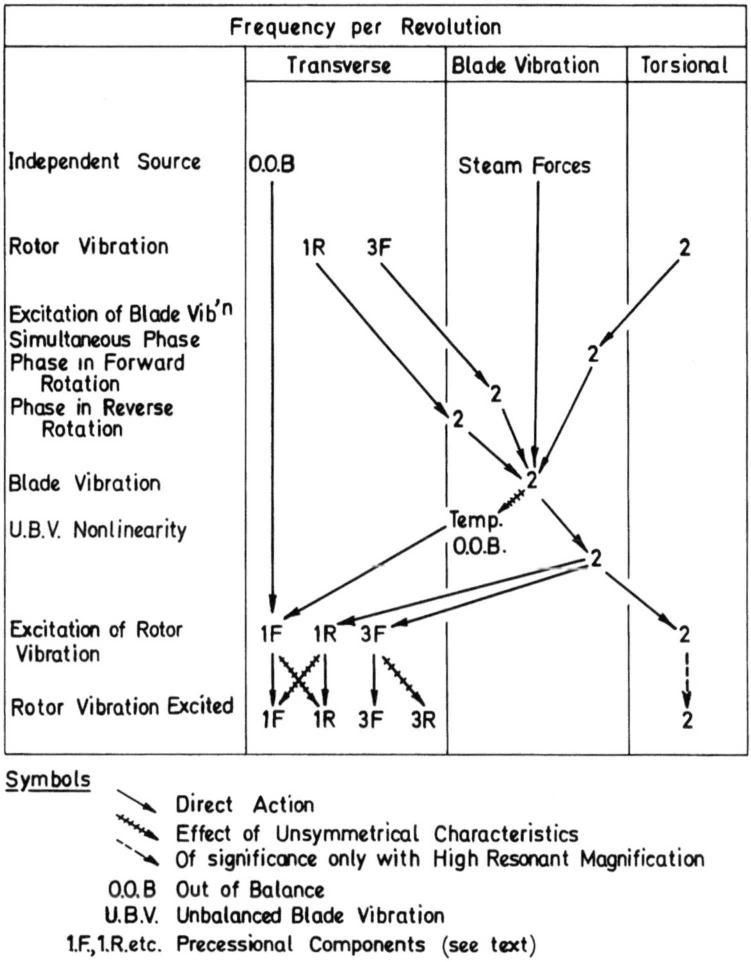

Fig. 5. Principal Actions in Later Build up of Synchronous Whirl.

causes torsional vibration to build up when conditions of resonance permit.

In the latter stages of build-up, as shown in Fig. 5, the vibration is dominated by self-excitation. Different sources of excitation of blade vibration, interacting on long heavy blades, set up U.B.V. due to nonlinearity of blade response characteristics. This U.B.V. excites rotor vibration, both transverse and torsional. The transverse vibration is drastically modified as amplitude increases by the nonlinear dynamic characteristics of the journal bearings.

The growth of vibration is slow owing to the limited rate of energy interchange between the various modes of vibration. Stabilisation of vibration may take place at an acceptable amplitude. This stabilisation is affected by a multiplicity of causes, some of which have been used as empirical means of controlling synchronous whirl.

4. SYNCHRONOUS WHIRL IN GENERAL

Other cases of synchronous whirl. The foregoing explanation of the dynamics of synchronous whirl is established only for the group of steam turbines described. Synchronous whirl has been observed on a wider range of plant, including generators running uncoupled. Interaction involving unbalanced blade vibration cannot apply in a component which has no blading.

In cases known to the writer of generator vibration of synchronous whirl type, the vibration phenomena have been so similar to those described in 1 . that he is satisfied that the vibration was due to transverse-torsional-interaction. The mechanism of interaction must have been different. The generators were of two-pole type, and there may have been a source of excitation of 2/rev. torsional vibration which does not occur in turbine rotors.

Engineers experienced in steam turbine operation have sometimes mistaken synchronous whirl for vibration of a different type, caused either by wander in out-of-balance or by running too close to a critical speed. This suggests that synchronous whirl may have played a greater part in earlier turbine vibration troubles than was realised at the time.

The problem of transverse-torsional interaction. The writer's approach to the synchronous whirl problem was that of an engineer searching for the explanation of a novel type of vibration experienced in service. The approaches of TONDL (1965) and of KONONENKO (1969) were analytical investigations of rotor motion with transverse-torsional interaction. The work of both authors was of great value to the writer in correlating observed symptoms of synchronous whirl with phenomena characteristic of transverse-torsional vibration, notwithstanding the writer's different conclusion as to principal mechanism of interaction.

On any future occurence of synchronous whirl, it would be of great interest to make experimental observations of torsional vibration and of blade vibration in conjunction with transverse vibration.

ACKNOWLEDGEMENTS - The writer gratefully acknowledges permission from G.E.C. Turbine Generators Limited (in which A.E.I. Turbine-Generator Division is incorporated) to publish the design information and operating experience on which this paper is based. He would particularly express his thanks to Dr. D. Kalderon, Mr. F.R. Harris, and Mr. D.E. Wood.

REFERENCES

[1] KER WILSON, W.: Practical Solution of Torsional Vibration Problems, Vol. 2, 1963, London: Chapman and Hall.

[2] KONONENKO, V.O.: Vibrating Systems with a Limited Power Supply, 1969, London: Iliffe Books.

[3] SMITH, D.M.: The Motion of a Rotor carried by a Flexible Shaft in Flexible Bearings, 1933, Proc. R. Soc. London: 142, 92.

[4] SMITH, D.M.: Journal Bearings in Turbomachinery, 1969, London: Chapman and Hall.

[5] TONDL, A.: Some Problems of Rotor Dynamics, 1965, London: Chapman and Hall.

On the Stability of Periodic Systems in Connection with Rotating Shafts

Ch. WEHRLI

Eidgenössische Technische Hochschule Zürich, Schweiz

1. INTRODUCTION

Critical angular velocities of rotating shafts play an important role in the field of technical stability. Even in simple cases the physical system is of such great complexity that one is faced with difficulties concerning the mathematical formulation and treatment which can hardly be overcome. In the attempt to solve a problem of this kind, the question of an adequate model for the system arises. Once the model is chosen, the mathematical form of the problem is determined and the corresponding results depend essentially on the definition of stability.

The idea of this investigation is to emphasize the importance of the general aspects; we will not be concerned, therefore, with mere technical details. For that reason we restrict ourselves to an elastic shaft equipped with a single disk and subjected to torsion. Moreover, we assume that the mass of the shaft is small compared to the disk, and we disregard the gyroscopic moments. To discuss the stability of the shaft rotating with constant angular velocity ω we use the method of Liapunov and the concept of total stability [1].

As we have already mentioned, a torque is acting on the shaft. Let us consider the three cases shown in Fig. 1.1 and let us assume that in each of these cases, the torque is transmitted from the disk to the end. If the torque is axial, the equilibrium of the system is unstable, in certain cases, for every angular velocity [2]. However, in a detailed analysis of the forces it turns out that in all prac-

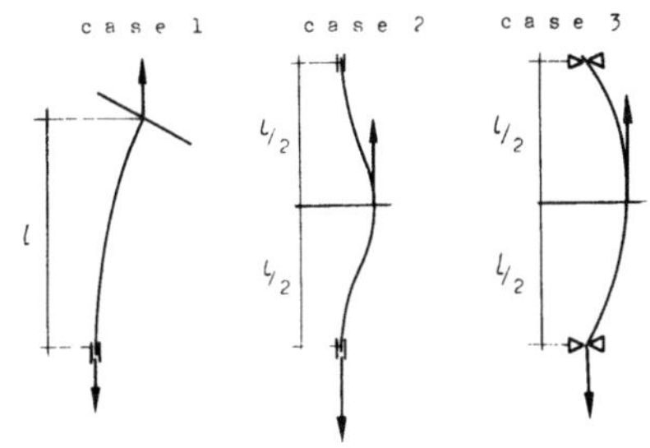

Fig. 1.1

tically conceivable cases the torque is always conservative [3]. We have to distinguish between the quasitangential torque

$$\underline{M} = M(v_x, 0, 1)^T, \quad M = 2Pa$$

(a couple acting on the disk, Fig. 1.2) and the semitangential torque

$$\underline{M} = \tfrac{1}{2}M(v_x, v_y, 2)^T, \quad M = nPa$$

($n > 2$ forces P uniformly distributed around the disk, Fig. 1.3)

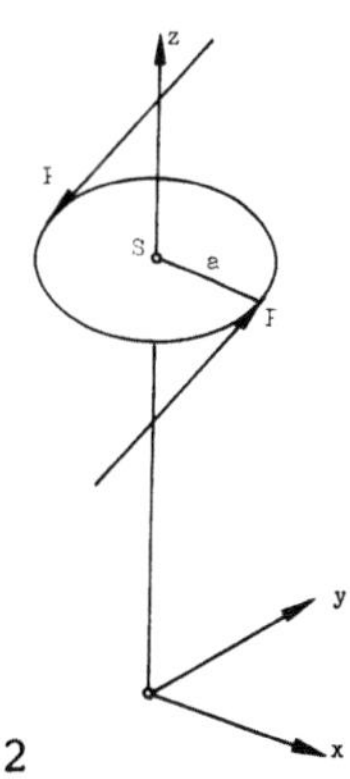

Fig. 1.2

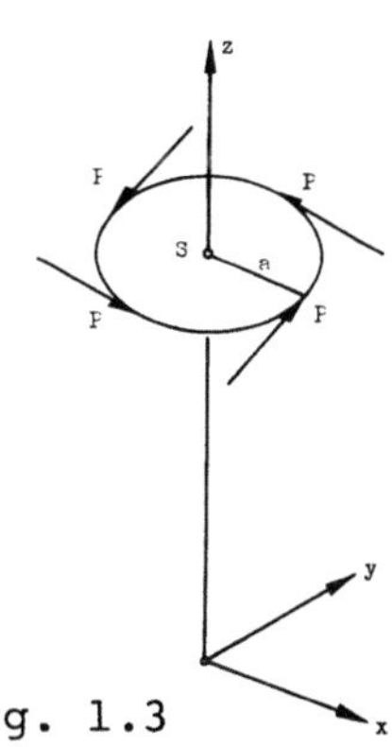

Fig. 1.3

$\underline{v} = (\frac{dx}{dz}, \frac{dy}{dz}, 1)^T$ being a tangential vector to the deflection curve at S. In case 3 the shaft is supposed to have Cardan links and as consequence a rotating quasitangential torque transmitted at the lower end. Calculations in all cases will be carried out on this bases and are subject to the condition that the torque M be small compared with the buckling moment.

In mathematical terms the principal aim of the paper is a study of the stability for periodic systems associated with our model in cases 1 to 3. According to the assumptions in case 1 (distinct flexural rigidities, quasitangential torque) and in case 3 (quasitangential torque) the mathematical structure of the system is of this type, while in case 1 (semitangential torque) and in case 2 a pulsating torque induces the periodicity.

It is of some importance to note that no instabilities in the sense of Liapunov occur in the first two cases for a shaft with circular cross section as long as the moment M is constant. The critical velocities indicated in [2] for these and similar cases have to be interpreted as resonance phenomena. However, with the results obtained in this context a completely different explanation of these critical speeds seems quite reasonable.

2. THE DIFFERENTIAL EQUATIONS OF MOTION

In principle, the motion of the disk can be described using a coordinate frame at rest or rotating with the constant angular velocity ω of the shaft. For our purpose it is convenient to start with the formulation of the differential equations of motion in the rotating frame of Fig. 2.1. If we denote by $p_x(x,y,t)$, $p_y(x,y,t)$ and $d(\dot{x},\dot{y})$ the terms arising from the nonlinear restoring and the external damping forces, respectively, and by $r_x(x,y,\dot{x},\dot{y},t)$, $r_y(x,y,\dot{x},\dot{y},t)$ the socalled permanent disturbances, including in particular the influence of an excentricity of the disk, Newtons law has the form.

$$(2.1)\qquad \begin{aligned} &\ddot{x}- 2\omega\dot{y} - \omega^2 x + d[\dot{x}-\omega y] - p_x = r_x \quad , \\ &\ddot{y}+ 2\omega\dot{x} - \omega^2 y + d[\dot{y}+\omega x] - p_y = r_y \quad . \end{aligned}$$

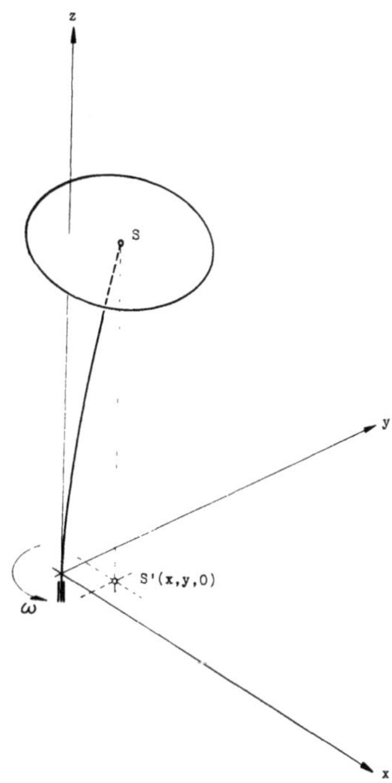

Fig. 2.1

Assuming that the nonlinear function on the left side can be expanded and putting

$$\underline{x}=\begin{bmatrix} x_1 \\ x_2 \\ x_3 \\ x_3 \end{bmatrix} := \begin{bmatrix} x \\ \dot{x} \\ \dot{y} \\ y \end{bmatrix}, \quad P(t,\varepsilon) := \begin{bmatrix} 0 & 1 & 0 & 0 \\ \omega^2-c_{11} & -2\gamma & 2\gamma\omega-c_{12} & 2\omega \\ 0 & 0 & 0 & 1 \\ -2\gamma-c_{21} & -2\omega & \omega^2-c_{22} & -2\gamma \end{bmatrix}, \quad \underline{r} := \begin{bmatrix} 0 \\ r_x \\ 0 \\ r_y \end{bmatrix},$$

(2.1) is transformed into

$$(2.2) \quad \dot{\underline{x}} = P(t,\varepsilon)\underline{x} + \underline{g}(\underline{x},t,\varepsilon) + \underline{r}(\underline{x},t) \ , \quad (\underline{r}(\underline{0},t) \not\equiv \underline{0}) \ ,$$

where $\underline{g}(\underline{x},t,\varepsilon)$ represents the nonlinear parts in the expansions. The matrix $P(t,\varepsilon)$ has the period $\frac{2\pi}{k\omega} = \frac{2\pi}{\Omega}$ $(k\in N)$ and depends on a loading parameter ε . The coefficients

$c_{ij}(t,\varepsilon) = c_{ji}(t,\varepsilon)$ are related to the influence numbers $\alpha_{ij}(t,\varepsilon) = \alpha_{ji}(t,\varepsilon)$ of the shaft by $\sum_{j=1}^{2} c_{ij}\alpha_{jk} = \frac{1}{m}\delta_{ik}$ (m mass of the disk).

Since α_{ij} can be written as power series

$$\alpha_{ij} = \sum_{0}^{\infty} \alpha_{ij}^{(k)} \varepsilon^k$$

with time-independent $\alpha_{11}^{(0)}$, $\alpha_{22}^{(0)}$ and $\alpha_{12}^{(0)} = 0$ we have

$$c_{ij} = \sum_{0}^{\infty} c_{ij}^{(k)}(t)\varepsilon^k \quad \text{and} \quad P(t,\varepsilon) = \sum_{0}^{\infty} P_k \varepsilon^k , \tag{2.3}$$

where

$$P_0 = \begin{bmatrix} 0 & 1 & 0 & 0 \\ \omega^2 - c_{11}^{(0)} & -2\gamma & 2\gamma\omega & 2\omega \\ 0 & 0 & 0 & 0 \\ -2\gamma\omega & -2\omega & \omega^2 - c_{22}^{(0)} & -2\gamma \end{bmatrix}, \quad P_k(t) = \begin{bmatrix} 0 & 0 & 0 & 0 \\ c_{11}^{(k)}(t) & 0 & c_{12}^{(k)}(t) & 0 \\ 0 & 0 & 0 & 0 \\ c_{12}^{(k)}(t) & 0 & c_{22}^{(k)}(t) & 0 \end{bmatrix}$$

$(k \geqq 1)$.

3. PRELIMINARY STABILITY CONSIDERATIONS

On account of the presence of persistant disturbances $\underline{r}(\underline{x},t)$ in (2.2) we have to apply the concept of total stability [1]. This means that the unperturbed systems which is obtained from (2.2) omitting $\underline{r}(\underline{x},t)$ is not essentially affected by the additional forces, provided they remain sufficiently small. The equilibrium of

$$\dot{\underline{x}} = P(t,\varepsilon)\underline{x} + \underline{g}(x,t,\varepsilon) \tag{3.1}$$

is totally stable if it is exponentially stable. In our case asymptotic stability of the solution $\underline{x} \equiv \underline{0}$ of the linearized equation (3.1) implies exponential stability of the equilibrium of (3.1) because $P(t,\varepsilon)$ is periodic. Total stability requires sufficiently small disturbances which in case of a rotating shaft are caused by imperfections as excentricity, vibrations of the bearings etc. We assume for our further investigations that $\underline{r}(\underline{x},t)$ has the necessary properties for total stability.

To solve the stability problem we have to deal with the linear system

$$\dot{\underline{x}} = P(t,\varepsilon)\underline{x} \ . \tag{3.2}$$

The behaviour of its solutions can be described by means of the characteristic multipliers and exponents [4]. In order to calculate these numbers and to establish conditions for stability it is convenient to transform (3.2) by means of

$$\underline{x} = e^{-\gamma t}\begin{bmatrix} 1 & 0 & 0 & 0 \\ -\gamma & 1 & 0 & 0 \\ 0 & 0 & 1 & 0 \\ 0 & 0 & -\gamma & 1 \end{bmatrix}\underline{y}$$

into

$$\dot{\underline{y}} = \begin{bmatrix} 0 & 1 & 0 & 0 \\ \omega^2-d_{11} & 0 & -d_{12} & 2\omega \\ 0 & 0 & 0 & 1 \\ -d_{12} & -2\omega & \omega^2-d_{22} & 0 \end{bmatrix}\underline{y} = Q(t,\varepsilon)\underline{y} \quad , \tag{3.3}$$

where $d_{ij}(t,\varepsilon) = c_{ij}(t,\varepsilon) - \gamma^2\delta_{ij}$. According to (2.3) we write

$$Q(t,\varepsilon) = \sum_{0}^{\infty} Q_k\varepsilon^k . \tag{3.4}$$

For the matrix $Y(t,\varepsilon)$ of a fundamental system of solutions of (3.3) satisfying $Y(0,\varepsilon) = E$ (E unity matrix) there exists a constant matrix $C(\varepsilon)$ such that $Y(t + \frac{2\pi}{\Omega},\varepsilon) = Y(t,\varepsilon)\, C(\varepsilon)$. The characteristic multipliers are the eigenvalues ρ_k of $C(\varepsilon)$ and the characteristic exponents are defined by $\mu_k := \frac{\Omega}{2\pi}\log \rho_k$. On the other hand, the μ_k's turn out to be the eigenvalues of the matrix $R(\varepsilon) := \frac{\Omega}{2\pi}\log C(\varepsilon)$. Floquet's theorem can now be expressed in the form

$$Y(t,\varepsilon) = N(t,\varepsilon)\exp[R(\varepsilon)t] \tag{3.5}$$

where the matrix $N(t,\varepsilon)$ has the period $\frac{2\pi}{\Omega}$. It is important to mention that the calculation of $R(\varepsilon)$ and $N(t,\varepsilon)$ is not elementary.

Using the notations of this paragraph we have the following stability criterion: if $\mathrm{Re}\mu_k < \gamma$, $(k = 1,2,3,4)$, (Re means real part), the equilibrium of (3.2) is asymptotically (in fact exponentially) stable and therefore the equilibrium of (3.1) is totally stable.

4. SHAFTS WITH A SINGLE FLEXURAL RIGIDITY

Here we have $\alpha_{11}^{(0)} = \alpha_{22}^{(0)}$ and consequently $c_{11}^{(0)} = c_{22}^{(0)} = {=:}\ \omega_o^2$. Since these quantities are independent of the directions of the axes x and y we choose a coordinate frame at rest. The governing differential equation is obtained from (2.2) by cancelling the ω-terms in $P(t,\varepsilon)$. In the three cases listed in section 1 the coefficients c_{ij} can be characterized as follows [5]:

case 1 with semitangential and case 2 with semi or quasi-tangential pulsating torque:

$$c_{11} = c_{22} = \frac{\alpha}{m\ell^3}\,(k_0 + k_1(\Omega t)\varepsilon + \ldots),\quad c_{12} = 0(\varepsilon^2)\;; \tag{4.1}$$

case 1 with quasitangential pulsating torque:

$$c_{11} = c_{22} = \frac{3\alpha}{m\ell^3} + 0(\varepsilon^2),\quad c_{12} = \frac{\alpha}{m\ell^3}\,k_2(\Omega t)\varepsilon + \ldots\;, \tag{4.2}$$

case 3 with semi- or quasitangential time -independant torque:

$$c_{11} = \frac{48\alpha}{m\ell^3}(1 + \frac{3}{32}\sin 2\omega t\varepsilon + \ldots),\quad c_{22} = \frac{48\alpha}{m\ell^3}(1 - \frac{3}{32}\sin 2\omega t\varepsilon + \ldots),$$
$$c_{12} = -\frac{9\alpha}{2m\ell^3}\cos 2\omega t\varepsilon + \ldots\;; \tag{4.3}$$

(α flexural rigidity, k_0, k_1, k_2 depending on case).

We first study some properties of the characteristic multipliers ρ_k. For this purpose we notice that if $\underline{y}$ is a solution of (3.3) then $\underline{z} = (y_2, -y_1, y_4, -y_3)^T$ is a solution of the adjoint system $\dot{\underline{z}} = -Q^T(t.\varepsilon)\underline{y}$. Choosing a $\underline{y}$ satisfying $\underline{y}(t + \frac{2\pi}{\Omega}) = \rho_k\underline{y}(t)$ we conclude that this system has the same characteristic multipliers as (3.3). On the other hand the characteristic multipliers of the adjoint system are ρ_k^{-1} [1]. Therefore the characteristic equation $\det[C(\varepsilon) - \rho E] = 0$ is reciprocal. As a consequence we have stable equilibrium of (3.3) as long as its roots are simple and of magnitude 1.

Since all ρ_k are continous functions of the parameters Ω and ε the parameter plane Ω,ε can be subdivided into stable and unstable domains.

On the boundary between these domains the characteristic equation has either conjugate complex double roots or multiple roots ± 1. From the roots in case $\gamma = 0$ on the axis Ω given $\rho_{1,2} = e^{\frac{2\pi i\omega_o}{\Omega}}$, $\rho_{3,4} = e^{-\frac{2\pi i\omega_o}{\Omega}}$ we infer that unstable domains are emanating from the points $(\frac{2\omega_o}{n},0)$ with $n \in N$. Unstable domains starting from other points of the axis Ω are in principle possible but it will be shown later that this is not the case. Provided $\gamma > 0$ is subject to $\gamma << 1$ the situation in the parameter plane concerning the equilibrium of (3.2) changes in so far as the unstable domains do no longer reach the points $(\frac{2\omega_o}{n},0)$ but approach them. For pairs Ω,ε from these domains the stability condition cited at the end of last section does not hold.

The usual way to analyse the distribution of stable and unstable points in the neighborhood of the axis Ω for the equilibrium of (3.2) makes use of μ_k [6]. A more efficient approach without speciallization of (3.2) but taking in account (2.3) is the application of the asymptotic method of Krylov and Bogoljubov [7]. Elimination of x_2 and x_4 in (3.2) leads to equations of the form

$$\text{(4.4)}\quad \begin{aligned} \ddot{x}_1 + \omega_o^2 x_1 &= -\varepsilon(c_{11}^{(1)}x_1 + c_{12}^{(1)}x_3 + 2c\dot{x}_1) + 0(\varepsilon^2)\,, \\ \ddot{x}_3 + \omega_o^2 x_3 &= -\varepsilon(c_{12}^{(1)}x_1 + c_{22}^{(1)}x_3 + 2c\dot{x}_3) + 0(\varepsilon^2) \end{aligned} \quad (\gamma = c\varepsilon)$$

with coefficients $c_{ij}^{(1)} = \sum_{o}^{\infty}(a_\nu^{ij}\cos\nu\Omega t + b_\nu^{ij}\sin\nu\Omega t)$.

In order to determine the unstable domains briefly discussed above we put

$$\Omega = \frac{2\omega_o}{n} + \eta = \frac{2\omega_o}{n} + \beta\varepsilon, \quad (\eta = 0(\varepsilon)) .$$

Then (4.4) becomes

$$\text{(4.5)}\quad \begin{aligned} \ddot{x}_1 + (\tfrac{n\Omega}{2})^2 x_1 &= -\varepsilon\{(c_{11}^{(1)} - n\omega_o\beta)x_1 + c_{12}^{(1)}x_3 + 2c\dot{x}_1\} + 0(\varepsilon^2)\,, \\ x_3 + (\tfrac{n\Omega}{2})^2 x_3 &= -\varepsilon\{c_{12}^{(1)}x_1 + (c_{22}^{(1)} - n\omega_o\beta)x_3 + 2c\dot{x}_3\} + 0(\varepsilon^2) . \end{aligned}$$

These equation are transformed by means of the substitution

$$(4.6)\quad \begin{aligned} x_k &= a_k\cos(\tfrac{n\Omega}{2}\,t+\theta_k)\ , \\ \dot{a}_k\cos(\tfrac{n\Omega}{2}\,t+\theta_k) &- a_k\dot{\theta}_k\sin(\tfrac{n\Omega}{2}\,t+\theta_k) = 0,(k=1,2) \end{aligned}$$

into the standard form

$$\dot{a}_1 = -\frac{\varepsilon}{\omega_o}\, f_{11}(a_1,a_3,\theta_1,\theta_3,t)\sin(\tfrac{n\Omega}{2}\,t+\theta_1) + 0(\varepsilon^2)\ ,$$

$$\dot{\theta}_1 = -\frac{\varepsilon}{a_1\omega_o}\, f_{11}(a_1,a_3,\theta_1,\theta_3,t)\cos(\tfrac{n\Omega}{2}\,t+\theta_1) + 0(\varepsilon^2)\ ,$$

$$\dot{a}_3 = -\frac{\varepsilon}{\omega_o}\, f_{31}(a_1,a_3,\theta_1,\theta_3,t)\sin(\tfrac{n\Omega}{2}\,t+\theta_3) + 0(\varepsilon^2)\ ,$$

$$\dot{\theta}_3 = -\frac{\varepsilon}{a_3\omega_o}\, f_{31}(a_1,a_3,\theta_1,\theta_3,t)\cos(\tfrac{n\Omega}{2}\,t+\theta_3) + 0(\varepsilon^2)\ .$$

(4.7)

Averaging the right sides of (4.7) over t yields the equation of the first approximation

$$\begin{aligned}
\dot{a}_1 &= \{\frac{\varepsilon}{4\omega_o}(a_n^{11}\sin2\theta_1+b_n^{11}\cos2\theta_1)-\gamma\}a_1 \\
&+ \frac{\varepsilon}{4\omega_o}\{2a_o^{12}\sin(\theta_1-\theta_3)+a_n^{12}\sin(\theta_1+\theta_3) + b_n^{12}\cos(\theta_1+\theta_3)\}a_3\ , \\
a_1\dot{\theta}_1 &= \frac{\varepsilon}{4\omega_o}\ \{2a_o^{11}+a_n^{11}\cos2\theta_1-b_n^{11}\sin2\theta_1+2n\omega_o\beta\}a_1 \\
&+ \frac{\varepsilon}{4\omega_o}\ \{2a_o^{12}\cos(\theta_1-\theta_3)+a_n^{12}\cos(\theta_1+\theta_3)-b_n^{12}\sin(\theta_1-\theta_3)\}a_3\ , \\
\dot{a}_3 &= \frac{\varepsilon}{4\omega_o}\ \{-2a_o^{12}\sin(\theta_1-\theta_3)+a_n^{12}\sin(\theta_1+\theta_3)+b_n^{12}\cos(\theta_1+\theta_3)\}a_1 \\
&+ \{\frac{\varepsilon}{4\omega_o}\ (a_n^{22}\sin2\theta_3+b_n^{22}\cos2\theta_3)-\gamma\}a_3\ , \\
a_3\dot{\theta}_3 &= \frac{\varepsilon}{4\omega_o}\{2a_o^{12}\cos(\theta_1-\theta_3)+a_n^{12}\cos(\theta_1+\theta_3)-b_n^{12}\sin(\theta_1+\theta_3)\}a_1 \\
&+ \frac{\varepsilon}{4\omega_o}\{2a_o^{22}+a_n^{22}\cos2\theta_3-b_n^{22}\sin2\theta_3+2n\omega_o\beta\}a_3\ .
\end{aligned}\qquad (4.8)$$

Let us now investigate the stability of the trivial solution $a_1 = 0$, $a_2 = 0$. If we introduce the substitution

$$\xi_1 = a_1\cos\theta_1, \quad \xi_2 = a_3\cos\theta_3, \quad \xi_3 = a_1\sin\theta_1, \quad \xi_4 = a_3\sin\theta_3$$

we obtain a linear system from (4.8):

$$\dot{\xi} = \begin{bmatrix} \frac{\varepsilon b_n^{11}}{4\omega_o} - \gamma & \frac{\varepsilon b_n^{12}}{4\omega_o} & \frac{\varepsilon}{4\omega_o}(-2a_o^{11}+a_n^{11}) + \frac{n}{2}\eta & \frac{\varepsilon}{4\omega_o}(-2a_o^{12}+a_n^{12}) \\ \frac{\varepsilon b_n^{12}}{4\omega_o} & \frac{\varepsilon b_n^{22}}{4\omega_o} - \gamma & \frac{\varepsilon}{4\omega_o}(-2a_o^{12}+a_n^{12}) & \frac{\varepsilon}{4\omega_o}(-2a_o^{22}+a_n^{22}) - \frac{n}{2}\eta \\ \frac{\varepsilon}{4\omega_o}(2a_o^{11}+a_n^{11}) + \frac{n}{2}\eta & \frac{\varepsilon}{4\omega_o}(2a_o^{12}+a_n^{12}) & -\frac{\varepsilon b_n^{11}}{4\omega_o} - \gamma & -\frac{\varepsilon b_n^{12}}{4\omega_o} \\ \frac{\varepsilon}{4\omega_o}(2a_o^{12}+a_n^{12}) & \frac{\varepsilon}{4\omega_o}(2a_o^{22}+a_n^{22}) + \frac{n}{2}\eta & -\frac{\varepsilon b_n^{12}}{4\omega_o} & \frac{\varepsilon b_n^{22}}{4\omega_o} - \gamma \end{bmatrix} \xi \, .$$

(4.9)

Instead of discussing the rather involved general situation described by (4.9)we confine ourselves to some examples leading to information about the stability behaviour of the shafts considered.

1. All a_ν^{ij}, b_ν^{ij} zero except $b_1^{11} = a_1^{12} = -b_1^{22} = a$,

   $\Omega = 2\omega$:

   boundary of stability if $\gamma = 0$ given by

   $\omega = \omega_o \pm \frac{a}{\omega_o}\varepsilon$, $(n = 1)$, $\omega = \frac{\omega_o}{n} + 0(\varepsilon^2)$, $(n > 1)$;

   conclusion: unstable domains for $n > 1$ neglible if $\gamma \neq 0$; boundary of stability if $\gamma \neq 0$ determined by

$$\eta^2 - \frac{a}{\omega_o^2}\varepsilon^2 = -4\gamma^2 \tag{4.10}$$

   in case 3 from (4.3): $\omega_o^2 = \frac{48\alpha}{m\ell^3}$, $a = \frac{9\alpha}{2m\ell^3}$, $\varepsilon = \frac{M\ell}{\alpha}$

   (Fig. 4.1).

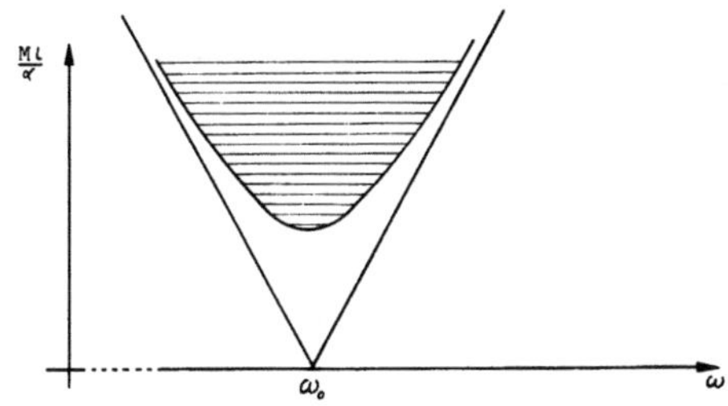

Fig. 4.1

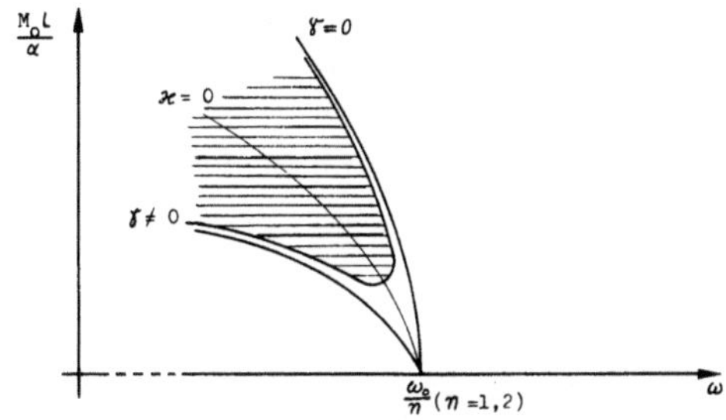

Fig. 4.2

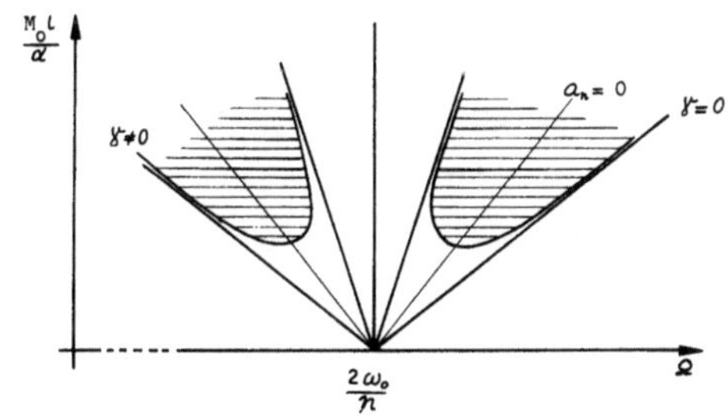

Fig. 4.3

2. $a_\nu^{11} = a_\nu^{22}$, $b_\nu^{11} = b_\nu^{22} = a_\nu^{12} = b_\nu^{12} = 0$:

boundary of stability given by

$$\Omega = \frac{2\omega_o}{n} - \frac{\varepsilon}{2n\omega_o}(2a_o^{11} \pm a_n^{11}), \quad (\gamma=0,\ n=1,2,\dots) ,$$

(4.11)

$$\left(n\eta + \frac{\varepsilon a_o^{11}}{\omega_o}\right)^2 - \left(\frac{\varepsilon a_n^{11}}{2\omega_o}\right)^2 = -4\gamma^2, \quad (\gamma \neq 0,\ n=1,2,\dots) ;$$

in case 2 from (4.1) taking $M = M_o(1+\kappa\cos 2\omega t)$:

$$\omega_o^2 = \frac{192\alpha}{m\ell^3}, \quad a_o^{11} = \frac{13\alpha}{40m\ell^3}(2+\kappa^2), \quad a_1^{11} = \frac{13\alpha\kappa}{10m\ell^3}, \quad a_2^{11} = \frac{13\alpha\kappa^2}{40m\ell^3},$$

$$\varepsilon = \left(\frac{M_o\ell}{\alpha}\right)^2 \quad \text{(Fig. 4.2)}.$$

3. $a_\nu^{11} = a_\nu^{22} = b_\nu^{11} = b_\nu^{22} = b_\nu^{12} = 0$:

boundary of stability determined by

$$\Omega = \frac{2\omega_o}{n} \pm \frac{\varepsilon}{2n\omega_o}(2a_o^{12}\ a_n^{12}),\ (\gamma=0,\ n=1,2,\ldots)$$

(4.12)

$$\frac{n^2}{4}\eta^2\ \frac{na_o^{12}}{2\omega_o}\varepsilon\eta+\{(a_n^{12})^2-4(a_o^{12})^2\}(\frac{\varepsilon}{4\omega_o})^2 = -\gamma^2,(\gamma\neq0,n=1,2,\ldots)$$

in case 1 with quasitangential torque from (4.2) taking

$$M = M_o(1+\sum_o^\infty a_\nu\cos\nu\Omega t)\ :$$

$$\omega_o^2 = \frac{3\alpha}{m\ell^3},\ a_o^{12} = \frac{9\alpha}{8m\ell^3},\ a_\nu^{12}= \frac{9\alpha a_\nu}{8m\ell^3},\ \varepsilon = \frac{M_o\ell}{\alpha} \qquad \text{(Fig. 4.3)}.$$

Finally we have to deal with pairs Ω,ε in the vicinity of $\varepsilon = 0$ in the intervals $\frac{2\omega_o}{n} > \Omega > \frac{2\omega_o}{n+1}$. With the same procedure as before starting with (4.4) and substituting ω_o for $\frac{n\Omega}{\alpha}$ in (4.6) one readily ends up with a rather simple system corresponding to (4.9) for which no unstable equilibrium arises. This remark completes the discussion in the first approximation.

5. SHAFTS WITH DISTINCT FLEXURAL RIGIDITIES

Under these circumstances the terms with c_{ij} in the differential equation only reduce to the simple form indicated in section 2 provided we use a rotating coordinate frame with axes x and y parallel to the principal directions of the cross section in the unloaded state of the shaft. Since now centrifugal and Coriolis terms are present it is no longer possible to proceed as in the last section.

Typical for this class of problems is the form α_{ij} illustrated in the following cases [8]:

case 1 and 2 with semitangential pulsating torque:

$$\alpha_{11} = \frac{k_o \ell^3}{\alpha_2} + k_{11}(\Omega t)\varepsilon^2 + \dots ,$$

(5.1)

$$\alpha_{22} = \frac{k_o \ell^3}{\alpha_1} + k_{22}(\Omega t)\varepsilon^2 + \dots, \quad \alpha_{12} = k_{12}(\Omega t)\varepsilon + \dots ;$$

case 1 with quasitangential time-independent torque:

$$\alpha_{11} = \frac{\ell^3}{3\alpha_2} - \frac{\alpha_3 \ell^4}{8\alpha_2^2}\sin 2\omega t + \dots, \quad \alpha_{22} = \frac{\ell^3}{3\alpha_1} + \frac{\alpha_3 \ell^4}{8\alpha_1^2}\sin 2\omega t\varepsilon + \dots,$$

(5.2)

$$\alpha_{12} = \frac{\ell^4}{\alpha_1\alpha_2}\left|\frac{1}{12}(\alpha_1-\alpha_2) - \frac{1}{8}\alpha_3\cos 2\omega t\right|\varepsilon + \dots, \quad \varepsilon = \frac{M}{\alpha_3} \quad ;$$

($\alpha_1 \neq \alpha_2$ flexural rigidities, α_3 torsional rigidity, k_o, k_{11}, k_{22}, k_{12} depending on kind of torque).

In order to establish the condition for asymptotic stability for the equilibrium of (3.2) we apply a recursive technique. If we diagonalize Q_o by means of a constant matrix A into $\tilde{Q}_o = AQ_oA^{-1} =: J$ [8] (3.3) becomes

(5.3) $$\dot{\tilde{Y}}(t,\varepsilon) = (\sum_0^\infty \tilde{Q}_k \varepsilon^K)\tilde{Y}(t,\varepsilon) = (J + \sum_1^\infty \tilde{Q}_k\varepsilon^k)\tilde{Y}(t,\varepsilon) \quad ;$$

the elements of J are $\delta_{jk}\Lambda_j$ with

$$\Lambda_j = \pm\frac{1}{\sqrt{2}}\sqrt{-(d_{11}^{(0)}+d_{22}^{(0)}+2\omega^2) \pm \sqrt{(d_{11}^{(0)}+d_{22}^{(0)}+2\omega^2)^2 - 4(d_{11}^{(0)}-\omega^2)(d_{22}^{(0)}-\omega^2)}}.$$

(5.4)

It is well-known that the characteristic exponents of (3.3) remain unchanged if we replace $Q(t,\varepsilon)$ by the similar matrix $\tilde{Q}(t,\varepsilon) = AQA^{-1}$. Thus the equilibrium of (3.2) is asymptotically stable if the eigenvalues λ_j of $\tilde{R}(\varepsilon) - \gamma E$ satisfy the condition $\mathrm{Re}\lambda_j < 0$, $(j=1,2,3,4)$ (see section 3).

Because of (3.4) we assume expansions

(5.5) $$\tilde{R}(\varepsilon) = \sum_0^\infty \tilde{R}_k\varepsilon^k , \quad \tilde{N}(t,\varepsilon) = \sum_0^\infty \tilde{N}_k(t)\varepsilon^k \quad .$$

From (3.5), written in terms of $\tilde{Y}, \tilde{N}, \tilde{R}$ and from (5.3) it follows that

$$(\dot{\tilde{N}} + \tilde{N}\,\tilde{R})\exp[\tilde{R}t] = \tilde{Q}\,\tilde{N}\,\exp[\tilde{R}t] \;.$$

Inserting (5.5) we get

$$(5.6)\quad \dot{\tilde{N}}_o = J\,\tilde{N}_o - \tilde{N}_o\tilde{R}_o,\quad \dot{\tilde{N}}_k = J\,\tilde{N}_{k-1} - \tilde{N}_k\tilde{R}_o + F_k - \tilde{N}_o\tilde{R}_k,\quad (k \geqq 1)$$

where $$F_k := \sum_{j=1}^{k} \tilde{Q}_j\,\tilde{N}_{k-j} - \sum_{j=1}^{k-1} \tilde{N}_j\,\tilde{R}_{k-j} \;.$$

It can be proved that the series (5.5) converge provided the series (3.4) converges [8], [9].
Starting with $\varepsilon = 0$ we have

$$\tilde{Y}(t,0) = \tilde{N}_o\,\exp[\tilde{R}_o t] \;;$$

and from the transformed equation (3.3) we get

$$\tilde{Y}(t,0) = \exp[\tilde{Q}_o t] = \exp[Jt],\quad (\text{if } \tilde{Y}(0,0) = E) \;;$$

consequently

$$\tilde{N}_o = E\;,\quad \tilde{R}_o = J\;.$$

Integration of the linear differential equation (5.6) for $\tilde{N}_k(t)$ subject to the condition $\tilde{N}_k(0) = 0$ yields

$$\tilde{N}_k(t) = \int_0^t \exp[-J\tau]\{F_k(\tau) - \tilde{R}_k\}\exp[J\tau]d\tau \;.$$

In this relation only $\tilde{R}_k$ is not known but can be calculated, since $\tilde{N}_k(t + \frac{2\pi}{\Omega}) = \tilde{N}_k(t)$, from

$$\int_0^{\frac{2\pi}{\Omega}} e^{-\Lambda_j t}\{f_{ij}^{(k)}(t) - \tilde{r}_{ij}^{(k)}\}e^{\Lambda_j t}dt = 0 \;.$$

All problems of the category treated here have the property $\tilde{r}_{ij}^{(1)} = 0$ whereas, in general,

$$r_{ij}^{(1)} = \frac{\int_0^{\frac{2\pi}{\Omega}} e^{(\Lambda_j - \Lambda_i)t} q_{ij}^{(1)}(t)dt}{\int_0^{\frac{2\pi}{\Omega}} e^{(\Lambda_j - \Lambda_i)t} dt} \quad , \quad (\Lambda_i - \Lambda_j \neq n, n \in N)$$

does not vanish ($q_{ij}^{(1)}(t)$ sums or series of $\cos\nu\Omega t$ and $\sin\nu\Omega t$) . This implies that in the first approximation $\lambda_j = \Lambda_j - \gamma$ or in other words that in the first approximation the stability behaviour is the same as for the unloaded shaft.

In the second approximation already the influence numbers α_{ij} are much more complicated and in addition $\tilde{R}$ is complex such that the criterion of Bilhanz and Schur has to be formulated to ensure that all λ_j have negative real parts. Explicit results can be obtained only numerically and by use of the computer.

Example:

Second order approximation in case 1 with quasitangential torque (see (5.2)), [8]; results shown in Figures 5.1 and 5.2:

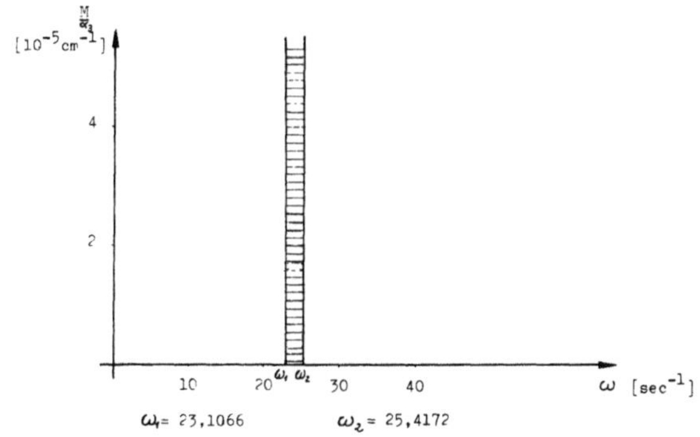

Fig. 5.1. Unstable domain of the first kind given by $\omega_1 \leqq \omega < \omega_2$.

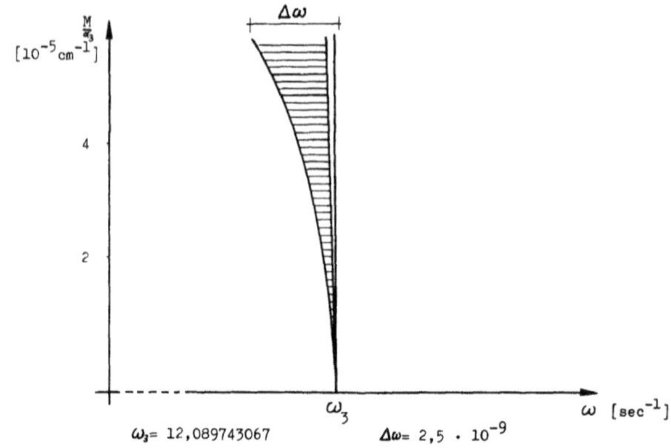

Fig. 5.2. Extremely narrow unstability domains of the second kind emanating from the point $(\omega_3, 0)$.

6. CONCLUDING REMARKS

Pulsating torsion might appear rather unrealistic. However, time-dependent torques occur more frequently than one would expect, in particular if the variation from the mean value M_O of the torque is small. We are faced with a similar situation in the case of distinct flexural rigidities since as a rule most shafts will have a slightly noncircular cross section.

All cases considered in this paper are characterized by the fact that for fixed values of the parameter ε intervals of critical ω's occur. This is due to the fact that in all of these cases the mechanical properties give rise, at least in the coordinate system at rest, to differential equations of motion with parametric exatation. For our model under these circumstances alone instabilities in the sense of Liapunov are possible.

Let us return to the question how the results presented here are related to former ones [2](see end of section 1). First we look at the applications of the theory to cases

1 and 2 in section 4, putting $\Omega = 2\omega$. The center lines of the unstable domains emanating from the point $(\omega_o, 0)$ are given by

$$\omega = \omega_o - \frac{13\alpha}{40\omega_o m\ell^3}\varepsilon \quad \text{(case 2, } \kappa \text{ assumed small)} ,$$

(6.1)

$$\omega = \omega_o \quad \frac{9\alpha}{16\omega_o m\ell^3}\varepsilon \text{ (case 1, quasitangential torque).}$$

These equations correspond exactly to the formulas for the critical angular velocities given in [2]. The coincidence is quite obvious if we consider (6.1) as limiting cases with $\kappa \to 0$, $a_1 \to 0$ and $\gamma \to 0$. The unstable domain represented in Fig. 4.3 by the perpendicular to the axis Ω at $\frac{2\omega_o}{n}$ is not contained in the results of [2]. Certainly this instability is of some significance because it ought to be interpreted as a limiting case of the unstable domain of Fig. 5.1 for $\alpha_2 \to \alpha_1$.

To get an idea of the motion for pairs Ω, ε in an unstable domain, the nonlinear vibrations of the mechanical system can be investigated by taking into account the first nonlinear term of the restoring force and neglecting the permanent disturbances. By means of the method used in section 4 it is possible to show that in the critical intervals for ω stationary oscillations exist (results to be published). Their amplitudes turn out to be stable.

REFERENCES

[1] HAHN, W.: Stability of Motion, Berlin: Springer 1967.

[2] ZIEGLER, H.: Principles of Structural Stability. Waltham/Massachusetts 1968.

[3] ZIEGLER, H.: ZAMP 3, 1952.

[4] KNOBLOCH, H.W. & KAPPEL, F.: Gewöhnliche Differentialgleichungen. Stuttgart: Teubner 1974.

[5] WEHRLI, Ch.: Ing.Arch. 47, 1967.

[6] WEHRLI, Ch.: ZAMP 18, 1967.

[7] BOGOLJUBOV, N.N. & MITROPOLSKI, J.A.: Asymptotische Methoden in der Theorie der nichtlinearen Schwingungen. Berlin: Akademie-Verlag 1965.

[8] BUCHSER, B.: Doctor's Theses ETH Zürich Nr. 4623, 1971.

[9] ERUGIN, N.P.: Linear Systems of Ordinary differential Equations, Academic Press 1966.

MIX
Papier aus verantwortungsvollen Quellen
Paper from responsible sources
FSC® C105338

If you have any concerns about our products,
you can contact us on
ProductSafety@springernature.com

In case Publisher is established outside the EU,
the EU authorized representative is:
Springer Nature Customer Service Center GmbH
Europaplatz 3, 69115 Heidelberg, Germany

Printed by Libri Plureos GmbH
in Hamburg, Germany